SYMMETRIES IN SCIENCE

SYMMETRIES IN SCIENCE

Edited by

Bruno Gruber
Richard S. Millman

Southern Illinois University
Carbondale, Illinois

Published in cooperation with the Illinois Academy of Science

PLENUM PRESS · NEW YORK AND LONDON

Library of Congress Cataloging in Publication Data

Main entry under title:

Symmetries in science.

"Proceedings of the Einstein centennial celebration science symposium on symmetries in science, held at Southern Illinois University, Carbondale, Illinois, February 23–March 2, 1979."
Includes index.
1. Symmetry–Congresses. 2. Einstein, Albert, 1879–1955–Anniversaries, etc. –Congresses. I. Gruber, Bruno, 1936-   II. Millman, Richard S., 1945-
III. Southern Illinois University.
Q172.5.S95S92   501   80-18665
ISBN 0-306-40541-5

Proceedings of the Einstein Centennial Celebration Science Symposium on Symmetries in Science, held at Southern Illinois University, Carbondale, Illinois, February 23–March 2, 1979.

© 1980 Plenum Press, New York
A Division of Plenum Publishing Corporation
227 West 17th Street, New York, N.Y. 10011

Printed in the United States of America

PREFACE

        Southern Illinois University at Carbondale undertook to
honor Albert Einstein as scientist and as humanitarian in commemo-
ration of his 100th birthday during an "Albert Einstein Centennial
Week", February 23 - March 2, 1979. During the course of this
week two Symposia were held, entitled "Symmetries in Science" and
"Einstein: Humanities Conscience", in addition to cultural and
social activities honoring Einstein. This volume presents the
Symposium "Symmetries in Science". It reflects the outstanding
response that was given to our "Albert Einstein Centennial Week"
by the international community of scientists.

        The motivation to have a celebration honoring Albert
Einstein at Southern Illinois University at Carbondale was supplied
by Dr. Paul A. Schilpp, the editor of the "Library of Living Philosophers". Albert Einstein has contributed to this series with his
autobiographical notes, a kind of autobiography of his scientific
life, in a volume entitled "Einstein: Scientist-Philosopher", the
most popular among all the outstanding volumes of this series.
Dr. Paul A. Schilpp's presence at Southern Illinois University at
Carbondale provided a natural link for an Einstein Celebration as
a kind of a continuation of the contribution he made to mankind
through the Einstein volume of his "Library of Living Philosophers".

        As far as the Symposium on "Symmetries in Science" is
concerned, I wish to express my deep gratitude to all members of
the "Einstein Centennial Committee" at Southern Illinois University
at Carbondale, for the encouragement and the support they gave me
in this undertaking. From among the members of the committee I
wish to thank in particular Joseph N. Goodman, Executive Director
of the Southern Illinois University Foundation. I also want to
express my deep gratitude to the authorities of Southern Illinois
University at Carbondale. Here I wish to thank in particular
former President Warren W. Brandt, Vice President for Academic
Affairs and Research Frank E. Horton, and the Dean of the College
of Science Norman J. Doorenbos. My greatest thanks, however, go
to Charles J. Lerner, Co-Chairman of the "Einstein Centennial

Committee" and Michael R. Dingerson, Director and Associate Dean
of the Graduate School. It was truly due to their day to day
cooperation and their personal efforts that the Symposium "Symmetries
in Science" became the success that it was.

                        Bruno Gruber
                        Chairman of Symposium
                        "Symmetries in Science"

CONTENTS

CONTENTS

ix

WHY WE BELIEVE IN THE EINSTEIN THEORY

P.A.M. Dirac

Physics Department
Florida State University
Tallahassee, Florida 32306

I am very happy to be invited to Carbondale and to have this
opportunity of paying tribute to Einstein. His influence on the
whole of modern physics is simply enormous and probably not always
appreciated and I will try to give you a better understanding of it.

Relativity was first introduced to the world in 1918, just at
the end of the first world war. Of course, the special theory of
relativity was then quite old. It was discovered in 1905 but it
was quite unknown except to a few specialists in universities, and
no one had heard of Einstein apart from that. Then, at the end of
the first world war, relativity just burst on the world with a tremendous impact.

The reason for this is that it was just at the psychological
moment when a terrible war had at last come to an end. Everyone
was quite sick of it, whichever side they were on, and people
wanted something new, something to enable them to forget about the
war and to start off on a new line of thinking. Relativity provided just that.

At that time, I was a student of engineering at Bristol University in England, just one of the undergraduates there. We were
caught up in this storm of relativity. Everyone was discussing it.
People had no really definite information to go on. Students and
professors just were discussing it from the point of view of hear-
say. Newspapers were continually writing articles about it and
all the magazines were full of articles about it. The people who
wrote the articles understood very little also, but they felt more
or less competent to try to explain things.

As engineering students we had been working all the time with Newton. Newton was our god and everything in engineering depended on Newton. Then we were told that in some mysterious way Newton was wrong. We had to adjust ourselves to new ideas.

Why should we believe in this new theory? There were two reasons that were given. One reason was that it was supported by experiemental evidence and another reason, given by the philosophers, was that philosophy demanded it. From general philosophical arguments they thought that it was necessary to have relativity and to get away from absolutism. I want to discuss these ideas more thoroughly. Neither of them is the true reason for believing in Einstein and appreciating the greatness of his ideas.

Special Relativity

If you just think of velocities, then in the first place it is quite obvious that the velocity of a body can have a meaning only with respect to the velocity of something else. It is only the difference of two velocities which is a well defined concept. But the question arises - is there some absolute zero to which all other bodies can be referred to give us an absolute velocity for a body. That is a question that cannot be decided by philosophy. It can only be decided by experiment, by observation. One must see if one can find this absolute zero of velocity.

Now experiments had been done by Michaelson and Morley to see if there is such an absolute zero in velocity. All physical theory at that time was based on the idea that there is an absolute ether that had to be used as a reference system. So one could talk about light moving with a definite velocity through the ether. Now the question arises - can one determine the velocity of the ether? More precisely, can one find out the velocity with which the earth is moving through the ether? That is just what Michaelson and Morley tried to do.

They did some careful experiments involving sending beams of light to and fro and making accurate interference observations. The result was that they were unable to find any velocity which could be counted as absolute zero. They were unable to determine what the velocity of the earth was through the ether. They did the experiment at different times of the year when the velocity of the earth would be different because of the motion around the sun, but their results were always negative. How could one understand that?

It was a great mystery to the people at that time. It was studied in particular by Lorentz and Fitzgerald and they supposed that one had to set up new ideas about how rigid bodies behave. The rigid bodies had to undergo a strange kind of contraction which was adjusted in such a way as just to neutralize the effects that

would otherwise be produced by the motion of the earth through the
ether.

At that time, the best physical theory we had was the electromagnetic theory, based on Maxwell's equations. Lorentz worked on
these equations a great deal and made a rather remarkable discovery.
He showed that from these equations, combined with suitable assumptions about how material bodies behaved when they were in motion,
one could set up different frames of reference in space and time
such that the equations appear the same with respect to all these
frames of reference.

From this discovery of Lorentz you get an immediate explanation
of the null result of the Michaelson and Morley experiment. It is
just that, as the earth moves with different velocities, you have to
pass from one Lorentz frame of reference to another, and then there
won't be anything observable to show up with the different velocity
of the earth at the different times. Lorentz found out that there
are these different frames of reference, and he worked out the
equations that transform from one frame of reference to another,
the Lorentz transformation. As a result, one could see immediately
that with any experiments just involving electromagnetic processes,
you could not get the velocity of the ether to show up in any way.
The results that you get will always be the same. The proof just
involves passing from one Lorentz frame of reference to another.

This was all done before Einstein came on the scene. Then
Einstein made a very bold assumption: he said that all these different Lorentz frames were equally good and you had to adopt a new
picture of space and time which treated all these Lorentz frames
symmetrically. Then you would never be able to find out the velocity
of the ether, because it was something which just didn't exist. That
was really the start of relativity.

At that time Lorentz did not accept it. Lorentz had really
done the hard mathematical work. He had discovered the transformations, but he did not accept the view that all the different frames
of references were equally good. Lorentz thought that one of these
frames of references was the really correct physical frame and all
the others were just mathematical fictions. That was the point of
view that Lorentz held. It was in direct contradiction to Einstein's.
The disagreement remained for quite a number of years. Poincaré had
also worked on the problem and held a similar point of view to
Lorentz.

It turned out that it was quite impossible to find out which
was the correct frame of reference. Under those conditions one
should, of course, give up the idea that there is just one frame
of reference which is correct. One then goes over to the idea that
all the frames of reference are equally good, which is just
Einstein's view.

In order to appreciate what Einstein's assumption involves you must see that it really goes a long way beyond the conclusions which Lorentz had obtained. Lorentz had established that so long as one keeps to electromagnetic forces it would be impossible to find an absolute zero of velocity. Einstein went beyond that and said it would never be possible to find an absolute zero in velocity, that there would not be other physical processes that would show up an absolute zero of velocity. It was something inherant in space and time that this absolute zero does not exist at all. You have to adopt a new picture of space and time.

This new picture was very much brought into prominence by Minkowski, the great geometer of that time. He set up the basic geometry. You have to describe physical events in a four dimensional world with this geometry, in which you don't have Lorentz frames of reference such that one is more basic than the others. It is called Minkowski space. You might say that Einstein's fundamental assumption was that the whole of physics has to be put into Minkowski space.

The Microwave Radiation

In one sense, Lorentz was correct and Einstein was wrong, because all Einstein should have said was that with the physics of that time it was impossible for an absolute zero in velocity to show up. But to say that it would never be possible for an absolute zero in velocity to show up was going a bit too far. An absolute zero in velocity has shown up with the more advanced technology which we have at the present day. This refers to the natural microwave radiation.

If one has suitable telescopes capable of observing microwaves of just a few centimeters wavelength, and one points these telescopes to the sky in various directions, one observes some weak radiation coming in. This radiation, called the natural microwave radiation, does not come from the sun, it does not come from the galaxy, it comes from all directions in space. It must be something of cosmological importance. People explain it by saying that it is the remains of the radiation which existed at a time close to the time when the universe was created. There was a lot of hot radiation then, which has cooled down and left some cold radiation which can be observed now by suitable telescopes.

This radiation is coming in equally from all directions for a suitable observer. If you take another observer who is moving relatively to that first observer, he will see it coming more strongly in the direction to which he is proceeding and less strongly from behind him. So it will only be symmetrical with respect to one observer. There is thus one preferred observer for which the microwave radiation is symmetrical. You may say this preferred

observer is at rest in some absolute sense, maybe he is at rest with
respect to an ether. That is just contradicting the Einstein view.

It is possible to observe the velocity of the earth through the
ether as defined by the microwave radiation. One finds that the
earth and the whole solar system are moving very rapidly, with a
speed that can be observed. The only reason why Michaelson and
Morley got a null result, why they failed to observe the motion of
the earth in an absolute sense, was because their technology was
inadequate. Present day technology can do much more than could be
done in those days, nearly one hundred years ago. With the more
modern technology, there is an absolute zero of velocity.

You might say that, with the microwave radiation showing Einstein was wrong, that would destroy relativity. But it has not
destroyed the importance of Einstein's work. The importance lies
in another respect. You shouldn't say that Einstein's theory rests
solely on its agreement with observation. There is agreement with
observation only provided that you don't use a sufficiently advanced
technology. In the absolute sense the agreement no longer holds.
The real importance of Einstein's work was that he introduced
Lorentz transformations as something fundamental in physics. The
whole of physics has to be expressed in Minkowski space, a space
which is subject to Lorentz transformations. That I would say is
the most important of the new ideas introduced by Einstein, and it
is of tremendous importance and is not disturbed by the more advanced
technology which cuts away the basis that Einstein had in proposing
his theory.

I should try to give you some idea of the immense importance
of our having to express all our physical theory in Minkowski space
which is subject to the Lorentz transformations. There are many
examples I could give. One of them, concerns de Broglie, a French
physicist. Just by studying the relationship between particles and
waves and using the Lorentz transformations he found that one could
set up a relationship between particles and waves which was invariant
under Lorentz transformations. That led him to postulate the existence of waves associated with particles. That was a fundamental
idea in atomic physics. It was taken up by Schrodinger and developed
by him. It has proved to be right at the basis of the whole of
modern atomic theory.

Further developments of the theory of Lorentz transformations
show that when you try to set up an equation for the motion of
electrons agreeing with Lorentz transformations and the fundamental
laws of quantum mechanics, you are led to a theory which provides
an explanation for the spin of the electron. If you go a little
farther, you are led to the existence of antimatter. These are all
consequences of the Lorentz transformations. They all follow
smoothly from Einstein's assumption that Lorentz transformations

dominate physics. This dominance of the Lorentz transformation is something which is excessively important and has affected the whole of physics since it was introduced at the beginning of the century.

I might put things a little differently and say that the Lorentz transformations are beautiful transformations from the mathematical point of view, and Einstein introduced the idea that something which is beautiful mathematically is very likely to be valuable in describing fundamental physics. This is really a more fundamental idea than any previous idea. I think we owe it to Einstein more than to anyone else, that one needs to have beauty in mathematical equations which describe fundamental physical theories.

That is certainly the situation with regard to the special theory of relativity. I would say the reason why we believe in special relativity is because it puts importance on these Lorentz transformations which are beautiful mathematically. There is certainly no general philosophical basis for it, and we cannot say that it is supported by experiment if we allow experiments involving the most advanced technology. We can say that it dominates atomic theory and that the examples in which a special zero velocity shows up refer to cosmological questions which should be left out of consideration in the development of atomic theory.

General Relativity

That is the situation with regard to the special theory of relativity. How does it go with the general theory? The general theory was introduced in 1916 by Einstein and it came from his attempts to make gravitation fit in with the ideas of the special theory with Lorentz transformations. It was a very difficult problem to satisfy. It took Einstein many years and he got the solution only in 1916 in the middle of the war.

At that time people in England knew nothing about Einstein's work except for one man, Eddington, the great astronomer. Eddington kept in touch with de Sitter in neutral Holland. De Sitter kept in touch with Einstein, so that indirectly Eddington was in touch with Einstein. Eddington was extremely interested in this theory of Einstein, the general relativity, and wanted to know whether it would be in agreement with observation. This could be tested because the theory of Einstein predicted certain effects.

There were three effects which were immediately predicted. One of them concerned the motion of the planet, Mercury. It had been known for a long time that the motion of the planet Mercury was not adequately described by Newton's Laws. The perihelion of the orbit of Mercury was seen to be advancing in a way that could not be explained by Newton. It advanced by 42 seconds per century, a very small amount, but still something which is quite large enough for astronomers to detect and measure. It was soon found that Einstein's theory did provide for this advance of the perihelion of Mercury, and just by the right amount, 42 seconds per century. That was the first success of the Einstein theory.

Another way of testing the Einstein theory was that according to the theory light passing close by the sun should be deflected. There would also be a deflection according to the old Newtonion theory, but the Einstein deflection would have been twice the Newtonion one. That was something that one could check by observation at the time of a total eclipse. It was impossible to observe it at any other time, because the light of the sun was too strong and would have obscured the light of any stars whose light passed close by the sun. So one had to wait for a suitable total eclipse.

Eddington noticed that there would be a total eclipse occurring in May 1919 which would be very suitable for checking on this effect. He made preparations for sending expeditions out to observe this eclipse. Of course, he knew very well that it would be quite impossible to send out such expeditions as long as the war was still going on, but he was hoping that the war would be finished in time, and it so happened that it was finished in time. Eddington sent out two expeditions, one of which he led himself. He made observations of the deflections of the stars just behind the sun at the time of total eclipse, and the results supported the Einstein theory.

There was tremendous excitement when these results were announced in November 1919. I doubt if there has ever been any other occasion when a scientific discovery has produced such a tremendous effect on the public. Here was Einstein's theory, which everyone had been talking about for so long, several months, being actually confirmed by observations.

The results were not so very accurate because of the great difficulty of making the observations. But they were as good as one could have expected under the circumstances. Many other eclipse expeditions have been sent out more recently to check on this Einstein effect. And the results have supported Einstein in every case, with a greater or less accuracy; always with as much accuracy as one could expect depending on how good the observing conditions were.

Then there was a third effect which the Einstein theory predicted at that time, namely that there should be a red shift in the spectral lines of light that is emitted in a strong gravitational field. The natural case to look for this red shift was in the light from the sun. One should examine light from the surface of the sun and compare the spectral lines with similar lines produced on earth. This turned out to be not a very good way of checking on the Einstein theory, because the motion of the atmosphere of the sun is quite

large and produces a Doppler effect, which disturbs the Einstein
red shift and makes the interpretation of the results rather uncertain. Still, there was some rough support for the Einstein theory
from this third effect.

Several years later Eddington noticed that this effect could
be checked more accurately from certain stars, called white dwarfs,
where the matter is extremely condensed and extremely compact. The
light from the surface of a white dwarf exhibits the Einstein red
shifting quite strongly, much more strongly than the light from the
sun. When we know enough about the white dwarf so that we can estimate its size and mass roughly, we can make the calculations and we
then get good support for the Einstein theory.

Much more recently it has been found that this effect can be
checked just from terrestrial experiments, if one uses suitable
radiation. It could be gamma rays. One takes the rays emitted
from a source at the top of a tower and observes them lower down.
Then these rays, while moving downward, get their spectral lines
shifted from the difference in the gravitational potentials at the
top and bottom of the tower. Just from these terrestrial experiments one can get a confirmation of this Einstein effect.

This effect can also be checked astronomically with the use of
radio waves instead of light. If there is a radio star, a source
of radio waves, behind the sun, then the light from this source,
in passing close by the sun, should also get deflected. That is
an observation that you can make at any time, because the sun does
not emit very strong radio waves. You don't have to wait for a
total eclipse of the sun to do that. This provides an independent
way of checking on this effect.

There is a complication coming in with this way of checking
the effect in that the sun's corona gives rise to a deflection of
radio waves. But this deflection is different for different frequencies. So by comparing the deflection for two different frequencies, you can separate the effect of the corona from the
Einstein effect. You get very good confirmation of the Einstein
theory in this way.

Now I am going to speak about a fourth test for the Einstein
theory; namely, that waves passing close by the sun get delayed.
Not only are they deflected, but they are delayed. If you send
radar waves to a planet lying behind the sun and then observe the
reflected waves that come back and measure how long it takes to
make the journey to and fro, then you can check to see whether
there is such a delay. That is work that has been done in the
last few years by Shapiro. Again, as you work with radio waves,
the sun's corona has an effect, but you can separate out the effect

of the corona from the Einstein effect by using two different wave lengths. The result is again a confirmation of the Einstein theory.

Another check has been provided in recent times by a binary pulsar. A pulsar is a star which emits radio waves in pulses with a very definite periodicity, extremely constant. Now, if this pulsar is a part of a binary system and moves around another star, the periodicity of the pulses will be changed by the motion, and that is something which can be observed. With the pulsar passing close by its companion, you get a very big effect, like the motion of the perihelion of the planet Mercury, but very much larger. This doesn't provide a very accurate test of the Einstein theory, because we do not know enough about the different parameters describing the binary system to be able to calculate just what the effect ought to be. But with reasonable estimates, it does come out approximately right and so it does provide a further rough check of the Einstein theory.

The Need for Mathematical Beauty

We have all these observations which have been made since Einstein first proposed his general theory of relativity. Every time the Einstein theory is confirmed; it has passed all the tests with flying colors. With all these observations you could say that there is very strong support for the Einstein theory of gravitation.

Still one should face the question: Suppose a discrepancy did turn up, how should one react toward it? How would Einstein himself have reacted toward it?

I don't think one should say that the whole foundation of the Einstein theory would be destroyed. Not even if the discrepancy is very well substantiated. One should interpret it rather by saying that there is some new effect that is not adequately explained. Our theory, at any time, should be looked at as in a temporary state and it is always liable to be improved upon. Any discrepancy which should show up should not be looked upon as fatal to the theory, but just as indicating that there is some further work to be done. It should stimulate people to look for further changes which could be made to account for the discrepancy.

I feel that the situation here is very similar to what it is with the special theory of relativity, where with modern technology applied to the observation of the microwaves one finds that there is an absolute zero of velocity. That doesn't spoil the Einstein theory, it just shows an inadequacy. It might very well turn up that there is an inadequacy with regard to the general theory of relativity, but so far it has not yet showed up. We shouldn't be too disturbed if it does show up in the future. It is not something that one should consider as destroying the foundations of the theory.

The foundations of the theory are, I believe, stronger than what one could get simply from the support of experimental evidence. The real foundations come from the great beauty of the theory. They come from the circumstance that Einstein has introduced new ideas of space which are extremely exciting, very elegant, and these ideas will survive no matter what the future brings before us. These ideas are based on the possibility that one can describe a physical force like gravitation just in terms of the properties of space. They lead one to look upon a physical field in general as just a disturbance of space, like a curvature.

Curvature of space is a bit difficult to understand, for a non-mathematician. Mathematicians have now gotten very used to it. Einstein introduced the simplest kind of curvature in space, a curvature which was first studied mathematically by Riemann a hundred years previously. Riemann worked out the basis of the mathematics which was used by Einstein for his general theory of relativity just like Lorentz had worked out the basis of the mathematics for special relativity.

It is the essential beauty of the theory which I feel is the real reason in believing in it. This must dominate the whole future development of physics. It is something which cannot be destroyed, even if there are experimental discrepancies turning up in the future. These experimental discrepancies must be looked upon merely as inadequacies in our present theory.

The Model of the Universe

There is one respect in which the present Einstein theory is clearly inadequate. It is a theory which gives equations describing the gravitational field, but does not give sufficiently complete equations for one to be able to get answers to definite problems. In order to be able to get definite answers, we have to have the field equations supplemented by boundary conditions. We have to know something about the conditions at very great distances. People up to the present have been working on the assumption that to study the gravitational field of particular masses, one can assume that at great distances space is just the flat space of Minkowski. That is certainly not a correct assumption, even though it has worked in describing phenomena in the solar system.

In order to understand the conditions at great distances, one has to have a model of the universe. One has to have some appreciation of what the universe is like when the local irregularities produced by the stars and galaxies being scattered about in it are smoothed out.

Einstein himself realized the need for these boundary conditions and thus the need for a model of the universe. Einstein proposed a

certain static model, Einstein's cylindrical model, it is called.
It was soon realized that this model would not do, because it was
observed that the matter at great distances from us, the galaxies,
are all receding from us. They are all moving away from us and
from each other. This contradicts Einstein's model, so Einstein
gave it up.

Another model was soon afterwards introduced by de Sitter.
De Sitter proposed a model which gave correctly the matter at
great distances moving away from us. It was a good model in that
respect. But it failed in another respect; namely, the de Sitter
model requires the average density of matter to be zero. That
plainly won't do. So, the de Sitter model had to be abandoned.

People then set to work to discuss other models and a great
many were p posed. General theories were set up by F edmann and
Lemaitre. Among all these other models, there is one hich I would
like to call to your attention. It was proposed jointly by Einstein
and de Sitter. They joined forces and produced a new model, the
Einstein - de Sitter model, in 1932. This model gave distant matter
to be receding from us in the way that it ought to and also gave a
non-zero average density for matter, an average density roughly of
the correct value. This Einstein - de Sitter model is the simplest
model which is acceptable, which does not have some obvious flaw.

I would like to bring this Einstein - de Sitter model into
general consideration. I believe it is a very good model and that
it should govern the cosmological development of the future. It is
certainly the simplest acceptable model. It gives the universe
starting off at a definite time, the Big Bang, as it is often called.
It was a terrific explosion. According to the model, the universe
will go on expanding forever. World without end, as they say in
religion.

Many of the other models which have been proposed would require
that the universe expands up to a definite limit and then contracts
again. This is an unnecessary complication and I believe there is
no justification for it. I would like people just to stick to the
Einstein - de Sitter model, where you have an expansion which goes
on forever, although it is continually slowing up. It will get
slower as time goes on, but never actually stops.

The law of expansion fits in very well with the properties of
the microwave radiation. I could talk a great deal about this
development, but I don't want to go into technical details. I have
been working for a good many years on developing these ideas, using
a theory on the Einstein - de Sitter model, which I feel very satisfied with, although it is not yet definitely proved. I am hoping
that soon the proof will be obtained. I would like to stop at this
point and I shall be very happy to answer any questions.

THE ROLE AND VALUE OF THE SYMMETRY PRINCIPLES AND

EINSTEIN'S CONTRIBUTION TO THEIR RECOGNITION

Eugene P. Wigner

Princeton University
Princeton, NJ   08540

It is not good to receive applause ahead of the speech because
it indicates that a good speech is expected. But it is a pleasure
to speak to you and it is a particular pleasure to speak after
dinner, because people are less critical after dinner than before.

A Few Words About Einstein

Let me say a few words about Einstein. In his early youth, he
was interested both in very special and in very general problems.
He wrote an excellent paper on Brownian motion, that is the motion
of tiny particles suspended in water or in other liquids, and bom-
barded on every side by the atoms and molecules that are around it.
The analysis of this phenomenon permits one to obtain the size of
the atoms. And both he and Smoluchowski did that. He also wrote
a wonderful article on the photoelectric effect. Well, many people
who are not physicists do not know about the photoelectric effect,
but Einstein knew it, and he wrote such a good paper that he received
the Nobel Prize for that paper. These were specialized papers --
but his most general papers were, of course, on relativity theory -
the theory which he founded. It is interesting that the Nobel Foundation gave him the prize for his paper on the photoelectric effect
and not for his founding the theory of relativity. The reason for
that was, it is believed, that the latter was so revolutionary and
of entirely unusual scope and content, that not everybody believed
in it. And the Nobel Foundation wanted to be careful and not give
a prize for an accomplishment which may turn out to be incorrect
later. All this was in the early youth of Einstein. In later years,
Einstein's interest centered on general problems, general problems
of physics. And I will speak on those today.

Let me mention another characteristic of Einstein. He was ready to absorb knowledge, to absorb new ideas, to believe in them whether they constituted his own invention or the invention of others. He believed in the existence of quanta more firmly than the inventor of quantum theory, Max Planck. This is actually what enabled him to discover the law of the photoelectric effect – the discovery which helped to make him a recipient of the Nobel prize. And in addition to his paper on the photoelectric effect, he wrote several other interesting articles on the subject of quantum theory, some of them of a statistical nature. And, when Bose's article on a new kind of statistics appeared, he accepted it at once, fully.

Let me mention one more thing about Einstein. This is that his philosophical convictions changed enormously in the course of his life. In his early youth, when he founded relativity theory, he was a positivist. He felt that all that counted was what one can observe and experience in some ways. In later years, he had a firm belief in the existence of world lines, of definite positions of all objects, including particles, at definite times. And this kept him away from the part of physics which most of his colleagues, both in Berlin and in Princeton, were most impressed by. However, his friendliness, his cooperation, his cordiality were not influenced by his beliefs. They overwhelmed everybody because his friendliness and cordiality and desire for equality and simplicity could not be surpassed by anybody.

Three Basic Concepts of Present-day Physics

I will begin with something that is very general, and for physicists, very obvious. The three categories which physics used to describe the world and its events are initial conditions, laws of nature, and symmetries. The initial conditions describe the situation as it now exists. And about this, physics does say virtually nothing. The fact that the moon is, let us say there, and is now in a certain state motion, and the fact that there <u>is</u> a moon, are entirely outside the scope of physics. The present state of the world is not "explained" by physics. Our introducer is interested in geography. He knows how geography describes where is Carbondale; what is the vegetation around Carbondale; how are the hills around Carbondale. To physics, none of this is part of its science. There is one exception to that which I will admit; no, two exceptions: first, and this we often forget to mention, that all electrons have the same charge, that all electrons have the same mass. And that same applies for protons. In other words, there is some part of the structure of the world which has a high regularity. But, these are about the only "initial condition" regularities. Physics assumes, in fact, that the other initial conditions, the present structures of the world, is as irregular as conceivable except for what one can really view, and see, and experience. In other words, the atoms in me and the molecules in me are in as

irregular motion as possible. And, as the physicists among you
know, this is the basis for the derivation of the law of the increase of entropy and is the basis of the kinetic theory of gases,
for instance. But, except for these two principles, the initial
conditions are entirely outside the area of physics.

On the other hand, the next category, the laws of nature, are
supposed to describe the future if the initial conditions are given.
And Einstein believed that they completely describe it, he believed
that just as firmly as he believed in those other things which I
mentioned. He believed that if I knew the structure of the world
now I, or somebody who is perfect at calculating, could predict
all the future. Of course, the world is a very complicated system
and its full description is beyond human ability, but if you had a
relatively simple system and if it is an isolated system so that
it doesn't get disturbed by other systems, or by people, then its
future behavior is, he believed, completely determined. The laws
of nature which permit this prediction are supposed to be simple and,
in the words of Einstein, mathematically beautiful. And this is
wonderful. Well, let me say that, in my opinion, Newton's greatest
merit was to have separated the two ideas: initial conditions and
laws of nature. His theory implies that the initial positions and
initial velocities of the planets of the sun are outside the scope
of physics, but given their initial positions and velocities, their
future positions can be calculated. And he did that. This was a
great discovery. His predecessor, Kepler, still seems to have believed that he could get as good regularities for the size of the
orbits of the planets as he could get for the motion of the planets.
And he could get some approximate rules which tend to derive from
an earlier irregularity reasonably well. But the separation of the
two, of initial conditions and laws of nature, is really Newton's
accomplishment.

Similarly I would say that one of Einstein's greatest merits,
and it is again something that is not much emphasized, is to have
pointed to the importance of invariances.

What are invariances? The most important invariances, and the
invariances which Einstein recognized as such, are easily described.
Namely, first of all, that the laws of nature are the same everywhere. An example for this is that if I dropped something, such as
this key, here, it would fall the same way as if I dropped it in
another corner. This invariance seems easy to verify but is
actually less easy than it appears. We must remember that the laws
of nature apply only to isolated systems and the object I drop here
is not isolated - it is strongly under the influence of the Earth
which attracts it. However, the key will fall in the same way
wherever I drop it in this neighborhood because it is in the same
relation to the Earth all over this place. On the North Pole, which
is a bit closer to the center of the Earth, it would fall slightly

differently. This example also shows that it is not always easy to
verify the invariances. They only postulate that the laws of nature
are the same everywhere but the laws of nature apply, strictly
speaking, only to isolated systems and it is not always easy to
ascertain whether a system is isolated. Of course, the same applies
to the difficulty to verify laws of nature - in both cases quite
often one applies corrections for some small outside influences.
But in principle we all believe that the laws of nature are the same
everywhere in the world - actually, if they were not it would have
been virtually impossible for us to discover any.

The second invariance, perhaps even more natural, and more easy
to verify than the first one, is that of the time independence of
the laws of nature. If these laws changed from today to tomorrow,
we would not know how to influence events, how to aim a rock at an
enemy, or a ball toward a friend. They would proceed differently
from the way we learned yesterday. Our life would not at all be the
same - it would not be worthwhile learning anything. Of course, if
the change would be only a billionth of a millionth, as Dr. Dirac
suggests, it would not be so bad.

The third invariance is what is called rotational invariance:
that is, that if the succession of two or three or any other number
of events is consistent with the laws of nature, then the same
events subjected all to the same rotation around any axis in space
are also consistent with the laws of nature. This invariance is
also evident to us if the axis of rotation is vertical. If I see,
for instance, a rock's path in the air looking to the North, I will
take it for granted that another path of the rock, which has the
same form if viewed in any other direction, is also possible. The
direction of rotation must be vertical because the Earth is under
us and if we want to prove the invariance with respect to other rotations, we must move to another part of the Earth so that its
position has the same relation to the path of the rock as it had in
the original experiment.

All these invariances appear natural to all of us, particularly
if we do not think of experiments carried out here on the Earth,
which by its attraction influences all common events, but in empty
space. Surely, if a succession of events by parts of an isolated
system is possible at one place, it is possible at all other places
and the same applies to the succession of events obtained by rotating
all objects of the original experiment about any definite axis by any
angle. The same applies to any geometrical figure: these can be
displaced in space (and also time) and rotated without changing any
length of the figure or any angle.

The fourth type of invariance is the one which Einstein re-
established in his special theory of relativity and modified somewhat in his general theory. It is very much less natural to most

people than the first three types and was, in fact, for a period
repudiated even by the physicists. It postualtes that if a set of
events is consistent with the laws of nature, then a similar set
of events is also possible if all objects forming the isolated system which produces the original set of events are put into a uniform
straight-line motion in any direction and at any velocity. The
events of this latter moving system will appear to be the same to
an observer moving along with it as the events of the original system appear to the observer at rest. Thus, the relation between a
uniformly moving system to one "at rest" is just the same as the
relation between two systems rotated with respect to each other or
displaced in space or time. The lack of belief in this invariance
was responsible for a good deal of the opposition to railroads - it
was feared that people inside the cars would feel terribly uncomfort-
able. The cause of the skepticism of the physicists had an entirely
different reason: they believed in the existence of an "ether" which
carries the electromagnetic radiation, the light, and the motion with
respect to it should be observable. And indeed, the transformation
of the space and time coordinates between observers moving with
respect to each other is (as given by the Lorentz transformation)
much more complicated than was anticipated. It was Einstein's
desert to have fully accepted the implications of these transformations and to have agreed to the fact that (in general) two events
which appear to be simultaneous from the point of view of a moving
observer but should not be considered to be simultaneous by him.
He should interpret his observations in exactly the same way as does
the observer at rest. In fact, since there is no difference between
the laws of nature as they appear to the two observers, there is
little point to call one as being at rest, the other one to be moving. It is more appropriate to say that they are moving with respect
to each other and not to say about either that he is "at rest". Of
course, the concept to be at rest with respect to a definite object
is a valid one - all of us here are at rest with respect to this
part of the Earth but there is no absolute "at rest" as the assumed
(and now denied) existence of the all-pervading ether would imply.

But, having recognized all this and having accepted the Lorentz
transformations also as conceptually valid, in other words to have
created the special theory of relativity, is only one of the fundamental realizations of Einstein in this connection. The fact to
have recognized the general nature and the importance of invariance
principles (that is, of symmetry laws) ranks, in my opinion, about
equally high. It can be compared with Newton's recognition of the
fundamental difference between initial conditions and laws of nature
- the importance of which was mentioned and emphasized before. I
may mention in this connection that the first physics book I read
did not give the invariance principles explicitly even though their
existence was evident from what the book did bring out. It gave
only two pages to one of the most important consequences of the
invariance with respect to rotation, to the conservation law of

angular momentum, without mentioning the symmetry principle from
which it can be derived. We learned about the basic importance of
these principles largely from Einstein. The recognition of the
connection between symmetry (that is, invariance) principles and
conservation laws is usually attributed to Klein and Noether. I
believe that Hamel recognized it also, even before.

Extensions of the Area of Physics

There were three changes in the history of physics which I consider of utmost importance. The first one, as was mentioned before,
was largely due to Newton. The Newtonian theory described the motion
of the planets, once the initial conditions were given, with amazing
accuracy. Newton also recognized the existence of all the invariances, including the one which was for some time abandoned because
of the assumed existence of the ether. The "events" in his theory
were the positions and velocities of the objects, in particular also
of falling objects here on our Earth.

One great accomplishment of Newton's theory was that it gave
two entirely different phenomena the same basis. These were the
motions of the planets, in particular of the moon around the Earth,
which is essentially a circle, and the objects falling here on the
Earth. It should be admitted, though, that his theory gave a close
description only to those phenomena in which only gravitational
forces play an important role.

The next great discovery which I wish to mention culminated in
Maxwell's theory of the electromagnetic forces. His equations also
give a common basis to apparently very different phenomena, such as
the electric attraction and repulsion (Coulomb's law), the magnetic
attraction and, most importantly, the phenomenon of light and its
propagation. His equations constituted an enormous extension of
the area of physics - they enabled us to describe a wealth of phenomena in which electromagnetic forces also play a role - in many,
if not most cases, a dominating role.

In the time of Maxwell, in fact up to the end of the 19th
century, physics was concerned almost solely with macroscopic bodies
and phenomena - as I often mention, the first physics book I read
said that "atoms and molecules may exist, but this is irrelevant from
the point of view of physics." And this was true at that time and,
as a consequence, the properties of material bodies were not the
subjects of physics. If one wanted to know the density of aluminum,
one looked it up in a handbook but this, and the properties of
materials in general, were not parts of physics. This has changed
drastically as a result of the next great development in the
history of physics, the establishment of quantum theory. This permits us to calculate the properties of atoms and molecules, and
hence also the structures of macroscopic bodies, though the

calculation is not easy and needs insights and also mathematical skill. But, as Dirac said, the properties and behavior of ordinary materials, under normal conditions, are implicitly contained in the now well-known equations of quantum theory. The establishment of this theory was the third significant extension of the area of physics.

Circumstances in Our World Which Made
the Development of Physics Possible

That laws of nature could be discovered by man is truly a miracle. It is a miracle in two ways. It is a miracle that we can think so deeply and abstractly that we could, for instance, create a geometry. This, created largely by the Greeks, was, I believe, the first result of abstract and deep thinking. If one tries to prove, without knowing how to do it, that the three altitudes of any triangle have a point in common, one needs very great skill to do it. That man can put one argument after another, and a third argument after the second one, and he is still correct - this is simply wonderful. This, in fact, distinguishes him from all animals. And it is not clear how we acquired this skill - we do not need it to maintain our life. But, we can do it.

The second miracle is that situations were presented to us in which the consequences of the laws of nature are amazingly simple. The first such situation is when only gravitational forces play a role as in the case of the motion of the planets. On the motion of the planets the electromagnetic forces, and all microscopic forces, have virtually no effect. And that made it possible for Newton to discover the gravitational law because the planets' motion, influenced only by gravitational forces, obeys simple regularities. It is a wonderful thing that we were given such a situation, a situation in which only gravitational forces play a role.

The next miracle was the possibility of a totally macroscopic physics which was certainly fostered by Maxwell's electromagnetic theory. And, as I mentioned, when this happened the first physics book I read said that in physics we don't want to consider microscopic objects, atoms or molecules.

The fourth was, of course, the discovery of the microscopic events, which are not, as I have mentioned and that is very important to realize, are not influenced by macroscopic systems. If somebody should send an electron with a reasonable velocity in vacuum in a certain direction, it is essentially uninfluenced by anything, except by the electromagnetic force which accelerates or deflects it. And this is, as I mentioned, also a miracle which made physics and the physical sciences possible: the existence of situations in which the laws of nature manifest themselves terribly simply. And the knowledge of these laws, as we all know very well, has changed the life style of man fundamentally.

Are we close to having discovered all laws of nature? This is
the next subject I wish to discuss. But before doing so, let me re-
emphasize the important role which symmetry principles played in the
discovery of the laws - surely all approximate - in the past. As I
mentioned, if they were not present, if the laws of nature would
change from day to day, we could not have discovered them. And the
other symmetry principles were - and continue to be - also of great
help. Not only those which we take for granted, the first three
which were mentioned, but also the fourth one which Einstein reestab-
lished. It is basic for the attempts toward the first and most popular exploration to be mentioned in the next section, and we believe
firmly in its validity, at least on the non-cosmological scale.

On the other hand, we must admit that we believe in Lorentz
invariance - and also in the other invariances - only on the non-
cosmological scale. The general relativity theory has its own invariances which are not in conformity with those discussed above.
The conflict of our symmetry principles is even more acute with
Dirac's new theory of the change of the gravitational constant on
the cosmic time scale. Even the invariance principles seem to be
epoch-dependent. Mach, in a sense, anticipated much of this.

I have emphasized the enormous help which the invariance principles have provided and do provide in our efforts to extend the
area of physics. Have they misled us in the past, even on the non-
cosmological scale? I know of only one example. About 20 years ago
it was firmly believed that if a set of events is possible, that is
compatible with the laws of nature, its mirror image is also possible. Now we know, as a result of the observations of Lee and Yang
and subsequent experimental confirmation of these observations, that
this is not true. But, when the principle was still accepted, it was
used to refute some experimental results, i.e., those of Cox, which
later turned out to have been correct. Cox observed some beta dis-
integrations and people pointed out that if the disintegration has
those characteristics, it is not reflection invariant, it contradicts
a symmetry principle and the observation must be incorrect. As a
result, Cox withdrew his observation. And it was a correct observation. In other words, the symmetry principles can have also un-
favorable effects, but so far, usually, they were very, very helpful.

What Areas Remain Unexplored?
Which Should Be Explored?

As Dirac so well said, the properties and behavior of materials
under ordinary conditions are well described by quantum mechanics -
which implies that the constituent particles are those present in
ordinary matter and their energies are not too high, surely not
large as compared with their rest energies. And a great deal of
effort of physicists is now directed toward the goal of overcoming

these limitations, in particular also toward a full union of quantum
and relativity theories. Much has been accomplished in this direction, yet, in my opinion, we are still far from its full achievement.

Are there phenomena which we encounter in everyday life which
are also outside the area of present-day physics? Indeed there are:
present-day physics does not describe the fact of our thinking, the
existence of pleasure and pain, our desires, that we know, we love,
we feel. It does not describe life. In fact, and I won't do it,
but it is easy to prove on the basis of present quantum theory in
about five lines, that the existence of consciousness is contradictory to present-day physics. I won't prove it as some people know
the five lines and I don't want to put down now any elaborate
equations.

Will man ever be able to establish a theory which gives a
description of these phenomena together with those which form the
subject of our present physics? I do not know. But earlier in our
century many questioned man's ability to understand microscopic
phenomena even of inanimate systems - and this we have managed.

Are high energy phenomena and life the only two areas not fully,
or not at all, incorporated into our physics? Again, I believe, we
do not know. It is quite possible that entirely new phenomena will
be discovered, perhaps a large number of them, and man will try to
extend science further, so that it encompasses the description of
these phenomena, just as it was extended in the past to the description of electromagnetic forces, microscopic events, strong interactions. We do not know what the future will bring and this is good.
And I am convinced Einstein was in agreement with this view.

STATISTICAL CONCEPTS IN EINSTEIN'S PHYSICS

E. C. G. Sudarshan

Center for Particle Theory
University of Texas at Austin
Austin, Texas   78712

It is particularly appropriate that we are gathered together
not so much to analyze the work of Albert Einstein, but to undertake
a critical appreciation.

In the Indian tradition it is said that in fact if there is any
ultimate entity, that ultimate entity is the Master: master in relation to disciple. And if there is a master in physics, Einstein
certainly commands that place: there are so many areas of physics
that he has contributed to so decisively and with such clarity.
In fact it is almost impossible to believe that one human being
could have done all that he did.

I have chosen to talk about a part of Einstein's work which
has fascinated me for a long time. My teachers taught me two things
about Einstein's contributions in this area: (1) that he had contributed significantly and systematically to statistical physics
and (2) that he simply did not agree with the contemporary inter-
pretation--statistical interpretation--of quantum physics. It
interested me greatly to examine why it is that the person who has
contributed so significantly to statistical and to quantum physics
is unwilling to accept a statistical interpretation of quantum
physics.

But in the course of these studies, thanks to the help of many
of my distinguished colleagues who know a great deal more about the
history of science and about statistical physics in particular, I
have come to realize that in fact much of what Einstein has done
was done with the motivation which was always very clear, namely to
elucidate the unknown through statistical assessment of the physical situation.

Albert Einstein's statistical physics contributions were of fundamental importance in the elucidation of atomic and quantum phenomena. It is still curious that Einstein, who contributed so definitively to the foundations of quantum theory, took exception to the generally held interpretation of quantum mechanics.

The first group of statistical papers of Einstein began at the turn of the century, 1902, with a paper on the quantum theory of statistical equilibrium and the second law of thermodynamics. This paper dealt with a method of trying to relate statistical concepts, and quantitative statistical mechanics to thermodynamics. Of course this general idea was underlying the derivation of the Maxwell distribution of velocities: a description of a statistical state, an imprecise (mixed) state of a collection of particles being put in correspondence with a state which was described in an entirely different language in terms of temperature and pressure. But this correspondence was always made with regard to a very specific and particular system, not with regard to dynamical systems in general. Einstein's treatment here is very curiously parallel to the work of the great Josiah Willard Gibbs who had done this work about a year earlier; and obviously the two men worked quite independently of each other. The general idea here was that instead of dealing with the mechanical system as it was, you dealt with many, many identical copies of this entity--the so-called ensemble--and the ensemble, that is the system with many, many identical copies, this surrealistic entity, became a substitute for the original system. This new system was then put into one-to-one correspondence with the system which was the object of the thermodynamic description. We now have a connection with the mechanics of the system: the evolution of the system was completely governed by the equations of mechanics. On the other hand, we had something that conventional mechanical systems did not possess, namely, a variety of physical states--variety of mechanical states--<u>all at the same time</u>. It is not one after another one, but all of them at the same time. So that the thermodynamic system was imaged by <u>not one system but many</u> copies of the same system.

This was followed the next year by another paper which is also now standard knowledge--so standard in fact that at the present time it is a shock to think that there was a time when this was not known--called the theory of the foundations of thermodynamics. In it the notion of ensembles of states was reexamined and one considered the possibility that an ensemble could be thought of for certain purposes as the same system in passage through time. Since the system was in statistical equilibrium there was going to be no long-range change, but there could be short-range fluctuations. The assertion was that, taking the time average of a (sufficiently complicated) mechanical system over a long period of time was more or less equivalent to taking the ensemble average by taking many copies of the same system.

But all these papers were, in a sense, the prelude to the real work, and the culmination was in the beautiful and celebrated paper on Brownian motion in 1905, that extraordinarily miraculous volume of the Annals of Physics in which there were the three famous papers by Einstein.

This paper and its sequel, called "On the Theory of Brownian Motion," in the following year are of particular interest to the physicists who are gathered here because one might say that this was one of the earliest examples of the direct application of symmetry principles to dynamical evolution and dynamical configurations. In talking about Brownian motion, unlike usual mechanical systems where the forces acting on a mechanical system have a definite direction and a definite magnitude, here we have a situation in which particles (pollen grains or other particles) are suspended in a liquid; the liquid is supposed to consist of a lot of restless molecules, and therefore they are jostled around all over the place. But no direction is preferred to any other direction, and therefore, the forces on the suspended particles due to the molecules--the net force--should have no preferred direction.

Now in a pure mechanical situation, it is impossible to deal with such a treatment. If you don't know anything, then you don't know anything further. That is too bad, but it cannot be helped. But when you come down to dealing with ensembles, the situation is quite different: If you don't know which way things go, you make one to go each way! To be on the safe side, you take out insurance. You deal with every direction--send one member of the ensemble in each possible direction.

While it looks like a very wasteful and very expensive way of doing things, in the long run such a system eventually reaches a certain equilibrium in which the ensembles do not change any further--and at that time we say an equilibrium has been reached. So the theory of Brownian motion may be said to be a step in the direction of application of symmetry to a mechanical system.

You will hear from Professor Wigner and Professor Dirac about symmetry with much more authority and much more elegance. We do know that in quantum mechanics symmetry considerations play a more fundamental role because a (pure) state can be symmetric without having to resort to ensembles. It was a fortunate thing that we didn't discover quantum mechanics too early: we may not have discovered the proper treatment of Brownian motion as an example in which two apparently contradictory aspects were reconciled.

One aspect which was the aspect of symmetry said that for the forces that are to be exerted on a particle buffetted by myriad molecules of a liquid no direction is preferred to any other direction. The other aspect is the law of mechanics which states that the change of the momentum is in the direction of the applied force and proportional to it. How are we going to talk about the motions in mechanical systems with a force in which the direction is unspecified? The answer is: Let us invent a mechanics which goes outside the usual framework. When two natural principles cannot be reconciled, it is a signal that something is missing from our perception of the system. We must enlarge our model so that a new model may come about.

Probably people were more courageous and inventive in the 1900's, because it appears that Einstein had no hesitation in inventing an entirely new model, the new model being one in which you had both the symmetry and the mechanical equations satisfied at the same time.

If symmetry (which has the popular association with Einstein's works in relation to what we now call the Poincaré group or the theory of special relativity and the theory of general relativity) could be related to statistical concepts I'm certainly not going to let you go without trying to tell you that all the other important work that Einstein did at that particular time was also very closely related with <u>statistical</u> concepts.

Boltzmann had found some decades earlier that there was a relation between the possible arrangements that were associated with a certain system (what is called the statistical probability, a statistical mechanical concept about configurations in a mechanical system) and the thermodynamic concept of entropy or disorder--that the entropy was, apart from a certain unit of measurement, (which we now call Boltzmann's constant), proportional to the logarithm of the statistical probability. There is a bridge between the two: if you knew the statistical configuration and the number of possible complexes that you could make, then you could find out how much disorder there could be in the system--how much entropy there could be in the system.

That's of course not very different from our conviction that if there are ten ways in which your books could be disarranged, then anyone who tries to clean up your books, obviously, disarranges them in all those possible ways.

We could now ask: Instead of starting with the statistical mechanical end, and then trying to find out the thermodynamic correlates like the entropy of the mechanical system couldn't we invert the order? After all an equation could be read from left to right, but it could also be read from right to left. If we do that it suggests a means of finding out something about the mechanical aspects of the system from the thermodynamic properties of the system.

In the case of a collection of gas atoms of the kind that we had talked about earlier (Maxwell distribution of velocities, etc.) we know what the mechanical system is and therefore we are anxious to find out about the thermodynamics of the system. But there are other systems in which we know more about the thermodynamics and we did not know or do not know much about its mechanical structure. And honorable mention must be made of the blackbody (a heated cavity from an enclosed area of space which is kept at a steady temperature). You peep into it through a small hole or allow the light to come out through a small hole. The blackbody has a familiar pattern of radiation, of light and heat and other electromagnetic waves inside it, which is characteristic of the temperature and independent of the kind of cavity. Max Planck, at the turn of the century, had at last solved the problem of the distribution of energy--what energies were available, what was the energy in each special range of wavelengths and frequencies and also how it changes as a function of temperature.

In developing his theory Planck introduced us to the concept of the quantum. The discovery of the correct law of blackbody radiation is synonymous with the quantum revolution. That blackbody radiation could be explained is an interesting, but not very compelling news item! But the birth of quantum theory, the introduction of the quantum: that was of course big news!!

Planck had taken the first step in this direction by observing that you had to get a radiation law, which was intermediate between two laws which were known to be approximately true at two ends of the spectrum--the law of Rayleigh and Jeans which was derived from the equipartition law and the Wien radiation law which was derived from entirely different considerations: both of them are approximately true in the two extreme ends of the spectrum. Planck had tried to find something in between, and he had made use of thermodynamics as a means of generating insight into this particular structure.

Einstein made use of a similar path to find out what exactly are the quanta of Planck. Are they real entities, or are they a means of deriving certain results? Are they only manifesting results when there is an exchange of energy between the oscillators and the field, or were the quanta substantial entities in the sense in which we said atoms and electrons were there, or was it only there in the sense of somebody writing down some equations. Einstein set out to find out. If it consisted of a collection of particles, there would be certain properties of the system that one could calculate; he calculated the probability of the radiation contained inside a volume being compressed to a fraction of the volume and found that this particular probability was proportional to a power of the fraction of the volume. If the volume was 1/2,

it was a certain power of 1/2, and the power was given by the total
energy divided by the energy of a single quantum. So if there were
quanta, this ratio was the number of quanta.

Now that sounds as if these quanta are really there. Because
each quantum could have been anywhere, with 50% probability of being
in one half and 50% in the other half, if you have two of them, then
it would be the square of 50%, that is 25%. If there were three of
them, it would be 12½%, etc. Having found this result, he was more
or less convinced that these quanta were there. Therefore it was
only a natural step to say, if the quanta are there, not only with
regard to the exchange of radiation with oscillators and giving rise
to the Planck radiation law, but also with regard to the fluctuations of energy density inside the cavity, then these quanta are
really there and when they come out of the cavity, they come out as
quanta. In that case we must find manifestations of these quanta
doing something in the interaction of radiation with matter.

It is very natural that these particles of light must manifest
themselves in the photoelectric effect, and this led to the very
celebrated paper of the application of quantum theory to the photoelectric effect. It's also in the same issue of Annals of Physics
in 1905.

Einstein did not stop at this point. Planck's quanta were for
radiation, they were manifesting themselves in various aspects of
radiation, but there was a question: Is this a quirk of radiation,
or is the appearance of quanta valid for all kinds of matter?

If it were something which is manifesting itself in phenomena
which involve radiation, as well as in entities which did not involve radiation, then it was a general principle of physics, and
therefore something which was going to revolutionize all physics.
So it was natural to search for a process in which radiation was
not directly important: a good clue to fruitful areas would be
to look for unusual and unexplained phenomena.

One such was the behavior of specific heat of solids. The
Rayleigh-Jeans law was good for sufficiently long wavelengths for
the blackbody radiation but not for short wavelengths, and was based
on the equipartition law. If you applied corresponding ideas one
deduced that the specific heat of all solids, when expressed in
molar units had to be a constant, because the number of degrees of
freedom that were associated with the solid could be easily counted
and if you took the same number of particles of different solids,
they had the same number of degrees of freedom. Since the energy
associated with each degree of freedom at any particular temperature
was fixed by thermodynamic considerations, when you had the same
number of particles, same number of molecules, in the solid, you

should have the same specific heat--the so-called Dulong-Petit law.
All solids seem to obey this particular requirement at room temperatures or higher. When the temperature came down, the solids seemed
to show a remarkable decrease of the specific heat, as if certain
degrees of freedom were being frozen. How do they freeze? What
was the mechanism of this freezing of the degrees of freedom? If
the quantum had anything to do with this one, then the quantum
better appear. So Einstein applied Planck's theory of radiation
to the theory of specific heats of solids in a paper published two
years later.

In this paper, Einstein showed that the specific heats decrease
rapidly with temperatures below a certain value even though the
actual result that Einstein derived for the behavior of specific
heats of solids showed too fast a decrease; but for the first time
it showed that the quantum was something that cut across the whole
domain of physical phenomena, not just for radiation.

We see a revival of this particular idea--this particular
theme--of recognizing that if the quantum is there, it better be
relevant for everything. (Very much like the simple but powerful
point of view that we would like to propagate at the present time,
namely that physics being the all-encompassing science there should
be no domain of human experience in which physics is not relevant!)
Accordingly Einstein decided that it should be possible to be able
to apply the notion of the quantum to study the induced and spontaneous emission of radiation by atoms. If atomic systems were
excited and in the process of excitation they de-excited, and emitted
radiation, the same physical process of course could be reversed so
that the atom in the ground state together with a little bit of
radiation could get excited and go to the higher state. If both
these processes are continually taking place eventually we would
reach thermal equilibrium; and then we not only find out what is
the probability of the atom being excited and how much of it being
in the ground state, but also how much radiation there is. But we
already have a theory of the radiation, the Planck distribution
law. Question: How shall we reconcile the Planck distribution
law with this particular dynamic mechanism? Planck's method was
a static method, a method in which you considered the equilibrium
state, but not the dynamic processes which are going on. Here you
ask the question: How do we derive the Planck law from considerations about the dynamic equilibrium? Einstein discovered that for
a two-level atom one had to include, in addition to the process of
emission and absorption stimulated by electromagnetic radiation,
also a spontaneous emission process enabling the atom in the
excited state to decay spontaneously. Therefore an excited atom
has two ways of decaying: (1) All by itself without paying any
attention to anybody else (spontaneous emission) or (2) stimulated
emission in which there is already existing radiation and this

radiation makes the excited energy come down as part of the interaction of the radiation with the atom.

Classical theory of course was quite good at considering the
second process: a charged particle could change from one of its
configurations to another one of its configurations. And that
process has the reverse, namely that if it is in the ground state--
the lower energy state of the atom--the atom can pick up energy from
the radiation that is existing and go to the upper excited state.
But in addition to this one there was the new process of spontaneous
emission, and Einstein was able to derive Planck's law of radiation
by deducing the ratio of the probabilities for spontaneous emission
and induced emission. Though this demonstration was carried out
for a very limited case, and though we knew the Planck law already,
it was important to demonstrate that the quantum principle was
applicable not only to radiation or to solids, but also to atoms in
interaction with radiation.

This work was completed about a decade or so later by S. N.
Bose who generalized it to dealing with an arbitrary number of
material bodies, arbitrary number of collisions, and arbitrary
spectrum of energy levels. Thus the Boltzmann method of dealing
with dynamic equilibrium could be generalized and adapted to the
case of radiation interacting with matter. The result is the
Einstein derivation (as a special case of the Bose derivation of
general radiation equilibrium).

But the important thing by this time was the fact that the
quantum had been shown to be playing the central role not only in
the process of blackbody radiation, but in several domains of
physics.

Lest you think that I am simply being carried away by the
auspicious occasion of the Centennial, and think that all the
things that could possibly be done and some more were done by
Einstein, allow me to mention that I do notice a curious lacuna
in the developments around this time. It is curious that the
applications of the specific heat of solids which Einstein started
was completed only a decade or so later by Debye, who introduced
what we would today call the phonon picture of a solid (namely that
the solid really could be effectively replaced for many purposes
by a collection of sound waves and that these sound waves were the
elementary low-lying excitations). I find it very strange that
none of the physicists at that particular time seemed to look at
the very close similarities between the blackbody spectrum which
had an energy density which is proportional to the 4th power of the
temperature (and therefore a specific heat or a heat capacity--
thermal capacity--which is proportional to the 3rd power of the
temperature); and the behavior of solids which also had the same

temperature dependence. If you look at Debye's theory, which was done in the 1920's, and Planck's theory, which was done in the 1900's, you see that in fact the mechanisms are quite similar. In fact, people very often write the same letter "c" for the velocity of light and velocity of sound so that they look even more similar. I am puzzled why this close connection was not recognized by anyone in the first two decades of the twentieth century!

A major group of contributions to statistical physics was the theory of the ideal Bose gas. The fundamental idea for dealing with the ideal Bose gas came from the Indian physicist Satyen Bose, in a very sketchy but remarkably personal letter to Einstein. Bose pointed out that if one were to think of the photons (the light quanta) as particles, they were not ordinary particles, but particles which were so strictly identical that they were indistinguishable; we must pay proper attention to the indistinguishability by saying that only complexions (arrangements) of photons were important, not which photon was where. If two identical photons exchanged places, it would not be a different state. This method of counting is somewhat different from the counting that we normally use; but Bose showed that if we used such a method of counting distinct states we could derive the Planck law of radiation.

Going back to the analogy that I mentioned about books being disarranged, somebody who did not know what you are working on or what your habits are or what ideas are important, would see a lot of sheets of paper distributed on the table and collect them all nicely together and arrange them all in one neat pile, little knowing that in fact there is information on those sheets: not only on the writing on the sheets, but the manner in which they are displayed. They are supposed to inspire you by being an ensemble of sheets, rather than a pile of sheets. For a person who does not see the distinction, there is in fact no difference. It would be foolish for anybody to have a collection of blank sheets of paper or sheets which are identical--almost everything written the same way-- which were displayed on the table and then complaining that by someone tidying his desk by piling them up you have lost the order in which they were placed. If on the other hand they were distinguishable in that some had more interesting doodles than the others, some had formulae or references or jokes written on them: then of course we can say that there has been a disturbance of the order of the papers; but if the sheets are strictly identical, the order of them should not make any difference, because nobody can tell them apart. This was the idea that Bose had brought up.

Einstein immediately recognized not only that this was a new method of derivation of Planck's law but that it was also a new property of light quanta--that light quanta were particles but they were not ordinary particles because there was a new law of counting the statistical probability of any possible configuration. By

recognizing this difference, one finds a new order of matter, not
in the composition of the components, but in the interrelationship
between the components. In a true sense of the term one may say
that this is the beginning of the appreciation of the notion of
correlation between things which do not interact with each other.
What could be more uncorrelated and independent than a collection
of particles?

Previously whenever we had a collection of particles we recognized that if they are interrelated then they must be interacting--
they must be exerting forces on each other. But the quantum particles are doing something quite different. They are correlated, but
they are not interacting. Their interrelationship was not by inter-
action--not by forces--but by a certain bond of affinity, a certain
kind of complexion, a certain kind of correlation between the particles. Einstein recognized that this particular affinity could be
made the beginning of a general theory applicable not only to light
quanta, but also for all matter. Because the quantum was not restricted to radiation, the quantum was cutting across all of physics;
if identical particles of light were to be treated in a certain
fashion, we should try to treat other identical particles also in
the same fashion.

The generalization from the treatment of light quanta to the
treatment of a general gas, required two more identifications. One
was that light quantum being light quantum was very light--it has
no mass, and therefore it had a certain relation between energy and
momentum. The energy was equal to the magnitude of the momentum
multiplied by the velocity of light. But if you wanted to apply it
to some other Bose gas--helium molecules--then the law that you use
should recognize that the helium atoms were rather heavy and moving
rather slowly, and that the relation between energy and momentum for
such a system was different.

The second thing was that light quanta had the property that
there could be creation and destruction of photons. There was no
particular reason in arranging for equilibrium to make sure that the
number of all the particles was preserved, but when you deal with
something like helium gas, the number of helium atoms are to be
preserved. You may rearrange their energy, but you have no business
to make them disappear. Otherwise it would be like cleaning up a
desk by throwing some papers away. You are told, no, no, you can't
throw anything away, you can't hide it, you must keep all of them,
you must really rearrange them!

This modification is introduced by a technical device called
the chemical potential. Whenever we consider variations, we have
to say that if some particles with high energy are moved out to
lower energy, then the number of lower energy particles would

increase, but the total number of particles remaining constant. Introducing the chemical potential and introducing the finite mass and the corresponding relation between momentum and energy of these particles, we may generalize Bose's idea to construct a theory of the ideal gas. This theory is now of course quite naturally known as the theory of the ideal Bose gas; and the particles as Bosons.

In all these things, the probabilities that one dealt with were classical probabilities, even so it was the harbinger of quantum theory. Much of the notion of probabilities, except the idea just mentioned which Einstein adopted, enlarged and generalized, nevertheless were still classical probabilities. By this I mean that they obeyed all the things that probabilities were supposed to obey: Probabilities by their very nature must be nonnegative quantities, usually nonzero quantities, the sum of all the probabilities should sum up to 1, and no probability could be larger than 1, nothing can be smaller than 0.

This meant that of course any bifurcation, any uncertainty with regard to the lack of determinacy with regard to the forces acting on the system or the dynamical processes was going to make the ensembles, the collection of realizations, larger and larger. I could make them split into many possibilities, but I could not recombine. So all the statistical processes had a certain direction; and being sensible people we always take the directions from the past to the future that disorder and disarray always increases. A very disheartening way of looking at things, if we thought that all human experience is subject to this law that the total disorder always increases!

It appears in all cases in which the probability was applied to a complex system which has many, many component parts for which the underlying dynamical law of the pure mechanical system was purely deterministic, it is only as a simplification of the complex dynamics that we use statistical methods. It appears to me that Einstein believed that physical law and physical states were respectively causal and precise. Physical law was precise; physical states were precisely specified, and the statistical mechanics that was applied was therefore a device for dealing with this particular complex system. Therefore statistical methods were to be used to deduce properties of the complex system which were hidden from us, which we could not directly perceive by looking for the regularities of the large entity that we have. It seems to me that Einstein would not have indulged either in the idea of intrinsic probability of all processes nor the notion of quantum probabilistic dynamics. Both would appear to him to be playing dice.

This mistrust of quantum probabilistic interpretation has been with him apparently since the early days. For example at the Solvay

conference of 1927, his remarks seem to indicate that he was deeply
disturbed by this. But the clearest exposition and the pinpointing
of this critique of the standard interpretation, the statistical
interpretation of quantum theory is in the paper titled "Can quantum
mechanical description of physical reality be considered complete?"
by Einstein, Podolsky and Rosen. It is written somewhat late according to this calendar of events that I am describing to you, namely
in 1935. The questions raised there are fundamental and demonstrate
the nonlocal character of quantum mechanics and have been restudied
extensively during the past few years. It seems to me that there is
a great revival of interest in questions of this kind, because there
is a suspicion that one had misread what Einstein had written. He
was not objecting to the existing theory, but was pointing out simply
that we had not drawn enough conclusions from it. It would be as if
we had been given a Christmas gift but we simply looked at the
wrapping paper and did not bother to open the present and find out
what was inside.

For those of you who are not familiar with physics of the
quantum systems let me just make a two-minute presentation of what
is the problem with the statistical interpretation. It is believed
that a quantum mechanical system is a system in which you cannot
simultaneously specify all attributes that it potentially possesses.
For example, in ordinary physics when we talk about a particle which
is moving it is almost axiomatic that we must be able to measure
both its position and its velocity, both its position and its momentum. If we cannot measure them, it is because we are either not
very good at measuring things or that we don't have the right kind
of apparatus (or we don't have enough research funds to deal with
it), but that in fact if all these things were available there should
be no difficulty of measuring it to arbitrary degree of precision:
that in substance there exists a thing to which we are approximat-
ing: that the system could, in principle, be measured.

Quantum mechanics seems to introduce a system where this is no
longer possible, that the dynamical attributes of position and momenta are there for the quantum particles but that you could not
make both of them measurable at the same time. If you tried to
measure the position very accurately, then the experimental arrange-
ment, the interaction with the system and the method of measurement
make it difficult to measure the momentum. If you measure the momentum precisely, you cannot measure the position accurately.

If you want to say that you have a certain amount of fog, we
must have the fog extend over a certain region. We cannot say the
fog is only at one particular point. By the very nature of fog,
it is rather foggy. It must extend. If you talk about a wave, or
a traffic jam, they too involve an extended object. And it is not
that you cannot measure this more precisely, but that in the very

nature of things you could not make the entity more precise. It
is not like a treasure mark in which there is a cross saying the
treasure is here. A fog is not here or there, a fog is all over
the place. The traffic jam is not at this particular traffic light,
but all around it. The quantum system seems to be something of this
kind; therefore the natural conclusion is that a quantum system is
an extended system. But when you try to detect the quantum particle,
say an electron, or alpha particle, or a light quantum, by a suitable
apparatus, you can find it at one point and one point only. So if
we were to think that it is extended, we are wrong: It is at one
point. When you think it is as if it were at one point we are wrong:
It is extended. You can't win; either way you are going to lose.

Therefore we say that a quantum mechanical system is intrinsi-
cally uncertain, the uncertainty quantified by Heisenberg's principle
of indeterminacy. Very often we can think of quantum experiments
which possibly have only spectacular outcomes which can depend not
only on the initial conditions that we set up, but also on chance.
Sometimes it may happen; sometimes it may not happen.

There is a very deep difference with regard to the quantum
probability: they come about not through the mechanism that we had
in classical theory of taking an ensemble, seven of them going this
way and three of them going that way, but by introducing the notion
of what may be called the probability amplitude; you square the
amplitude to get the probability. But the great advantage of a
square root of probability is that it does not have to be positive.
If you take a positive number and square it, you get a positive
number; if you take a negative number and square it you still get
a positive number. Therefore a square root can be positive or negative. Actually in quantum theory it turns out that the square is
really the absolute value squared of a complex number; the probability amplitude needs not even be real. But being a complex quantity, you can have two probability amplitudes, either one of which
by itself would have given you a distribution, but the two of them
added together produce a probability in which everything is concen-
trated at one point. And therefore the quantum probability is a
new kind of probability. It is this kind of probability that
Einstein was unwilling to accept, because it appeared to him that
either the lessons of the quantum were not fully understood or
quantum theory was a provisional theory, a theory which was describing only a limited level of the description of the system. In statistical systems when we employ the probability densities, etc., we
were making a simplified description of a complex system, with an
aim to elucidate some properties of the complex system which we
could not handle in its entirety. But quantum theory as we understand at the present time is not a simplified but incomplete
description; Einstein said that completeness of description and
probability do not go together. Either the quantum is not a final

and complete theory--but only an approximate and provisional description, or we must not talk about probabilities in the manner in which we talk about it.

It is interesting to note that the use of probability amplitudes furnishes new possibilities and problems of interpretation. In classical probability it was always the density of ensembles which was averaged over; for example, taking forces which are completely symmetric was not possible except by employing ensembles. So you had to get a nondeterministic situation. You had to get a distribution before you could get a group average which is rotationally invariant, something which is the same as seen by all different people with their different orientations. The price of democracy in this case was impurity. But quantum theory with its probability amplitude provides us with a new possibility. We could have a state which was invariant under rotation, which was the same looked at from all possible directions, or a state which was rotationally invariant, but which was nevertheless a pure state. For example, to the extent that you can neglect the spin of the electron, the ground state of the atom is one which is rotationally invariant, the same when viewed by differently oriented observers. Yet it is not analyzable into simpler entities. In fact it is the simplest possible entity. The state is rotationally invariant.

So group integration, the summing over contributions over all possible directions, the averaging is now done for the probability amplitude. There is a new probability amplitude, but the state is now a pure state, it is the most deterministic state of the system. It is both tantalizing and curious. It appears to be a mixture if you try to produce a classical picture of it, to try to see it as if it is a classical system.

Another aspect of the vector space nature of pure states in quantum theory is that if you took a composite system, a system which had many parts, either two particles or one million photons, for a classical system it means that you know the motion of every particle. Every particle had a definite position and a definite momentum. Even if you decided to look at a subsystem and concentrate your attention on the subsystem, the subsystem would be in a pure state. Therefore, every pure state of the composite system viewed as the state of a subsystem continued to be pure. But in a quantum system described by probability amplitudes, this is no longer true. One could think of pure states of the composite system in which the states of the subsystem were not pure states. The reason is the same old notion of correlation, the interrelationship between things not describable in terms of forces, but describable in terms of a connection, in terms of a phase connection between the particles. It is an understanding between the particles, not an imposition upon each other. Correlation is lost when you fragment the total system into subsystems.

This direct contrast with classical physics creates lots of problems and the paper of Einstein, Podolsky and Rosen talks about the appearance of an independent correlated probability distribution when you would have been led to expect uncorrelated behavior for spatially separated subsystems. We find that we don't have consistency.

This correlation has nothing to do with how far the particles are, because correlation is not a force which is acting from one entity to another one, not an interaction which is propagated; a correlation is preserved however far you go. So two particles which constituted part of one system in a pure state, however far they go, still belong to the same system, and the correlation is something that is lost if you talk only about the subsystem. It is something which is inherent in both of those subsystems taken together and something which is not lost when they move apart.

I would like to believe that Einstein was a very sophisticated person, a very clear thinker, a person who read what other people were writing and talking about. His objections to the standard interpretation of quantum theory in this question of the separation between two particles, was couched very carefully: He said that either certain ideas about quantum theory are wrong or a subsystem cannot be considered as separable from the rest of the system. I would therefore like to suggest that perhaps he was saying that we must look at the holistic nature of the entire system; that perhaps quantum theory is suggesting to us that correlation cannot be ignored. In classical theory we could ignore correlations: out of sight, out of mind. But in the quantum theory this was not so; the total perception of the system was essential. The total perception was not something that was expressible in terms of the fragmented components.

There is a consistent view that underlies all his statistical work including the photoelectric effect, the Brownian motion, the radiation equilibrium, the specific heat, and finally the work criticizing the standard interpretation of quantum theory. Whenever you see a statistical theory and you deduce conclusions from it, it should be the way of uncovering something which is covered up, rather than a way of covering up things which are embarrassing! He seemed to feel that in quantum theory we were not entirely honest in uncovering something, and he seemed to be bent on calling our attention to it, even after he is no longer amongst us in life: calling our attention to the fact that in fact there are lessons from quantum theory which have to be uncovered.

To Einstein statistical theories were tools to elucidate the simple but important features of dynamics of complex systems, and it is not difficult to share with him the conviction that probabilities, even in quantum theory, should lead to the elucidation of a new picture of physical reality.

Professor Wigner has for many years tried to continue in the same vein, to remind us that perhaps we have left something out. He reminds us that maybe we should remember that all of us are conscious, and consciousness is a very crucial part of our human experience, and if we consistently study scientifically all of our human experience, consciousness cannot be left behind. In a situation like the quantum theory where the machinery works perfectly, but somehow or other we cannot agree what it is that works, maybe what we have left out is the role of consciousness with regard to it. Obviously, it is a very difficult problem.

Allow me to conclude this tribute by saying that it is very difficult for us to conceive of a person with the same degree of versatility, clarity, productivity and abiding impact on physics as Albert Einstein.

WHAT ARE THE TRUE BUILDING BLOCKS OF MATTER

A. O. Barut*

International Centre for Theoretical Physics
Trieste, Italy

I. INTRODUCTION

"Can high-energy physics be too easy?" asked a recent editorial
in "Nature."[1)] At present, the picture mostly used in high-energy
phenomenology is becoming admittedly very complicated. Besides
leptons (which we see), one introduces families of "quarks," each
with different colours, then the so-called "gluons," which are the
gauge vector mesons binding the quarks, then there are the so-
called "Higgs particles," which give masses to some of the vector
mesons (all of which are not seen in the laboratory). One is
already beginning to talk about a second generation of more fundamental and simpler objects for these quarks and gluons etc., even
though these first generations of "basic" objects have not been
seen. This type of framework seems to create more problems than
it solves.[2)]

Against this background of recent developments, we wish to
expand here a very intuitive and simple physical theory, along the
traditions of atomic and nuclear structure theories, from which a
unified picture of high-energy phenomena can be deduced. High-
energy physics is very expensive. One must have alternative views,
if only to test better the inevitability of the orthodox picture.
Furthermore, physical phenomena must be explainable in a simple
intuitive form in terms of already verified definite primary concepts, and continuous with the existing physics.

II. THE PHYSICAL PRINCIPLES

Atoms and molecules are best described as built from electrons
and nuclei bound by Coulomb forces because they disintegrate

into electrons and nuclei, which we detect, and because these constituents are stable as far as atomic processes are concerned. In
turn, nuclei and all the hadrons eventually decay into the absolutely stable particles: protons, electrons, neutrinos and photons
(electromagnetic field). We present here a theory in which all
matter is made up of these stable constituents, bound again by
electromagnetic forces. One can of course ask questions about the
nature of the absolutely stable particles themselves. This is
another level of enquiry. In this paper we shall take these as
given and elementary.

At first such an idea might seem impossible or outrageous,
because electromagnetic froces between p, e and ν (and their antiparticles) cannot possibly, one would think, give the necessary
strong binding and strong interactions between hadrons, nor the
so-called weak forces. On the other hand, the idea that stable
particles are the constituents of hadrons is probably very old as
a general idea, if not carried out in specific details. For
example, with the hypothesis of neutrino in β decay, Pauli's model
of the neutron was a bound state of proton, electron and antinuetri-
no.[3] This model was soon abandoned (to be revived much later[4]) for
one did not know how to suppress the large magnetic moment of the
electron (on nuclear scale) inside the nucleus, and one did not know
any deep enough well to contain or confine the electron inside the
nucleus.

What is new, however, is the recognition that magnetic forces
between the stable particles, when treated non-perturbatively,
become very strong at short distances (short ranged), provide a
deep enough well to give rise to high mass narrow resonances, have
saturation property and give rise, by magnetic pairing, to the com-
pensation of the large magnetic moment of the electron. In the
construction of atoms and molecules we make use only of the electric
(Coulomb) part of the electromagnetic forces and treat magnetic
forces as small perturbations. There is, however, another regime
of energies and distances in which magnetic forces play the dominant
role and the electric forces are small perturbations. We shall show
this duality with explicit calculations. It would have been
strange if Nature provided magnetic forces just to be tiny corrections to the building principle of atoms and molecules (which could
exist without them) and not to play an equally important role in
the structure of matter. Clearly, a model of this type also
automatically provides a dynamical theory of nuclear forces.

There are two main immediate questions or objections to our
propositions. Why do we not see in the laboratory strong forces
between proton and electron, electron and positron, or electron and
neutrino etc., whereas we see strong forces between pions and
protons, or protons and neutrons etc.? How can we obtain the rich
world of hadrons just starting from the three stable particles

p, e, ν (and their antiparticles), the multitude of internal quantum numbers like isospin, strangeness, charm etc., the multiplet structures and symmetries?

Correspondingly, this work has two parts. A kinematical part showing the composition of all hadrons and their multiplet structures, hence the meaning of internal quantum numbers in terms of the stable particles, p, e, ν. This by itself is a remarkable mapping of hadron states onto the combinations of stable particles, the eventual final products of all unstable matter, and of hadron quantum numbers into those of three stable particles, p, e, ν.

The second part is dynamical showing that ordinary magnetic spin-spin and spin-orbit forces, when treated non-perturbatively, have the correct strength and shape to give hadronic and nuclear states.

We begin with the second part in order to answer immediately the problems raised above.

A number of models, with increasing complexity, have been studied in recent years, and we have a good understanding of the spin-spin and spin-orbit potentials at short distances.[4]-[7] Consider, for example, a relativistic charged spinless particle m in the field of a fixed (quantum) magnetic momentum $\mu\vec{\sigma}$,[8] or alternatively, a charged spin $\frac{1}{2}$ particle of mass m and magnetic moment $\vec{\mu}$, in the field of a fixed charge.[9] In both cases, the effective radial equation can be written, in appropriate co-ordinates, as

$$\left[-\frac{d^2}{dy^2} + V(j,\ell,r) \right] u = \lambda^2 u, \tag{1}$$

where the effective potential is given, apart from the Coulomb potential $\frac{a}{y}$, by

$$V(j,\ell,r) = \frac{\ell(\ell+1) - \alpha^2}{y^2} + \varepsilon \frac{2c(j,\ell)}{y^3} + \frac{1}{y^4} \tag{2}$$

with $\varepsilon = \pm 1$ (relative sign of the charge and magnetic moment); $c(j,\ell)$ is equal to $-(\ell+1)$ for $\ell = j + \frac{1}{2}$ and equal to ℓ for $\ell = j - \frac{1}{2}$. Furthermore (in units $c = \hbar = 1$), $r = \mu e y = \mu_0 \frac{\alpha}{2M} y$ (M is the mass of fixed magnetic moment--in the second case put M = m), and the eigenvalue λ^2 is

$$\lambda^2 = (E^2 - m^2) \mu^2 \alpha^2 = (E^2 - m^2) \frac{\mu_0^2 \alpha^2}{4M^2} . \tag{3}$$

If we solve the same problem with a Dirac equation and give also an
anomalous magnetic moment a to the particle, then additional terms
are added to Eq.(2).[5] Further models also treat the magnetic
moments of both of the particles.

The potential (2) is treated in atomic phenomena (lately also
in the quark model) as a perturbation. This is justified if the
energies are of the order of Coulomb energies and for Coulombic
bound state wave functions. New phenomena occur, however, if the
magnetic potential is treated non-perturbatively. Fig. 1 shows the
schematic form of the potential at two different energies and
angular momenta in the case when the anomalous magnetic moment terms
are also included. We see three distinct regions of potential
wells: The Coulomb region at distances $r \cong \frac{1}{\alpha m}$ (Bohr radius), hence
momenta of the order of αm or non-relativistic energies of the
order of $\alpha^2 m$, the nuclear region at $r = \frac{1}{m}$, (relativistic) energies
$\frac{m}{\alpha}$ ($\sim$70 MeV); and the supernuclear region of $r \simeq \frac{\alpha^2}{m}$ and energies $\frac{m}{\alpha^2}$
(10 GeV).

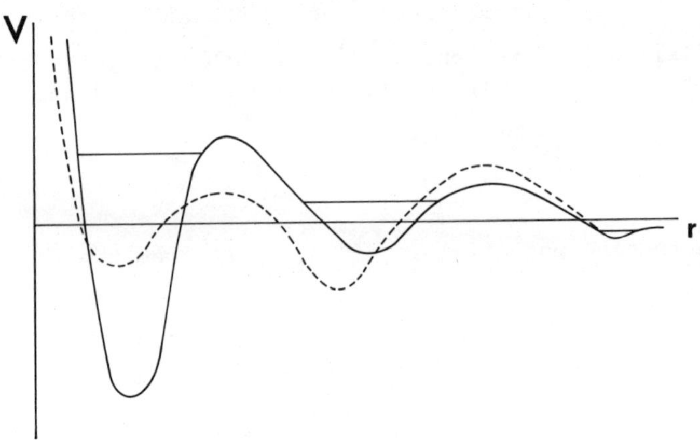

<u>Fig. 1</u>   Schematic form of the effective radial magnetic potential
          V as a function of the radial distance r for two different
          fixed values of energy and angular momentum.

The form of the potential at very short distances is still
quite uncertain in these models. Furthermore, the potentials are
modified by form factors. Form factors must also be calculated

non-perturbatively, and self-consistently from the wave functions
which are localized around each well, respectively, in Fig. 1.[6],[7]
Form factors can easily be incorporated into the model (1)-(2) by
taking $\mu = \mu(r)$. At intermediate distances the form of the potential
is essentially correct. Unfortunately, quantum electrodynamics
cannot tell us anything about the non-perturbative short distance
behavior of the potential between two particles.

Zero-mass limit

It is important for our model later to remark that Eqs. (1) and
(2) also hold for a massless particle in the field of a magnetic
moment, or for a massless particle with an anomalous magnetic moment
(or with only an anomalous form factor) in the field of a charge.[10]
Note that mass m appears only in Eq.(3).

We can now answer the question as to why we apparently do not
see strong interactions in the laboratory between the stable particles p, e, ν.

Scattering against a barrier

The effect of large repulsive potential barriers as in Fig. 1
on the scattering of two fermions (say e^+, e^-) can be evaluated
numerically (and sometimes analytically). The cross-section of
penetration to the attractive region is very small except at the
sharp energy and angular momentum of the resonance, when "resonance
penetration"[11] takes place. The partial phase shift, shown in
Fig. 2 shows a sharp jump of about π near the resonance energy
anomalous scattering). The sharper the resonance, the steeper is
the jump of the phase shift. The effect of this behavior on the
total cross-section is, however, only a small bump, its width
being proportional to the width of the resonance (Fig. 2). Indeed
most hadron resonances are experimentally seen as such small bumps
in cross-sections on a large background. Some predictions based
on this phenomenon will be made after we present the model of
hadrons.

On the other hand, a pion, being itself a spin-zero resonance
state of stable particles (see following sections), can penetrate
much more easily into the region of strong magnetic forces of
other hadron constituents, because of the absence of the spin-orbit
barrier.

An important property of magnetic potentials (Fig. 1) is that
the scattering amplitude is analytic in the whole of the angular
momentum plane, hence is a sum of Regge pole contributions only.
This has many applications in the analysis of scattering processes.

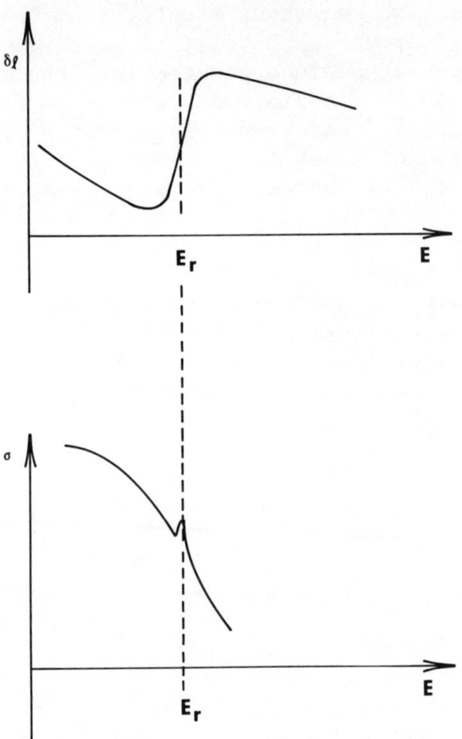

<u>Fig. 2.</u>   The effect of a repulsive barrier on the cross-section σ
around the resonance energy E_r.

III.   ORDINARY AND STRANGE MATTER

Ordinary matter can be built up from p, e and ν (and their
antiparticles) according to the rules that we shall state explicitly.
These are pions, neutron and Δ resonances, hence also   nuclear
matter, atoms and molecules. In order to describe the building-up
principle in a more general way to include "strange" particles, we
must first talk about the μ meson. The μ meson can be thought of
as a magnetic excitation of the electron due to the interaction
of its anomalous magnetic moment with its own field. These arguments are at present semiclassical.[12],[13] Another (perhaps
equivalent) way from our point of view, is to consider μ as a

magnetic resonance state of $(e\nu\bar{\nu})$ into which it decays. We shall
see that the pairs of the type $(e\bar{\nu})$ are identified with pions.
Thus, in order to obtain a spin-$\frac{1}{2}$ state we need three stable particles, and $(e\nu\bar{\nu})$ should be then dynamically a little more stable
than the $(e\bar{\nu})$ states.

The magnetic three-body problem $(e\nu\bar{\nu})$ can be approximated by an
equivalent two-body problem $(e\bar{\nu})\nu$ and considerations similar to
Eqs.(1)-(3) may be applied. The charge-magnetic moment system gives
in the Bohr-Sommerfeld quantization a quantized energy spectrum of
the form $E = \lambda n^4$, $n = 1,2,3,\ldots$ Adding this to the rest mass, one
obtains a leptonic mass spectrum

$$M_N = m_e + \frac{3}{2}\frac{1}{\alpha} m_e \sum_{n=0}^{N} n^4 \tag{4}$$

For electron $(N = 0)$, muon $(N = 1)$, $\tau(N = 2),\ldots$ The predictions
for muon (105.55 MeV) and τ(1786.08 MeV) work very well and the next
lepton predicted is δ(10.293 GeV). The coefficient $\lambda = \frac{3}{2}\frac{1}{\alpha} m_e$ can
also be derived by semi-classical arguments.[12] These results
should only be considered as a beginning of a dynamical theory of
heavy leptons. Nevertheless, they are interesting, because we have
no other hints or ideas concerning the repetitions of leptons in
the series e, μ, $\tau,\ldots$, which is one of the most fundamental open
problems of particle physics.[14]

The ν resonances are inferred from the m = 0 limit of the
Dirac equation in models similar to Eqs.(1)-(3). Hence an interacting ν is necessarily a four-component neutrino. Only in the
asymptotic region can the free Dirac equation be split into two
two-component equations. We shall make the hypothesis that the
neutrino has an anomalous magnetic moment, or at least a magnetic
form factor, even if its magnetic moment is zero (on the mass
shell). We also do not make, at this stage, a difference between
ν_e and ν_μ.

The μ meson, behaving very much like the electron, can in
turn form magnetic pairings and resonances with the stable particles, forming the so-called "strange" hadrons. In fact, it will
turn out that the number of $\mu^{\pm}$ mesons in hadrons is exactly equal
to the "strangeness" quantum number of hadrons. These apparently
new types of hadrons are more unstable and decay into ordinary
hadrons if the μ inside the hadron decays. During strong interactions, μ is stable, hence strangeness is conserved (see also
next section). The μ meson, rather than being a "redundant" particle ("the world would be the same if μ did not exist"(!)) now
plays an essential role in building up the hadrons. This process

is then continued with the τ-excitations, etc.

IV. CONSTRUCTION OF HADRON STATES AND BUILT-IN CONSERVATION LAWS

There is a very simple relationship between lepton quantum numbers and quark quantum numbers. If we compare the triplet $\ell = (\nu, e^-, \mu^-)$ with the quark triplet $q = (u, d, s)$, we have

$$Q_q = Q_\ell + \frac{2}{3} B_\ell, \; B_q = B_\ell - \frac{2}{3} B_\ell \; , \tag{5}$$

where B_ℓ stands for the lepton number and B_q for the baryon number. This we have called the "shifting principle": shifting two-thirds of the lepton number into the electric charge. Hence

$$Q_\ell + B_\ell = Q_q + B_q \; .$$

It is then straightforward to construct the meson quantum numbers as $(\ell\bar{\ell})$-states, both pseudoscalar and vector mesons.

In the case of baryons, the proton is always a final constituent of all baryons. The baryons cannot be constructed as $(\ell\ell\ell)$-states because then L would be equal to 3 and B = 0 but as $(p\ell\ell)$-states giving total baryon number B = 1 and lepton number L zero.

The conservation of lepton and baryon numbers and charge are automatically built-in in this model, because p, e and ν are absolutely stable. The only dynamical process is the pair production of constituents which conserves Q, B and L.

A physical interpretation of the mysterious internal quantum numbers, like isospin and strangeness, emerges from the model. As we have noted, the μ number is equal to the strangeness number S. Hence the number of all quantum numbers is reduced by 1:

$$S = N_{\mu^+} - N_{\mu^-}.$$

The isotopic spin quantum number essentially counts the number of stable constitutents (p, e and ν). In order to see this more precisely, we first define the third component of isospin and the isospin creation and annihiliation operators

$$I_3 = \frac{1}{2} (N_p - N_{\bar{p}} + N_{e^+} - N_{e^-} + N_\nu - N_{\bar{\nu}}) \; ,$$

$$I_+ = \frac{1}{\sqrt{2}} (a_\nu^+ a_{e^-} + a_{e^-}^+ a_{\bar{\nu}}), \; I_- = (I_+)^\dagger \; . \tag{6}$$

The empirical Gell-Mann-Nishijima formula is now derived and automatically also built in the model:

$$Q = N_p - N_{\bar{p}} + N_{e^+} - N_{e^-} + N_{\mu^+} - N_{\mu^-} = I_3 + \frac{1}{2}(N_p - N_{\bar{p}} + S), \quad (7)$$

because $\sum_\ell N_\ell = \sum_\ell N_{\bar{\ell}}$ for all states (i.e. $N_{e^+} + N_{\mu^+} + N_{\bar{\nu}} = N_{e^-} + N_{\mu^-} + N_\nu$).

Figs. 3, 4 and 5 show the hadron multiplets in minimal realization.[15] We can of course add to each hadron a lepton pair ($\ell\bar{\ell}$) <u>of the same species</u> without changing the quantum numbers. For example, the physical proton can be thought of as having a π^0 cloud:

$$P_{\text{physical}} = p\left[\frac{1}{\sqrt{2}}(e^-e^+ - \nu\bar{\nu})\right], \quad (8)$$

as can be seen by applying I_- to it or I_+ to the neutron state.

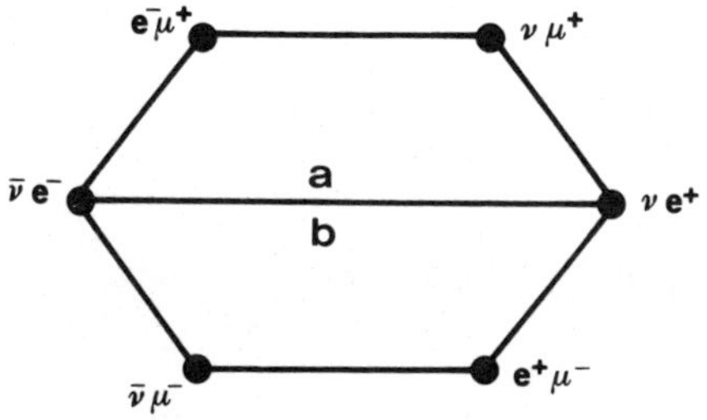

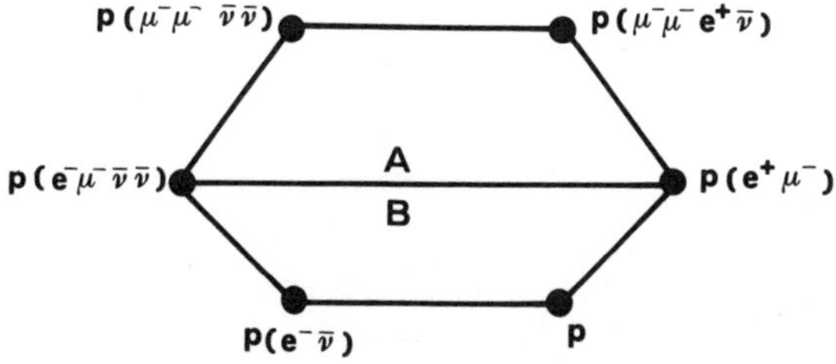

$$a = \frac{1}{\sqrt{2}}(\bar{\nu}\nu - e^+e^-), \quad b = \frac{1}{\sqrt{6}}(\bar{\nu}\nu + e^+e^- - 2\mu^+\mu^-)$$

<u>Fig. 3</u>   The meson octet.

$$A = p(\mu^-\bar{\nu}), \quad B = p(\mu^-\bar{\nu}\nu\bar{\nu})$$

<u>Fig. 4</u>   The baryon octet.

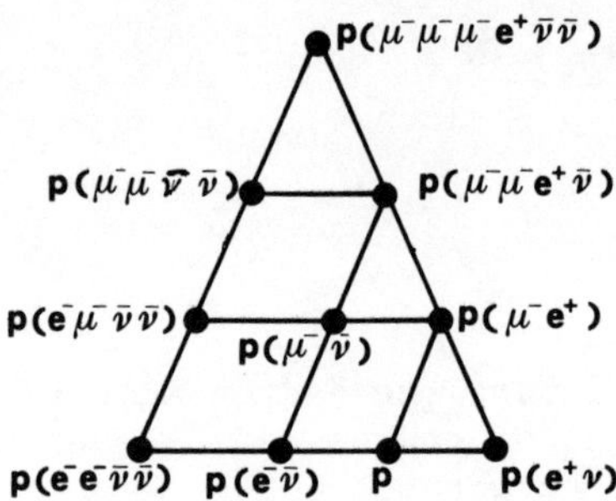

Fig. 5   The baryon decouplet.   The nearly linear mass formula of about the μ mass is a consequence of nearly zero-energy bound states in the magnetic potential well.

A full physical interpretation can be given to the concept of _isospin_ as the quantum-mechanical exchange process of the lepton pair $(e^-\bar{\nu})$ between two systems, exactly like the exchange effects in H_2^+ molecule.   To see these we go to the two-nucleon problem, where the notion of isospin has historically origninated.   The states of definite isospin are

$$pp, \quad \frac{1}{\sqrt{2}}\,(pn + np), \quad nn \quad (I = 1), \quad \text{and} \quad \frac{1}{\sqrt{2}}\,(pn - np) \quad (I = 0)\ .$$

In the $I_3 = 0$ state, $(e\bar{\nu})$ is exchanged between the two protons and we have the symmetric $(I = 1)$ and antisymmetric $(I = 0)$ states with respect to the exchange, which are eigenstates of total Hamiltonian.   We could make a similar isospin triplet and singlet in atomic physics with

$$pp, \quad \frac{1}{\sqrt{2}}\,(Hp + pH) \equiv H_2^- \text{ sym.} \quad , \; H_2; \quad \frac{1}{\sqrt{2}}\,(Hp - pH) \equiv H_2^- \text{ antisym.} \quad \cdot$$

Here (p,H) is an isospin-doublet $(I_3 = +\tfrac{1}{2}$ and $-\tfrac{1}{2})$ and $Q = I_3 + \tfrac{1}{2}$.   Also $I_+ = a_p\, a_H^+$.   Similarly, if we look at two-pion states of definite isospin

$$|\pi^{\pm}\pi^{\pm}\rangle\ , \quad \frac{1}{\sqrt{2}}\{|\pi^{\pm},\pi^0\rangle + |\pi^0,\pi^{\pm}\rangle\}, \quad \frac{1}{\sqrt{6}}\{2|\pi^0\pi^0\rangle + |\pi^+\pi^-\rangle +$$
$$|\pi^-\pi^+\rangle\}$$

$$\frac{1}{\sqrt{2}}\{|\pi^{\pm},\pi^0\rangle - |\pi^0,\pi^{\pm}\rangle\}, \quad \frac{1}{\sqrt{2}}\{|\pi^+\pi^-\rangle - |\pi^-\pi^+\rangle\}$$

$$\frac{1}{\sqrt{3}} \{ |\pi^+,\pi^-> + |\pi^-\pi^+> - |\pi^0\pi^0> \} \quad ,$$

or, pion-nucleon states of definite isospin

$$p\pi^+ \ , \ \frac{1}{\sqrt{3}} \{2|p\pi^0> + |n\pi^+> \}, \ \frac{1}{\sqrt{3}} \{2|n\pi^0> + |p\pi^-> \}, \ n\pi^-$$

$$\frac{1}{\sqrt{3}} \{ |p\pi^0> - 2|n\pi^+> \}, \ \frac{1}{\sqrt{3}} \{-|n\pi^0> + 2|p\pi^-> \} \quad ,$$

we see that the isospin is identical to the symmetric and anti-symmetric <u>exchange</u> or rearrangement of constitutents. Isospin conservation is always used or tested in the reactions of two or more hadrons when stable constituents can be exchanged between the two hadrons, as between two atoms. It is not necessary to assign an isospin to individual hadrons, let alone to the constituents of hadrons, although the third component of isospin can be assigned to the constituents via the Gell-Mann-Nishijima formula. The con-servation of the third component of isospin is equivalent to the conservation of the number of stable constituents, because the only processes occurring in nature, according to the present model, are the rearrangement of constituents when two hadrons interact and pair production and annihiliation of stable particles. The conservation of $\vec{I}$ or I^2, <u>in strong interactions</u>, on the other hand, is the con-servation of symmetry properties of stable leptons $(e\bar{\nu})$ <u>under exchange</u> between the hadrons.

The physical intuitive meanings given to the abstract internal quantum numbers of hadrons is an important feature of the present theory: The constituents no longer carry mysterious properties such as strangeness, isospin, charm etc. The only charge is the electric charge.

<u>Relation to quark assignments</u>

The relation of our constituents to quark constituents is very simple. For mesons: $\ell\bar{\ell} \to q\bar{q}$, and for baryons: if we think of p as (uud) then our assignments become the same as the $q_1q_2q_3$ assignment with additional definite $(q\bar{q})$ terms <u>of the same species</u> (so-called $q\bar{q}$ sea terms). Such terms are introduced into the quark model anyway.

If we continue this correspondence or shift between quarks and leptons, then the next "excited" neutrino with the quantum numbers of ν_μ would correspond precisely to the so-called "charmed" quark and the next leptons τ and ν_τ to the other two new quarks, b and t. It is not known at present if ν_μ or ν_τ are massless or absolutely stable. According to the experimental limit so far, ν_μ

is heavier than the electron!

It is important to remark that from deep inelastic electron-nucleon scattering experiments one can infer two solutions for the charges of constituents (assumed to be point-like at high energies)[16]. One solution gives for proton constituents the charges +1, -1, 0. This is in agreement in our model with the physical proton being pe^+e^- and neutron being $pe^-\bar{\nu}$. The second solution gives the fractional quark charges. The _additivity_ assumption of the magnetic moments and equal additive quark masses then selects quark assignments. However, in a dynamical physical bound state model, magnetic moments also have orbital contributions and constituent masses are unequal.

V.   STRONG AND WEAK INTERACTIONS

All strong interactions including nuclear forces are, according to the present theory, of magnetic type and are further determined by the composite structure of the hadrons. Specifically there are two fundamental processes at short distances when hadrons collide: i) Rearrangement of constituent stable particles, ii) pair production (or annihilation) of leptons (and subsequent rearrangement). It is possible to give diagrams for every strong process using i) and ii). The ideas of the old meson theory, the many models of meson exchanges or Regge-pole exchanges emerge as approximate schemes from this theory, as well as the ideas of the S-matrix theory and nuclear democracy: different rearrangements of constituents with real or virtual lepton pairs obviously imply that hadrons can be thought to be built of other hadrons. In particular, the meson cloud around the nucleon is an immediate approximation here, but not in the quark model.

We propose here a new _model of the nucleus_, which seems to combine two apparently contradictory features of the nucleus. On the one hand, the nucleus consists of closely packed large nucleons with an occupancy between 60 and 90%, or may even have a crystalline structure. On the other hand, the nucleons seem to be moving freely inside the nucleus, as the shell model or other Fermi gas models are implying. These two features are reconciled in the present theory as follows. The stable protons form the closed packing or even the crystalline skeleton of the nucleus. On top of it the stable lepton pairs $(e^-\bar{\nu})$ acting like a boson are hopping from one proton to another. When an $(e^-\bar{\nu})$ is attached to a proton, it then becomes a neutron. Thus moving $(e^-\bar{\nu})$'s will appear exactly as moving neutrons, or moving protons in the opposite direction. We can then study the motion of $(e^-\bar{\nu})$ pairs in the periodic potential of the lattice of protons.

The weak interactions of the β-decay type are due to barrier

penetration, e.g. $n(pe^-\bar\nu)$ decay or $\mu(e\nu\bar\nu)$ decay. In fact, a theory
of the neutron with an equation of type (1)-(2) correlates (in this
approximation) the lifetime of the neutron, the n-p mass difference
(which is positive and can be estimated as the excess magnetic
energy of $(e^-\bar\nu)$ bound to the proton) and the magnetic moment of
the neutron.[8] Hence, indirectly, the Fermi constant G is related
to the fine-structure constant α. All other decay modes of hadrons
can be understood as a barrier penetration between two wells of
the potential (see Fig. 1), μ decay inside the hadron (suppressed by
the Cabibbo angle as compared with the free μ decay) and barrier
penetration with or without μ decay. Different decay channels
result in different rearrangements of the constitutents. Finally,
a weak scattering process such as $e\nu \to e\nu$ should be related to the
anomalous magnetic moment of the neutrino. This remains to be seen
when we shall have more experimental data on the angular and
energy dependence of this process.

VI. SOME FURTHER APPLICATIONS: K^0 PHYSICS AND CP VIOLATION

As an example of the intuitive value of the model we consider
its application to the remarkable <u>physics of the K^0 mesons</u>.

According to Fig. 3, K^0 and $\bar K^0$ mesons are $(e^-\mu^+)$ and $(e^+\mu^-)$,
respectively, i.e. the magnetic analogues of muonium and anti-
muonium. (Such states have also been called superpositronium
(e^+e^-) or supermuonium $(e^-\mu^+)$.) They are obviously charge con-
jugates of each other. If one of the states is produced, say
$e^-\mu^+$, and we view μ^+ as $(e^+\nu\bar\nu)$, then $(\nu\bar\nu)$ pair can oscillate between
e^- and e^+ in a magnetic potential as shown in Fig. 6. When $(\nu\bar\nu)$ is
attached to e^+ we have a $\bar K^0$, when it is attached to e^- we have a
K^0. Under these circumstances, we know from general quantum
mechanics that the observed eigenstates of the energy are the symmetric and antisymmetric combinations with respect to the $(\nu\bar\nu)$
exchange, namely $K_{\substack{S\\AS}} = K^0 \pm \bar K^0$, which are also eigenstates of CP.

In fact the problem is exactly the same quantum-mechanically as in
the ammonium (NH_3) laser,[17] where N oscillates between two positions in a potential as in Fig. 6. We therefore have the unam-
biguous prediction that the antisymmetric states is heavier than
the symmetric one. In our case $m(K_L) > m(K_S)$. This is, to my
knowledge, the first theory of the sign of the K_L – K_S mass difference. Moreover, the Dennison-Uhlenbeck mass formula[18] gives
for the mass difference $\dfrac{\Delta m}{m} = \dfrac{1}{\pi A^2}$, where A is the barrier penetration factor in the potential (Fig. 6). We do not know A, but we
can obtain it from the decay rate Γ_S of K_S into $\pi^- + \pi^+$ $(e^-\bar\nu + e^+\nu)$,
which uses the same potential barrier. This gives $\Delta m = \dfrac{1}{2}\,\Gamma_S$.
Experimentally we have for the K_L – K_S mass difference
$\Delta m = 0.477\,\Gamma_S$.

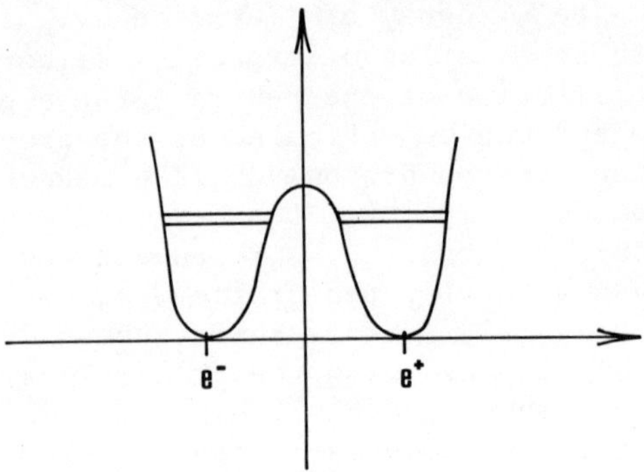

<u>Fig. 6</u>   The effective magnetic potential barrier for ν and $\bar{\nu}$
exchange and oscillations between e^- and e^+ in the
$K^0 - \bar{K}^0$ system.

The two decay modes of K_S are given by two ways of rearranging
the constituents. K_L cannot decay in this way because of CP
invariance. But an additional lepton pair production gives all
the decay channels of K_L. The rate is down by $\pi\alpha$ due to this pair
production, which agrees with experiment.

Finally we discuss a <u>mechanism of CP violation</u> which occurs
only in the K^0 mesons. CP violation in our picture means a small
violation of the symmetric and antisymmetric combinations. There is,
in fact, a feature in the model, which brings an asymmetry. In the
above discussion we have not made a distinction between ν_e and ν_μ.
If we do make a distinction, then we have $(e^- \bar{\nu}_e \nu_\mu e^+)$ combination
for $\bar{K}^0$ and $(e^- \bar{\nu}_\mu \nu_e e^+)$ combination for K^0. Hence an extra interaction
must convert $\bar{\nu}_e \nu_\mu$ into $\bar{\nu}_\mu \nu_e$, which provides a further asymmetry
between K_1 and K_2 leading to K_L and K_S. We can further predict
that CP violation should also occur in the neutral mesons built
from $(e^- \tau^+$ and $e^+ \tau^-)$ and $(\mu^- \tau^+$ and $\mu^+ \tau^-)$.

VII.   CONCLUSIONS

High-energy physics according to the present theory can be
considered as an extension of atomic and molecular physics. The
Coulomb forces being replaced by the short-ranged strong magnetic
forces. The only additional particle not present in atomic physics
is the neutrino, which is in fact a limiting case of the electron.
There is then a welcome continuity and simplicity in the physics,
which was perhaps lost by the abstract concepts and free inventive-
ness of particle physics. No <u>new</u> particles, or no new interactions

or forces are introduced[19] except the stable ones and the electromagnetic field. In this sense it is a truly already-unified theory
with one coupling constant e. The only parameter so far, in
principle, is the neutrino magnetic moment. All other "particles"
are transitory; they come as resonances and eventually decay into
the absolutely stable particles. The division of forces in nature
into strong, weak and elementary was a temporary one; there is no
need for such a division.

Although much detailed quantitative work must be done, and is
being done, we have shown that, conceptually and logically, it is
possible to understand the world of fundamental particles and their
interactions from the very simple framework of stable particles and
stable electromagnetic forces. Our guiding principle has been the
same as that of Lord Kelvin under similar circumstances: "I want
to understand light as well as I can, without introducing things
that we can understand even less of."

REFERENCES

1) Nature editorial, $\underline{273}$, 479 (1978).
2) V.F. Weisskopf, Physikalische Blätter $\underline{35}$, 3-12 (1979), cf.
    Sec. 5, Unsolved problems;
    M. Gell-Mann [Summary talk, Jerusalem Einstein Centennial
    Symposium, March 1979.]
3) See the historical account, L.M. Brown, "The idea of neutrino,"
    Phys. Today, $\underline{31}$, 23 (September 1978).
4) A.O. Barut, in Structure of Matter, Proc. Rutherford Centennial
    Conference, 1971, Ed. B.G. Wybourne (Univ. of Canterbury Press,
    1972), p. 69.
5) A.O. Barut and J. Kraus, Phys. Letters $\underline{59B}$, 175 (1975); J. Math.
    Phys. $\underline{17}$, 506 (1976).
6) A.O. Barut and J. Kraus, Phys. Rev. $\underline{D16}$, 161 (1977).
7) A.O. Barut and R. Raczka, Acta Phys. Polon., $\underline{B10}$, 687-703
    (1979).
8) A.O. Barut and G. Strobel, to be published.
9) C. Piron and F. Reuse, Helv. Phys. Acta $\underline{51}$, 146 (1978);
    F. Reuse, Helv. Phys. Acta $\underline{51}$, 157 (1978).
10) A.O. Barut, "Magnetic resonances between massive and massless
    spin-$\frac{1}{2}$ particles," ICTP Trieste, preprint IC/79/51. J. Math.
    Phys. (in press).
11) R.W. Gurney, Nature $\underline{123}$, 565 (1929).
12) A.O. Barut, Phys. Letters $\underline{73B}$, 310 (1978).
13) A.O. Barut and J.P. Crawford, Phys. Letters $\underline{82B}$, 233 (1979).
14) S.L. Glashow, Comments Nucl. Part. Phys. $\underline{8}$, 105 (1978).
15) A.O. Barut, "Leptons as 'quarks'," University of Geneva pre-
    print UGVA-DPT 1978/08-175, and in Proceedings of the Texas
    Conference on Group Theory and Mathematical Physics, September
    1978, Eds. A. Bohm and E. Takasugi (Springer Lecture Notes in

Physics, 1979), Vol. <u>94</u>, 490-98 (1979); Surveys in High Energy
Physic, Vol. 1, No. 2, Jan. 1980 issue.

16) L. Berkelman, in <u>Orbis Scientiae Proceedings,</u> Coral Gables (1979).

17) C.H. Townes and A.L. Schawlov, in <u>Microwave Spectroscopy</u>
(McGraw Hill, 1955), p. 300.

18) D.M. Dennison and G.E. Uhlenbeck, Phys. Rev. <u>41</u>, 313 (1932).

19) "When one thinks back to these days, one finds that it is
really remarkable how unwilling people were to postulate a
new particle. This applies both to theoretical and experimental workers. It seems that they would look for any
explanation rather than postulate a new particle. It needed
the most obvious and unassailable evidence to be presented
before them before they were reluctantly forced to postulate
a new theory. The climate has completely changed since these
early days. New particles are now being postulated and proposed
continually, in large numbers. There are a hundred or more
in current use today. People are only too keen to publish
evidence for a new particle, whether this evidence comes from
experiment or from ill-established theoretical ideas." (P.A.M.
Dirac. in <u>The Development of Quantum Theory</u> (Gordon and Breach,
1971), p. 60.)

*Permanent Address: Department of Physics, University of Colorado,
Boulder, Colorado 80308, USA.

CLASSIFICATION OF WIGNER OPERATORS BY A NEW TYPE

OF WEIGHT SPACE DIAGRAM[+]

L.C. Biedenharn* and J.D. Louck**

*Department of Physics, Duke University
 Durham, North Carolina
**Los Alamos Scientific Laboratory
 Los Alamos, New Mexico 87545

1. *Introduction*

It is quite appropriate that we commemorate Einstein with a
conference devoted to symmetry in physics, since emphasis on the
importance of symmetry concepts in theoretical physics is indeed
one of Einstein's many legacies. I would like to begin by emphasizing that the practical applications of symmetry techniques in
physics have, in the final analysis, depended (explicitly or implicitly) upon the use of the Wigner coefficients of the symmetry
group. Physically interesting operators are, in this approach,
classified by the symmetry as instances of generic tensor operators
and the matrix elements of the generic operators define the relevant Wigner coefficients. This is a concept whose depth and importance is common knowledge among the participants of this conference.

We would like to review today the status of this tensor operator approach to the family of unitary groups U(n), n=1,2,3... .
Since this family obeys the inclusion $U_n \supset U_{n-1}$ for all n, the inverse limit sequence defines U_∞, and the operator approach thus
aims ultimately at nothing less than classifying operators in Hilbert space itself. Clearly this is a 'large' program, and current-
ly available results are *very* far from possessing such sweeping
generality. Complete results are, of course, available for U(2)--
from the work of Giulio Racah and Eugene Wigner. Complete results
are available also for U(3), at least in principle and to a considerable extent in practice. For the general U(n) problem there has
recently been very substantial progress (see below).

Our plan today is to sketch the main ideas and techniques, the
known results, and recent new results that hint at possible definitive progress. As a particular result, we will describe in detail

the totally unexpected way in which weight patterns for SU(3) appear as the classifying concept for a new family of polynomials defined by SU(3) canonical tensor operators. This result constitutes the new space of our title.

Progress on the tensor operator problem for various U(n) has been due to the efforts of many researchers. Besides Wigner and Racah already mentioned, let us note the contributions of S.J. Alisauskas[1], G. Baird[2], V. Bargmann[3], E. Chacon[4], M. Ciftan[5], M. Hamermesh, K.T. Hecht, C.J. Hinrich[6], A.P. Jucys[1], M. Lohe[14], and M. Moshinsky[7], as well as ourselves. For lack of space and time we shall not attempt a more adequate referencing for these varied contributions; apologies are offered in advance for this lack of scholarship.

2. *Notational Preliminaries*

The unirreps (linear <u>unitary</u> <u>irreducible</u> <u>rep</u>resentations) of the unitary group U(n) are well known to be classified uniquely and exhaustively by the set of all standard Young frames having n or less rows; each such irrep (we assume all representations to be linear and unitary henceforth) is characterized by a lexical array of integers (positive, negative or zero) $[m] \equiv [m_{1n}, m_{2n}, \ldots, m_{nn}]$. (Lexicality means that $m_{in} \geqslant m_{i+1n}$ for all i.) Irreps of SU(n) have $m_{nn} = 0$.

A given vector belonging to the irrep m is specified by the *Weyl branching law* applied to a chain of subgroups: $U(n) \supset SU(n-1) \times U(1), \ldots$ ending with U(1). (This is also known as a flag.) Fixing a specific chain, we may label a unique vector (one dimensional subspace) of the representation [m] by a Gel'fand pattern:

$$
(m) \equiv \begin{pmatrix} m_{1n} & & m_{2n} & & \cdots & & m_{nn} \\ & m_{1n-1} & & m_{2n-1} & & m_{n-1\ n-1} & \\ & & & \cdots & & & \\ & & & m_{11} & & & \end{pmatrix}
$$

where $m_{ij} \geqslant m_{i+1,j+1} \geqslant m_{i+1,j}$ (betweenness condition).

It is remarkable that from the betweenness condition alone one can determine the Weyl dimensionality formula (by direct summation over all allowed patterns of a fixed irrep [m]).

The vector belonging to the irrep [m] having the Gel'fand pattern (m) is denoted as the ket vector $|(m)>$. The set $\{|(m)>\}$ is an orthonormal basis for the irreps [m].

Let us now construct a particular Hilbert space -- call it $\mathcal{H}$ -- whose vectors are a direct sum of the vectors of each irrep [m], each distinct irrep of SU(n) occurring once and only once. The *operators* carrying $\mathcal{H}$ into itself themselves define representa-

tions of U(n) and can thus be classified by irrep labels [M] and
Gel'fand labels (M). This is the *tensor operator classification,*
but the classification is far from unique, failing in two ways:
(a) multiplying a given operator by an invariant operator leaves
the irrep labels of the operator unchanged and (b) there exists
in general an *intrinsic multiplicity* -- distinct operators exist,
not related by (a), having the same labels [M].

Difficulty (a) is resolved by focussing attention on the *unit*
tensor operators admitting as "scalars" the ring of invariant
operators.

Difficulty (b) is the so-called "multiplicity problem"; what
is required for its resolution is *the analog for operators of the
Weyl branching law classification for states.* This problem is
trivial for U(2) but non-trivial for U(3); but here it has been
completely resolved, in a canonical fashion. (Canonical means no
free choices to within equivalence.) How far it has been canonically resolved for U(n) will be discussed below.

It is remarkble, however, that there *does exist* a natural way
to label unit tensor operators! To see this, note first that a
unit tensor operator <[M]> acting on a given irrep [m] of $\mathcal{H}$ can
be classified by the irrep labels [m'] of the irrep produced by
the action. (A unit tensor operator is defined to have the intertwining number (I) equal to 1 or 0 when acting on any (and hence
all) irreps [m] of $\mathcal{H}$.) Thus we can associate to the operator
<[M]> the *shift labels* [Δ] $\equiv$ ($m'_{1n}-m_{1n},\ldots,m'_{nn}-m_{nn}$). (Cf. Ftn. (16).)

Let us define a given unit tensor operator [M] to have the
same shift [Δ] on all irreps of $\mathcal{H}$ for which $I=1$. This removes
completely any multiplicity for U(2) (unit tensor) operators --
but not for U(n) operators, n>2.

If we consider the shift labels [Δ] to be the "operator space
analog" of Cartan's weights for irrep vectors, then it is natural
to ascribe to the operator <[M]> an *operator pattern:*

$$
(\Gamma) = \begin{pmatrix}
 & & \Gamma_{11} & & \\
 & & \cdots & & \\
 & \Gamma_{1n-1} & \Gamma_{2n-1} & \Gamma_{n\,n-1} & \\
M_{1n} & M_{2n} & \cdots & & M_{nn}
\end{pmatrix}
$$

where the entries in the pattern are integers obeying the betweenness relations. (Observe that the betweenness relations are unchanged for an inverted triangle.)

The shift labels [Δ] are the *weights of the operator pattern*
(Γ) -- denoted [$\Delta(\Gamma)$] -- and defined by: $\Delta_{1n}=\Gamma_{11}$, $\Delta_{2n}=\Gamma_{12}+\Gamma_{22}-\Gamma_{11}$,
..., (difference of sums of adjacent rows).

We now assert: *the multiplicity of distinct operator pat-
terns (Γ) having the same shift labels [Δ], is precisely the multiplicity of unit tensor operators <[M]> with shift [Δ] when acting on an irrep of $\mathcal{H}$ in general position.* ("General position"

means that the $\{m_{in}\}$ are sufficiently large and far apart.)

A basis for the unit tensor operators may accordingly be labelled by the set of operators: $\{<\left\{\begin{smallmatrix}\Gamma\\(m)\end{smallmatrix}\right\}>\}$ or, better, with the amended notation $<\begin{smallmatrix}(\Gamma)\\[M]\\(m)\end{smallmatrix}>$, as we wish to call attention to the fact that the irrep labels [M] are common *both* to the operator pattern (Γ) *and* to the Gel'fand pattern (m). (The n'th row is thus deleted in the patterns (Γ) and (m) henceforth.)

Operator patterns[8] implicitly define a new space -- completely distinct from the space of Gel'fand patterns -- and it is of considerable interest to explicate and elucidate the properties and meaning of this new space.

3. *The Construction of a Canonical Labelling*

The fact that the Gel'fand pattern (m) is a useful labelling of the state vectors of an irrep [m] is a consequence of the Weyl branching law which ascribes a *group theoretic significance*[15] to each label m_{ij}.

The Gel'fand labelling *does* involve free choices, namely the chosen specific sequence of subgroups. This freedom of choice is precisely that afforded by the Weyl group of U(n) -- the permutation group S(n) -- and a given Gel'fand labelling is the choice of a specific Weyl chamber. Since all Weyl chambers are abstractly equivalent, the Gel'fand labelling is accordingly canonical.

What significance can be ascribed to the operator pattern labels?

To answer this, let us begin by considering that specific Weyl chamber defined by lexical ordering. To a given irrep of U(n) specified by [m] we associate the new variables: $p_{in} \equiv m_{in} + n - i$ ($i = 1,\ldots,n$). It is easily seen -- but no less remarkable -- that the variables $\{p_{in}\}$, $i = 1,\ldots,n$ are the proper *symmetric* variables with which to discuss the matrix elements of unit tensor operators in $\mathcal{H}$. In fact, these matrix elements function as the carrier space of an S(n) group *which must be carefully distinguished from the Weyl group of* U(n).[11]

It is essential now to impose the unimodular condition on the set of all irrep labels of U(n) without prejudicing the S(n) symmetry in operator pattern space. To do so we project the n-dimensional orthogonal space labelled by the $\{p_{in}\}$ onto the sub-space perpendicular to the unit vector $n^{-\frac{1}{2}}$ $(1,1,\ldots,1)$. This defines an n-simplex in (n-1 dimensions) with the barycentric coordinates

$$y_i \equiv p_{in} - (\sum_{i=1}^{n} p_{in})/n.$$

The variables $\{y_i\}$ $i = 1,\ldots,n$ obey the rule: $\Sigma y_i = 0$, so that the labels $\{y_i\}$ specify a unique irrep of SU(n) associated to the set of U(n) irreps $\{[m]\}$ by $m_{in} \to m_{in} - m_{nn}$.

We now define a specific realization of the space of SU(n) irrep labels. To the n-simplex having equal edge lengths we asso-

ciate the n-dimensional barycentric coordinate frame whose axes
are the n lines through the center of the simplex perpendicular to
the n (n-1 dimensional) hyperplanes. The lattice points inside
the cone $y_i > 0$, i $\leqslant$ n-1, define the lexical region, and every lattice point inside the cone corresponds to a distinct SU(n) irrep
and conversely.

Let us introduce a new diagram to be realized on the lexical
region: *the operator intertwining number - null space diagram*[9,10]
For the unit tensor operators labelled by <[M]> and [Δ] we associate to the point (y_i) (which corresponds to irrep [m]) the intertwining number I ([m][M]$\rightarrow$[m+Δ]). Define *level surfaces* by join-
ing points having the same value of I.

This construction defines the intertwining number - null
space diagram.

Let us sketch how the null space of an operator is to be
used to effect a canonical labelling.

The level surfaces of the intertwining number - null space
diagram divide the lexical chamber into disjoint regions; because
the intertwining number is non-decreasing for all directions within the lexical cone, the level surfaces are nested. Because of
the nesting property, we may *dissect* the level surfaces into a
family of characteristic regions, each region to be associated to
a specific operator such that this operator has I=1 inside the
characteristic region and 0 outside. A specific operator is to be
canonically associated to the null space of the operator (that
part of the lexical region having I=0).

The assignment of specific operator patterns to the canonically identified operators is carried out by inducing the labels from
limit properties of the matrix elements. This is a separate problem which (since fully resolved) can be safely put aside. Suffice
it to say that maximal null space is assigned to the stretched pattern, denoted (Γ_s). (Among the (n-1) independent adjoint operators <2$\dot{1}$0> with Δ=(11...1) in U(n) minimal null space identifies
the generators.)

*The remarks sketched in the above two paragraphs require con-
siderable verification and proof; only for U(3) has the complete
proof been published.* Partial results[3] have been published for
general U(n).

For U(3) the intertwining number - null space diagram, and its
dissection into characteristic sets, is easily grasped since the
barycentric coordinate system is that of a plane. In Section 5,
below, we discuss an example which illustrates these concepts in
some detail (see Fig. 1 especially).

4. *Boson Operators and the Pattern Calculus*

In the preceding sections we have discussed the global concepts underlying the canonical resolution of the multiplicity,
but when it comes to explicit details, as physicists we always
calculate with boson operator techniques. These calculational

techniques include the *pattern calculus*[11], a symbolic calculational device of quite surprising efficacy. Let us sketch the principal ideas and results.

The basic underlying idea is the *Jordan map:* given a set of
nxn matrices $\{\Lambda\}$ over $\mathbb{R}$, define the mapping:

$$J: \quad \Lambda = (\Lambda_{ij}) \to \Lambda_{op} = \sum_{i,j} (A_{ij} \, a_i \bar{a}_j) \; ,$$

where the $\{a_i\}$, $\{\bar{a}_i\}$ for i=1,...,n are *boson operators* obeying the
defining relations:

$$[a_i, a_j] = [\bar{a}_i, \bar{a}_j] = 0 \; ,$$

$$[\bar{a}_i, a_j] = \delta_{ij}, \quad i=1,2,\ldots,n \quad .$$

The mapping J has the property that it preserves commutation relations $[\Lambda_{op}^{(1)}, \Lambda_{op}^{(2)}] = [\Lambda^{(1)}, \Lambda^{(2)}]_{op}$. This step is quite familiar in physics, for it coincides with Schwinger's realization of the general
SU2 generators as the Jordan map of the Pauli matrices.

We can extend this concept by considering in place of the
n component boson $(a_1, a_2, \ldots, a_n)$ the nxn boson operator a_j^i, or
matrix boson operator $A \equiv (a_j^i)$. These n^2 operators obey:
$[a_j^i, a_{j'}^{i'}] = [\bar{a}_j^i, \bar{a}_{j'}^{i'}] = 0$, $[\bar{a}_j^i, a_{j'}^{i'}] = \delta_i^{i'} \delta_{j'}$. Two different Jordan maps
now defined:

$$J_u \equiv J_{lower}: \quad (\Lambda)=(\Lambda_{ij}) \to J_\ell(\Lambda) = \sum_{k=1}^{n} \Lambda_{ij} a_i^k \bar{a}_j^k$$

$$J_\ell \equiv J_{upper}: \quad (\Lambda)=(\Lambda_{ij}) \to J_u(\Lambda) = \sum_{k=1}^{n} \Lambda_{ij} a_i^i \bar{a}_k^j \quad .$$

Both maps preserve commutation relations, but now *all* $J_u(\Lambda)$
commute with all $J_\ell(\Lambda)$. Hence if the $\{\Lambda\}$ are the generators of a
fundamental irrep of U(n), the extended Jordan map yields generators of the group U(n)xU(n) for the *general case*. Letting (Λ)
include the $n^2 \times n^2$ matrices of a fundamental irrep of $U(n^2)$ we obtain generators for all totally symmetric irreps of $U(n^2)$ *adapted
to the group decomposition:* $U(n^2) \supset U(n) \times U(n)$.

Let us now determine more explicitly the irreps defined by
this procedure. Using the elements of the matrix boson A as indeterminates, it is known (from Schur's thesis) that we might as
well consider homogeneous polynomials as the basis. These are the
boson polynomials,[12] denoted by:

$$B \begin{pmatrix} (m') \\ [m] \\ (m) \end{pmatrix} (A) \quad ,$$

where A is the matrix boson (a_j^i), and [m] denotes the Young pattern labels $[m_{1n}, m_{2n}, \ldots, m_{nn}]$ with (m) and (m') being Gel'fand
patterns of n-1 rows. The operators $J(\Lambda)$ act on the B(...) by

commutation and the boson polynomials are, by construction, the vectors denoted by the respective Gel'fand patterns of the group $U(n) \times U(n)$.

The explicit form of these boson polynomials is given by:

$$B \begin{pmatrix} (m') \\ [m] \\ (m) \end{pmatrix} (A) = M^{\frac{1}{2}}([m]) \sum_{\boxed{\alpha}} C \begin{pmatrix} (m') \\ [m] \\ (m) \end{pmatrix} (\alpha)$$

$$\times \prod_{i,j=1}^{n} (a_i^j)^{\alpha_i^j} / [(\alpha_i^j)!]^{\frac{1}{2}} ,$$

where $C(...)$ is an *explicitly known coefficient* (cf. Ref. 12).

In the summation $\boxed{\alpha}$ denotes the following square matrix of nonnegative integers with constraints on the sums of the entries in the rows and columns:

$$\boxed{\alpha} = \begin{array}{|cccc|l} \alpha_1^1 & \alpha_1^2 & \cdots & \alpha_1^n & W_1 \\ \alpha_2^1 & \alpha_2^2 & \cdots & \alpha_2^n & W_2 \\ & \vdots & & & \\ \alpha_n^1 & \alpha_n^2 & & \alpha_n^n & W_n \\ \hline W_1' & W_2' & & W_n' & \end{array}$$

The symbols $W_i (W_j')$ written to the right of row i (below column j) designate that the entries in row i (column j) are constrained to add to $W_i (W_j')$.

(Here $M([m])$ denotes an explicitly known normalization (see p. 68 below).)

For totally symmetric irreps $[m] = [\dot{p}0]$, the coefficients $C(..)$ above all have the value unity.

Example: If (M') and (M) are maximal then

$$B \begin{pmatrix} (max) \\ [M] \\ (max) \end{pmatrix} (A) = \prod_{k=1}^{n} (a_{12...k}^{12...k})^{M_{kn} - M_{k+1n}} ,$$

where $a_{1...k}^{1...k}$ is the determinant formed from the k bosons a_j^i, $i,j \leq k$.

Remarks:

(1) By construction the boson polynomials obey the product law:

$$B\begin{pmatrix}(\mu')\\[m]\\(\mu)\end{pmatrix}(\tilde{X}AY) = \sum_{(m)(m')} B\begin{pmatrix}(\mu)\\[m]\\(m)\end{pmatrix}(X)B\begin{pmatrix}(m')\\[m]\\(m)\end{pmatrix}(A)\begin{pmatrix}(\mu')\\[m]\\(m')\end{pmatrix}(Y) .$$

(Here the tilde denotes matrix transposition.)

The boson variables A appearing in B(A) are indeterminates. If we specialize $A \to U$ with U an $n \times n$ unitary matrix we see (from the product law) that the B([m])(U) are finite dimensional unitary matrix representations of U(n). (That they define irreps labelled by [m] follows from the action of the operators $J(\Lambda)$.)

(2) We are, however, free to interpret the indeterminates A as elements of an arbitrary non-singular complex matrix. Just as in remark (1) we can conclude that *these same polynomials are matrix irreps of* GL(n,C).

(3) It is helpful to translate these results for the boson polynomials into the language used by mathematicians. This is particularly useful since our results are quite explicit and involve a specific choice of basis.

To do so we first remark that the boson variables (adjoining unity) are what are called a "Weyl algebra" [cf., for example, Kirillov's book]; the boson variables viewed as operators are a "Heisenberg group." The extension (noted above) from unitary matrices to GL(n,C) is an instance of what is called "holomorphic induction."

To cast our results into basis-free language, we first use Gel'fand's decomposition of U(n): $g \epsilon U(n)$ $g = Z\Delta Z$, where Δ is diagonal, Z lower triangular (with 1's along diagonal) and Z upper triangular (with 1's along the diagonal).

By construction, the boson polynomials specialize to irreps of the subgroup Δ.

Now consider the boson polynomials of the form:

$$B\begin{pmatrix}(max)\\[m]\\(m')\end{pmatrix}(A).$$
(Here "max" denotes a maximal Gel'fand pattern i.e., a pattern whose weights are [m] itself.)

Choosing A to be of the form NA' when $N = Z\Delta$ and Z' is general we find

$$B\begin{pmatrix}(max)\\[m]\\(m')\end{pmatrix}(NA') = \Pi(N) \cdot B\begin{pmatrix}(max)\\[m]\\(m')\end{pmatrix}(A') , \qquad\qquad (*)$$

where $\Pi(N) \equiv \prod_{i=1}^{n} [\mu_{ii}(N)]^{m_{in}-m_{i+1,n}}$ with $\mu_{ii}(N)$ the minor of N formed from rows 1...i and columns 1...i. (A similar result obtains for $B\begin{pmatrix}(m)\\[m]\\(max)\end{pmatrix}$ when one uses: $A \to A'\Delta Z$.)

By the use of standard theorems, one recognizes that this property (*) characterizes the set of vectors $\{B\begin{pmatrix}(max)\\ [m]\\ (m')\end{pmatrix}(A)\}$ for $(m') \in [m]$ as the carrier space of an irrep $[m]$ of $U(n)$. This result is *independent of the basis* $\{(m')\}$.[13] A similar remark holds for $B\begin{pmatrix}(m)\\ [m]\\ max\end{pmatrix}$.

The factorization lemma:

We have seen that this matrix boson realization involves the direct product group $U(n) \times U(n)$. One sees in fact that this boson realization really involves the group $U(n^2)$ and all totally symmetric irreps thereof. This defines an imbedding of $U(n)$ in the sequence of groups $U(n^2) \supset U(n) \times U(n) \supset U(n)$, in which, moreover, the irrep labels of the two $U(n)$ groups in $U(n) \times U(n)$ coincide [we denote this by $U(n)*U(n)$]. This structure is the analog to that exhibited by the tensor operators of $U(n)$, and we exploit this formal analogy to discuss the factorization lemma.[8]

Let

$$\left|\begin{pmatrix}(M')\\ [M]\\ (M)\end{pmatrix}\right\rangle\!\!\!\rangle$$

denote a normalized basis vector in an irrep space of $U(n)*U(n)$. In this notation, the first $U(n)$ refers to the $U(n)$ group with generators $J_u(\Lambda)$, the second to the $U(n)$ group with generators $J_\ell(\Lambda)$. These two $U(n)$ groups are isomorphic but distinct (and commuting); the placement of the indices is merely a reminder as to which group is which ("upper" vs. "lower") – there is no other implication.

The star signifies that the Casimir invariants of the irreps of these two groups coincide. Hence, both

$$(M) = \begin{pmatrix}[M]\\ (M)\end{pmatrix} \quad \text{and} \quad (M') = \begin{pmatrix}(M')\\ [M]\end{pmatrix}$$

are Gel'fand patterns, the second one being inverted. The basis vectors may also be written in the form

$$\left|\begin{pmatrix}(M')\\ [M]\\ (M)\end{pmatrix}\right\rangle\!\!\!\rangle = M([M])^{-\frac{1}{2}} B\begin{pmatrix}(M')\\ [M]\\ (M)\end{pmatrix}(A)|0\rangle , \text{ where } B\begin{pmatrix}(M')\\ [M]\\ (M)\end{pmatrix}(A)$$

is an operator-valued polynomial in the set of boson operators $A = \{a^i_j\}$, the symbol $|0\rangle$ denotes the vacuum ket, and $M([M])$ is the measure of the highest weight tableau associated with $[M]$:

$$M([M]) \equiv \frac{\prod\limits_{i=1}^{n} (M_{in} + n - i)!}{\prod\limits_{i<j=1}^{n} (M_{in} - M_{jn} + j - i)} \ .$$

The introduction of $M^{-\frac{1}{2}}$ defines the manner in which the boson polynomials are normalized.

The boson polynomials are clearly tensor operators with respect
to transformations in the respective U(n) subgroup of U(n)*U(n),
denoted by the lower or upper Gel'fand pattern. As such, it must
be bilinear in the canonical Wigner operators which are defined,
respectively, on the two U(n) groups. *The factorization lemma
asserts that the precise form of this bilinear relation is:*

$$B \begin{pmatrix} (M') \\ [M] \\ (M) \end{pmatrix} (A) = M^{\frac{1}{2}} \sum_{(\Gamma)} \left\langle \begin{matrix} (\Gamma) \\ [M] \\ (M) \end{matrix} \right\rangle_{\ell} \left\langle \begin{matrix} (\Gamma) \\ [M] \\ (M') \end{matrix} \right\rangle_{u} M^{-\frac{1}{2}} \ , \qquad (**)$$

where M is an invariant operator of U(n)*U(n) which has eigenvalue
(defined above) equal to $M([m])$ for an arbitrary vector with labels
$[m]$. The indices ℓ and u designate the fact that the Wigner operators act, respectively, on the lower and upper Gel'fand patterns of
an arbitrary vector of U(n)*U(n). (Note that the two Wigner operators *commute* since they act in different spaces.)

Remarks:

(1)   There is an important special case of (**):

$$a_i^j = M^{\frac{1}{2}} \left(\sum_{\tau=1}^{n} \left\langle [1 \ \overset{\tau}{\ } \dot{0}]_i \right\rangle_{\ell} \left\langle [1 \ \overset{\tau}{\ } \dot{0}]_j \right\rangle_{u} \right) M^{-\frac{1}{2}} \ ,$$

where the $\langle 1\dot{0} \rangle$ denote the (unique) fundamental Wigner operators.
It is this special case that accounts for the term "boson factorization."

(2)   The result in (**) is important not only in affording a
method to calculate Wigner operators but also in showing that, at
this stage, all resolutions of the multiplicity are equivalent
(since an arbitrary orthogonal transformation leaves the (Γ) sum
invariant).

(3)   When we use the factorization lemma below we will take
the labels (Γ) to be the canonical labelling defined by null space
(cf. Sect. 3).

To simplify the discussion of the unit tensor operators in $U(n)$ it is useful to employ recursive methods and consider that all $U(n-1)$ operators are known.

We can then consider *projective operators*, $\begin{bmatrix}(\Gamma)\\ [M]\\ (\Gamma')\end{bmatrix}$, which are scalar products in $U(n-1)$ of unit tensor operators $\left\langle\begin{matrix}(\Gamma)\\ [M]\end{matrix}\right\rangle$ in $U(n)$ and $\left\langle\begin{matrix}(\Gamma')\\ [M]\end{matrix}\right\rangle_{n-1}$ in $U(n-1)$.

Note that both patterns (Γ) and (Γ') in a projective operator are operator patterns.

It is a remarkable fact that the explicit matrix elements of all extremal[17] unit $U(n):U(n-1)$ projective operators can be calculated from a few simple rules of the pattern calculus (cf. ref. 11). In particular, this class of explicitly known projective operators includes all *elementary* operators of the form $[\dot{1}_k \dot{0}_{n-k}]$ and $[\dot{0}_{n-k}\dot{1}_k]$ (a dot over a numeral implies that the numeral is repeated a number of times equal to the subscript), which themselves are *a basis for constructing all* $U(n)$ *tensor operators* (using Cartan multiplication).

Applications:

(1) The pattern calculus constitutes an effective technique *to calculate explicitly bases for all tensor operators in all* $U(n)$.

(2) For $U(3)$ there exists a simple way to resolve the multiplicity. Let us consider the projective operator $\begin{bmatrix}(\Gamma)\\ [m_{13}m_{23}m_{33}]\\ (\Gamma')\end{bmatrix}$.

If we specify (Γ') *to effect maximal shift, i.e.,* $\Delta = (m_{13}m_{33})$, *in the* $U(2)$ *subgroup* $[m_{13}\ m_{33}]$*, then the operator vanishes unless* $(\Gamma)=(\Gamma_{Stretched})$. In other words, the choice of this splitting configuration projects onto the unique operator having maximal null space (denoted by Γ_S).

Using the factorization lemma, one then sees that the sum reduces to a single term and that moreover choosing the splitting configuration in the lower $U(2)$ group *determines the operator* $\left\langle\begin{matrix}\Gamma_s\\ [M]\end{matrix}\right\rangle$ *on arbitrary irrep labels.*

(2) Let us now sketch, by an example, exactly how the factorization lemma serves to give an elegant definition of the denominator function for the operator $\left\langle\begin{matrix}\Gamma_s\\ [M]\end{matrix}\right\rangle$. We consider the operator $\langle[6\ 3\ 0]\rangle$ as an example.

The boson polynomial corresponding to extremal states is given by:

$$B\begin{pmatrix}\text{extr.}\\ [6\ 3\ 0]\\ \text{extr.}\end{pmatrix} = (a_1^1)^3\ (a_{13}^{13})^3$$

Let us choose the initial state to be:

$$|\text{init.}\rangle = \left| \begin{matrix} & & m_{12} & & \\ & m_{12} & & m_{22} & \\ m_{13} & & m_{23} & & m_{33} \\ & m_{12} & & m_{22} & \\ & & m_{12} & & \end{matrix} \right\rangle$$

and the final state to be:

$$|\text{final}\rangle = \left| \begin{matrix} & & m_{12}+6 & & \\ & m_{12}+6 & & m_{22} & \\ m_{13}+3 & & m_{23}+3 & & m_{33}+3 \\ & m_{12}+6 & & m_{22} & \\ & & m_{12}+6 & & \end{matrix} \right\rangle .$$

(Note that we have chosen maximal SU2 shifts to effect the splitting.)

From the factorization lemma we have:

$$\langle\text{final}|B(\ldots)|\text{init.}\rangle = \left(\left\langle (m)+\Delta=\begin{pmatrix} 333 \\ 6\ 0 \end{pmatrix} \right| \begin{bmatrix} & \Gamma_s & \\ 6 & 3 & 0 \\ & 6 & 0 \\ & & 6 \end{bmatrix} | (m) \rangle \right)^2 .$$

(The matrix element on the right is squared since the same matrix element enters for both the upper and lower U(3) groups. We omit the operator M which will drop out. See below.)

Next we apply the factorization lemma to the boson operators $a^\dagger$ and a^{13}_{13} themselves. Noting that the operator M drops out, we find:

$$\left(\left\langle (m)+\Delta \left| \begin{bmatrix} & 3 & \\ 6 & 0 & \\ 6 & 3 & 0 \\ & 6 & 0 \\ & & 6 \end{bmatrix} \right| (m) \right\rangle \right)^2$$

$$= \left\langle \text{final} \left| \left(\sum_\gamma \begin{bmatrix} & \gamma & \\ 1. & 0 & 0 \\ & 1 & 0 \\ & & 1 \end{bmatrix}_u \begin{bmatrix} & \gamma' & \\ 1 & 0 & 0 \\ & 1 & 0 \\ & & 1 \end{bmatrix}_\ell \right)^3 \right. \right.$$

$$\left. \left. \left(\sum_{\gamma'} \begin{bmatrix} & \gamma' & \\ 1 & 1 & 0 \\ & 1 & 0 \\ & & 1 \end{bmatrix}_u \begin{bmatrix} & \gamma' & \\ 1 & 1 & 0 \\ & 1 & 0 \\ & & 1 \end{bmatrix}_\ell \right)^3 \right| \text{initial} \right\rangle , \quad (+)$$

where $\Delta(\gamma)+\Delta(\gamma') = (333)=\Delta(\Gamma_s)$.

Now we make use of the standard form[12] of stretched projective operators:

$$
\begin{bmatrix} & \Gamma_s & \\ M_{13} & M_{23} & M_{33} \\ & M_{13} & M_{33} \\ & M_{13} & \end{bmatrix} = \frac{\text{Numerator Pattern Calculus Factor}}{\text{Denominator} \left\langle \begin{matrix} \Gamma_s \\ [M] \end{matrix} \right\rangle \cdot (U(2) \text{ Denominator})}
$$

Inserting this in (+) we find that both the numerator pattern calculus factor and the U(2) denominator function *cancel on
both sides of Eq. (+) -- regardless of the order of the* [100] *and*
[110] *operators*. Moreover, the U(3) denominator functions enter
as *squares*.

These changes simplify the resulting formula for the denominator function in a most remarkable way. Let us denote the denominators of the elementary operators [100] and [110] by $d(\lambda)$ where
λ is one of the six possible shifts: (100), (010) or (001) for
[100] or (110), (101), (011) for [100]. Then we find:

$$
\frac{3!3!}{6!} \cdot (\text{Denom} \left\langle \begin{matrix} \Gamma_s \\ 630 \end{matrix} \right\rangle)^{-2} = \sum_{\lambda_1 \cdots \lambda_6} (\prod_{i=1}^{6} d^2(\lambda_i))^{-1}
$$

where the sum is over all possible values of the λ_i such that the
total shift is (333).

Within the last few weeks we have found a sweeping generalization of this formula. Let us first recognize that any unit tensor operator in U(n) may be written in terms of the Cartan multiplication of elementary operators:

$$
\langle [M] \rangle \propto \prod_{i=1}^{n-1} (\langle \dot{1}_i \; \dot{0}_{n-i} \rangle)^{M_{in} - M_{i+1,n}} .
$$

Using this form for the general U(n) unit tensor operator, we
have shown that *the denominator function of the operator* $\langle [M] \rangle$
having maximal null space (Γ_s) *is given by the sum-over-paths
formula*:

$$
(\text{Denom} \langle [M] \rangle)^{-2} = (\text{numerical constant}) \cdot
$$

$$
\cdot \sum_{\text{paths}} [\prod (\text{elementary denominators})^2]^{-1}
$$

where the sum-over-paths is constrained by:

(a) the elementary denominators induce the total shift
$\Delta(\Gamma_s) = (\Delta_1 \Delta_2 \ldots \Delta_n)$;

(b) all possible paths are admitted with equal weight, pro-
vided the number of elementary denominators of type $\langle \dot{1}_i \dot{0}_{n-i} \rangle$ is
precisely $M_{in} - M_{i+1,n}$.

*This result implies an elegantly simple canonical definition
of the unit tensor operator in U(n) having maximal null space.*
Moreover, it can be shown that this is the first step in an itera-

tive procedure that resolves all multiplicity.

5. *The Existence of a New Type of Weight Pattern Space*

The use of null space to classify operators is a remarkably powerful concept but there are great difficulties in actually calculating with this concept, since it is non-constructive. For U(3) we were very fortunate in finding a splitting configuration that projected uniquely onto that operator in any given multiplicity set having maximal null space. This procedure can be iterated to determine all U(3) operators canonically.

The important lesson to be learned from these explicit results is that *all information on null space is encoded in the* U(3) *denominator function*. Complete knowledge of the unit tensor operator is -- for null space considerations -- inessential; one need only determine the denominator function.

Stated differently: *The canonical resolution of the multiplicity is given by the existence of Denominator functions which effect the dissection of the intertwining number-null space diagram into characteristic sets.*

The denominator functions (Denom $(_{[M]}^{\Gamma}))^2$ determine the characteristic functions by the association (in the lexical region):

$$(\text{Denom})^2 > 0 \Rightarrow X(\Gamma) = 1$$

$$(\text{Denom})^2 = \cdot 0 \Rightarrow X(\Gamma) = 0 \text{ , where } X(\Gamma) \text{ is the characteristic}$$
$$\text{function.}$$

The resulting "nesting property" of the characteristic sets then yields: $X(\Gamma_i)X(\Gamma_j) = X(\Gamma_i) \Longleftrightarrow N(\Gamma_i) \subset N(\Gamma_j)$, where $N(\Gamma_i)$ denotes the null space of the operator $\langle[M]\rangle$ having operator pattern Γ_i.

The required denominator functions [12] have been constructed in U(3) and have these properties:

(1) They are invariant under the S(n) group acting on the variables $\{p_{in}\}$;

(2) They have the form:
$$(\text{denom fcn})^2 = (\text{Product of linear functions}) \times$$
$$\times (\text{ratio of two polynomials}).$$

(3) The ploynomials, $G_q^{(t)}$,entering this ratio in (2) have remarkable symmetry properties.

In order to make these results intuitively understandable, let us consider in detail the tensor operator $\langle 6\ 3\ 0 \rangle$ (the self-conjugate "64-plet" operator) which induces the shifts $(3\ 3\ 3)$ on an arbitrary irrep $[m]$. There are four distinct tensor operators with operator patterns (Γ) explicitly given by:

$$\Gamma_1 = \begin{pmatrix} & 3 & \\ & 6 & 0 \\ 6 & 3 & 0 \end{pmatrix} \qquad \text{largest null space ("stretched operator pattern")}$$

$$\Gamma_2 = \begin{pmatrix} & 3 & \\ & 5 & 1 \\ 6 & 3 & 0 \end{pmatrix}$$

$$\Gamma_3 = \begin{pmatrix} & 3 & \\ & 4 & 2 \\ 6 & 3 & 0 \end{pmatrix}$$

$$\Gamma_4 = \begin{pmatrix} & 3 & \\ & 3 & 3 \\ 6 & 3 & 0 \end{pmatrix} \qquad \text{minimal null space}$$

The general intertwining number–null space diagram for U(3) has the form:

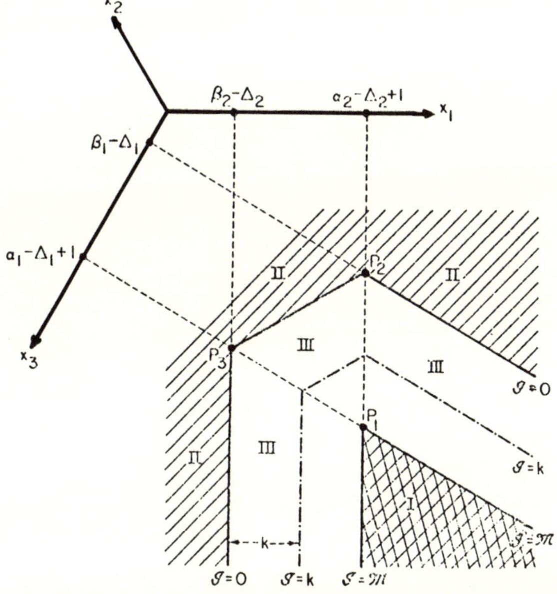

Fig.1. The intertwining number–null space diagram. The intertwining number $\mathcal{J}$ is defined at each lattice point (points having integral coordinates) of the barycentric plane. At each lattice point in the cross-hatched region I, including the bent solid line, the value of $\mathcal{J}$ is zero; at each lattice point in the region III between the two bent solid lines, the value of $\mathcal{J}$ is 1,2,...,M–1, its value being k at the lattice points on the bent dash-dot line designated by $\mathcal{J}$ = k.

How shall we dissect this diagram into characteristic regions? We have no choice in this dissection, since the separation into characteristic sets is fully determined by the resolution of the multiplicity.

From explicit knowledge of the actual tensor operators we find:

$$(\text{Denom } (\Gamma_t, x))^{-2} = G_k^t(x)/G_k^{t-1}(x) \cdot \prod_{i=1}^{3} \prod_{\lambda=1}^{k-t+1} (x_i^2 - \lambda^2) \qquad (++)$$

The first characteristic feature to observe in Eq. (++) is the *existence of lines of zeroes*, that is, explicit linear factors in the (denom)^2.

Let us remove these linear factors. We obtain $G_k^t(x)/G_k^{t-1}(x)$, that is, a ratio of polynomials. In the lexical region (the region interior to $I=0$ in Fig. 1) we find explicitly that this ratio of polynomials has only first order poles and zeroes. For each of the four cases we find:

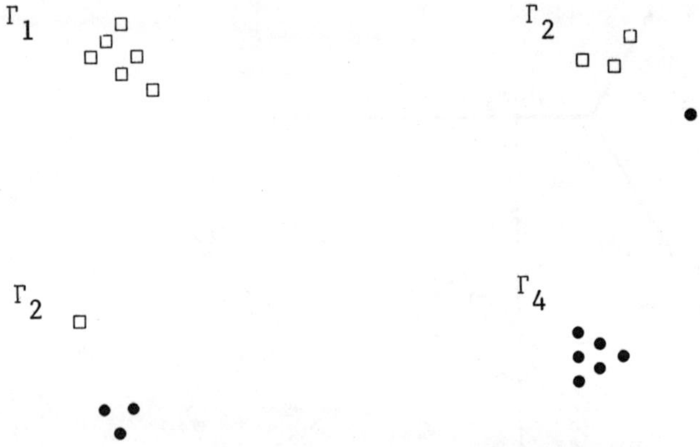

Γ_1

Γ_2

Figure 2

Γ_2

Γ_4

Note: □ = poles,   ● = zero.

To explain this curious assortment of poles and zeroes (all first order) one looks again at the denominator function defined in (++).

The G_q^t functions entering into (++) are explicitly known[14].

If we examine these G-functions we find they are highly symmetric functions which are non-negative in the lexical region and *whose zeroes have weight space patterns*.

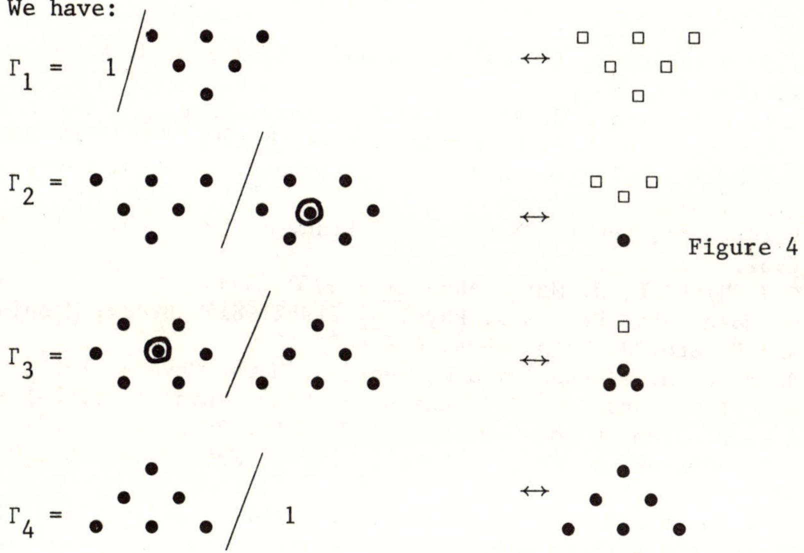

$G^{(1)}$: $\leftrightarrow$ $[\,2\ \ 0\ \ 0\,]$

$G^{(2)}$: $\leftrightarrow$ $[\,1\ \ 0\ -1\,]$ Figure 3

$G^{(3)}$: $\leftrightarrow$ $[\,0\ \ 0\ -2\,]$

*These are the new weight space patterns which were promised
in the title of this talk.* It is a general property implied by
the canonical resolution of the multiplicity using null space.

One detail remains to show how the zeroes and poles arise
from these patterns.

We have:

$\Gamma_1 = $ $\leftrightarrow$

$\Gamma_2 = $ $\leftrightarrow$ Figure 4

$\Gamma_3 = $ $\leftrightarrow$

$\Gamma_4 = $ $\leftrightarrow$

Let us remark that the family of functions $G^t_q(x,\Delta)$ have some
most remarkable properties, and have proved to be of interest to
mathematicians concerned with special functions.

It is sometimes asserted that the tensor operator classification problem -- just as for the Wigner coefficients in angular
momentum theory -- are not really of basic interest because these
operators are "basis dependent". That this view is invalid can be

seen from the $G_q^{(t)}$ functions which implement the canonical split:
the information contained in these symmetry patterns is invariant
under the operator-space S_n group and is obviously *basis free.*

Concluding Remarks

The most natural question at this point is probably this:
Do these weight space properties generalize to arbitrary $U(n)$?
Stephen Milne has recently shown that the $G(\Delta,x)$ functions do have
a generalization to all n preserving the known weight space
properties. This is a major advance and the open question now is
whether the null space dissection in $U(n)$ implies this particular
generalization. This question is (as might be expected!) under
current study.

References and Footnotes

1. S.J. Alisauskas, A.-A.A. Jucys and A.P. Jucys, J. Math. Phys.
    13 (1972) 1349.
2. G.E. Baird and L.C. Biedenharn, J. Math. Phys. 5 (1964) 1730.
3. V.Bargmann and M.Moshinsky,Nucl.Phys.18(1960)697;23(1961)177.
4. E. Chacon, M. Ciftan and L.C. Biedenharn, J. Math. Phys. 13
    (1972) 577.
5. Biedenharn, Louck, Chacon and Ciftan, J. Math. Phys. 13 (1972)
    1957.
6. C.J. Hinrich, J. Math. Phys. 16 (1975) 2271.
7. M. Moshinsky, Rev. Mod. Phys. 34 (1962) 813; Brody, Moshinsky
    and Renero, J. Math. Phys. 6 (1965) 1540
8. Biedenharn, Giovannini and Louck, J. Math. Phys 8 (1967) 691.
9. J.D. Louck and L.C. Biedenharn, J. Math. Phys. 11 (1970) 2368.
10. L.C. Biedenharn and J.D. Louck, J. Math. Phys. 13 (1972) 1985.
11. L.C. Biedenharn and J.C. Louck, Commun. Math. Phys. 8 (1968)
    89
12. J.D. Louck and L.C. Biedenharn, J. Math. Phys. 14 (1973) 1336.
13. One of the authors (JDL) wishes to thank Professor W.H. Klink
    for discussions related to this point.
14. L.C. Biedenharn, M.A. Lohe and J.D. Louck, in "Group Theoreti-
    cal Methods in Physics", edited by A. Janner, T. Janssen and
    M. Boon (Springer Verlag, Berlin 1976) p. 395.
15. Operators whose eigenvalues are the individual m_{ij} are con-
    structed in ref. 9 (cf. p. 2384)
16. It is numerically convenient to use n shift labels (for $U(n)$
    operators) even though the operator action is defined to
    carry $SU(n)$ irreps into $SU(n)$ irreps. The unimodular re-
    striction can always be imposed on the shift labels subse-
    quently. (This situation has an analogy in defining uni-
    modular weights.)
17. Extremal means that the shifts induced by the operator are a
    permutation of the (operator) irrep labels.
+   Supported in part by the National Science Foundation
    and performed under the auspices of the USERDA.

DE SITTER FIBERS AND SO(3,2) SPECTRUM GENERATING GROUP FOR HADRONS

A. Bohm

Physics Department
The University of Texas at Austin
Austin, Texas 78712

1. INTRODUCTION

In order to convince you of the relevance of my topic for an
Einstein Centennial celebration let me start with a quote.[1]

> Suppose that the fundamental idea of the general
> relativity theory is not only limited to gravitational
> processes but also applicable to the processes of strong
> interaction, then one is led to the conclusion that in
> the small domains, in which the process is relevant,
> the space is curved. In the same way as the gravita-
> tional interactions might lead in a certain state of
> equilibrium to the de Sitter universe (the most symmetri-
> cal curved space), the strong interactions might cause--
> if the equilibrium is not disturbed by the presence of
> other nearby systems--a de Sitter world of much smaller
> size because the gravitational interactions are much
> weaker than the strong interactions. The group of
> motion in this small curved domain is no longer the
> Poincaré group, but the de Sitter group, in particular
> the (4+1)-de Sitter group, if we assume that our physi-
> cal model system is finite in space and infinite in
> time direction. ... As the dynamics in general rela-
> tivity are a property of the space-time-metric, it
> should also be inherent in the group of motion of the
> curved space.

This idea of connecting hadrons with miniature de Sitter spaces
attached to the space-time points of the Minkowski space has come
about in two independent ways: (1) The phenomenological-

mathematical way[1] started from the desire to obtain the following
discrete mass spectrum for hadrons

$$m^2 = m_0^2 + \lambda^2 s(s + 1) \ . \tag{1}$$

This mass spectrum (which we will call the relativistic rotator
spectrum) seems to be supported by experimental evidence and has
an analogy in the non-relativistic quantum mechanical rotator which
is realized by numerous physical systems in molecular and nuclear
physics. The experimental evidence for (1) was earlier not very
strong and it disagrees with the prevailing belief of linear rising
Regge trajectories. Later experimental results, however, strongly
favor (1). To obtain the relativistic rotator spectrum required a
very particular construction of the SO(4,1) de Sitter algebra in
terms of the Poincaré algebra and it was this construction that
lead to the above geometrical interpretation.

(2) The philosophical-mathematical way[2] is based on the intention to understand elementary particles without asking what they
"consist of" and clearly shows Heisenberg's influence. To formulate
internal degrees of freedom differential geometrically led to the
Cartan type fiber bundles introduced into mathematics in 1922. The
simplest possibility for such fiber bundles having the Minkowski
space as base space is a de Sitter bundle, which again leads to the
above geometric interpretation.

This talk is divided into five parts. First I shall give a
vague description of Drechsler's de Sitter fiber bundle.[2] Then I
give a brief review of Biedenharn and van Dam's "dynamical stability
group"[5,6] which functions as a spectrum generating group. The theory
that I shall present can be understood as a combination of the two
above proposals: the system which this theory describes I like to
call the rigid relativistic rotator--the subject of part four. I
shall slightly generalize this idea by (a) allowing for a discrete
number of radii of the de Sitter fibers instead of one radius,
which can be interpreted as taking for the de Sitter radius an
operator with discrete spectrum instead of a number and (b) by
taking instead of Biedenharn and van Dam's remarkable representations[6,7] of SO(3,2) a more complicated representation which I
think is even more remarkable. This generalization is the general
relativistic rotator. The particle spectrum of this de Sitter
rotator I shall then compare with the experimental data in the last
part of my talk on experimental evidence for the de Sitter spectrum.

2. DRECHSLER'S DE SITTER FIBER BUNDLE

Drechsler's idea originates from Cartan's 1922 and 1924 papers[3]
and is based on the intention to understand elementary particles
without asking the question of what they consist of. According to

this idea the space in which the physical phenomena take place is not just the (3+1) dimensional space but an eight dimensional differentiable manifold. This is a (4+1) de Sitter bundle over the space-time for which we take the Minkowski space (we ignore gravitational effects).

A fiber bundle is in a certain sense a union of fibers F_x:

$$E = \bigcup_{x \in B} F_x \tag{2}$$

where F_x for every $x \in B$ (the "base space") is related to a "standard fiber" F by a map (diffeomorphism, homeomorphism):

$$F_x \to F \; . \tag{3}$$

Base space B and fiber F are (usually) differentiable manifolds and two maps

$$F_x \to F$$

$$F_x' \to F$$

differ from each other by an element of a group, the structure group of the fiber bundle, or in physical terms, the gauge group G.

A Cartan type bundle is a fiber bundle in which the fiber is "soldered" to the base space in such a way that the fiber over $x \in B$ is tangent to B for every $x \in B$. This can be done when certain conditions are fulfilled for B and F (e.g., dim B = dim F) and is achieved by the identification:

$$T_{\overset{\circ}{\xi}}(F_x) \equiv T_x(B) \tag{4}$$

where $T_x(B)$ is the tangent space to B at x and $T_{\overset{\circ}{\xi}}(F_x)$ is the tangent space to F_x at the point of contact $\overset{\circ}{\xi} \in F_x$, $\overset{\circ}{\xi} \equiv x$. See Fig. 1.

The (4,1)-de Sitter bundle $T^R(V_4)$ is a special Cartan type bundle[2]:

$$T^R(V_4) = \bigcup_{x \in V_4} V_4'(x) \tag{5}$$

where V_4 is the Minkowski space and $V_4'(x) \to V_4' = SO(4,1)/SO(3,1)$ is the (4,1)-de Sitter space, and where the structure group or gauge group G is SO(4, 1). Thus the de Sitter bundle is an expansion of the usual Minkowski space V_4 by attaching one copy of the de Sitter space V_4' to each point x of V_4 as shown in Fig. 2.

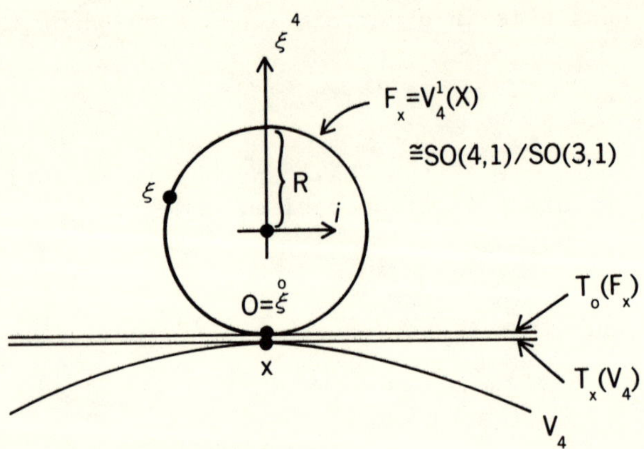

Fig. 1.   de Sitter fiber soldered to the base space, V_4 (drawn here
as curved). Of the base space, only one space dimension,
x, is drawn. For the time dimension the circle goes over
into a hyperbola.

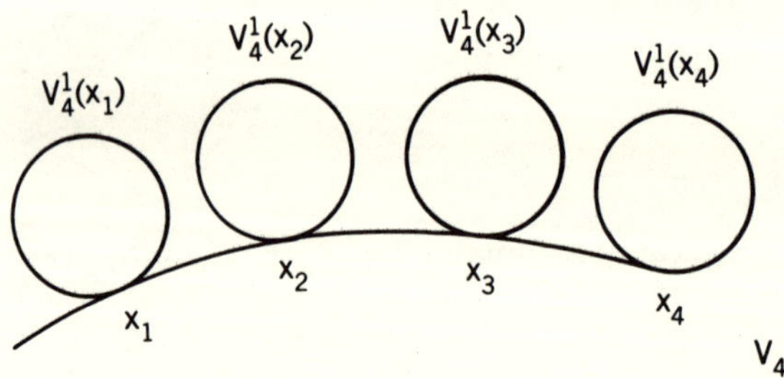

Fig. 2.   Illustration showing how the de Sitter fiber bundle has
a de Sitter space V_4^1 attached to each point of the
Minkowski space V_4.

It is not quite the direct product of base space V_4 and de Sitter space V_4' but sort of the direct product modulo a transformation of SO(4, 1) acting on V_4': going from one point in the base space to another point one also makes a transformation in the fiber.

The de Sitter space can be imbedded into a 5-dimensional pseudo-Euclidean space with signature (+----) and is then given by a 4-dimensional hyperboloid which is infinite in time and finite in space direction. In the two-dimensional picture of Fig. 1, one of the space directions x is represented together with the auxiliary 4-direction orthogonal to the Minkowski space (for the t-directions one does not have a circle but a hyperbola). In the coordinates ξ^a of the 5-dimensional pseudo-Euclidean space the hyperboloid is given by

$$\xi^a \xi_a = \xi^a \xi^b \eta_{ab} = -R^2 \qquad \eta_{ab} = (+----) \qquad (6)$$

$$a,b = 4,\mu,\nu \qquad \mu,\nu = 0,1,2,3$$

where R is the radius.

Originally a distance has only a physical meaning in the Minkowski space, where it is measured in cm, and the radius R of an ordinary fiber bundle would have no meaning. However, since in the Cartan type bundle one identifies the Minkowski space $V_4 = T_x(V_4)$ with the tangent space to the fiber over x:

$$V_4 = T_\xi(V_4'(x)) , \qquad (4')$$

the meaning of distance is transferred to the de Sitter fiber, so that R is again measured in cm. Later we will see that experimental data require R to be of the order of 10^{-13} cm.

The de Sitter group SO(4,1) is the group of transformations $\xi^{a'} = A^a_b \xi^b$ that leaves $\xi^a \xi_a = \xi^{a'} \xi_{a'} = -R^2$ invariant, i.e., the group of motions in the de Sitter space. The generators are given by the intrinsic differential operators

$$\lambda_{ab} = -i\left(\xi_a \frac{\partial}{\partial \xi^b} - \xi_b \frac{\partial}{\partial \xi^a}\right) \qquad a,b = 0,1,2,3,4$$

where $\lambda_{4\mu}$, $\mu = 0,1,2,3$ are the generator of the rotation (pseudo-rotation for $\mu = 0$) in the 4-μ plane and $\lambda_{\mu\nu}$, $\mu,\nu = 0,1,2,3$, are SO(3,1)-generators that generate rotations (pseudo-rotations) around the ξ^4 axis leaving the point $\overset{\circ}{\xi}$ fixed. If one introduces $b_\mu = \frac{1}{R} \lambda_{4\mu}$ then b_μ is the generator of a translation (in cm) along the de Sitter fiber.

The Poincaré group is the group of motions in the Minkowski space, $p_\mu = -i\partial_\mu$ are the generators of translation along the space-time coordinates and $\ell_{\mu\nu} = -i\left(x_\mu \frac{\partial}{\partial x^\nu} - x_\nu \frac{\partial}{\partial x^\mu}\right)$ are the generators of transformations in V_4 leaving the point x fixed. From the identification (4') follows

$$\lambda_{\mu\nu} = \ell_{\mu\nu} \, . \tag{7}$$

The gauge group SO(4, 1) has the Lorentz group as a subgroup. If one fixes the gauge[4] by choosing $\xi^\mu(x) = 0$, then the residual structure group is the Lorentz group. Gauge transformations of this subgroup are symmetry transformations; transformations generated by the b_μ, which change the points ξ^μ along the fiber, may not be symmetry transformations.

So far we have described the de Sitter group SO(4, 1) and the Poincaré group P as they arise in a geometrical picture. A quantum physical system is not described by a geometrical picture (though such a picture may be helpful) but by an algebra of operators in the space of physical states. To the group of motion in Minkowski space correspond physical systems described by the irreducible unitary representation spaces $H(m,s)$ of the Poincaré group. The geometrical operators p_μ, $\ell_{\mu\nu}$, are represented by Hermitian operators $P_\mu = U(p_\mu)$, $L_{\mu\nu} = U(\ell_{\mu\nu})$. And, as I am a follower of Wigner, I shall use these representation spaces and the algebra of observables generated by P_μ, $L_{\mu\nu}$ as the basis for my description of hadrons: The quantum physical system "elementary particle" has as its mathematical image the irreducible representation space $H(m, s)$ of the Poincaré group P.

To introduce some interaction one may try to use the substitution in analogy to the introduction of the electromagnetic interaction ("minimal coupling"). In classical mechanics the electromagnetic interaction is introduced by the substitution

$$\vec{p} \to \vec{\pi} = \vec{p} + e\vec{A} \, . \tag{8}$$

In quantum mechanics the numbers should be replaced by operators so that

$$\vec{P} \to \vec{\Pi}^{op} = \vec{P} + e\vec{A}^{op} \, . \tag{9}$$

In the position representation of quantum mechanics where

$$\langle \vec{x}|P_i|\phi\rangle = \frac{1}{i}\frac{\partial}{\partial x_i}\langle \vec{x}|\phi\rangle = \frac{1}{i}\frac{\partial}{\partial x_i}\phi(x) \qquad i = 1,2,3 \, . \tag{10}$$

This leads to

$$\langle \vec{x} | \Pi_i^{OP} | \phi \rangle = \frac{1}{i} \frac{\partial}{\partial x_i} \langle \vec{x} | \phi \rangle + e \langle \vec{x} | \vec{A}^{OP}(Q_j) | \phi \rangle \tag{11}$$

or

$$\frac{1}{i} D_i \langle \vec{x} | \phi \rangle = \left\{ \frac{1}{i} \frac{\partial}{\partial x_i} + e A_i(x_j) \right\} \langle \vec{x} | \phi \rangle \tag{12}$$

where it was assumed that the operator $A_i^{OP}(Q_j)$ is only a function of the position operators Q: $Q_i | \vec{x} \rangle = x_i | \vec{x} \rangle$. In this form the description of the interaction is generalized and leads to the general definition of the covariant derivative for non-abelian gauge groups

$$D_\mu = \partial_\mu + i e A_\mu^a(x_\nu) U(I_a) \qquad \mu = 0,1,2,3 \tag{13}$$

where I_a are the generators of the non-abelian gauge group G and $U(I_a)$ are their hermitian representations in a unitary representation: $G \to g \to U(g)$. Here the $A_\mu^a(x_\nu)$ are numbers, which are functions of the numbers x_ν. (13) is the relativistic non-abelian analogue of (12) which is only identical with (9) in the position representation, i.e. when applied to the "wave function in the position representation." Independent of the question what the meaning of a position representation should be in a relativistic quantum theory, (13) is certainly not the full analogue of (9) if one wants to use it in a more general way than just applying it to functions $\phi(x_\nu)$ of the coordinates x_ν. In general, the analogue of (9) for the introduction of interaction in a relativistic theory with a non-abelian gauge group would be the substitution:

$$P_\mu \to B_\mu = P_\mu + \frac{1}{2}\{b_\mu^a, U(I_a)\} \tag{14}$$

where the b_μ^a are not numbers but operators and are the quantum mechanical analogues of the numbers $e A_\mu^a$. This general form of the "minimal coupling" would then hold not only when applied to fields (functions of x_ν) but also as a general operator expression that can be applied to general vectors of the space of physical states.

In our case we choose for the $U(I_a)$ the representatives $L_{\rho\sigma}$ of the generators for the residual gauge group. The problem is then to find for a particular interaction the operators b_μ^a. We make the ansatz:

$$b_\mu^{\rho\sigma} = \lambda \eta^\rho_\mu \hat{P}_\sigma \qquad \text{where} \qquad \hat{P}_\sigma = M^{-1} P_\sigma , \qquad M = (P_\mu P^\mu)^{\frac{1}{2}}$$

and λ is a constant of the dimension MeV.

Then the substitution (14) has the form

$$P_\mu \to B_\mu = P_\mu + \frac{\lambda}{2} M^{-1} \{P^\sigma, L_{\mu\sigma}\} . \tag{15}$$

From this it follows that

$$[B_\mu, B_\nu] = i\lambda^2 L_{\mu\nu} \tag{16}$$

so that the B_μ, $L_{\mu\nu}$ generate a representation of $SO(4,1)$, which is the group of motion in a de Sitter space with radius $R = \frac{1}{\lambda}$. By the Inönü-Wigner contraction[8a]:

$$B_\mu \xrightarrow[\lambda \to 0]{} P_\mu$$

the de Sitter group goes into the Poincaré group[8b] and the de Sitter bundle goes into the affine tangent bundle.[8c]

Mathematical formulas similar to (15) have reappeared several times in the literature over the past three decades, however the operator corresponding to the B_μ had the entirely different meaning of a center-position[9] and when it was realized that they are related to $SO(4,1)$, the de Sitter space was chosen to have the radius of the universe.[10]

3. A SPECTRUM-GENERATING GROUP AS THE DYNAMICAL STABILITY GROUP OF BIEDENHARN AND VAN DAM

If one wants to describe not a single hadron but a tower of hadrons, each hadron being considered as a different state of the system "hadron-tower," one has to take a reducible representation space of the Poincaré group P and introduce operators which describe transitions between the different irreps $H(m, s)$ of P. A scheme which does this is the spectrum generating group $SO(3,2)$[6] or the dynamical stability group of Biedenharn and van Dam.[5] They choose a Lorentz vector operator Γ_μ. (They call it V_μ for vector current, in Ref. 5. I call it Γ_μ because in a particular "4-dimensional" version, in which the representation space is identical with the space of solution of the Dirac equation, the matrix elements $\langle|\Gamma_\mu|\rangle = \frac{1}{2}\gamma_\mu$, the Dirac matrices.) This vector operator together with the part of the Lorentz generators $L_{\mu\nu}$ which in the spinor basis act only on the indices and which I call $S_{\mu\nu}$ form an $SO(3,2)_{S_{\mu\nu}\Gamma_\mu}$ algebra. (The matrix elements $\langle|S_{\mu\nu}|\rangle = \frac{1}{2}\sigma_{\mu\nu}$ in the "4-dimensional" version.) This $SO(3,2)$ is the anti-de Sitter group which leaves $\eta^{5^2} + \eta^{0^2} - \eta^{1^2} - \eta^{2^2} - \eta^{3^2}$ invariant. However unlike the $SO(4, 1)_{B_\mu, L_{\mu\nu}}$ de Sitter group there is no known geometrical meaning for the $SO(3, 2)_{\Gamma_\mu S_{\mu\nu}}$. It just serves as the spectrum generating group.

Due to their vector operator character the Γ_i $i = 1,2,3$ will change the spin. Biedenharn and van Dam choose a very particular representation of this SO(3, 2), which Dirac[7] called the "remarkable representation" but which is equivalent to the representation which Majorana found decades ago[7] and which I therefore call the Majorana representation. The reduction of the integer spin Majorana representation into irreducible representations of the Poincaré group is given by

$$H(\text{Majorana}) \doteq \sum_{\substack{s=0,1,2\ldots \\ |n|=s+1}} \oplus\, H^n(m_{(n,s)}, s) \ . \tag{17}$$

n is the eigenvalue of $\Gamma_\mu P^\mu M^{-1}$ (or Γ_0 in the rest frame) and is a label which is redundant in the Majorana representation because $|n| = s + \frac{1}{2}$. (The reduction in the particular "4-dimensional case is given by $H \doteq H^{n=-\frac{1}{2}}(m, s = \frac{1}{2}) \oplus H^{n=\frac{1}{2}}(m, s = \frac{1}{2})$ which is the familiar space of solutions of the Dirac equation.)

Whether Γ_i will change the mass and in what way it will change the mass depends upon what Biedenharn and van Dam call the postulated constraint-relation. This constraint can be arbitrarily imposed: $P_\mu P^\mu = f(s)$, and Biedenharn and van Dam favor the one which leads to hadron Regge bands $m^2 = \alpha + \beta s$. If one assumes the constraint $[P_\mu P^\mu, \Gamma_\gamma] = 0$ one obtains a degenerate mass spectrum. And if one postulates the constraint $P_\mu \Gamma^\mu = \kappa = \text{const}$, one obtains the equally unrealistic Majorana spectrum $m = \dfrac{\kappa}{s + \frac{1}{2}}$.

4. COMBINATION OF BIEDENHARN-VAN DAM'S DYNAMICAL STABILITY GROUP
    WITH DRECHSLER'S DE SITTER FIBER BUNDLE: THE RELATIVISTIC
    ROTATOR

If--according to Drechsler's idea--the internal motion of the hadron is viewed as an excitation along the de Sitter fiber, then one should not take Biedenharn-van Dam's Regge band constraint or the elementary particle constraint $[P_\mu P^\mu, \Gamma_\nu] = 0$, but the constraint

$$[Q, \Gamma_\nu] = 0 \tag{18}$$

where

$$Q = \frac{1}{\lambda^2} B_\mu B^\mu - \frac{1}{2} L_{\mu\nu} L^{\mu\nu} = \text{Casimir operator of SO(4, 1).}$$

As Q commutes already with all the other generators except the Γ_μ, the constraint (18) makes the invariant operator of the de Sitter space Q into the invariant operator of the whole algebraic structure

which describes the de Sitter bundle. Its eigenvalue α^2 will then
characterize the physical system (in the same way as the eigenvalue
of $M^2 = P_\mu P^\mu$ characterizes the elementary particle). From this constraint relation it follows after a lengthy calculation

$$m^2_{(n,s)} = \lambda^2(\alpha^2 - 9/4) + \lambda^2 s(s + 1) \qquad \lambda = \frac{1}{R} . \tag{19}$$

If one chooses as an example the tower of $I = 0$ mesons, one obtains
from the experimental values of $m_\omega (s = 1^-)$ and $m_U (s = 4^+)$ the empirical value for λ and R:

$$\lambda^2 \approx 0.29 \; \text{Gev}^2 \Rightarrow R \approx \frac{1}{3} \cdot 10^{-13} \; \text{cm} \left(R = \frac{\hbar c}{\lambda} \right) . \tag{20}$$

Thus the experimental data provide the radius of Drechsler's de
Sitter fiber. (19) is a simple relation, and I have often been
asked whether this simple result (which is not even the full SU(6)
mass formula) warrants the enormous mathematical apparatus that was
necessary to derive it. The answer is of course that this is an
entirely different type of result from the SU(3) of SU(6) "mass"
formulas. These are obtained from the assertion that an operator
with a certain transformation property is the mass operator. That
this "mass operator" has something to do with the observable $P_\mu P^\mu$
which is measured as the mass-square and that its expectation values
(and what is called spin in the SU(6) mass formula) are the quantities that characterize the elementary particles, is never even
attempted to be shown. Here, the quantities m and s in (19) are
exactly the quantities that characterize (according to Wigner) the
elementary particles and (19) is, therefore, of much greater depth
than those mass formulas that provide a relation between two quantities which are just called "mass" and "spin." Therefore, the derivation of (19) requires more mathematics but in return also provides
a much deeper insight leading to the picture of hadrons as micro-
de Sitter spaces attached to the space-time points.

5. GENERALIZATION OF THE "REMARKABLE" REPRESENTATION AND GENERALI-
    ZATION OF THE DE SITTER FIBER BUNDLE--THE GENERAL RELATIVISTIC
    ROTATOR

The mass spectrum (19) is that of a rigid rotator.   Non-relativistic rotators are realized in nature by numerous molecules and
nuclei. From these examples one knows that the model of a rigid
rotator is really of limited applicability and that non-rigidity
has to be taken into account if one wants a more accurate description, especially for problems that involve higher angular momenta.
In the classical picture of the molecule, the origin of this non-
rigidity can be explained as the effect of centrifugal forces which
change the moment of inertia (or radius). In the quantum mechanical
description of the rotating molecule this can be described as replacing the number for the inverse moment of inertia by an operator.

For the relativistic rotator this would mean classically, that the radius of the de Sitter fiber is not constant[11] and quantum mechanically, that the number λ should be replaced by an operator Λ.

Therefore, in order to obtain a quantum mechanical description of de Sitter fibers with varying radii, we have to construct a non-trivial Lorentz invariant operator Λ, which commutes with the observables that specify the states of a single hadron (otherwise the radius would depend upon, e.g. the polarization of the hadron). In the "remarkable" representation used by van Dam and Biedenharn this does not seem to be possible.

However, there exist even more remarkable representations[12] than the Majorana representation which have the additional advantage that they are not only representations of the quantum mechanical Poincaré group but also representations of the quantum mechanical Poincaré group extended by P, T and C.[13] I call these representations Generalized Dirac Representations (GDR), because for the half-integer spin case the 4×4 upper corners of the infinite matrices of Γ_μ are equivalent to the Dirac γ-matrices. Therefore, these representations appear to be ideally suited for the description of an infinite baryon tower whose lowest states are proton and anti-proton.

There is also an integer spin GDR with the Poincaré group reduction[12]

$$H(\text{GDR}) \doteq \sum_{\substack{s=0,1,2,\ldots \\ |n|=s,s-2,\ldots 0 \text{ or } 1}} \oplus H^n(m_{(n,s)}, s) \tag{21}$$

which therefore describes a meson tower containing

$$(n, s^{PC}) = (0,0),(1,1^{--}),(0,2^{++}),(2,2^{++}),(1,3^{--}),(3,3^{--}),$$

$$(0,4^{++}),(2,4^{++}),(4,4^{++}),(1,5^{--}) \ , \tag{22}$$

In these GDR one can form a non-trivial Lorentz invariant operator Λ defined by

$$\Lambda^2 = \lambda_1^2 - \lambda_2^2(\hat{W} - (\Gamma_\mu \hat{P}^\mu)^2)$$

where $\tag{23}$

$$\hat{W} = -\hat{W}_\mu \hat{W}^\mu \qquad \hat{W}_\mu = \tfrac{1}{2} \varepsilon_{\mu\nu\rho\sigma} L^{\nu\rho}\hat{P}^\sigma \qquad \hat{P}_\mu = P_\mu M^{-1}$$

λ_1^2 and λ_2^2 are constants with dimension MeV^2. In the Majorana representation the operator Λ^2 is the constant $\Lambda^2 = \lambda_1^2 + \tfrac{1}{4}\lambda_2^2$. In the GDR the operator Λ^2 has the following spectrum:

$$\lambda^2_{(s,n)} = \lambda^2_1 - \lambda^2_2(s^2 + s - n^2) \ . \tag{24}$$

Thus the radius of the de Sitter fiber $\frac{1}{\lambda_{(s,n)}}$ increases with increasing spin, as one would intuitively expect for a rotator. With this operator Λ one can define the operator B_μ (the generator of translation along the fiber) now by

$$B_\mu = P_\mu + \Lambda M^{-1} \frac{1}{2}\{P_\nu, L^\nu_{\cdot\mu}\} \ . \tag{25}$$

And because of the Lorentz invariance of Λ the $L_{4\mu} = \Lambda^{-1} B_\mu$ fulfill together with the $L_{\mu\nu}$ the commutation relation of SO(4, 1). If one now again considers the constraint relation

$$[\Gamma_\mu, Q] = 0 \ ; \qquad Q = \Lambda^{-1} B_\mu \Lambda^{-1} B^\mu - \frac{1}{2} L_{\mu\nu} L^{\mu\nu} \tag{26}$$

then one obtains the mass spectrum

$$m^2_{(n,s)} = \lambda^2_{(s,n)} (\alpha^2 - 9/4) + \lambda^2_{(s,n)} s(s + 1) \tag{27}$$

where α^2 is again the eigenvalue of the invariant operator Q and, therefore, characterizes the physical system described by the algebraic structure.

6.   COMPARISON WITH EXPERIMENT

The mass formula (27) expresses the masses in the tower in terms of two empirical constants λ_1 and λ_2--which should be the same for all towers--and the value α which is in general different for different towers. If one believes that the rotator is rather rigid, which is supported by the fact that the rigid rotator spectrum (19) gives already a good approximation of the experimental values, then $\lambda^2_1 \ll \lambda^2_2$. An experimental verification of the mass formula provides not only a check of our model but determines also the radius of the de Sitter fibers.

A preliminary comparison of (27)--using two suitably picked sets of values for (λ_1,λ_2)--is given in the Table. In the comparison in the table both the I = 1 and I = 0 resonances are given, because of their almost equal masses. It is not clear whether the transition operator V^β_μ has an internal (SU(3)) property $\beta = 1,2, \dots 8$ so that Γ_μ is only its restriction to H(GDR). In the case that it did, the isospin could change with spin.

The radii of the de Sitter fibers calculated from the fitted values of λ_1 and λ_2 are approximately $R_{(n,s)} \approx 0.36 \times 10^{-13}$ cm and differ very little for small values of s because $\lambda^2_1 \gg \lambda^2_2$. For large values of s however $\lambda^2_2(s^2 + s - n^2)$ can become large, even larger than λ^2_1 and as one would expect in analogy to the centrifugal

Table: Comparison of (27) with experimental data.

n,s^{PC}	$m_{(n,s)}$ [McV]		Experimental data from Particle Data Table or later References
	for $\lambda_1^2=0.298$ for $\lambda_2^2=0.005$	for $\lambda_1^2=.1300$ for $\lambda_2^2=0.0061$	
$1,1^{--}$	765	762	$\rho(770)$ $\omega(783)$
$0,2^{++}$	1269	1260	$f(1270)$
$2,2^{++}$	1313	1318	$A_2(1310)$
$1,3^{--}$	1705	1670	$g(1680)$ $\omega(1670)$
$3,3^{--}$	1840	1850	a $\tilde{p}n(1890)$ has been seen in the D-wave,[14a] with C = -1 follows sP = 3^-.
$0,4^{++}$	1990	1925	$S(1940)$
$2,4^{++}$	2090	2050	$h(2040)$
$4,4^{++}$	2360	2350	$(p\tilde{p}\rightarrow\pi\eta)(2320)$[14c,g] $U(2350)$[14c,d]
$1,5^{--}$	2140	1940	
$3,5^{--}$	2440	2280	$N\tilde{N}(2480)$ with $s^{PC} = 5^{--}$ [14d]
$5,5^{--}$	2860	2850	$N\tilde{N}(2850)$[14e] 2950[14f] s^{PC} unknown

forces for the non-relativistic rotator (rotating molecule) the relativistic rotator ceases to exist.

ACKNOWLEDGEMENTS

I am grateful to P. Moylan for his help with the writing of these lecture notes.

I would like to thank L. C. Biedenharn, H. van Dam, and W. Drechsler for reading the manuscript and suggesting changes for the final version.

REFERENCES

1. A. Bohm, Lectures in Theoretical Physics, Boulder Lectures
        (1966), Vol. 9B, p. 327, Gordon and Breach, A. O. Barut,
        editor. The idea to picture hadrons as micro-de Sitter
        spaces of infinite time and finite space extension was
        suggested in A. O. Barut and A. Bohm, Phys, Rev. $\underline{139}$, B1107
        (1965); A. Bohm, Phys. Rev. $\underline{145}$ 1212 (1966). The general
        idea that the "dynamics" of hadrons leads to a "dynamical
        symmetry group," which is larger than the space-time symme-
        try group, has been proposed earlier, see A. O. Barut,
        Phys. Rev. $\underline{135}$ B839 (1964). de Sitter momentum spaces
        have been introduced by A. D. Donkov, V. G. Kadyshevsky,
        M. D. Mateev, and R. M. Mir-Kasimov, Bulgar. Journ. of
        Physics $\underline{1}$, 58, 150, 233 (1974); $\underline{2}$, 3 (1975); see also V.
        G. Kadyshevsky, p. 114, in Group Theoretical Methods in
        Physics, Lecture Notes in Physics, Vol. 94, Springer-
        Verlag, New York (1979). That this (4+1) de Sitter group
        be a gauge group of strong interaction was suggested by
        W. Drechsler using Cartan type fiber bundles in Ref. 2.
2. W. Drechsler, Fortschritte der Physik $\underline{23}$, 607 (1975).
    W. Drechsler, Found. Phys. $\underline{7}$, 629 (1977).
    W. Drechsler, "Gauge Theory of Strong and Electromagnetic Inter-
        actions Formulated on a Fiber Bundle of Cartan Type,"
        Texas Lectures, Springer Lecture Notes in Physics, Volume
        67 (1977).
3. E. Cartan, Ann. E. Norm. $\underline{40}$, 325 (1922); $\underline{41}$, 1 (1924); Bull. Sc.
        Math. $\underline{48}$, 294 (1924). For the general definition of a
        solder form see S. Kobayashi, Can. J. Math. $\underline{8}$, 145 (1956).
        For a comparison of Cartan's viewpoint with the convention-
        al approach to connections, see Comment 5 of S. Sternberg
        in Differential Geometrical Methods in Mathematical Physics
        II, K. Bleuler et al., editors, p. 2, Springer Verlag,
        1978.
4. This is called the unitary gauge choice in K. S. Stelle and
        P. C. West, Imperical College London preprint ICTP/78-
        79/19.
5. H. van Dam, L. C. Biedenharn, Phys. Rev. $\underline{D14}$, 405 (1976);
        Physics Letters $\underline{62B}$, 190 (1976).
6. A. Bohm, Phys. Rev. $\underline{175}$, 1767 (1968). This was based upon an
        idea by P. Budini and C. Fronsdal, Phys. Rev. Lett. $\underline{14}$,
        968 (1965); V. Ottoson, A. Kihlberg, J. S. Nilsson, Phys.
        Rev. $\underline{137}$, B658 (1965).
7. P. A. M. Dirac, J. Math. Phys. $\underline{4}$, 901 (1963).
    E. Majorana, Nuovo Cimento $\underline{9}$, 335 (1932).
8. a) E. Inönü, E. P. Wigner, Proc. N.A.S. $\underline{39}$, 50 (1953);
    b) A. Bohm, "The Connection Between Representations of SO(4, 1)
        and the Poincaré Group," p. 197, Studies in Mathematical
        Physics, A. O. Barut, ed. (1973);
    c) W. Drechsler, J. Math. Phys. $\underline{18}$, 1358 (1976).

9. F. Rohrlich, Nucl. Phys. B112, 177 (1978); L. P. Staunton,
        Phys. Rev. D13, 3269 (1976); H. Bacry, J. Math. Phys. 5,
        109 (1964); R. J. Finkelstein, Phys. Rev. 75, 1079 (1949);
        H. S. Snyder, Phys. Rev. 71, 38 (1947).
10. F. Gürsey, Group Theoretical Concepts and Methods in Elementary
        Particle Physics (1964), p. 365.
     C. Fronsdal, Rev. Mod. Phys. 37, 221 (1965).
11. Gauge groups whose structure constants change over space-time,
        A. N. Lesnov, V. I. Manko, Lebedev Inst. preprint No. 22,
        Moscow, 1978 and "bundles" with fibers of different curva-
        tures are generalizations of the conventional differential
        geometrical concepts whose mathematical foundations have
        not yet been worked out.
12. L. Jaffe, J. Math. Phys. 12, 882 (1971).
13. E. P. Wigner, Unitary Representations of the Inhomogeneous
        Lorentz Group Including Reflections, Istanbul Lectures
        (1964). F. Gürsey, ed.
14. a) T. E. Kalogeropoulos, G. S. Tzanakos, Proc. of 3rd European
         Symposium on $\bar{N}N$ interactions, Stockholm (1976), and private
         communication;
     b) L. Montanet, Proc. of XIII Rencontres de Moriond, Flaine,
         France (1978), CERN(EP) Phys. 77-22 (1977);
     c) R. S. Dulude et al., IV International Antiproton Symposium,
         Strasbourg (1978), Physics Letters, June 1978;
     d) A. A. Carter et al., Phys. Letters 67B, 117 (1977);
     e) A. Apostolakis et al., Phys. Letters 66B, 185 (1977);
     f) C. Evangelista et al., Phys. Letters 72B, 139 (1977);
     g) A. A. Carter, unpublished, May 1978;
     h) A. Carter, Rutherford Laboratory Preprint RL/78/032 (1978);
     i) R. Baldi et al., Phys. Lett. 74B, 413 (1978).

THE WIGNER-RACAH ALGEBRA FOR FINITE AND COMPACT CONTINUOUS GROUPS

Philip H. Butler

Physics Department
University of Canterbury
Christchurch, New Zealand

The algebra of vector coupling coefficients as developed
by Wigner for simply reducible groups, and by Racah for the groups
appropriate to fractional parentage, has found extensive applications
in many areas of the physics and chemistry. We review the generalization of Wigner's original treatment of simply reducible groups.
For compact groups the generalization is now known to be a straightforward inclusion of multiplicity labels and complex conjugation
labels.

Many authors have looked for special symmetries of 3jm and 6j
symbols, that is relationships particular to a specific group or
class of groups. A number of their results are discussed.

The question of greatest interest to the applications is where
is a table, or formula, or computer program for a group in the
required basis? Failing this, how can a table be most easily
generated? Basis projection operators, ladder operators, boson
polynomials, integration over group operations, integration over
group classes have all been used but what is the minimum that we
need to know about the irreps of a group before the Wigner-Racah
algebra can be constructed?

1.   INTRODUCTION

Ten years ago we needed some coefficients of fractional
parentage (cfps) for some work on the possible SO_4 symmetry of the
electronic structure of light atoms (Butler and Wybourne 1970). By
following Judd (1963) it was easy enough to obtain formulas for the
magnitudes of many of the coefficients required. One knew that cfps

were coupling coefficients for certain groups, and that coupling
coefficients had various factorization and symmetry properties. One
also knew that certain choices of phase and choices of multiplicity
separation existed in the coupling coefficients of a given group
so it should have been a simple matter to complete our calculation.
However it was not, for although Racah (1949) made essentially
arbitrary choices in his cfp calculations, there was a suggestion
that certain special "canonical" choices might exist (Baird and
Biedenharn 1964). Also Derome and Sharp (1965) and Derome (1966)
had proved that if phases were chosen appropriately the Wigner
Racah algebra for any compact group would have many of the symmetries
of the angular momentum case. (See the collected papers in
Biedenharn and van Dam 1963).

Sections 2 to 8 of this article review the known general properties of the Wigner-Racah algebra. Detailed results are to be found
elsewhere (Butler 1979, 1980). The earliest attempt at a systematic
treatment was by Wigner (1940). We discuss the generalization of
his results, which were for a single simply reducible group, to any
compact group. Klimyk has shown that much of the Wigner-Racah
algebra carries over to certain representations of locally compact
groups. (See his article in this volume).

The calculation of numerical values or formulas for the values
of the coefficients, especially recoupling coefficients (6j symbols)
and isoscalar factors (3jm factors) has been the subject of many
papers over the past forty years. It continues to be an important
area of research. Sections 9 and 10 show that the values of the
coefficients may be obtained directly from the general equations
of the Wigner-Racah algebra and the global properties of the irreps
of the groups under study. No use need be made of knowledge of
matrix elements of generators of the groups, instead the familiar
procedures of using ladder operators or of using generators to obtain
coupling coefficients can be reversed. This has proved useful for
the exceptional Lie group E_7 where the matrix elements of generators
were not known in any basis.

We conclude our review by giving (section 12) some indication
of the wealth of additional structure in the Wigner-Racah algebra
of certain classes of groups.

2.   GROUP BASES FOR REPRESENTATION SPACES

The Wigner-Racah algebra is very much concerned with basis
choices and changes of basis (coupling and recoupling). The Wigner-
Eckart theorem can also be looked at in this light. It is very much
to our advantage therefore, to use a notation in which distinct
bases are easily handled.

We denote the elements of a group G by R,S, ... and write
$|\Lambda i>$, i = 1,2, ... $|\Lambda|$ for the basis vectors of a vector space $V(\Lambda)$
of dimension $|\Lambda|$. The vector space $V(\Lambda)$ is said to be a represen-
tation space of the group G if there exist a set of linear
operators $O_R : V(\Lambda) \rightarrow V(\Lambda)$ for all R $\in$ G such that

$$O_R O_S = O_{RS} \tag{2.1}$$

In terms of the basis vectors we have the action of the group as

$$O_R |\Lambda i> = \sum_j |\Lambda j><\Lambda j|O_R|\Lambda i> \tag{2.3}$$

It is convenient to write the matrix elements of the group operators
as

$$\Lambda(R)_{ji} = <\Lambda j|O_R|\Lambda i> \tag{2.3}$$

It follows from the linearity of the various elements, together with
(2.1), that the matrices $\Lambda(R)$ form a matrix representation of the
group.

All groups under present consideration are compact (including
finite) groups. Hence we may without loss of generality assume that
the group operators O_R are unitary and that all bases are orthonormal.
In these circumstances the representation space $V(\Lambda)$ may be decomposed as a direct sum of irreducible representation spaces (irrep
spaces)

$$V(\Lambda) = \bigoplus_{x\lambda} V(x\lambda) \tag{2.4}$$

If $|x\lambda\ell>$ is a typical basis vector (a partner of $V(x\lambda)$ then the
above decomposition means that the action of O_R on $|x\lambda\ell>$ gives a
vector within $V(x\lambda)$ so that (2.2) becomes

$$O_R |x\lambda\ell> = \sum_{\ell'} |x\lambda\ell'><x\lambda\ell'|O_R|x\lambda\ell> \tag{2.5}$$

It is a result of representation theory that the irrep spaces
fall into equivalence classes (which we label by greek letters).
Different spaces of the same equivalence class are distinguished
by latin letters. Irrep spaces V(xλ) and V(yλ) are said to be
equivalent if and only if there exist bases of the spaces so that
the corresponding matrix elements

$$<x\lambda\ell'|O_R|x\lambda\ell> \text{ and } <y\lambda\ell'|O_R|y\lambda\ell>$$

are equal. We say that $|x\lambda\ell>$ are elements of a G-basis of representation space $\underset{\sim}{V}(\Lambda)$ if the matrix elements $\langle x\lambda\ell'|O_R|x\lambda\ell\rangle$ depend only on the equivalence class of irreps of G to which the kets belong. We write

$$\lambda(R)_{\ell'\ell} = \langle x\lambda\ell'|O_R|x\lambda\ell\rangle \qquad (2.6)$$

We say that two G-bases of a representation space $\underset{\sim}{V}(R)$ are equivalent if the various irrep matrices $\underset{\sim}{\lambda}(R)$ are the same in the two bases. Unless otherwise specified all our G-bases will be equivalent.

A G-basis of a representation space is not unique. The uniqueness properties are given by Schur's lemma.

3.   SCHUR'S LEMMA

Schur's lemma relates the concept of irreducibility to comutatability. Proofs of his lemma are to be found in most books on group theory. Schur's lemma concerns the matrix elements of an operator, A, which maps from one irrep space $\underset{\sim}{V}(x\lambda)$ to another $\underset{\sim}{V}(y\mu)$, and which commutes with the group action

$$A:\underset{\sim}{V}(x\lambda) \to \underset{\sim}{V}(y\mu) \qquad (3.1)$$

$$AO_R = O_R A \qquad (3.2)$$

The fact that all matrix elements $\langle y\mu m|A|x\lambda\ell\rangle$ are zero unless $\mu = \lambda$ follows from the definition of irreducibility, and their ℓm independence follow from the G-basis choice, (2.6).

$$\langle y\mu m|A|x\lambda\ell\rangle = \delta_{\lambda\mu}\,\delta_{\ell m}\langle y\mu\|A\|x\lambda\rangle \qquad (3.3)$$

Extension of this result (by linearity) shows that any linear map between any representation spaces of a group $\underset{\sim}{G}$, if it also satisfies (3.2), will have matrix elements of the form (3.3).

Schur (1905) bases his treatment of group representations on this lemma. One can set up the entire generalized Wigner-Racah algebra using (3.3) (Butler 1979). Such an approach has the advantage of not needing concepts of group integration, etc. As such, the approach is more direct than that of Wigner (1940) and Racah (see especially his 1949 paper) and other standard texts.

The importance of Schur's lemma lies, for us, solely in the conclusions we may draw about two different but equivalent G-bases of a representation space $\underset{\sim}{V}(\Lambda)$. Label the two sets of basis kets as $|\,\rangle_{one}$ and $|\,\rangle_{two}$. The matrix elements $U(\Lambda)_{y\mu m,x\lambda\ell} = \langle y\mu m|U|x\lambda\ell\rangle$ of the transformation between bases form a unitary matrix $\underset{\sim}{U}(\Lambda)$ and satisfy

$$|x\lambda\ell\rangle_{two} = \sum_{y\mu m} |y\mu m\rangle_{one} \langle y\mu m|U|x\lambda\ell\rangle \qquad (3.4)$$

A few short steps show that $\underset{\sim}{U}(\Lambda)$ commutes with all O_R so satisfying (3.3). Hence

$$U(\Lambda)_{y\mu m\ x\lambda\ell} = \langle y\mu m|U|x\lambda\ell\rangle = \delta_{\lambda\mu}\ \delta_{\ell m}\langle y\mu\|U\|x\lambda\rangle$$

$$= \delta_{\lambda\mu}\ \delta_{\ell m}\underset{\sim}{U}(\lambda)_{yx} \qquad (3.5)$$

The submatrices $\underset{\sim}{U}(\lambda)$ are unitary matrices, of dimension the multiplicity of irrep $\tilde{\lambda}$ in representation Λ.

We have shown that any equivalent G-bases of $\underset{\sim}{V}(\Lambda)$ are related by submatrices $\underset{\sim}{U}(\lambda)$. The converse is also true, any transformation of the form (3.5), with $\underset{\sim}{U}(\lambda)$ unitary, gives rise to equivalent G-bases. This freedom was explicitly used by Derome (1966) to find limits as to when the phases and multiplicity separations of 3jm symbols could be chosen to give 3jm symbols with simple symmetry properties. Various extensions of Derome's symmetries are to be found in Butler (1975), Butler & Wybourne (1976a) and Butler and Ford (1979).

4. COUPLING COEFFICIENTS

Coupling coefficients are matrix elements of the change of basis from the G-bases $\{|x\Lambda\ell\rangle\},\{|y\mu m\rangle\}$ of two irrep spaces $\underset{\sim}{V}(x\lambda)$, $\underset{\sim}{V}(y\mu)$ to a G-basis of the tensor product space $\underset{\sim}{V}(x\lambda) \otimes \underset{\sim}{V}(y\mu)$. The canonical basis for the product space is

$$|x\lambda\ell,y\mu m\rangle = |x\lambda\ell\rangle|y\mu m\rangle \qquad (4.1)$$

and it is simple to show it is a representation space, containing various irreps ν of $\underset{\sim}{G}$ with a multiplicity $m_{\lambda\mu}^{\nu}$, obtainable say by character theory (Hamermesh 1962, Wybourne 1970). We write the G-basis of the product space as $|(x\lambda,y\mu)r\nu n\rangle$ where r distinguishes members of the class ν of irreps. The matrix elements of the change of basis (the coupling coefficients) $\langle(x\lambda,y\mu)r\nu n|x\lambda\ell,y\mu m\rangle$ may, in the spirit of eq (2.6), be chosen to be independent of the labels x,y

$$\langle r\nu n|\lambda\ell,\mu m\rangle = \langle(x\lambda,y\mu)r\nu n|x\lambda\ell,y\mu m\rangle \qquad (4.2)$$

The G-basis of the product space is only fixed up to equivalence. bases related by (3.4) will give rise to different sets of coupling coefficients. This freedom is often called the "outer multiplicity problem". We prefer to call the unitary matrix $U(\lambda\mu\nu)$ satisfying

$$\langle\lambda\ell,\mu m|r\nu n\rangle_{two} = \sum_{r'} \langle\lambda\ell,\mu m|r'\nu n\rangle_{one}\ U(\lambda\mu\nu)_{r'r} \qquad (4.3)$$

a "free phase matrix", for it is a simple generalization of the free
phase which occurs in the simply reducible case (Wigner 1940).

5.   jm SYMBOLS

Derome (1966) used the freedom in the coupling coefficients
expressed by $\underset{\sim}{U}(\lambda\mu\nu)$ to show that there exists at least one choice of
product multiplicity separations and phases, so that certain general-
izations of the 3j symbols of angular momentum have known (and
reasonably simple) permutational symmetries. The term 3jm symbol
is often used instead of 3j symbol because the coefficient depends
on both group labels (j in angular momentum theory), and partner
labels (m).

The 3jm symbol may be defined in terms of coupling coffiecients

$$\begin{pmatrix} \lambda_1 & \lambda_2 & \lambda_3 \\ \ell_1 & \ell_2 & \ell_3 \end{pmatrix} r = \sum_{\lambda\ell} \langle 000|\lambda\ell,\lambda_3\ell_3\rangle\langle r\lambda\ell|\lambda_1\ell_1,\lambda_2\ell_2\rangle \tag{5.1}$$

unless the phases of the coupling coefficient is fixed by other
considerations. For example, for angular momentum if Condon and
Shortly phases are used, a factor of $(-)^{\lambda_1-\lambda_2+\lambda}$ must be included.

It is a result of character theory that the only coupling in
which the identity irrep 0 occurs is when an irrep λ is coupled
with its complex conjugate, $\lambda*$. Such couplings are sufficiently
important to introduce a special notation, the 2jm (or 1j) symbol

$$\begin{pmatrix} \lambda & \lambda' \\ \ell & \ell' \end{pmatrix} = |\lambda|^{\frac{1}{2}} \langle 000|\lambda\ell,\lambda'\ell'\rangle \tag{5.2}$$

Butler and Wybourne (1976a) show that one may always choose one's
G-basis so that 2jm symbols are ± 1 or 0, leading to the simpler
notation (Butler 1979)

$$\begin{pmatrix} \lambda & \lambda' \\ \ell & \ell' \end{pmatrix} = \delta_{\lambda'\lambda*} \, \delta_{\ell*\ell'} \begin{pmatrix} \lambda \\ \ell \end{pmatrix} \tag{5.3}$$

where $\ell*$ is some unique "partner" to ℓ and where

$$\begin{pmatrix} \lambda \\ \ell \end{pmatrix} = \pm 1 \tag{5.4}$$

(For angular momentum $\begin{pmatrix} J \\ M \end{pmatrix} = (-)^{J-M}$) .

Derome and Sharp (1965) proved an important lemma concerning
the complex conjugation properties of 3jm symbols

$$\begin{pmatrix} \lambda_1 & \lambda_2 & \lambda_3 \\ \ell_1 & \ell_2 & \ell_3 \end{pmatrix} r^* = \sum_s A(\lambda_1\lambda_2\lambda_3)_{rs}^* \begin{pmatrix} \lambda_1 \\ \ell_1 \end{pmatrix} \begin{pmatrix} \lambda_2 \\ \ell_2 \end{pmatrix} \begin{pmatrix} \lambda_3 \\ \ell_3 \end{pmatrix} \begin{pmatrix} \lambda_1^* & \lambda_2^* & \lambda_3^* \\ \ell_1^* & \ell_2^* & \ell_3^* \end{pmatrix} r \qquad (5.5)$$

where A is a unitary matrix. Butler and King (1974) show that for
all compact Lie groups and all point groups, one can make choices
of phase so that A is the unit matrix, but Frame (see Butler 1973)
provided a finite group counterexample. We called such groups
quasiambivalent because the usual constraints such as a product
of an orthogonal irrep with a symplectic irrep contains symplectic
irreps or complex conjugate pairs of irreps, can be strengthened.
The complex irreps may be classified as quasi-orthogonal or quasisymplectic and the product of say, a (quasi)orthogonal irrep and a
(quasi)-symplectic irrep contains only symplectic or quasisymplectic
irreps. Our definition of quasiambivalence differs from that of
Sharp et al. (1973).

6.   THE G–H–BASIS

A major contribution of Racah's 1949 paper was his factorization
lemma. Although this paper stimulated much work in nuclear physics,
beginning immediately with a paper by Jahn (1950), much later work
does not use this particular result, (for example Griffith 1962,
Koster et al. 1963).

Consider a subgroup $\underset{\sim}{H}$ of $\underset{\sim}{G}$. The action of $\underset{\sim}{G}$ on a representation
space $\underset{\sim}{V}(\Lambda)$ leads, as we have seen, to irrep subspaces $\underset{\sim}{V}(x\lambda)$. Each
subspace $\underset{\sim}{V}(x\lambda)$ is a representation space of $\underset{\sim}{H}$, so any irrep of $\underset{\sim}{G}$
may be reduced into irreps of $\underset{\sim}{H}$

$$\underset{\sim}{V}(x\lambda) = \underset{a\rho}{\oplus} \underset{\sim}{V}(x\lambda a\rho) \qquad (6.1)$$

where ρ labels classes of equivalent irreps of $\underset{\sim}{H}$, and a enumerates
equivalent irreps contained in the one irrep λ of $\underset{\sim}{G}$. The label a
is known as a branching multiplicity label. Enumerate the partners
of the space $\underset{\sim}{V}(x\lambda a\rho)$ by an index i, to produce basis kets $|x\lambda a\rho i\rangle$
of the space $\underset{\sim}{V}(\Lambda)$. The three labels $a\rho i$ together replace the previous
index ℓ in enumerating the partners of the space $\underset{\sim}{V}(x\lambda)$.

The transformation from the product G–H–basis of a space
$\underset{\sim}{V}(x_1\lambda_1) \otimes \underset{\sim}{V}(x_2\lambda_2)$, to a G–H–basis consists of the previous G-
coupling coefficients

$$\langle x_1\lambda_1 a_1\rho_1 i_1, x_2\lambda_2 a_2\rho_2 i_2 | (x_1\lambda_1, x_2\lambda_2) r\lambda a\rho\rangle$$

$$= \langle \lambda_1 a_1\rho_1 i_1, \lambda_2 a_2\rho_2 i_2 | r\lambda a\rho i\rangle \qquad (6.2)$$

The tensor product may be looked at piecewise and each piece,
$\underset{\sim}{V}(x_1\lambda_1 a_1\rho_1) \otimes \underset{\sim}{V}(x_2\lambda_2 a_2\rho_2)$, being a representation space for $\underset{\sim}{H}$,

be transformed to an H-basis by H-coupling coefficients

$$\langle x_1\lambda_1 a_1\rho_1 i_1, x_2\lambda_2 a_2\rho_2 i_2 | (x_1\lambda_1 a_1\rho_1, x_2\lambda_2 a_2\rho_2)s\rho i\rangle$$

$$= \langle \rho_1 i_1, \rho_2 i_2 | s\rho i\rangle \tag{6.3}$$

The transformation from the H-basis to the G-H-basis commutes with the operations of $\underset{\sim}{H}$, and so by combining Schur's lemma (3.3) with (6.2) and (6.3) one has Racah's (1949) factorization lemma.

$$\langle \lambda_1 a_1\rho_1 i_1, \lambda_2 a_2\rho_2 i_2 | r\lambda a\rho i\rangle$$

$$= \sum_s \langle \lambda_1 a_1\rho_1, \lambda_2 a_2\rho_2 | r\lambda a\rho\rangle_s \langle \rho_1 i_1, \rho_2 i_2 | s\rho i\rangle \tag{6.4}$$

In terms of 3jm symbol notation one has

$$\begin{pmatrix} \lambda_1 & \lambda_2 & \lambda_3 \\ a_1 & a_2 & a_3 \\ \rho_1 & \rho_2 & \rho_3 \\ i_1 & i_2 & i_3 \end{pmatrix}rG = \sum_s \begin{pmatrix} \lambda_1 & \lambda_2 & \lambda_3 \\ a_1 & a_2 & a_3 \\ \rho_1 & \rho_2 & \rho_3 \end{pmatrix}\!\!{}_{sH}^{rG}\begin{pmatrix} \rho_1 & \rho_2 & \rho_3 \\ i_1 & i_2 & i_3 \end{pmatrix}sH \tag{6.5}$$

Namely a G-H-3jm symbol factorizes into a sum over multiplicity of G-H-3jm factors and H-3jm symbols.

Interestingly, several of the properties of the "3j symbol" of angular momentum depend strongly on the special nature of the basis chosen for SO_3 rotations, namely it is an SO_3-SO_2-basis, and the "3j symbol" is an SO_3-SO_2-3jm factor. Alternative SO_3-bases lead to quite different results (see Patera and Winternitz 1973, Bickerstaff and Wybourne 1976, Kibler and Grenet 1977, Butler and Reid 1979, Butler 1980).

7.   THE WIGNER-ECKART THEOREM

The Wigner-Eckart theorem is fundamental to most applications of the Wigner-Racah algebra. A simple proof of the theorem follows from a knowledge of basis operators and Schur's lemma applied to maps between operators (Butler 1975). A basis for the space of linear operators $\mathcal{L}:\underset{\sim}{V}(\Lambda) \to \underset{\sim}{V}(\Lambda)$, is provided by all ket-bra products of the form $|x_1\lambda_1\widetilde{\ell_1}\rangle\langle x_2\lambda_2\widetilde{\ell_2}|$. Such products transform under the action of the group according to the product representation $\lambda_1 \times \overset{*}{\lambda_2}$. The linear combinations

$$U(x_1\lambda_1, x_2\lambda_2)^{r\lambda}_{\ell} = \sum_{\ell_1\ell_2} |x_1\lambda_1\ell_1\rangle\langle x_2\lambda_2\ell_2| \begin{pmatrix} \lambda_1 \\ \ell_1 \end{pmatrix}\begin{pmatrix} \overset{*}{\lambda_1} & \lambda & \lambda_2 \\ \overset{*}{\ell_1} & \ell & \ell_2 \end{pmatrix}r \tag{7.1}$$

transform irreducibly as partner ℓ of irrep λ. Because of the unitary nature of jm symbols the operators of (7.1) form an alternative basis to the space of linear operators. Such a basis was first used for SO_3 by Elliott (1958).

A tensor operator set T^λ is defined as a set of linear operators T^λ_ℓ (on our space $\underset{\sim}{V}(\Lambda)$) which transforms as the irrep space $\underset{\sim}{V}(x\lambda)$, namely

$$O_R T^\lambda_\ell O_R^{-1} = \sum_{\ell'} T^\lambda_{\ell'} {}^\lambda(R)_{\ell'\ell} \tag{7.2}$$

Any operator is a linear combination of the basis operators (7.1) and on application of Schur's lemma one shows the combination coefficients are diagonal in λ and independent of ℓ.

$$T^\lambda_\ell = \sum_{x_1\lambda_1 x_2\lambda_2} U(x_1\lambda_1,x_2\lambda_2)^{r\lambda}_\ell \; \langle x_1\lambda_1\| T^\lambda \| x_2\lambda_2\rangle_r \tag{7.3}$$

The coefficients $\langle x_1\lambda_1\| T^\lambda \| x_2\lambda_2\rangle_r$ are known as reduced matrix elements.

Combining (7.1) with (7.3) proves the Wigner-Eckart theorem

$$\langle x_1\lambda_1\ell_1 | T^\lambda_\ell | x_2\lambda_2\ell_2\rangle = \sum_r \langle x_1\lambda_1\| T^\lambda \| x_2\lambda_2\rangle_r \begin{pmatrix} \lambda_1 \\ \ell_1 \end{pmatrix} \begin{pmatrix} \lambda_1^* & \lambda & \lambda_2 \\ \ell_1^* & \ell & \ell_2 \end{pmatrix}^r \tag{7.4}$$

A simple corollary is that a H-reduced matrix element is related to a G-reduced matrix element by G–H–jm factors

$$\langle x_1\lambda_1 a_1\rho_1\| T^{\lambda a\rho} \| x_2\lambda_2 a_2\rho_2\rangle^H_s = \sum_r \langle x_1\lambda_1\| T^\lambda \| x_2\lambda_2\rangle^G_r \begin{pmatrix} \lambda_1 \\ a_1 \\ \rho_1 \end{pmatrix} \begin{pmatrix} \lambda_1^* & \lambda & \lambda_2 \\ a_1^* & a & a_2 \\ \rho_1^* & \rho & \rho_2 \end{pmatrix}^{rG}_{sH}$$

$$\tag{7.5}$$

8.   THE j SYMBOLS

The jm symbols concern the coupling of two irreps to a third and contain information on basis choices. The various j symbols relate different ways of coupling and can be shown to be completely basis independent (using Schur's lemma of course).

Any number of irrep spaces $\underset{\sim}{V}(x_i\lambda_i)$, may be coupled together by successive use of coupling coefficients (or their equivalents, the 3jm symbols). Likewise the relationship between any pair of differently coupled products may be obtained by successive use of a single kind of recoupling coefficient, (or its equivalent, the 6j symbol).

Consider the triple tensor product space $\underset{\sim}{V}(x_1\lambda_1) \otimes \underset{\sim}{V}(x_2\lambda_2) \otimes \underset{\sim}{V}(x_3\lambda_3)$. A set coupling coefficients may be used to produce the partially reduced basis $|(x_1\lambda_1,x_2\lambda_2)r_{12}\lambda_{12}\ell_{12},x_3\lambda_3\ell_3\rangle$, and a further set used to complete the reduction into irrep spaces $\underset{\sim}{V}((x_1\lambda_1x_2\lambda_2)r_{12}\lambda_{12},x_3\lambda_3)r\lambda)$. It follows from (5.1) that the appearance of identity irreps $\lambda = 0$, corresponds to the 3jm symbol

$$\langle((x_1\lambda_1,x_2\lambda_2)r_{12}\lambda_{12},x_3\lambda_3)000|x_2\lambda_1\ell_1,x_2\lambda_2\ell_2,x_3\lambda_3\ell_3\rangle$$

$$= \begin{pmatrix} \lambda_1 & \lambda_2 & \lambda_3 \\ \ell_1 & \ell_2 & \ell_3 \end{pmatrix} r_{12} \quad \delta_{\lambda_{12}\lambda_3^*} \tag{8.1}$$

Schur's lemma may be applied to transformations between identity irreps obtained by different intermediate couplings, to derive from (8.1) the symmetries of 3jm symbols, which were obtained by Derome and Sharp (1965) using group integration methods. Such symmetries are a special case of those relations between 3jm symbols which may be obtained by applying Schur's lemma to the transformation

$$\langle((x_1\lambda_1,x_2\lambda_2)r_{12}\lambda_{12},x_3\lambda_3)r\lambda\ell|(x_1\lambda_1,(x_2\lambda_2,x_3\lambda_3)r_{23}\lambda_{23})r'\lambda'\ell'\rangle$$

This transformation is closely related to Derome and Sharp's (1965) 6j symbol (see Butler 1975). All known general relationships involving jm symbols and j symbols may be obtained in this manner. For example the Biedenharn (1953)-Elliott (1953) sum rule was proven independently, both authors using this technique.

9.   CALCULATION OF 6j SYMBOLS

The fact that 6j symbols are basis independent is worth emphasizing. Consider SO_3-6j symbols which were first calculated by Racah (1943) by extensive simplifications of a formula obtained as a sum over a product of known 3jm formulas. The result is independent of the J_z (or SO_2) basis used, any point group basis must give rise to the same SO_3-6j symbols (within the phase freedoms expressed by eq (4.3)).

Butler (1976) shows how SO_3-6j symbols may be calculated without knowledge of the basis, a calculation that has since been extended to all point groups (including the icosahedral group which has multiplicity 3 products, Butler 1980) and to various continuous groups (Butler, Haase and Wybourne 1978, 1979). The method has its origins in the recursive calculations familiar to nuclear physics and known by them as the building-up principle. (See also Fano and Racah, 1959, App.I).

Take a particular faithful irrep of the group in question, say the spin $\frac{1}{2}$ irrep of SO_3 or other point groups, or the irrep $\{1\}$

of U_n, and call it the primitive irrep η. Any other irrep will be contained in some power of $(\eta+\eta^*)$. The generalized Biedenharn-Elliott sum rule provides a sufficient recursion relation to obtain all 6j symbols not involving the primitive irrep η once all 6j symbols involving η are known. (Butler and Wybourne 1976a,b). No counterexample exists to disprove their hypothesis that the ortho-gonality, Racah back-coupling and Biedenharn-Elliott sum rules provide sufficient equations to solve for the primitive 6j symbols of any group.

10.   CALCULATION OF G–H–3jm FACTORS

As with 6j symbols, a relation exists between G–H–3jm factors (and 6j symbols of both groups), that allows a recursive calculation of 3jm factors not involving η in terms of those involving η (Butler and Wybourne 1976a). Recursive calculation of the 3jm factors in-volving η itself is usually quite straightforward. Consider the angular momentum or (SO_3-SO_2) case (Butler 1976). The phase we choose to be associated with the partners $|xJM\rangle$ of irrep $\underset{\sim}{V}(xJ)$ of SO_3 may be easily fixed. The 3jm $\begin{pmatrix} J & \frac{1}{2} & J-\frac{1}{2} \\ M & \frac{1}{2} & -M-\frac{1}{2} \end{pmatrix}$ is related to the 3jm $\begin{pmatrix} J-\frac{1}{2} & \frac{1}{2} & J-1 \\ -M-\frac{1}{2} & \frac{1}{2} & M \end{pmatrix}$ by orthogonality and a column interchange symmetry. The orthogonality equation gives

$$\begin{pmatrix} J & \frac{1}{2} & J-\frac{1}{2} \\ M & \frac{1}{2} & -M-\frac{1}{2} \end{pmatrix} = (-)^{J+M} \sqrt{\frac{J+M+1}{(2J+1)(2J+2)}} \qquad (10.2)$$

Butler, Haase and Wybourne (1979) have used this method to give a number of $E_7 - SU_3^{colour} \times SU_6^{flavour} - 3jm$ factors, and Butler and Reid (1979) and Butler (1980) have prepared tables of 6j symbols and 3jm factors for all possible point group chains. The latter tables are complete for the finite groups (except D_n and C_n for n > 6) and up to J=8 for SO_3.

The pure rotation symmetry group of the tetrahedron, the group T, provides several interesting test cases. It is a small, non-abelian, non-simply-reducible group. The group T has seven irreps. A E' T E_1' E_2' B_1 and B_2 (where $E_2' = E_1'^*$, $B_2 = B_1^*$). Its abelian sub-group D_2 has five irreps $0\ \frac{1}{2}\ \tilde{0}\ 1\ \tilde{1}$ (where all irreps are real). The norms of the various 3jm factors involved in the orthogonality

$$\begin{pmatrix} E' & T & E' \\ \frac{1}{2} & 1 & \frac{1}{2} \end{pmatrix} \begin{pmatrix} E' & T & E' \\ \frac{1}{2} & 1 & \frac{1}{2} \end{pmatrix}^* + \begin{pmatrix} E_1' & T & E' \\ \frac{1}{2} & 1 & \frac{1}{2} \end{pmatrix} \begin{pmatrix} E_1' & T & E' \\ \frac{1}{2} & \tilde{0} & \frac{1}{2} \end{pmatrix}^* + \begin{pmatrix} E_2' & T & E' \\ \frac{1}{2} & 1 & \frac{1}{2} \end{pmatrix} \begin{pmatrix} E_2' & T & E' \\ \frac{1}{2} & \tilde{0} & \frac{1}{2} \end{pmatrix} = 0$$

$$(10.3)$$

are quite readily obtained, in fact each has a norm $\dfrac{1}{\sqrt{3}}$. Writing the phase of each of the products as α, β and γ respectively e.g.

$$\begin{pmatrix} E' & T & E' \\ \tfrac{1}{2} & 1 & \tfrac{1}{2} \end{pmatrix} \begin{pmatrix} E' & T & E' \\ \tfrac{1}{2} & \tilde{0} & \tfrac{1}{2} \end{pmatrix} = \tfrac{1}{3}\alpha$$

shows

$$\alpha + \beta + \gamma = 0 \tag{10.4}$$

which implies that at least two phases are strictly complex, that is,
have both real and imaginary parts non-zero.

11. FORMULAS INVOLVING GROUP CHARACTERS

The 6j symbols are basis independent, likewise the G-H-3jm
factors are independent of the H-basis. Group characters are
other basis independent objects. Characters have formed both a
major tool and a major object of study in the pure mathematical
study of groups and their representations. These facts would seem
to indicate that relationships between characters and 6j symbols
should be of considerable importance. Derome and Sharp (1965)
generalized several of Wigner's (1940) results, Kibler (1977) contains
more relations, and other references. The only character theory
results used in our recursive calculations have been the simplest
possible, namely the integer relations which resolve a Kronecker
(coupled) product into irreps, and those which resolve an irrep of
a group into irreps a subgroup. The symmetry relations of the
algebra, and the phase choices which must be made, place sufficient
constraints on the values of j and jm symbols so that no additional
character relations have been needed.

12. SPECIAL PROPERTIES

The various special properties of a particular set of G-j symbols
or G-H-jm factors can be readily classified into two classes, a
class of properties which arise because the irrep structure of the
group is uncomplicated, and a class of non-trivial properties which
arise because of some special feature of the group.

In the first class we would place all the special properties of
simply reducible groups, such as the value of a 6j being invariant
under column interchanges. Other examples are that SU_3 and all
point groups are simple-phase, namely their 3jm symbols simply
change sign under column interchanges, or that $T-C_3$-3jm factors are
either real or imaginary whereas some $T-D_2$-3jm factors are strictly
complex.

The Regge (1958,1959) symmetries of SO_3-SO_2-3jms (equivalently
SU_2-U_1-3jms) and SO_3-6js (or SU_2-6js) are a prime example of the
second class of symmetries. Kramer and Seligman (1969) have shown
that they hold for all unitary group 6j symbols and for many unitary

group 3jm symbols. The symmetries result from the close interrelation of symmetric groups S_ℓ, and unitary groups U_n, together
with the properties of the alternating irrep $[1^\ell]$ of S_ℓ.

Hamermesh (1962) and Griffith (1962) showed that the existance
of one dimensional irreps of a group leads to symmetries of the
Wigner-Racah algebra . Such symmetries arise because the product
of a one dimensional irrep ε, with any irrep λ is again an irrep.
We define the irrep $\tilde{\lambda}$ by

$$\varepsilon \times \lambda* = \tilde{\lambda} \tag{12.1}$$

References will be found in Butler and Ford (1979) who connect this
tilde symmetry with the general permutation and complex conjugation
symmetries. In some cases, notably for the rotation group of the
octahedron, imposition of the tilde symmetry selects a multiplicity
separation which has advantages over the multiplicity separations
in the literature (Griffith 1962, Koster et al. 1963). The new
table of 6j symbols has no factors of 5 and several more zeros.

The symmetric groups and the unitary groups are of fundamental
importance, for they have as subgroups all other finite and compact
continuous groups. Schur was one of the first to recognize the
importance of the close relationship between the structures of their
irreps. The standard tool in this area of character theory, a
bialternant, or bideterminant, was renamed as S-function by Littlewood
in honour of Schur. The graphical techniques originated by Young
and developed principally by Littlewood and Robinson provide the
most powerful means available for analysing Kronecker products and
branching rules of the classical groups (see Littlewood, 1950;
Robinson, 1961; Wybourne, 1970; Butler and King, 1974; Wybourne and
Bowick, 1977; and references therein).

The work on the interrelation of the Wigner-Racah algebras of
U_n and S_ℓ has its origins in the Young symmetrizer techniques used
in early cfp calculations. There was considerable development
along these lines in the 1950's and 1960's. (See for example
Jahn, 1960; Kaplan, 1962; Horie, 1964; Moshinsky and Devi, 1969).

Kramer (1968) introduced the 6f symbol as a special case of the
transformation bracket of earlier work. Transformation brackets
are elements of the transformation between inequivalent G-bases,
say a G-H-H∩K basis and a G-K-H∩K-basis. If G is a symmetric group
$S_{\ell+m+n}$, H is $S_{\ell+m} \times S_n$ and K is $S_\ell \times S_{m+n}$ the transform is described
by a 6f symbol. If the third groups (H∩K) of the two bases are not
identical but merely conjugate in $S_{\ell+m+n}$, the transformations are
related to the matrix elements of a double coset representative.

A comparison of the effect of U_n - coupling coefficients and the
action of symmetric group operators, especially the Young symmetrizers

and the double coset representatives, gives many profound properties
of U_n-6j and 3jm symbols, and 6f symbols of the symmetric groups,
for example one shows, up to the ubiquitous free phase matrix of
(4.2) and certain dimension factors, that

$$1, \quad \text{a } U_n\text{-6j equals a 6f}$$
$$2, \quad \text{a } U_{n+m} - U_n \times U_m\text{-3jm factor also equals a 6f.}$$

Now a 6f is a symmetric group object and is absolutely independent
of the U_n involved. It also has various orthogonality and symmetry
relations of its own, in particular the special symmetry resulting
from the $[1^\ell]$ irrep. It follows that U_n-6js and $U_{n+m}-U_n \times U_m$-3jm
factors are (within dimension factors, etc) n independent, obey
special sum rules and have the tilde symmetry, Kramer and Seligman
(1969) indicate that this tilde symmetry becomes the SU_2-6j and
SU_2-U_1-3jm Regge symmetries. Sullivan (1973-78) and Patterson and
Harter (1976) have followed up many of these ideas.

13. SUMMARY

Wigner's (1940) paper on properties of the Wigner-Racah algebra
of simply-reducible groups and Racah's (1941-49) work on cfps have
been extended to all finite and compact groups. The principal
change is that certain free phases for the simply-reducible case
become free phase matrices, certain choices of which give rise to
"symmetries" of j symbols and jm factors, we have seen how Schur's
lemma (and finite dimensional irreps) lead to the Derome-Sharp
(1965) complex conjugation symmetry, the Derome (1966) permutational
symmetries, Racah (1949) factorization and the general Wigner-
Eckart theorem. Relations such as the Racah backcoupling and
Biedenharn-Elliott sum rules generalize. A summary of results was
given by Butler (1975).

Development since 1975 has been mainly in computational methods,
where it has been shown that usually, if not always, the numerical
values of coefficients follow from a recursive calculation within
the algebra. One has no need for ladder operators, explicit matrix
elements of generators, etc. Of particular note one now knows how to
calculate 6j symbols before any 3jms. Such calculations have lead
to various indications that few general results remain to be discovered.

The computational algorithms are simple enough for the computer
generation and tabulation of the many thousands of numbers required
for all point groups in all possible basis schemes. (Butler 1980).
They are also simple enough for hand calculation of tables for
groups such as E_7 where the smallest irrep is 56 dimensional.

The Wigner-Racah algebra of the unitary groups has been the
object of considerable study by Biedenharn and his coworkers (see
Biedenharn's article in these proceedings). The equivalent algebra
of the symmetric groups has had much less attention (Hamermesh
1962, Vanagas 1971). Studies of the duality of the two series of
groups is the basis of the S-functional methods of the character
theory of groups. Such methods bring together finite groups,
matrix groups, combinatorics and generating functions. The recent
studies on the duality of the Wigner-Racah algebra of these groups
indicate that a wealth of special results remain to be discovered.

REFERENCES

[BvD] indicates the paper is reprinted in Biedenharn and van Dam
(1965).

Baird, G.E., and Biedenharn, L.C., 1964, J. Math. Phys., 5: 1730-1747.
Bickerstaff, R.P. and Wybourne, B.G., 1976, J. Phys. A., 9:1051-1068.
Biedenharn, L.C., 1953, J. Math. Phys., 31:287-293 [BvD].
Biedenharn, L.C., 1979, Article in these proceedings.
Biedenharn, L.C., and van Dam H., 1965, "Quantum theory of angular
     momentum", Academic, New York.
Butler, P.H., 1973, J. Math. Phys., 14:540.
Butler, P.H., 1975, Trans Roy. Soc., (London), 277:545-598.
Butler, P.H., 1976, Int. J. Quantum Chem., 10:599-614.
Butler, P.H., 1979, "Properties and application of point group
     coupling coefficients",in Proceedings of NATO Advanced Study
     Institute:"Recent Advances in Group Theory and their Application
     to Spectroscopy" (Ed. J. Donini), Plenum, New York.
Butler, P.H., 1980, "Point Group Symmetry Applications: Methods and
     Tables", Plenum, New York. (in press)
Butler, P.H., and Ford, A.M., 1979, J. Phys. A., 12:1357-1366.
Butler, P.H., Haase, R., and Wybourne, B.G., 1978. Australian J.
     Phys., 31:131-135.
Butler, P.H., Haase, R., and Wybourne, B.G., 1979, Australian
     J. Phys. 32:137-154.
Butler, P.H., and King, R.C., 1974, Can. J. Math., 23:328-339.
Butler, P.H., and Reid, M.F., 1979, J. Phys. A, 12: 1655-1668.
Butler, P.H., and Wybourne, B.G., 1970, J. Math. Phys., 11:2517-2524.
Butler, P.H., and Wybourne, B.G., 1976, Int. J. Quant. Chem.,
     10:581-598; 615-628
Derome, J.R., 1966, J. Math. Phys., 7:612-615.
Derome, J.R., and Sharp, W.T., 1965, J. Math. Phys., 6:1584-1590
Elliott, J.P., 1953, Proc. Roy. Soc., (London) A218:370 [BvD]
Elliott, J.P., 1958, Proc. Roy. Soc., (London) A245:128-145
Fano, U., and Racah, G., 1959, "Irreducible Tensorial Sets",
     Academic, New York.
Griffith, J.S., 1962, "The irreducible tensor method for molecular
     symmetry groups", Prentice Hall, Englewood Cliffs, N.J.

Hamermesh, M., 1962, "Group theory and its application to physical
     problems", Addison Wesley, Reading Mass.
Horie, H., 1964, J. Phys. Soc., (Japan), 19:1783-1799.
Jahn, H.A., 1950, Proc. Roy. Soc., (London) A201:516-544 [BvD]
Jahn, H.A., 1960, Trans. Roy. Soc., (London), 523:27-53
Judd, B.R., 1963, "Operator techniques in atomic spectroscopy",
     McGraw Hill: New York.
Kaplan, I.G., 1962, Soviet Phys. JETP, 14:401-407.
Kibler, M.R., 1977, J. Phys. A., 10:2041-2052.
Kibler, M.R., and Grenet, G., 1977, Int. J. Quantum Chem., 11:359-379.
Klimyk, A.U., 1979, Contribution in this volume.
Koster, G.D., Dimmock, J.D., Wheeler, R.G., and Statz, H., 1963,
     "Properties of the thirty-two point groups", MIT Press, Cambridge,
     Mass.
Kramer, P., 1968, Z. Phys., 216-68-83.
Kramer, P., and Seligman, T.H., 1969, Z. Phys., 219:105-113.
Littlewood, D.E., 1950, "The theory of group characters and matrix
     representation of groups", 2nd ed. Oxford University Press.
Moshinsky, M., and V.S. Devi, 1969, J. Math. Phys., 10: 455-466.
Patera, J., and Winternitz, P., 1973, J. Math. Phys., 14:1130-1139
Patterson, C.W. and Harter, W.G., 1976, J. Math. Phys., 17:1125-1136;
     1137-1142.
Racah, G., 1942a, Phys. Rev., 61:186-197 [BvD]
Racah, G., 1942b, Phys. Rev., 62:438-462 [BvD]
Racah, G., 1943, Phys. Rev., 63:367-382 [BvD]
Racah, G., 1949, Phys. Rev., 76:1352-1365 [BvD]
Regge, T., 1958, Nuovo Cimento, 10:544-545 [BvD]
Regge, T., 1959, Nuovo Cimento, 11:116-117 [BvD]
Robinson, G de B., 1961, "Representations of the symmetric group",
     Edinburgh University Press.
Schur, I., 1905, Sitzungsber. Preuss. Akad., 406.
Sharp, W.T., Biedenharn, L.C., de Vries, E., and van Zanten, J., 1973,
     Canad. J. Math.
Sullivan, J.J., 1973, J. Math. Phys., 14:387-395.
Sullivan, J.J., 1975, J. Math. Phys., 16:756-760.
Sullivan, J.J., 1975, J. Math. Phys., 16:1707-1709.
Sullivan, J.J., 1978, J. Math. Phys., 19:1674-1680.
Sullivan, J.J., 1978, J. Math. Phys., 19:1681-1687.
Vanagas, V.V., 1971, "Algebraic methods in nuclear theory"
     (in Russian) Mintis, Vilnius, USSR.
Wigner, E.P., 1940, unpublished manuscript, now published in BvD.
Wybourne, B.G., 1970, "Symmetry principles in atomic spectroscopy",
     (with an appendix of tables of P.H. Butler), Wiley, New York.
Wybourne, B.G., and Bowick, M.J., (1977), Australian J. Phys.,
     30:259-286.

APPLICATIONS OF COHERENT STATES IN THERMODYNAMICS AND DYNAMICS

R. Gilmore

Physics Department
University of South Florida
Tampa, Florida 33620

Coherent states for an arbitrary Lie group G are defined and
their properties enumerated. These properties lead to a simple
algorithm for studying the thermodynamic critical properties of
Hamiltonians constructed from the generators of G. The algorithm
is:
1. Replace H by its "classical limit" and add an entropy term $-kTs(r)$.
2. Determine how the minima of the resulting function change as a
    function of T.

Step 1 is implemented using Group Theory. The classical limit is
taken using coherent states, and $s(r)$ is a group multiplicity factor.
Step 2 is implemented using Catastrophe Theory. This algorithm is
used to study the thermodynamics of extended Dicke models, general
bialgebraic models, and nuclear models. Ground state energy phase
transitions are determined by setting $T = 0$ and studying the bifurcations of the classical limit of H as a function of the interaction parameters. The "crossover theorem" is discussed. The
dynamical properties of H are also described simply in the coherent
state representation. Dynamical equations of motion are derived
when $\hat{H} \in \mathfrak{g}$ and TDHF equations are derived in the general case.
Phase transitions for nonequilibrium steady state systems are
determined by a "dissipation-transformation" theorem, analog of
the "fluctuation-transformation" theorem for equilibrium phase
transitions. The identification of the equilibrium and nonequilibrium steady state equations of state (for the laser) with a
(cusp) catastrophe manifold is shown.

1. History of Coherent States

Coherent states were first discussed by Schrödinger[1] in connection with the classical limit of the quantum-mechanical harmonic oscillator. They were subsequently used by Bloch and Nordsieck[2] to treat the "infrared catastrophe." The properties of these coherent states were then expounded formally by Schwinger,[3] who emphasized the relation between the properties of these states and the properties of the underlying Lie algebra. Finally, they were used by Glauber[4] in a very beautiful formulation of Quantum Optics. The harmonic oscillator coherent states are now a standard tool in any field in which the harmonic oscillator is a useful model for physical processes.

The realization came slowly that coherent states were more intimately associated with a Lie group rather than a Lie algebra, and that they could be associated with Lie groups more complicated than H(4) (harmonic oscillator group). Coherent states associated with the Lie algebra su(2) were discussed by Atkins and Dobson,[5] by Kutzner,[6] and by Radcliffe,[7] but the deep connection with the associated Lie group SU(2) was made by Arecchi, Courtens, Gilmore, and Thomas,[8] who showed that all of the properties of the "atomic coherent states" could be derived from the group SU(2) and its representations. The extension of coherent states to Lie groups more complicated than H(4) and SU(2) was made by Gilmore[9] for SU(r) and by Perelemov[10] for general Lie groups. A detailed study of the properties of coherent states for arbitrary groups, including disentangling theorems, semiclassical theorems, P- and Q- representations, operator bounds, and D-algebra mappings, was made by Gilmore.[11]

2. Definition of Coherent States

Coherent states may be defined in terms of the following mathematical structures:[11]
1. a Lie group G with Lie algebra $\mathcal{G}$;
2. an invariant subspace V^Λ which carries an irreducible square-integrable representation Γ^Λ;
3. a highest (or extremal) weight Λ with corresponding basis vector $|\Lambda,\Lambda>$ (or $|\text{ext}>$) $\varepsilon\ V^\Lambda$;
4. a maximal stability subgroup $H \subset G$ with the property
$h\ \varepsilon\ H\ \Leftrightarrow\ h|\Lambda,\Lambda> = |\Lambda,\Lambda>\ e^{i\phi(h)}$.

Every element g ε G can be written uniquely in the form

$$g = \Omega h \qquad\qquad \begin{array}{l} g\ \varepsilon\ G \\ h\ \varepsilon\ H \\ \Omega\ \varepsilon\ G/H \end{array} \qquad\qquad (2.1)$$

The effect of g on the vector $|\Lambda,\Lambda\rangle$ is

$$g\left|^{\Lambda}_{\Lambda}\right\rangle = \Omega h\left|^{\Lambda}_{\Lambda}\right\rangle = \Omega\left|^{\Lambda}_{\Lambda}\right\rangle e^{i\Phi(h)} = \left|^{\Lambda}_{\Omega}\right\rangle e^{i\Phi(h)} \tag{2.2}$$

Definition: The states $|\Lambda,\Omega\rangle$

$$\left|^{\Lambda}_{\Omega}\right\rangle = \Omega\left|^{\Lambda}_{\Lambda}\right\rangle = \left|^{\Lambda}_{"M"}\right\rangle \Gamma^{\Lambda}_{"M",\Lambda}(\Omega) \tag{2.3}$$

are called coherent states of G in the invariant space V^{Λ} with respect to the state $|\Lambda,\Lambda\rangle$, or simply coherent states. Here "M" is a Gel'fand-Tsetlein pattern which indexes all the orthonormal basis states in V^{Λ}.[12,13]

Coherent states exist in 1-1 correspondence with the coset representatives $\Omega \in G/H$.[14] Since G/H has a natural geometric strucuture, including Haar metric and measure, the coherent states come endowed with a natural geometric interpretation. The coherent states $|\Lambda,\Omega\rangle$ are nondenumerable in the Hilbert space over V^{Λ}, so they must be nonorthogonal and over-complete. Many useful properties of coherent states derive from their geometric interpretation and their over-completeness.
Example 1.   G = H(4), $|\Lambda,\Lambda\rangle = |0\rangle$ ($a|0\rangle = 0$, $[a,a^{+}] = I$), H = U(1) x U(1), G/H = H(4)/U(1) x U(1) $\simeq \mathbb{R}^2$. The familiar harmonic oscillator coherent states[1-4] exist in 1-1 correspondence with points in the plane.
Example 2.   G = SU(2), $V^{\Lambda} \rightarrow V^{j}$, $\Gamma^{\Lambda} \rightarrow D^{j}$, $|\Lambda,\Lambda\rangle \rightarrow |j,j\rangle$, H = U(1), G/H = SU(2)/U(1) $\cong$ S^2 (sphere).[8]
Example 3.   G = SU(3), $V^{\Lambda} \rightarrow V^{(\lambda_1,\lambda_2)}$, $\Lambda = (\lambda_1-\lambda_2)(2/3,-1/3,-1/3) + \lambda_2(1/3,1/3,-2/3)$, H = U(2) for $\lambda_2 = 0$ and H = U(1) x U(1) for $\lambda_2 \neq 0$, so G/H = SU(3)/U(2) for symmetric representations and G/H = SU(3)/U(1) x U(1) for other representations.[9,11]

The definition of coherent states given above is not standard among all authors. For example, Perelomov[10] does not require the state $|\psi\rangle$ appearing in (2.3) to be an extremal state nor the representations Γ^{Λ} to be square-integrable, while Barut and Girardello[15] define coherent states as eigenstates of a maximal solvable subalgebra of $\mathfrak{g}$. We have adopted the above definition because it appears to be the most useful for physical applications.

3.   Properties of Coherent States[11]

1.   For any unitary irreducible representation Γ^{Λ} and extremal state $|ext\rangle \in V^{\Lambda}$, $\mathfrak{g}$ has a decomposition

$$\mathfrak{g} = \mathfrak{g}_{+} + \mathfrak{g}_{0} + \mathfrak{g}_{-}$$

where $S_{+}|ext\rangle = 0$ $S_{+} \in \mathfrak{g}_{+}$

and $\quad S_0 |ext> = (multiple)|ext>$ $\qquad S_0 \in \mathcal{G}_0$ $\hfill$ (3.1)

2. All Gel'fand-Tsetlein states $|\Lambda, "M">$ in V^Λ can be written as products of shift-down operators in $\mathcal{G}_-$ applied to $|\Lambda,\Lambda>$

$$|\Lambda,"M"> = E_\gamma \cdots E_\beta E_\alpha |\Lambda,\Lambda> \qquad (3.2)$$

3. The coset representatives can be written as exponentials

$$\Omega = EXP(S_+ + S_-) \qquad (3.3)$$

where $S_+ \in \mathcal{G}_+$, $S_- = -S_+^\dagger \in \mathcal{G}_-$. The coordinates of the basis vectors in $S_\pm$ may be used to parameterize the space G/H.[14]

4. Baker-Campbell-Hausdorff formulas[14,16] can be developed

$$e^{(S_+ + S_-)} = e^{S_-'} \, e^{S_0'} \, e^{S_+'} \qquad (3.4)$$

$(S_i' \in \mathcal{G}_i$, $i = \pm,0)$ and used to simplify calculations.

5. Coherent states may be expanded in terms of Gel'fand-Tsetlein states

$$|{}^\Lambda_\Omega> = \sum_{k=0}^{\infty} \frac{(S_+ + S_-)^k}{k!} |{}^\Lambda_\Lambda>$$

$$= e^{S_-'} |{}^\Lambda_\Lambda> \times <{}^\Lambda_\Lambda|e^{S_0'}|{}^\Lambda_\Lambda> \qquad (3.5)$$

6. Three types of "eigenvalue" equations can be constructed

$$\Omega \text{ Operator} \quad \Omega^{-1}|{}^\Lambda_\Omega> = \lambda|{}^\Lambda_\Omega> \qquad (3.6)$$

where Operator $|\Lambda,\Lambda> = \lambda|\Lambda,\Lambda>$. The three useful types of eigenvalue equation correspond to the following choices
(a) Operator belongs to the universal enveloping algebra of $\mathcal{G}$, but is not necessarily a Casimir operator;
(b) Operator $\in \mathcal{G}_0$;
(c) Operator $\in \mathcal{G}_+$, in which case $\lambda = 0$.

7. Uncertainty relations can be constructed for the hermitian operators Re $S_+' = (S_+' + S_+'^\dagger)/2$, Im $S_+' = (S_+' - S_+'^\dagger)/2i$, where $S_+' = \Omega S_+ \Omega^{-1}$ and $S_+ \in \mathcal{G}_+$:

$$<\Delta(\text{Re } S_+')^2> \, <\Delta(\text{Im } S_+')^2> \, \geq \, |<\tfrac{1}{2}[S_+',S_+'^\dagger]>|^2 \qquad (3.7)$$

This uncertainty is minimized in the coherent state $|\Lambda,\Omega>$, but states which minimize (3.7) are not necessarily coherent states.

8. The inner product of two coherent states is

$$\langle \Lambda_\Omega | \Lambda_{\Omega'} \rangle = \langle \Lambda_\Lambda | \Omega^{-1}\Omega' | \Lambda_\Lambda \rangle = \Gamma^\Lambda_{\Lambda\Lambda}(c)\ e^{i\Phi(h)}$$

where $\Omega^{-1}\ \Omega' = ch$ (3.8)

9. The overcompleteness of the coherent states allows the following resolution of the identity

$$I = \frac{\dim(\Lambda)}{\mathrm{Vol}(G/H)}\ \int |\Lambda_\Omega\rangle\ d\Omega\ \langle\Lambda_\Omega|$$ (3.9)

Here $d\Omega$ is the Haar measure on G/H and, if G is compact, $\mathrm{Vol}\ (G/H) =$ $\mathrm{Vol}(G)/\mathrm{Vol}(H)$ is the invariant volume of G/H and $\dim(\Lambda)$ is the dimension of the representation Γ^Λ.[14] If G is not compact, $\dim(\Lambda)/\mathrm{Vol}(G/H)$ is the Plancherel measure.

10. Generating functions of the form

$$f(A,\Omega) = \langle\Lambda_\Omega|\ e^{A_i X_i}\ |\Lambda_\Omega\rangle\ ,$$ (3.10)

where X_i span ⅂, can easily be constructed using disentangling theorems.

4. Operator Mappings and Bounds

Every linear operator $\hat\theta$ on V^Λ has a Q-representative defined by

$$Q_\Lambda\ (\hat\theta;\Omega) = \langle\Lambda_\Omega|\ \hat\theta\ |\Lambda_\Omega\rangle$$ (4.1Q)

If G is compact, $\hat\theta$ also has[8] a P-representative defined by

$$\hat\theta = \frac{\dim(\Lambda)}{\mathrm{Vol}(G/H)}\int |\Lambda_\Omega\rangle\ P_\Lambda\ (\hat\theta;\Omega)\ \langle\Lambda_\Omega|\ d\Omega$$ (4.1P)

The P- and Q-representatives are related by a group covolution integral[9,17]

$$Q_\Lambda(\hat\theta;\Omega) = \frac{\dim(\Lambda)}{\mathrm{Vol}(G/H)}\ \int |\langle\Lambda_{\Omega'}|\Lambda_\Omega\rangle|^2\ P_\Lambda(\hat\theta;\Omega')\ d\Omega'$$ (4.2)

If $\hat\theta$ is an irreducible tensor operator $T^{\Lambda'}_{"M"}$ so that

$$g\ T^{\Lambda'}_{"M"}\ g^{-1} = T^{\Lambda'}_{"N"}\Gamma^{\Lambda'}_{"N","M"}(g)$$ (4.3)

then

a) $Q_\Lambda(T^{\Lambda'}_{"M"};\Omega) = q(\Lambda,\Lambda')\ y^{\Lambda'}_{"M"}(\Omega)$

b) $P_\Lambda(T^{\Lambda'}_{"M"};\Omega) = p(\Lambda,\Lambda')\ y^{\Lambda'}_{"M"}(\Omega)$ (4.4)

Here $y^{\Lambda'}_{"M"}(\Omega)$ is the "M"th spherical function of $\Gamma^{\Lambda'}$ on G/H.[18,19]
The proportionality factors $q(\Lambda,\Lambda')$ can be computed from generating
functions of type (3.10) and the ratio $p(\Lambda,\Lambda')/q(\Lambda,\Lambda')$ determined
from the kernel appearing in the convolution in (4.2), which reduces
to the square of a Clebsch-Gordan coefficient.[8] For $SU(2)$[20]

$$Q_J(y^L_M(J);\Omega) = \frac{(2J)!}{(2J-L)!2^L}\, Y^L_M(\Omega)$$

$$P_J(y^L_M(J);\Omega) = \frac{(2J+1+L)!}{(2J+1)!2^L}\, Y^L_M(\Omega) \ . \tag{4.5}$$

If $\hat{\theta}$ is a hermitian operator, lower and upper bounds may be
placed on the trace of EXP $\hat{\theta}$ using the Q- and P- representatives
of $\hat{\theta}$

$$\frac{\dim(\Lambda)}{\mathrm{Vol}(G/H)} \int e^{Q_\Lambda(\hat{\theta};\Omega)}\, d\Omega \le \mathrm{Tr}_\Lambda e^{\hat{\theta}} \le Q \to P \ . \tag{4.6}$$

The lower bound is due to Bogoliubov and is valid for noncompact as
well as compact G. Lieb's proof of the upper bound for SU(2) extends
without difficulty to other compact G.[21]

Mappings $\hat{X} \to \mathcal{D}(\hat{X})$ of operators $\hat{X} \in \mathcal{G}$ into first order linear
differential operators $\mathcal{D}(X)$ on G/H can also be constructed and are
useful when G is compact. In this case any operator, for example,
the density operator $\hat{\rho}$, has a diagonal P- representation. Then Lie
algebraic operator products of the type $\hat{X}_i\hat{\rho}\hat{X}_j$ can be replaced by
ordinary differential operators of type $\mathcal{D}(\hat{X}_i)\mathcal{D}(\hat{X}_j)\, P_{}(\hat{\rho};\Omega)$. $\mathcal{D}$-opera-
tor algebras have been constructed for $H(4)$,[4] $SU(2)$,[22] and $SU(r)$.[17]

5. Classical Limits

It is well known[23] that when the SU(2) representation label J
is large the angular momentum operators $\underset{\sim}{J}$ can be replaced by c-num-
bers according to $J_z \to J\cos\theta$, $J_x = J\sin\theta\cos\phi$, $J_y = J\sin\theta\sin\phi$,
and that this approximation becomes better as J becomes larger. The
construction of this classical limit can be carried out by using
atomic coherent states, for example, by taking a limit in (4.5).
Classical limits analogous to the angular momentum limit can be
constructed for all compact groups.[24] For $X \in \mathcal{G}$

$$\underset{N\to\infty}{\mathrm{Lim}}\ \Gamma^\Lambda(X/N) = \sum_{i=1}^{r} s_i g(\underset{\sim}{f}_i,X,\Omega) \tag{5.1}$$

where r is the rank[14] of $\mathcal{G}$, $\underset{\sim}{f}_i$ is the highest weight of the ith
fundamental irreducible representation of $\mathcal{G}$,

$$\underset{\sim}{\Lambda} = \sum_{i=1}^{r} \mu_i \underset{\sim}{f}_i$$

$$s_i = \lim_{N\to\infty} \mu_i/N$$

$$g(\underset{\sim}{f}_i, X, \Omega) = \; <\underset{\Omega}{\overset{f_i}{\sim}}| \; X \; |\underset{\Omega}{\overset{f_i}{\sim}}> . \tag{5.2}$$

The proof of this result involves the construction of generating functions (3.10) and the application of Baker-Campbell-Hausdorff formulas (3.4).

6. Thermodynamics of Dicke Models

The Dicke model[25] has long been a useful tool in Quantum Optics. The Dicke model Hamiltonian describing the interaction of a single mode of the radiation field of energy $\hbar\omega$ with an ensemble of N-identical 2-level atoms is, in the long wavelength approximation,

$$\hat{H} = \hbar\omega a^\dagger a + \frac{\varepsilon}{2} \sum_{i=1}^{N} \sigma_i^z + \frac{\lambda}{\sqrt{N}} \sum_{i=1}^{N} a^\dagger \sigma_i^- + a\sigma_i^+ \tag{6.1}$$

where $\varepsilon = \varepsilon_2 - \varepsilon_1$ is the atomic energy level spacing, λ is the coupling constant, essentially an electric dipole matrix element, and σ_i^z, $\sigma_i^\pm$ are the Pauli spin operators for the ith atom.

The thermodynamic properties of (6.1) were first discussed by Hepp and Lieb,[26] who used involved estimates to prove that a second order phase transition would occur if $|\lambda|^2 > \varepsilon\hbar\omega$ at a critical temperature $T_c = 1/k\beta_c$ defined by

$$\frac{|\lambda|^2}{\varepsilon\hbar\omega} \tanh \tfrac{1}{2} \beta_c \varepsilon = 1 \; . \tag{6.2}$$

The description of this phase transition was greatly simplified by Wang and Hioe, who replaced the field operators $a^\dagger a \to \alpha^*\alpha$, $a \to \alpha$, $a^\dagger \to \alpha^*$ by their Q-representatives in the field coherent state representation, took the trace of the resulting function

$$\mathrm{Tr}\; e^{-\beta\hat{H}} \to e^{-\beta\hbar\omega\alpha^*\alpha} \left(\mathrm{Tr} \; \mathrm{EXP} - \beta \begin{bmatrix} \varepsilon/2 & \lambda\alpha/\sqrt{N} \\ \lambda\alpha^*/\sqrt{N} & -\varepsilon/2 \end{bmatrix} \right)^N \tag{6.3}$$

and minimized with respect to α, α^* to obtain an estimate for F/N, the free energy per particle. This estimate was not rigorous, although it was simple and heuristically useful.

A rigorous treatment using atomic coherent states was subsequently given by Hepp and Lieb.[28] They first decomposed the atomic Hilbert space of dimension 2^N into SU(2) irreducibles

$$(\mathbb{C}^2)^{\otimes N} \rightarrow \sum_{J=0 \text{ or } \frac{1}{2}}^{N/2} Y(N,J) \ V^J \tag{6.4}$$

where the multiplicity factor $Y(N,J) = N!(2J+1)/(\frac{1}{2}N+J+1)!(\frac{1}{2}N-J)!$.
Bounds on the partition function $Z_J = \mathrm{Tr}_J \ e^{-\beta\hat{H}}$ in the subspace V^J
were computed using the inequalities (4.6) and the sums estimated
using Stirlings approximation for

$$Y(N,J) = e^{Ns(r)}$$

$$s(r) = -\{(\tfrac{1}{2}+r) \ \ln(\tfrac{1}{2}+r) + (\tfrac{1}{2}-r) \ \ln(\tfrac{1}{2}-r)\} \tag{6.5}$$

$$\operatorname*{Lim}_{N\to\infty} r = J/N$$

and Laplace's method[29] for estimating integrals.

These two dual methods lead to algorithms, the latter rigorous,
the former not, for discussing the thermodynamics of Dicke Hamiltonians. These two approaches were synthesized[30] into a single
simple algorithm for computing the ground state energy per particle
E_g/N and the free energy per particle F/N in the thermodynamic limit
$N \to \infty$. First, define $h = h(u_3,u_+,u_-;v_3,v_+,v_-)$ and obtain from it
the operator h_Q and the function h_C through the substitutions

	$h \rightarrow h_Q$	$h \rightarrow h_C$
u_3	$a^\dagger a/N$	$\mu^*\mu = \alpha^*\alpha/N$
u_+	$a^\dagger/\sqrt{N}$	$\mu^* = \alpha^*/\sqrt{N}$
u_-	$a/\sqrt{N}$	$\mu = \alpha/\sqrt{N}$
v_3	$\frac{1}{2}\left(\sum_{i=1}^{N} \sigma_i^z\right)/N$	$r \cos\theta$
v_+	$\left(\sum_{i=1}^{N} \sigma_i^+\right)/N$	$v^* = r \sin\theta \ e^{i\phi}$
v_-	$\left(\sum_{i=1}^{N} \sigma_i^-\right)/N$	$v = r \sin\theta \ e^{-\phi}$

where μ, α are complex, $r \in [0,\tfrac{1}{2}]$, and $(\theta,\phi) = \Omega$ parameterize the
sphere surface $SU(2)/U(1) \simeq S^2$. Then

    1. If $\hat{H}/N = h_Q$

2. $\lim\limits_{N\to\infty} E_g/N = \min\limits_{\theta,\phi} h_C$

$\lim\limits_{N\to\infty} F/N = \min\limits_{r,\theta,\phi} \Phi$

where $\Phi(r,\theta,\phi;\beta) = h_C(r,\theta,\phi) - kT\, s(r)$.

This algorithm was used[30] to determine the critical properties of the following model systems

a. $h_D = u_3 + \epsilon v_3 + \lambda(u_+v_- + u_-v_+)$ [25-28]

b. $h_{CR} = h_D + \lambda'(u_+v_+ + u_-v_-)$ [27,31]

c. $h_{CR+A} = h_{CR} + \kappa(u_+ + u_-)^2$ [32]

d. $h = h_D + Q(\tfrac{1}{4} - v_3^2) + K(\tfrac{1}{2} + v_3)^2$ [33]

e. $h = u_3 + f[\epsilon v_3 + \lambda(u_+v_- + u_-v_+)]$ [34]

f. $h = u_3 + \epsilon v_3 + g[\lambda(u_+v_- + u_-v_+)]$ [34]

g. $h = h_D + c^*u_+ + c^*u_-$ [35,36]

h. $h = u_3 + \epsilon v_3 + \lambda(u_+^2 v_- + u_-^2 v_+)$ [30]

The existence of a second order phase transition is determined by the vanishing of the determinant of a 2×2 matrix, which is essentially the quadratic form obtained from $\Phi(r,\theta,\phi;\beta)$ along $\theta = \pi$.

7. Thermodynamics of Bialgebraic Models

The photon operators $a^\dagger a$, $a^\dagger$, a, I span the harmonic oscillator algebra $h(4)$ and the collective atomic operators $J_z = \sum\limits_{i=1}^{N} \tfrac{1}{2} \sigma_i^z$, $J_\pm = \sum\limits_{i=1}^{N} \sigma_i^\pm$ span the algebra $su(2)$. The Dicke model (6.1) is therefore a particularly simple example of a "bialgebraic Hamiltonian"[37]

$$\hat{H} = u_i H_i + v_j K_j + \frac{1}{n} \sum_{\alpha,\gamma} E_\alpha C_{\alpha\gamma} F_\gamma \; . \tag{7.1}$$

Here H_i span the Cartan subalgebra of a Lie algebra $\mathcal{G}_1$ and E_α are the shift operators in this algebra.[14] The operators K_j, F_β span $\mathcal{G}_2$ with a similar Cartan decomposition. More general Hamiltonians of bialgebraic type can be written

$$\hat{H}/N = h(X_i/N, \, Y_j/N) \tag{7.2}$$

where the operators X_i span $\mathcal{G}_1$ and Y_j span $\mathcal{G}_2$, and the factors $1/N$ are present for thermodynamic purposes.[34,38]

The form (7.1) of the bialgebraic Hamiltonian is particularly suited to a bifurcation analysis of the type originally carried out for Dicke Hamiltonians.[32] A second order phase transition can occur when a nontrivial primary branch bifurcates from the trivial branch $\langle E_\alpha \rangle = \langle F_\gamma \rangle = 0$. This in turn can occur when the matrix

$$\begin{bmatrix} u \cdot \alpha \, \delta_{\alpha\alpha'} & M_{\alpha\gamma'} \\ N_{\gamma\alpha'} & v \cdot \gamma \, \delta_{\gamma\gamma'} \end{bmatrix} \tag{7.3a}$$

becomes singular. The eigenvector $\mathrm{col}(\langle E_{\alpha'} \rangle, \langle F_{\gamma'} \rangle)$ of (7.3a) with zero eigenvalue indicates the initial direction in order-parameter space of the nontrivial bifurcating solution. The matrices M, N are given by

$$M_{\alpha\gamma'}(\beta) = \langle \alpha \cdot H \rangle_{1,0} \, C_{-\alpha,\gamma'}$$

$$N_{\gamma\alpha'}(\beta) = \langle \gamma \cdot K \rangle_{2,0} \, C_{\alpha',-\gamma} \tag{7.3b}$$

where the thermodynamic expectation value $\langle \cdot \rangle_{1,0}$ is taken with respect to $H_1 = u \cdot H$ and $\langle \cdot \rangle_{2,0}$ with respect to $H_2 = v \cdot K$. Since the matrices (7.3b) are closely related to fluctuations

$$\langle \alpha \cdot H \rangle_{1,0} = \langle [E_\alpha, E_{-\alpha}] \rangle_{1,0} = \langle \{E_\alpha, E_{-\alpha}\} \rangle \, \tanh \tfrac{1}{2} \beta u \cdot \alpha$$

$$\langle \gamma \cdot K \rangle_{2,0} = \langle [F_\gamma, F_{-\gamma}] \rangle_{2,0} = \langle \{F_\gamma, F_{-\gamma}\} \rangle \, \tanh \tfrac{1}{2} \beta v \cdot \gamma \tag{7.4}$$

the result (7.3) indicates that a transformation from disordered to ordered state ($\langle E_\alpha \rangle = 0 \to \langle E_\alpha \rangle \neq 0$ for some α, etc.) occurs when the temperature-weighted fluctuations become sufficiently large. The result (7.3) is therefore called the fluctuation-transformation theorem.[39]

More general bialgebraic Hamiltonians (7.2) can be treated following the steps indicated in §6:

1. Decompose the Hilbert space for G_1 into its invariant subspaces

$$V^{(1)} \to \sum Y_1(N,\Lambda) \, V^\Lambda \, . \tag{7.5}$$

2. Use inequalities (4.6) to put upper and lower bounds on $\mathrm{Tr}_\Lambda e^{-\beta \hat{H}}$.
3. Introduce an entropy function

$$s_1(r) = \lim_{N \to \infty} \frac{1}{N} \ln Y_1(N,\Lambda) \ . \tag{7.6}$$

4. Ditto for G_2.
5. Estimate the resulting upper and lower bounds by Laplace's method.

The difference between these bounds is of order $1/N$, so that

$$\lim_{N \to \infty} F/N = \min_{r_1,\Omega_1;r_2,\Omega_2} \Phi(r_1,\Omega_1,r_2,\Omega_2;\beta)$$

$$\Phi(r_1,\Omega_1,r_2,\Omega_2;\beta) = h(<X_i/N>_1, \ <Y_j/N>_2)$$

$$- kT(s_1(r_1) + s_2(r_2)) \ . \tag{7.7}$$

Here $<X_i/N>_1$ are the classical limits for G_1, as given by (5.1). When evaluated at the minimum, $s_1 + s_2$ gives the intensive entropy and, if $T = 0$, h gives the intensive ground state energy.

Under rather general conditions, involving a minimal symmetry assumption on h (7.2),[30,39] a "crossover theorem" can be proved.[40] This relates the existence of a ground state energy phase transition as a function of increasing interaction parameters ($C_{\alpha\gamma}$ in (7.1), λ in (5.1)) to the occurrence of a thermodynamic phase transition as a function of decreasing temperature.

Example 1. If we consider r-level Dicke models[32] with only one nonzero coupling constant between a single pair of levels and a resonant field mode, only one bifurcation can occur. If one of the two levels is the ground state, a second order thermodynamic phase transition will occur if the coupling constant λ exceeds a critical value λ_{cr}.[41] The bifurcation diagram is shown in Fig. 1a.

If the coupling does not involve the ground state there are four critical values of λ (Fig. 1b) with behavior as follows[41]

$\lambda < \lambda_1$. $<a/\sqrt{N}> = 0$ at all temperatures.

$\lambda_1 < \lambda < \lambda_2$ (curve A, $\lambda = \lambda_A$). At sufficiently low temperatures a metastable ordered state exists.

$\lambda_2 < \lambda < \lambda_3$ (curves B, C, D, $\lambda_B < \lambda_C < \lambda_D$). At low temperatures an ordered state exists. For λ_B the ordered state is metastable to the right of b and stable to the left. A first order phase transition occurs at b. These first-order phase transitions are not surrounded by spinodal lines.

$\lambda_3 < \lambda < \lambda_4$ (curve E, $\lambda = \lambda_E$). Two primary branches bifurcate from the disordered branch at temperatures determined by the fluctuation-transformation theorem.[39] The lower primary branch E_2 is

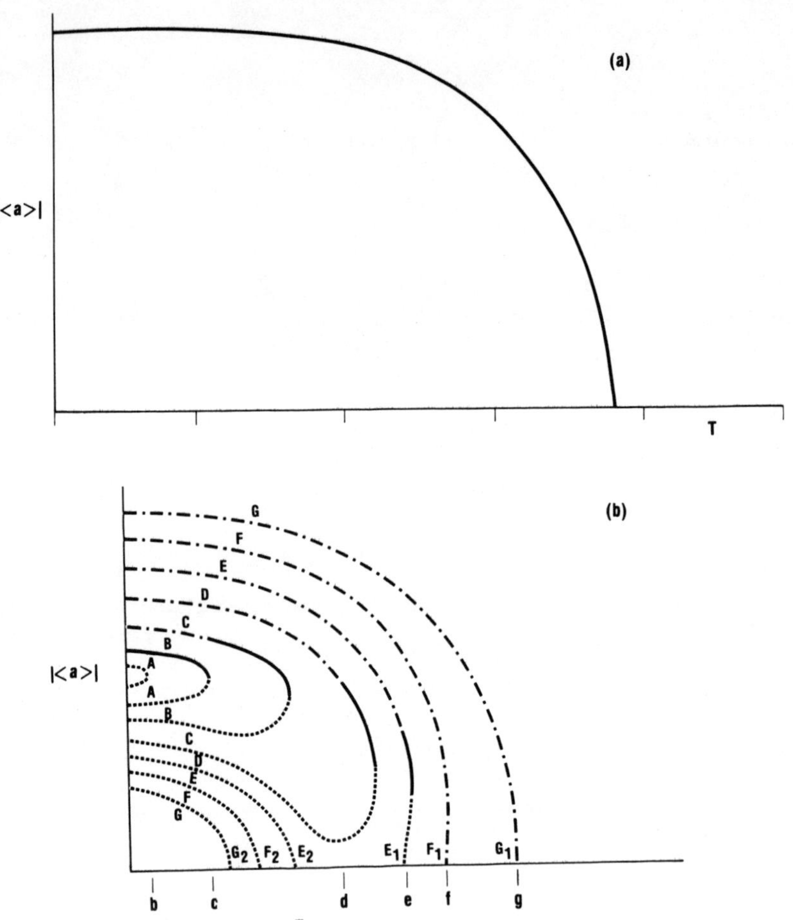

Fig. 1. The critical properties of the r-level extension of the
         Dicke model depend on the pair of levels between which
         the strong interaction occurs. 1a. If the ground state
         is involved a second order thermodynamic phase transition
         can occur. 1b. If two excited states are involved, the
         ground state energy critical properties depend on the valu
         of the coupling constant λ in relation to four critical
         values λ_i, i = 1,2,3,4.

unstable. The upper branch E_1 is initially unstable, becomes
locally stable at the point of vertical tangency, and globally
stable at e. The disordered branch is stable down to e, metastable
between e and the E_1 bifurcation point, unstable between E_1 and E_2,
and metastable below E_2. The first order transition at e is surrounded by spinodal lines.

$\lambda = \lambda_4$ (curve F, $\lambda = \lambda_F$). The F_1 bifurcation point is a tri-
critical point.

$\lambda_4 < \lambda$ (curve G, $\lambda = \lambda_G$). The bifurcating branch G_1 is stable
below g, and G_2 is always unstable. The disordered branch is stable
above g, unstable between the G_1 and G_2 bifurcation points, and
metastable below the G_2 bifurcation point.

Example 2. Primary, secondary, ..., k^{th} order branches can occur
in bialgebraic models (7.1), (7.2), where $k = \min (B_1, B_2)$ and B_i is
the bifurcation index[39] of $\mathcal{G}_i$. The bifurcation index of SU(r) is
r - 1, so r-level Dicke models can exhibit primary, ..., $(r - 1)^{ary}$
branches, but none higher. For 3-level Dicke models only primary
and secondary branches can occur. Typical temperature dependences
for the order parameters on the globally stable branches are shown
in Fig. 2.[42]

8. Thermodynamic Algorithm

The problem of studying the thermodynamic critical properties
associated with a Hamiltonian constructed from operators belonging
to Lie algebras has now been reduced to a simple algorithm for
estimating the free energy per particle. This algorithm becomes
increasingly accurate as N becomes large. The algorithm involves
two steps:
1. Replace the Hamiltonian $\hat{H}/N$ by its classical limit, and add an
    entropy term $-kT \, s(r)$;
2. Determine how the minimum of Φ changes as a function of changing
    temperature.
The first step is implemented using the machinery of coherent states.
The second step is implemented using the machinery of catastrophe
theory.[43-46]

To be more specific, assume that f(x;c) is a k-parameter family
of (potential) functions depending on n state-variables or order-
parameters $x \in \mathbb{R}^n$ and k control parameters $c \in \mathbb{R}^k$. Then under suitable conditions $\mathbb{R}^k$ is partitioned into disjoint open sets by the sets
S_B (bifurcation set) and S_M (Maxwell set) defined by

$$S_B: \quad \begin{array}{c} \nabla f = 0 \\[2mm] \det \dfrac{\partial^2 f}{\partial x_i \partial x_j} = 0 \end{array} \qquad\qquad (8.1)$$

$$S_M: \quad \begin{array}{c} \nabla f = 0 \\[2mm] f(x;c) - f(x';c) = 0 \; . \end{array} \qquad\qquad (8.2)$$

On the bifurcation set an equilibrium becomes degenerate. On the
Maxwell set two or more equilibria assume the same values. The
Maxwell set is determined by equations of Clausius-Clapeyron type.[46]

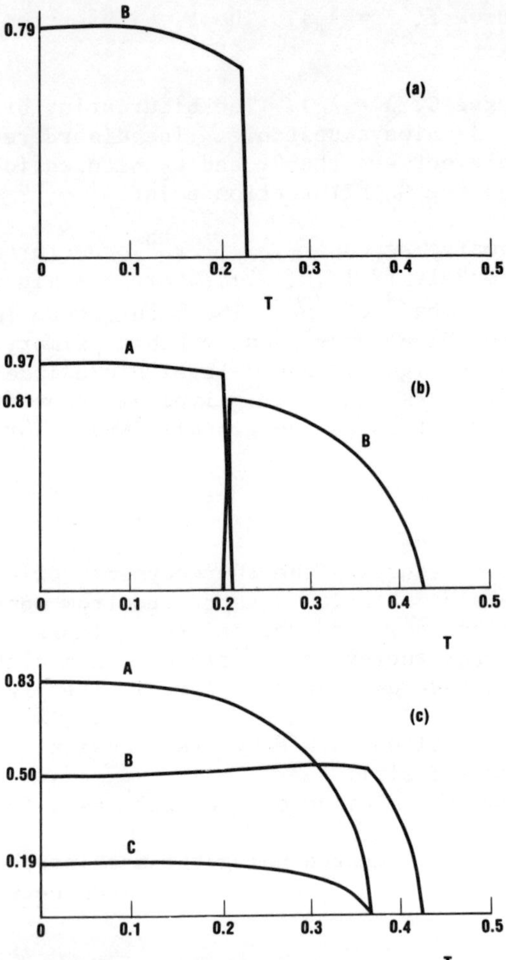

Fig. 2. The thermodynamic critical properties of the 3-level Dicke
model with resonant interaction $\hbar\omega_{ji} = \varepsilon_j - \varepsilon_i$ have been
studied as a function of the dimensionless coupling con-
stants $\Lambda_{ji} = \lambda_{ji}/(\varepsilon_j - \varepsilon_i)$. Here $r = \varepsilon_2 - \varepsilon_1/\varepsilon_3 - \varepsilon_1 = .2$
and $A = \langle E_{21}\rangle$, $B = \langle E_{32}\rangle$, $C = \langle E_{31}\rangle$, for the following
values of $(\Lambda_{12},\Lambda_{23},\Lambda_{13})$. a) (0.0,1.7,0.0) a first order
phase transition occurs. b) (2.0,1.8,0.7) a second order
phase transition occurs at high temperature and a first
order phase transition occurs at lower temperature. c)
(2.0,1.8,0.8) two phase transitions occur, both of second
order.

   As we follow some path $c(s) \in \mathbb{R}^k$, a second (first) order phase
transition will occur if the path crosses a component of S_B (S_M)
describing the global minimum.

For discussing ground state energy phase transitions, we identify the control parameters $c \in \mathbb{R}^k$ with the interaction parameters, such as $C_{\alpha\gamma}$ in (7.1). The order parameters can be chosen as the expectation values of the shift operators spanning $\mathfrak{g}_+ + \mathfrak{g}_-$. For discussing thermodynamic phase transitions we use the same order parameters but the single control variable T.

9. Thermodynamics of Nuclear Models

Nuclear systems are particularly difficult to treat because they contain too many particles for a useful description in terms of the individual particle coordinates and too few particles for a useful description in terms of statistical methods. Two general methods have emerged for the description of such systems: the group-theoretical and the liquid drop models.

Wigner initially exploited the approximate invariance of the nucleon-nucleon interaction under spin and isospin to provide the SU(4) model of nuclear Hamiltonians applicable to light (A $\leq$ 16) nuclei.[47] Although this model loses its usefulness when the protons and neutrons occupy different orbitals, or when spin-orbit interactions become important, it nevertheless retains some predictive capabilities for much heavier nuclei (30 $\leq$ A $\leq$ 110).[48] The SU(3) model of Elliott[49] is also an extremely useful approximation in the range 16 $\leq$ A $\leq$ 24. Recently serious attempts have been initiated[50] to understand heavy (A > 100) even-even nuclei on the basis of the dynamical symmetry group $\widetilde{SU(6)}$, which arises naturally under the assumptions that nucleons have a strong attractive interaction in the L = 0 channel, a somewhat weaker atrractive interaction in the L = 2 channel, and other channels can be neglected.

These approaches have been fairly successful at phenomenologically reproducing energy level spectra and some transition rates within their various regions of applicability. However, the methods described in the previous Section allow us to study the ground state energy and the thermodynamic critical properties of model Hamiltonians constructed from operators belonging to a Lie algebra.

We first tested the value of the coherent state methods in the laboratory called the Meshkov-Glick-Lipkin (MGL) pseudospin (= SU(2)) Hamiltonian.[51] The minimum value of the Q-representative of the MGL Hamiltonian gave a good approximation to the exact ground state energy computed by matrix diagonalization.[52] The error is about 4% for N = 14 and decreases as N increases. The ground state energy phase transition inherent in this model is particularly easy to visualize using coherent states. In the weak interaction regime the coherent state which is a best approximation to the exact ground state is represented by a point at the south pole of the sphere $S^2 \simeq SU(2)/U(1)$. When the quadrupole interaction strength exceeds a critical value the point moves off the south pole in a standard Ginzburg-Landau second-order phase transition.

The thermodynamic critical properties of the MGL Hamiltonian have also been studied.[53] The upper bound on the free energy per nucleon is a good approximation to F/N, which was computed numerically for N = 30, 50, 70. The approximation becomes better as either T decreases or N increases. Although the second order thermodynamic phase transition is not clearly signalled by matrix diagonalization, it is clearly indicated in the coherent state treatment, with a deformed state represented by a point off the south-polar axis and a spherical state by a point on the south-polar axis.

The phase transitions can also be studied by variational methods.[53,54] For the ground state case, coherent states are used as trial states to minimize the Hamiltonian operator $\hat{H}$. In the finite-temperature case Gibbs states, constructed as outer products of coherent states using (4.1P), are used as trial states to minimize the free energy operator $\hat{F} = \hat{H} - T\hat{S}$. In the former case the variational calculation leads to an upper bound identical to that constructed in Section 7, while the finite temperature variational calculation does not.

These methods were extended by Gilmore and Feng to any compact group.[55] For the r-level extension of the MGL model (G = SU(r)) with a quadrupole interaction between a single pair of levels, we find[56] a ground state energy phase transition as a function of increasing interaction strength. The transition is second order if the interaction involves the ground state, first order if it does not. The classical limit $\Phi(\theta_2,\theta_3)$ of this Hamiltonian is shown in Fig. 3 as a function of increasing quadrupole strength Q, where θ_2,θ_3 (r = 3) are the important SU(3) order parameters. We anticipate that the finite temperature critical properties of this model will be closely analogous to the finite-temperature critical properties of the r-level extension of the Dicke model.[41]

A second approach to the description of nuclear properties is based on the liquid drop model. Whereas the former approach is quantum-mechanical in spirit, this approach is primarily classical in spirit. The former approach is discrete in the sense that only a finite number of parameters are fitted in the Hamiltonian; the latter approach involves a nondenumerable number of degrees of freedom. It is a hope[57] that coherent states can bridge the gap between these two approaches. These states appear to have one foot in each school. They are constructed from a finite-dimensional Lie algebra and live in a finite dimensional space (G compact), but there is a nondenumerable number of such states.

10. Dynamics

Coherent states also provide a useful tool for treating problems of quantum dynamics. If $\hat{H}$ is constructed from operators belonging to a Lie algebra $\mathbf{g}$ and if the quantum state is initially in the

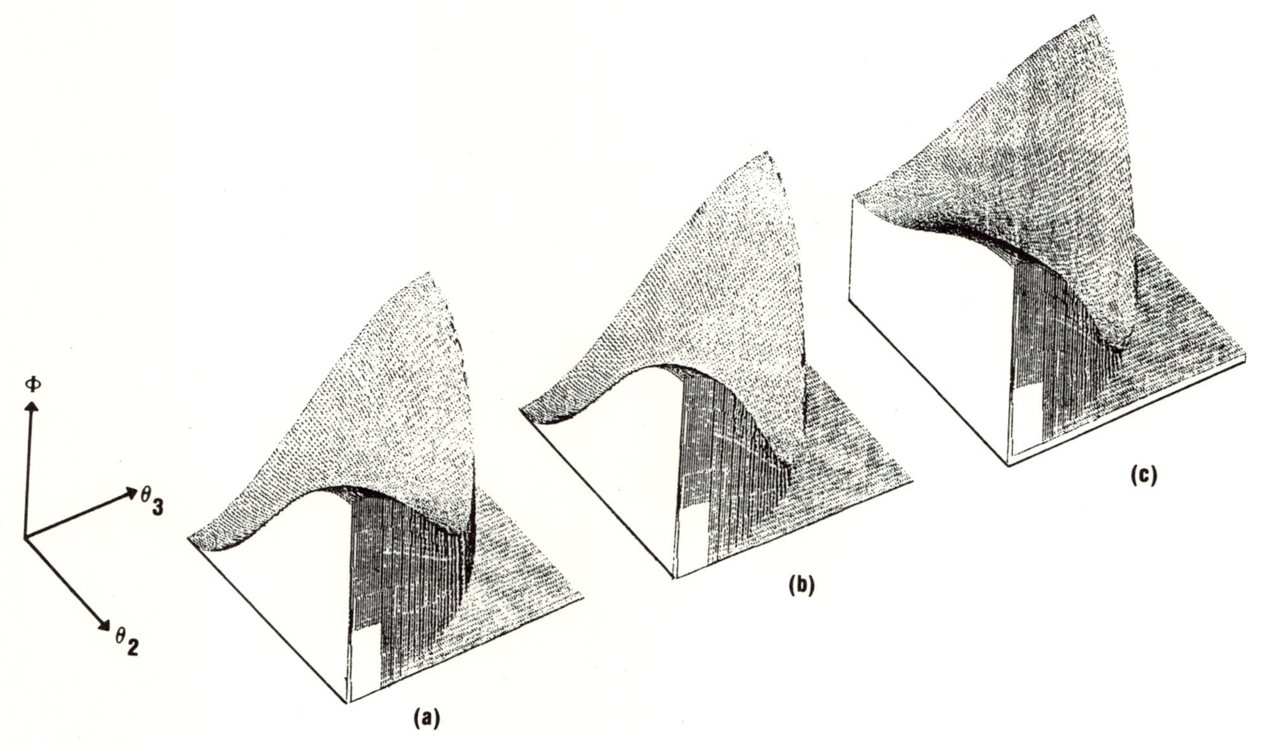

Fig. 3. The shape of $\Phi(\theta_2,\theta_3)$ at $T = 0$ is shown for the 3-level extension of the MGL Hamiltonian. Here θ_2, θ_3 are the relevant SU(3) order parameters, $r = \varepsilon_2-\varepsilon_1/\varepsilon_3-\varepsilon_1 = 0.8$, and the quadrupole interaction Q couples the two excited levels. The quadrant $0 \leq \theta \leq \pi$ is shown. A first order phase transition occurs at $Q = 3.59$. a) $Q = 2.5$; b) $Q = 3.59$; c) $Q = 6.0$.

invariant subspace V^Λ, it will remain in V^Λ for all later times.
To be specific, we can write

$$|\psi(t)> = \int \; |^\Lambda_\Omega > \; \psi(\Omega,t)d\Omega \tag{10.1}$$

where $\psi(\Omega,t) = <\Lambda,\Omega|\psi(t)>$. Then the Hamiltonian operator acting on
$|\psi>$ can be replaced by its $\mathcal{D}$ operator algebra equivalent to give a
c-number equation of the form

$$\mathcal{D}(\hat{H}) \; \psi(\Omega,t) = i\hbar \, \frac{\partial}{\partial t} \, \psi(\Omega,t) \; . \tag{10.2}$$

If $\hat{H}$ contains no higher than second degree operator products of the
basis vectors of $\mathfrak{g}$, (10.2) is an equation of Fokker-Planck type on
the (curved) surface G/H.

If the quantum system cannot be represented by a pure state $|\psi>$
but is rather a statistical mixture of states, then it can be represented by a density operator

$$\hat{\rho}(t) = \int \; |^\Lambda_\Omega> \; \rho(\Omega,t) \; <^\Lambda_\Omega| \; d\Omega \; . \tag{10.3}$$

A c-number equation for $\rho(\Omega,t)$ can be constructed using an appropriate $\mathcal{D}$-algebra mapping. Such equations have been particularly
useful in the description of equilibrium properties ($i\hbar t \to kT$,
Schrödinger equation $\to$ Bloch equation).

In the event that $\hat{H}$ is a linear superposition of the operators
spanning $\mathfrak{g}$, then $\hat{H} \in \mathfrak{g}$ and more powerful statements can be made.
For example, there is a powerful selection theorem: "once a coherent
state, always a coherent state."[4,8,11] The proof is simple. Since
$H \in \mathfrak{g}$, EXP $- i\hat{H}\Delta t/\hbar \in$ G, so that the S matrix is a group element
in G. Then

$$S(t_2,t_1) \; |^\Lambda_{\Omega_0}> = S(t_2,t_1)\Omega_0|^\Lambda_\Lambda>$$

$$= \Omega(t_2,t_1) \; h(t_2,t_1)|^\Lambda_\Lambda>$$

$$= |^\Lambda_{\Omega(t_2,t_1)}> \; e^{i\Phi[h(t_2,t_1)]} \tag{10.4}$$

where $\Omega(t_1,t_1) = \Omega_0$.

The equations of motion for the coherent state parameters can be
derived quite easily using Schur's formula (Appendix   ).[58,59] If
we write the S-matrix in a coset decomposition

$$S(t) = e^{X(t)} \; e^{Y(t)}$$

where

$$X(t) \in \mathfrak{g}_+ + \mathfrak{g}_-$$

$$Y(t) \in \mathfrak{g}_0 \tag{10.5}$$

then the equations of motion are determined from

$$S(t + \Delta t)$$

$$e^{X(t + \Delta t)} \, e^{Y(t + \Delta t)} = e^{-i\hat{H}\Delta t/\hbar} \, e^{X(t)} \, e^{Y(t)} \ . \tag{10.6}$$

If we write $X(t + \Delta t) = X + \Delta X$, $X = X(t)$ and $\Delta X = \Delta t \, (dX/dt)$, and similarly for $Y(t + \Delta t)$ then (10.6) can be written

$$e^X \{I + \frac{I - e^{-AdX}}{AdX} \Delta X\}\{I + \frac{e^{AdY} - I}{AdY} \Delta Y\} \, e^Y$$

$$= e^X \, e^{-X} \{I - \frac{i\Delta t}{h} H\} \, e^X \, e^Y \ . \tag{10.7}$$

The equations of motion are determined from the first order terms in (10.7). In particular, if we set

$$X(t) = b_i(t) \, X_i$$

$$Y(t) = \omega_\alpha(t) \, Y_\alpha$$

$$H(t) = h_i(t) \, X_i + h_\alpha(t) \, Y_\alpha \ ,$$

where the X_i span ($\mathfrak{g}_+ + \mathfrak{g}_-$) and the Y span $\mathfrak{g}_0$, then

$$i\hbar \begin{bmatrix} M_{ij}(t) & M_{i\beta}(t) \\ \\ M_{\alpha j}(t) & M_{\alpha\beta}(t) \end{bmatrix} \begin{bmatrix} \dfrac{db_j}{dt} \\ \\ \dfrac{d\omega_\beta}{dt} \end{bmatrix} = \begin{bmatrix} h_i(t) \\ \\ h_\alpha(t) \end{bmatrix} \tag{10.8}$$

where

$$M_{ij} = \left(\frac{e^{AdX} - I}{AdX}\right)_{ij} \ , \qquad M_{i\beta} = \left(e^{AdX} \frac{e^{AdY} - I}{AdY}\right)_{i\beta}$$

$$M_{\alpha j} = \left(\frac{e^{AdX} - I}{AdX}\right)_{\alpha j} \ , \qquad M_{\alpha\beta} = \left(e^{AdX} \frac{e^{AdY} - I}{AdY}\right)_{\alpha\beta} \ . \tag{10.9}$$

In all physical cases that have arisen so far, the decomposition of $\mathfrak{g}$ has the special property

$$[\mathfrak{g}_+ + \mathfrak{g}_-, \ \mathfrak{g}_+ + \mathfrak{g}_-] \subseteq \mathfrak{g}_0$$

$$[\mathfrak{g}_0, \ \mathfrak{g}_+ + \mathfrak{g}_-] \qquad \subseteq \mathfrak{g}_+ + \mathfrak{g}_-$$

$$[\mathfrak{g}_0, \ \mathfrak{g}_0] \qquad\qquad \subseteq \mathfrak{g}_0 \ . \tag{10.10}$$

In this case AdY is a block diagonal matrix and AdX is block off-diagonal, so that the matrices (10.9) simplify somewhat

$$M_{ij} = \left(\frac{\sinh\ AdX}{AdX}\right)_{ij}, \qquad M_{i\beta} = (\sinh\ AdX)_{i\beta'}\left(\frac{e^{AdY} - I}{AdY}\right)_{\beta'\beta}$$

$$M_{\alpha j} = \left(\frac{\cosh\ AdX - I}{AdX}\right)_{\alpha j}, \qquad M_{\alpha\beta} = (\cosh\ AdX)_{\alpha\beta'}\left(\frac{e^{AdY} - I}{AdY}\right)_{\beta'\beta}.$$

$$\tag{10.11}$$

These equations have been integrated explicitly in the cases $G = H(4)^4$ and $G = SU(2)$ where $(AdY)_{\alpha\beta} = 0$ because $\mathfrak{g}_0$ is commutative. Special cases of the equations of motion (10.8), well-known in nuclear magnetic resonance,[60] are the Bloch equations and the spin-echo equations [$G = SU(2)$]. For $G/H = SO(3,1)/SO(3)$, the equations for $d\omega/dt$ reduce to the Thomas precession equations[14,61] when $h_i(t)$ = constant, $h_\alpha(t) = 0$.

Equations (10.8) represent a (classical) dynamical system on the Lie group G, or a flow on the direct product space $G/H \otimes H$. If the time dependence of the source terms $h_i(t)$, $h_\alpha(t)$ dies out sufficiently rapidly as $t \to \pm\infty$, then the S matrix

$$S = \lim_{\substack{t_1 \to -\infty \\ t_2 \to \infty}} S(t_2, t_1) \tag{10.12}$$

can be determined by analyzing suitable $\alpha \to \omega$ flows. The S-matrices for several systems have been calculated by Perelomov.[62]

11.   TDHF Calculations

The Time Dependent Hartree Fock equations of motion are[63,64]

$$\delta < \psi| \ i\hbar\frac{\partial}{\partial t} - \hat{H}|\psi> = 0 \ . \tag{11.1}$$

If the Hamiltonian $\hat{H}$ is constructed from operators belonging to Lie algebra $\mathfrak{g}$, the most general state $|\psi>$ that can be obtained by applying a dynamical group transformation to the ground or other extremal state is a coherent state, up to phase factor. We can therefore study (11.1) by choosing $|\psi> = g|\Lambda,\Lambda>$, $g = \Omega h$. In this case

$$\langle\psi| \; \hat{H} \; |\psi\rangle = Q_\Lambda(H;\Omega) \; . \tag{11.2}$$

For finite N the Q-representative of $\hat{H}$ can be written down once $\hat{H}$ has been expressed in terms of irreducible tensor operators. For N large, the Q-representative is obtained to a good approximation by taking the classical limit of $\hat{H}$.

The time derivative may be taken exactly as in the previous Section. The result is

$$\langle\psi| \, i\hbar \frac{\partial}{\partial t} \, |\psi\rangle = (\langle X_i\rangle \, , \quad \langle Y_\alpha\rangle) \begin{bmatrix} M_{ij} & M_{i\beta} \\ M_{\alpha j} & M_{\alpha\beta} \end{bmatrix} \begin{pmatrix} \dot{b}_j \\ \dot{\omega}_\beta \end{pmatrix} \tag{11.3}$$

where

$$\langle X_i\rangle = \langle{}^\Lambda_\Omega| \; X_i |{}^\Lambda_\Omega\rangle$$

$$\langle Y_\alpha\rangle = \langle{}^\Lambda_\Omega| \; Y_\alpha |{}^\Lambda_\Omega\rangle \tag{11.4}$$

are simple to compute and the notation of the previous section has been used.

The TDHF equations are therefore

$$\delta \; (\; (11.3) - Q_\Lambda(\hat{H};\Omega) \;) = 0 \tag{11.5}$$

where the variation is over the coordinates in the group G, or the coordinates b_i, ω_α, where $b_i X_i \in (\mathfrak{g}_+ + \mathfrak{g}_-)$ and $\omega_\alpha Y_\alpha \in \mathfrak{g}_0$. Since $Q_\Lambda(\hat{H};\Omega)$ and Ad X are independent of the coordinates ω_α, these coordinates are easily seen to be cyclic. The equations (11.6) were explicitly derived for MGL models[55] using the coherent states $| \, J; \, \theta\phi \rangle$. The azimuthal coordinate ϕ is also cyclic.

12. Nonequilibrium Steady State Systems

The laser is an important example of an open physical system. Energy is pumped through this system. If the rate of dissipation is sufficiently great, a phase transition from disordered to ordered state ("off" to "on") can occur.[65]

We can describe the dynamics of open quantum-mechanical systems by computing the equations of motion

$$i\hbar \frac{d}{dt} \langle X_i\rangle = \langle[X_i,\hat{H}]\rangle + \text{more} \; . \tag{12.1}$$

The additional terms model the dissipative process. They can be taken into account by replacing the term on the left by

$$\frac{d}{dt} \langle X\rangle \rightarrow (\frac{d}{dt} + \gamma_i)(\langle X_i\rangle - \langle X_i\rangle_c) \; . \tag{12.2}$$

Here γ_i describes the dissipation process, i.e., how fast the expectation value $\langle X_i \rangle$ will relax to the value $\langle X_i \rangle_c$ which is "clamped" by forces causing the flow through the system.

It is useful to consider the nonequilibrium steady state $(d/dt \rightarrow 0)$ properties of the bialgebraic Hamiltonian (7.1). If $\langle E_\alpha \rangle_c = \langle F_\gamma \rangle_c = 0$, then the coupled nonlinear equations are

$$i\hbar\, \gamma_i (\langle H_i \rangle_s - \langle H_i \rangle_c) = \langle [H_i, E_{\alpha'}] \rangle_s\, C_{\alpha'\gamma'}\, \langle F_{\gamma'} \rangle_s \qquad \text{(a)}$$

$$i\hbar\, \gamma_\alpha\, \langle E_\alpha \rangle_s = \langle [E_\alpha, E_{\alpha'}] \rangle_s\, C_{\alpha'\gamma'}\, \langle F_{\gamma'} \rangle_s \qquad \text{(b)}$$

$$i\hbar\, \gamma_j (\langle K_j \rangle_s - \langle K_j \rangle_c) = \langle E_{\alpha'} \rangle_s\, C_{\alpha'\gamma'}\, \langle [K_j, F_{\gamma'}] \rangle_s \qquad \text{(c)}$$

$$i\hbar\, \gamma_\gamma\, \langle F_\gamma \rangle_s = \langle E_{\alpha'} \rangle_s\, C_{\alpha'\gamma'}\, \langle [F_\gamma, F_{\gamma'}] \rangle_s \ . \qquad \text{(d)}$$

$$\text{(12.3)}$$

The subscript s indicates the steady state expectation value.

Equations (12.3) may be studied by eliminating $\langle H_i \rangle_s$ between a and b and $\langle K_j \rangle_s$ between c and d. In the absence of dissipation and for low flow rates the only solution to these coupled nonlinear equations is $\langle E_\alpha \rangle_s = \langle F_\gamma \rangle_s = 0$. Bifurcations of nontrivial solutions from the trivial solution can be determined by the standard linearization method. The bifurcation condition is given by

$$\det \begin{bmatrix} -i\hbar\gamma_\alpha\, \delta_{\alpha\alpha'} & M_{\alpha\gamma'} \\ N_{\gamma\alpha'} & -i\hbar\gamma_\gamma\, \delta_{\gamma\gamma'} \end{bmatrix} = 0 \qquad \text{(12.4)}$$

$$M_{\alpha\gamma'} = C_{-\alpha\gamma'}\, \langle \alpha \cdot H \rangle_c$$

$$N_{\gamma\alpha'} = C_{\alpha', -\gamma}\, \langle \gamma \cdot K \rangle_c \ . \qquad \text{(12.5)}$$

Because of its close relationship with the fluctuation-transformation theorem (7.3), this result is called the dissipation-transformation theorem.

When applied to the Dicke model,[36] this calculation shows that there is a formal analogy between the thermodynamic phase transition (discussed earlier) and the nonequilibrium steady state phase transition. But is does more. Part of the thermodynamic algorithm involves the study of bifurcation properties using the methods of catastrophe theory. The Ginzburg-Landau phase transition responsible for both phase transitions is structurally unstable,[42-46] and requires one additional unfolding parameter for a universal unfolding. This may be introduced by adding a classical near-resonant coherent field to the system.[66] When this is done the laser equation of state is diffeomorphic with the cusp catastrophe manifold. Increasing the amplitude of this field when the pump rate is held below threshold can lead to a first order phase transition. Such transitions have been observed by Gibbs, McCall, and Venkatesan,[67] and have been investigated extensively.[68,69]

For the laser:[45,66]

1.   the nonequilibrium equations of (steady) state are ("typically") manifolds in the direct product group parameter space;

2.   the density operator factors ($\hat{\rho} = \hat{\rho}_1 \otimes \hat{\rho}_2$) where $\hat{\rho}_1(\hat{\rho}_2)$ is the reduced density operator for the subsystem with group $G_1(G_2)$;

3.   the reduced density operators can be written as exponentials: $\hat{\rho}_1 \simeq EXP(r_i X_i)$, $\hat{\rho}_2 \simeq EXP(s_j Y_j)$;

4.   the expectation values $\langle X_i \rangle$, $\langle Y_j \rangle$ determine the coefficients r_i, s_j up to overall multiplicative factor;

5.   this factor is determined by the variances $\langle X_i^\dagger X_i \rangle - \langle X_i^\dagger \rangle \langle X_i \rangle$, etc;

6.   the reduced density operators factor

$$\rho = [\hat{\rho} \ (Geometry)]^{Physics}$$

in the spirit of the Wigner-Eckart theorem. Here $\hat{\rho}$ (Geometry) is an operator defined by a point on some (catastrophe) manifold and "Physics" is a number characterizing noise;

7.   the density operator $\hat{\rho} = \hat{\rho}_1 \otimes \hat{\rho}_2 = EXP(r_i X_i + s_j Y_j)$ can be obtained by linearizing a Hamiltonian H'. This means the steady state dynamics of H is identifiable with the thermodynamics of H'.

It is our hope that this relation between dynamics and thermodynamics can be extended further.

13.   Summary and Conclusions

Coherent states may be defined with respect to the following structures: 1) a Lie group G with Lie algebra $\mathfrak{g}$; 2) a square-integrable unitary irreducible representation Γ^Λ on an invariant Hilbert space V^Λ; 3) an extremal state $|ext \rangle$ in V^Λ annihilated by a maximal solvable subalgebra in $\mathfrak{g}$; 4) a closed subgroup $H \subset G$ fixing $|ext \rangle$ up to phase. Coherent states are "the orbit of $|ext \rangle$ under G" and are parameterized by the projective space G/H. Properties derived from their rich geometric structure are summarized in Section 3.

They provide a particularly convenient set of states for constructing operator mappings (Section 4) and for constructing classical limits for Lie algebras more general than SU(2) (Section 5). The P- and Q-representatives can be used to put upper and lower bounds on partition functions. In the thermodynamic limit the bounds on F/N converge to the value obtained by taking the classical limit of $\hat{H}/N$, adding an entropy term $-kT \ N^{-1} \ln Y(N,\Lambda)$, and minimizing over (Ω,r), $\Omega \ \epsilon \ G/H$, $r \ \epsilon$ dual to H. This two-step algorithm exploits Lie Group Theory in the first step and Elementary Catastrophe Theory in the second.

This algorithm was used to study the ground state energy phase transition and thermodynamic phase transitions in models of Dicke type (Section 6), generalized bialgebraic models (Section 7), and nuclear models (Section 9). A bifurcation analysis leads to a "fluctuation-transformation" theorem, which determines bifurcations of ordered branches from the disordered branch.

The dynamical properties of systems described by Hamiltonian $\hat{H}$ can also be treated using coherent states. If $\hat{H} \in \mathfrak{g}$, the "semi-classical" theorem can be written explicitly as a set of first order ordinary coupled nonlinear differential equations over G/H (Section 10). If $\hat{H} \notin \mathfrak{g}$ but $\hat{H}$ is constructed from operators in $\mathfrak{g}$ ($H \in U(\mathfrak{g})$) then the semiclassical theorem fails, wave packets on G/H spread, but the TDHF equations (Section 11) are closely analogous to the semiclassical equations.

Nonequilibrium steady state systems were studied in the same way that equilibrium equations can be studied, through their equations of motion. These sets of equations are closely comparable. A bifurcation analysis leads in this case to a "dissipation-transformation" theorem (Section 12). These identifications are made explicit for Dicke models, where the "unfolded" Dicke Hamiltonian gives an equation of state diffeomorphic to the cusp catastrophe manifold under both equilibrium and nonequilibrium mappings. This suggests that it may be possible to study the steady state dynamics of H variationally by studying the thermodynamics of some associated Hamiltonian H' variationally.

APPENDIX

Schur's formula[58,59] for computing operator differentials may conveniently be derived by solving the differential equation[32]

$$\frac{d}{dt} e^{-tX} e^{t(X + \varepsilon Y)} = e^{-tX} \varepsilon Y e^{t(X + \varepsilon Y)} \tag{A.1}$$

formally by iteration in powers of the small parameter ε

$$e^{-tX} e^{t(X + \varepsilon Y)} - I = \varepsilon \int_0^t dt' \, e^{-t'X} Y e^{t'X} \{ I + \varepsilon \int_0^{t'} dt'' \ldots \tag{A.2}$$

By truncating beyond linear terms in ε, setting $t = 1$, and carrying out the integration formally, we obtain

$$e^{X + \varepsilon Y} = e^X \left\{ I + \frac{I - e^{-AdX}}{AdX} \varepsilon Y \right\} \tag{A.3}$$

where AdX is the regular representation of X, defined by

$AdX\ Y = [X,Y]$, $e^{AdX}\ Y = e^{X}\ Ye^{-X}$. The latter relation provides the alternative decomposition

$$e^{X+\varepsilon Y} = \{I + \frac{e^{AdX} - I}{AdX}\ \varepsilon\ Y\}\ e^{X}\ . \qquad (A.4)$$

REFERENCES

1. E. Schrödinger, Naturwiss. $\underline{14}$, 644 (1927).
2. F. Bloch and A. Nordsieck, Phys. Rev. $\underline{52}$, 54 (1937).
3. J. Schwinger, Phys. Rev. $\underline{91}$, 728 (1953).
4. R. J. Glauber, Phys. Rev. $\underline{130}$, 2529 (1963), $\underline{131}$, 2766 (1963).
5. P. W. Atkins and J. C. Dobson, Proc. Roy. Soc. (London) $\underline{A321}$, 321 (1971).
6. J. Kutzner, Phys. Lett. $\underline{41A}$, 475 (1972).
7. J. M. Radcliffe, J. Phys. $\underline{A4}$, 313 (1971).
8. F. T. Arecchi, E. Courtens, R. Gilmore, and H. Thomas, Phys. Rev. $\underline{A6}$, 2211 (1972).
9. R. Gilmore, Ann. Phys. (NY) $\underline{74}$, 391 (1972).
10. A. M. Perelomov, Commun. Math. Phys. $\underline{26}$, 222 (1972).
11. R. Gilmore, Rev. Mex. de Fisica $\underline{23}$, 143 (1974).
12. I. M. Gel'fand and M. L. Tsetlein, Dokl. Akad. Nauk SSSR $\underline{71}$, 825 (1950), $\underline{71}$, 1017 (1950).
13. R. Gilmore, J. Math. Phys. $\underline{11}$, 3420 (1970).
14. R. Gilmore, Lie Groups, Lie Algebras, and Some of Their Applications, NY: Wiley, 1974.
15. A. O. Barut and L. Girardello, Commun. Math. Phys. $\underline{21}$, 41 (1971).
16. R. Gilmore, J. Math. Phys. $\underline{15}$, 2090 (1974).
17. R. Gilmore, C. M. Bowden, and L. M. Narducci, Phys. Rev. $\underline{A12}$, 1019 (1975).
18. W. Miller, Jr., Lie Theory and Special Functions, NY: Academic, 1968.
19. N. Ja. Vilenkin, Special Functions and the Theory of Group Representations, Providence: American Mathematical Society, 1968.
20. R. Gilmore, J. Phys. $\underline{A9}$, L65 (1976).
21. E. H. Lieb, Commun. Math. Phys. $\underline{31}$, 327 (1973).
22. L. M. Narducci, C. M. Bowden, V. Bluemel, G. P. Carrazana, and R. A. Tuft, Phys. Rev. $\underline{A11}$, 973 (1975).
23. G. Herzberg, Atomic Spectra and Atomic Structure, NY: Dover, 1944.
24. R. Gilmore, J. Math. Phys. $\underline{20}$, 891 (1979).
25. R. H. Dicke, Phys. Rev. $\underline{93}$, 99 (1954).
26. K. Hepp and E. H. Lieb, Ann. Phys. (NY) $\underline{76}$, 360 (1973).
27. Y. K. Wang and F. T. Hioe, Phys. Rev. $\underline{A7}$, 831 (1973).
28. K. Hepp and E. H. Lieb, Phys. Rev. $\underline{A8}$, 2517 (1973).
29. D. V. Widder, An Introduction to Transform Theory, NY: Academic (1971).
30. R. Gilmore, J. Math. Phys. $\underline{18}$, 17 (1977).

31. H. J. Carmichael, C. W. Gardiner, and D. F. Walls, Phys. Lett. A46, 47 (1973).

32. R. Gilmore and C. M. Bowden, J. Math. Phys. 17, 1617 (1976), Phys. Rev. 13, 1898 (1976).

33. Y. A. Kudenko, A. P. Slivinsky, and G. M. Zaslavsky, Phys. Lett. A50, 411 (1975).

34. R. Gilmore, Physica 86A, 137 (1977).

35. J. P. Provost, F. Rocca, G. Vallee, and M. Sirugue, Physica 85A, 202 (1976).

36. R. Gilmore, Phys. Lett. A60, 387 (1977).

37. R. Gilmore, in: Journees Relativistes 1976, M. Cahan, R. Debener, and J. Geheniau, Eds., Brussels: Université Libre, 1976, p. 71.

38. D. Ruelle, Statistical Mechanics, NY: Benjamin, (1969).

39. R. Gilmore, Bialgebraic Models (unpublished).

40. R. Gilmore and C. M. Bowden, in Proceedings of the First Army Conference on High Energy Lasers, C. M. Bowden, D. W. Howgate, and H. R. Robl, Eds., NY: Plenum, 1978, p. 335.

41. R. Gilmore, J. Phys. A10, L131 (1977).

42. R. Gilmore, S. R. Deans, and D. H. Feng, to be published.

43. R. Thom, Structural Stability and Morphogenesis, Reading: Benjamin, 1975.

44. E. C. Zeeman, Catastrophe Theory, Selected Papers 1972-1977, Reading: Addison-Wesley, 1977.

45. T. Poston and I. N. Stewart, Catastrophe Theory and its Applications, London: Pitman, 1978.

46. R. Gilmore, Catastrophe Theory for Scientists and Engineers, NY: Wiley, 1980 (to appear).

47. E. P. Wigner, Phys. Rev. 51, 106 (1937).

48. P. Franzini and L. A. Radicati, Phys. Lett. 6, 322 (1963).

49. J. P. Elliott, Proc. Roy. Soc. (London) A245, 128 (1958).

50. A. Arima and F. Iachello, Phys. Lett. 53B, 309 (1974), Phys. Rev. Lett. 35, 1069 (1975) and 40, 385 (1978), Ann. Phys. (NY) 99, 253 (1976) and 111, 201 (1978).

51. H. J. Lipkin, N. Meshkov, and A. J. Glick, Nucl. Phys. 62, 188, 199, 211 (1965).

52. R. Gilmore and D. H. Feng, Phys. Lett. 76B, 26 (1978).

53. R. Gilmore and D. H. Feng, Nucl. Phys. A301, 189 (1978).

54. R. Gilmore and D. H. Feng, Phys. Rev. C19, 1119 (1978).

55. R. Gilmore and D. H. Feng, to be published.

56. R. Gilmore and D. H. Feng, Phys. Lett. (submitted).

57. F. Iachello, private communication.

58. I. Schur, Math. Ann. 35, 161 (1890); Leipz. Ber. 42, 1 (1890).

59. S. Helgason, Differential Geometry and Symmetry Spaces, NY: Academic, 1962.

60. A. Abragam, The Principles of Nuclear Magnetism, Oxford: Clarendon Press, 1961.

61. L. H. Thomas, Nature 117, 514 (1926).

62. A. M. Perelomov, Sov. Phys. Usp. 20, 703 (1977).

63. P. A. M. Dirac, Proc. Camb. Phil. Soc. $\underline{26}$, 376 (1930).
64. D. J. Rowe, <u>Nuclear Collective Motion - Models and Theory</u>,
         London: Methuen, 1970.
65. H. Haken, Rev. Mod. Phys. $\underline{47}$, 67 (1975).
66. R. Gilmore and L. M. Narducci, $\underline{A17}$, 1747 (1978).
67. H. M. Gibbs, S. L. McCall, and T. N. C. Venkatesan, Phys. Rev.
         Lett. $\underline{36}$, 1135 (1976).
68. R. Bonifacio and L. A. Lagiato, Opt. Commun. $\underline{19}$, 1972 (1976),
         Phys. Rev. Lett. $\underline{40}$, 1023 (1978).
69. G. S. Agarwal, L. M. Narducci, R. Gilmore, and D. H. Feng,
         Optics Letters $\underline{2}$, 88 (1978), Phys. Rev. $\underline{A18}$, 620 (1978).

THE APPLICATION OF IMBEDDING THEORY TO THE ATOMIC SHELL MODEL

Bruno Gruber*, F. A. Matsen** and M. S. Thomas***

*Physics Department, Southern Illinois University,
  Carbondale, Illinois 62901

**Departments of Physics and Chemistry, The University
  of Texas, Austin, Texas 78712

***Mathematics Department, Royal Military College of
  Canada, Kingston, Ontario K7L 2W3

INTRODUCTION

The atomic shell model has been shown by Racah (1), Judd (2),
and Wybourne (3) to exhibit a group theoretical structure
consisting of a network of interlocking chains of subgroups. In
the present article we apply formal imbedding theory to the
elucidation of the atomic shell network.

Exact atomic eigenvectors and eigenvalues can, in principle,
be computed by diagonalizing an N-electron Hamiltonian in the
antisymmetric subspace of its full Hilbert space. The antisymmetrized subspace is spanned by the antisymmetric projections
of the set of all Nth rank tensor products of fermion orbitals,
$\mid \eta >$ where $\eta = (n, \ell, m_\ell, m_s)$ or (n, ℓ, j, m_j) which may, in
principle, be computed by the Hartree-Fock procedure. These basis
vectors are eigenvectors to a zero order Hamiltonian for which the
complementary perturbation Hamiltonian contains the residual
electron-electron interaction. Finally, again in principle, the
matrix representation of the perturbation Hamiltonian is computed
and diagonalized to obtain exact eigenvectors and eigenvalues.

Of course the full antisymmetrized space is too large for
the implementation of this program and, in practice, the space is
truncated. A widely used truncation is the atomic shell model,
$(n\ell)^N$ in which the truncated space is spanned by zero order states
with closed inner shell configurations, from which orbitals outside

the $n\ell$ shell are excluded and which are degenerate to the zero
order Hamiltonian. Exact eigenvectors and eigenvalues can, in
principle, be computed from an effective Hamiltonian computed by
means of degenerate perturbation theory (2d, 4, 5). In practice
the effective Hamiltonian is chosen empirically (3a, 6) and the
values of the parameters of the effective Hamiltonian determine
its symmetry or near-symmetry. To a high level of approximation
the total angular momentum group $SU_J(2)$ is the group of the Hamiltonian and J is an "exact" quantum number. $SU_J(2)$ may, in consequence, be regarded as the tail group of a network of interlocking
chains of subgroups any one of which may have physical significance
and supply good quantum numbers.

The $(n\ell)^N$ antisymmetrized vector space can also be constructed
by applying fermion creation operators, $a^\dagger_\eta$ to the vacuum state
and the effective Hamiltonian expressed as an electron number
conserving polynomial in the creation and annihilation operators
with parameters selected as described above.

A conventional choice for the head group of the $n\ell$ shell
network is $SO(8\ell + 5)$ whose Lie algebra is composed of the set of
all linear and bilinear creation and annihilation operators (2c)
and which does not conserve particle-number and orbit, spin, and
total angular momentum. There are two reasons for this choice:
i) Higher degree combinations do not close on a Lie algebra.
ii) The Lie algebra is the algebra of quasi-particles and the
Bogoliubov transformation (8, 9, 10) and is physically significant
when the effective pairing interaction is strong. The network of
all chains of subgroups can be computed by imbedding theory as is
discussed in the subsequent sections of this article. Some of the
results of these computations are exhibited in Figures one through
four.

Figure one contains the network of chains which contain the
electron-number conserving group $U(4\ell + 2)$ and which is complete
up to $\ell \le 6$. The L − S chain contains $SO_L(3) \times SU_S(2)$ which is
physically significant for light atoms where spin-orbit interaction
is weak and L and S are good quantum numbers. If spin-orbit
interaction can be neglected there exists a spin-free atomic
theory (11, 12) based on the chain

$$U(2\ell + 1) \supset SO(2\ell + 1) \supset \ . \ . \ . \ SO_L(3).$$

The j−j chain contains $U(2\ell + 2) \times U(2\ell)$ and is physically
significant for heavy atoms where spin-orbit interaction is
strong and the individual j's are good quantum numbers. The remaining two chains contain $Sp(4\ell + 2)$ which is physically significant
when the pairing interaction is strong and seniority is a good quantum number. These chains are modified L − S and j−j chains.

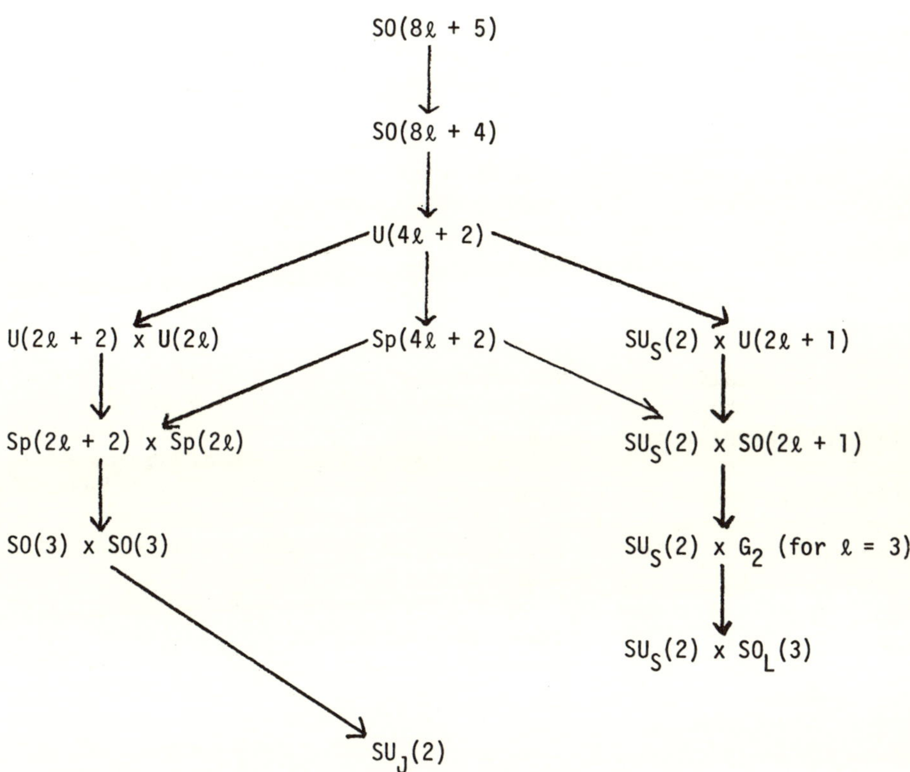

Figure 1. Symmetry chain for electronic configurations

Figure two contains figure one and additional chains which involve quasi spin groups $SU_Q(2)$ and symplectic groups. These chains introduce seniority ν as a good quantum number and they are useful in the treatment of the effective pairing interaction.

Figure three contains still another subnetwork for the head group $SO(8\ell + 5)$, a network in which spin is unconserved throughout and where the words "up" and "down" refer to spin orientation. The left hand chain is called the <u>quasi particle</u> chain because it contains $(SO(2\ell + 1) \times SO(2\ell + 1)) \times (SO(2\ell + 1) \times SO(2\ell + 1))$ a

$$\text{up} \qquad\qquad\qquad\qquad \text{down}$$

group, which like the head group, conserves neither spin nor particle number and whose tail group is $SO_L(3)$ rather than $SU_S(2)$. The physical significance of this and the other chains in the networks is not known at present.

Finally, figure four exhibits part of the network computed from the $(\ell_1 + \ell_2)^N$ mixed configuration space.

THE ATOMIC SHELL AND ITS ALGEBRA

The formalism of second quantization can be formally made use of to define the states for the atomic shell, and to derive a Lie algebra for which the states form basis vectors of an irreducible representation. In order to achieve this goal creation and annihilation operators a^+_η, a_ρ for single electrons are introduced. These operators are characterized by the four quantum numbers $\eta = (n, \ell, m, m_s)$ introduced in the preceding section, with the principal quantum number n and the orbital momentum ℓ being the same for all electrons of a given shell. These fermion creation-annihilation operators satisfy the anticommutation relation

$$\{a^+_\eta, a_\rho\} \equiv a^+_\eta a_\rho + a_\rho a^+_\eta = \delta_{\eta,\rho} \tag{2.1}$$

where $\delta_{\eta,\rho} = \delta_{m,m'} \delta_{m_s,m_s'}$ denotes a product of Kronecker

symbols. If $|0>$ denotes the vacuum state, then the states of an atomic shell are given by the collection of states

$$\begin{array}{c} |0> \\ a^+_\eta |0> \\ a^+_\eta a^+_\rho |0> \\ \cdot \\ \cdot \\ \cdot \\ a^+_\eta a^+_\rho \cdots a^+_\sigma |0> \end{array} \tag{2.2}$$

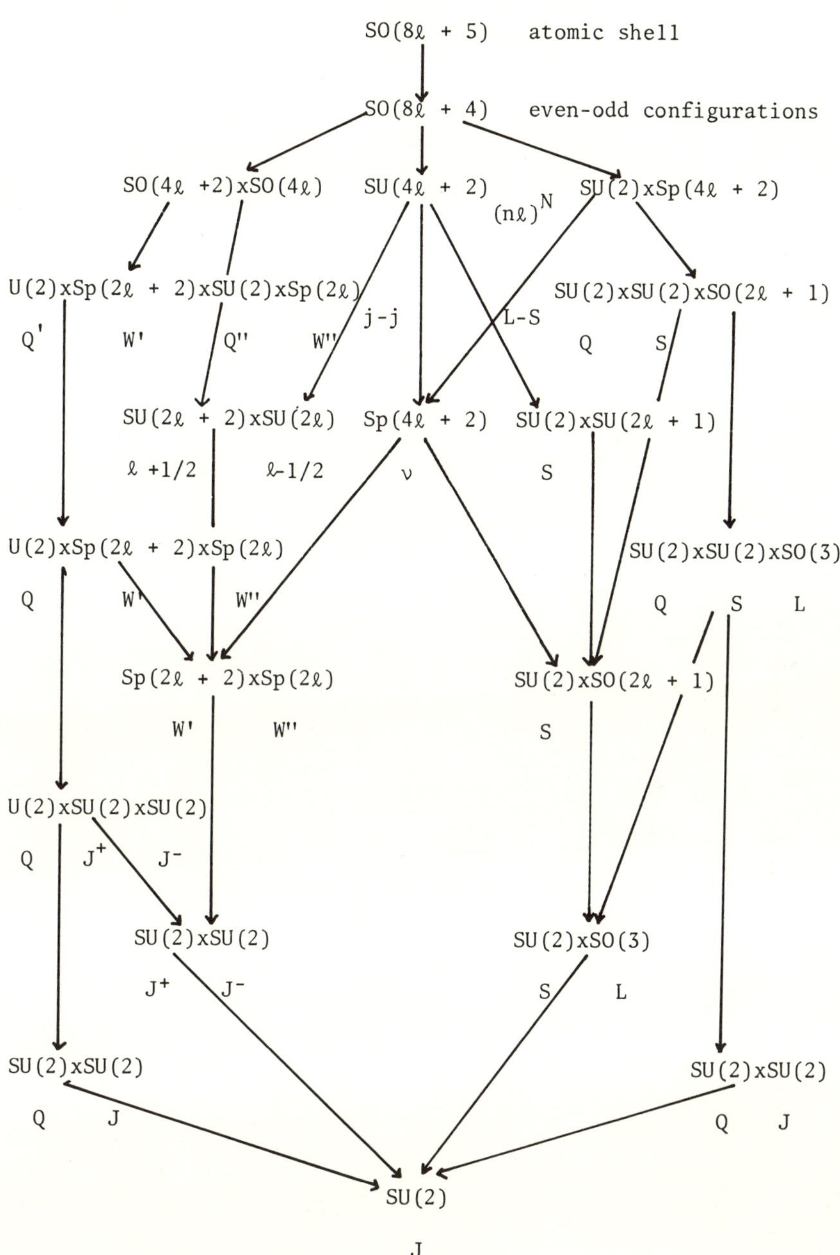

Figure 2. Symmetry chain for the atomic shell

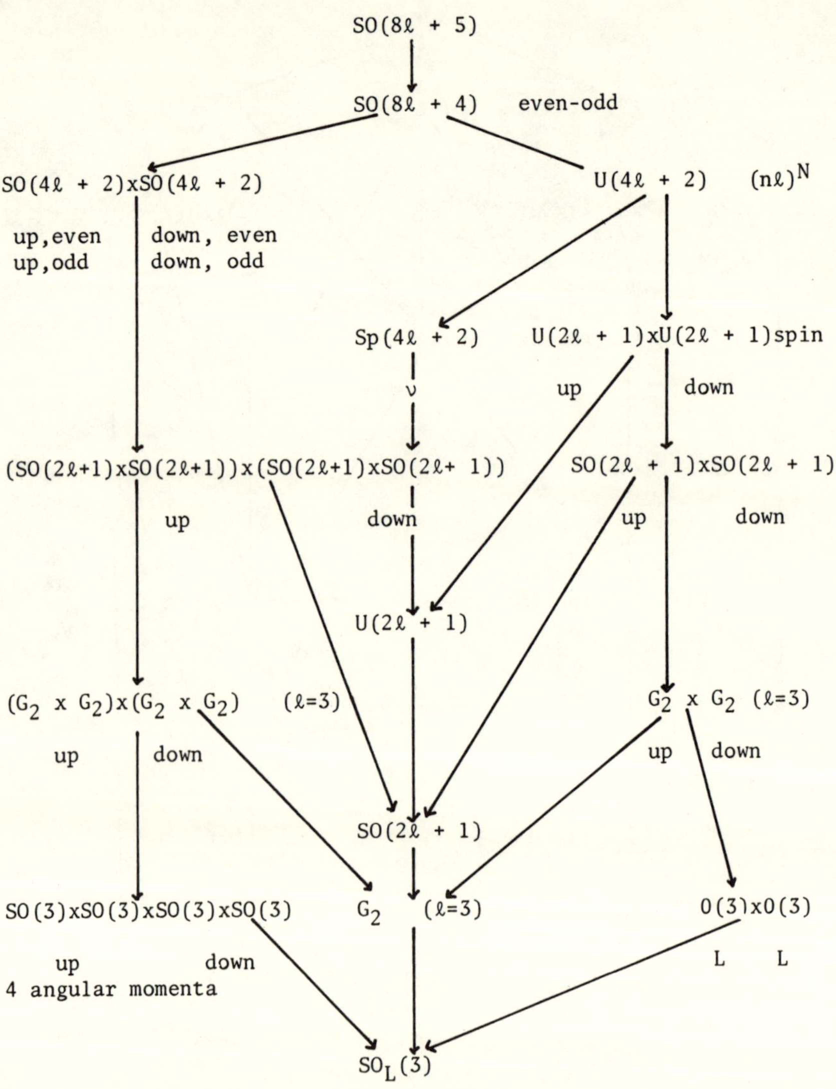

Figure 3.   Symmetry chain conserving total orbital angular momentum

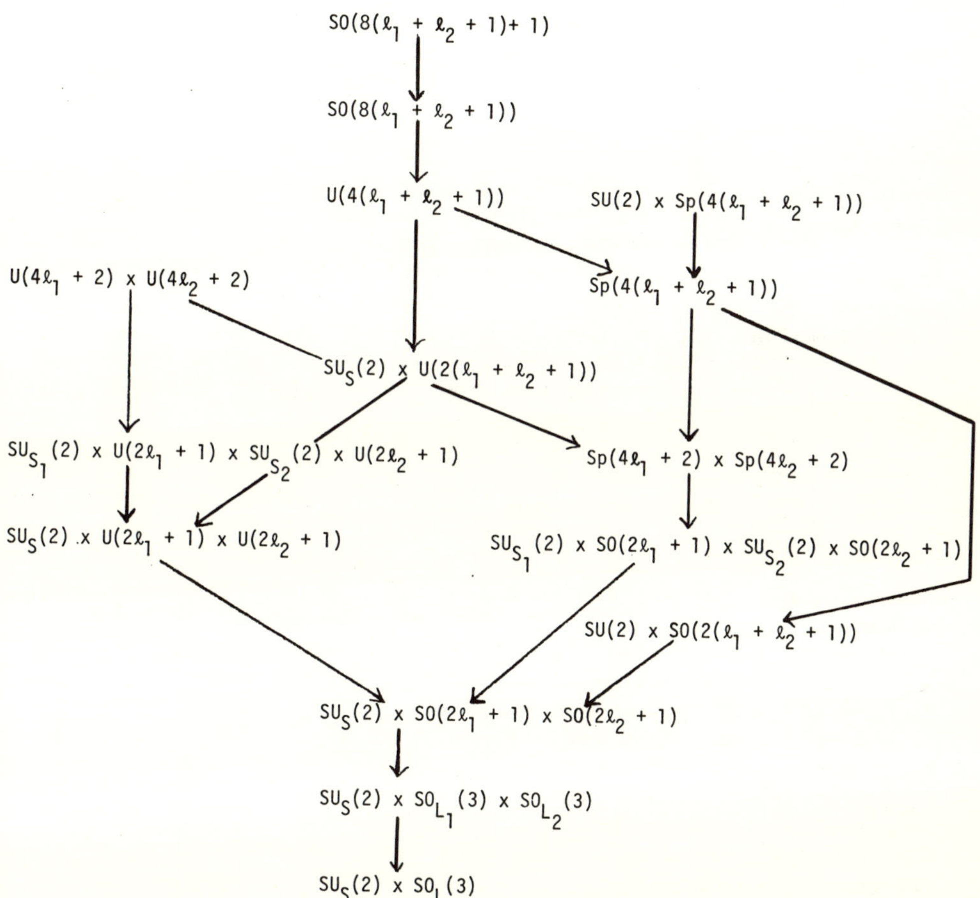

Figure 4. Symmetry chain for mixed atomic shells

with each line representing the states of a configuration $(n\ell)^N$ of
N electrons in succession, for N = 0, 1, 2, ..., $4\ell + 2$. The last
line presents the full shell, containing all the $4\ell + 2$ electrons
permitted by the Pauli exclusion principle. Also, due to the
anticommutation property eqn (2.1), these states automatically
satisfy the Pauli principle, i.e. they are antisymmetric with
respect to the exchange of pairs of electrons.

The elements of the (simple) Lie algebra L, for which the
states of eqn (2.2) form an orthonormal basis of an irreducible
representation, are in terms of the creation-annihilation operators
a^+_η, a_ρ given by

$$\{a^+_\eta; a_\rho; a^+_\eta a^+_\rho, \eta \neq \rho; a_\eta a_\rho, \eta \neq \rho; a^+_\eta a_\rho, \eta \neq \rho;$$

$$a^+_\eta a_\eta - 1/2\}. \tag{2.3}$$

The elements $H_\eta \equiv a^+_\eta a_\eta - 1/2$ commute and form the $4\ell + 2$ basis
elements of a Cartan subalgebra H of L. Forming all possible
commutators, and utilizing the anticommutation relation eqn (2.1),
it is easy to prove that the elements given by eqn (2.3) form a
basis for the simple (complex) Lie algebra $B_{4\ell + 2}$. Certain
linear combinations can then be chosen such as to form a basis
for the real (simple) Lie algebra of the compact group $SO(8\ell + 5)$.
An excellent review of this approach has been given by Judd[2c].

THE CARTAN CANONICAL FORM

The Cartan Canonical Form for a simple Lie algebra is the form
which is most frequently utilized in applications [13, 14]. The
reason for this is that in this form the elements of the Lie
algebra are, in a matrix representation, either _diagonal_ matrices,
or of the form of _simple_ shift operators (rather than linear
combinations of simple shift operators). The diagonal matrices
represent the commuting basis elements of the Cartan subalgebra.
And for the case of a Lie algebra of direct physical significance,
their (real) eigenvalues represent physically significant quantum
numbers. The (simple) shift operators then effect transitions from
one pure state to another pure state (simultaneous eigenvectors of
all commuting elements).

The Cartan Canonical Form is necessarily given for _complex_
Lie algebras. The diagonalization process requires an extension
of the field of real numbers $\mathbb{R}$ to the complex numbers $\mathbb{C}$. The
Cartan subalgebra H of L is however not affected by this change
and thus retains its original properties. That is, in a representation of L the matrices representing the elements H_i of H will

still be represented by hermitian matrices with real eigenvalues
(or antihermitian matrices with purely imaginary eigenvalues), while
the shift operators will no longer be represented by hermitian (or
antihermitian) matrices.

The Cartan Canonical Form for a simple complex Lie algebra L
is given by

$$[H_i, H_j] = 0$$

$$[H_i, E_\alpha] = \alpha_i E_\alpha \qquad\qquad (3.1)$$

$$[E_\alpha, E_\beta] = N_{\alpha,\beta} E_{\alpha+\beta} \text{ , if } \alpha + \beta \in \Sigma,$$

$$= 0, \text{ if } \alpha \neq -\beta, \quad \alpha + \beta \notin \Sigma,$$

$$= \alpha_i H_i \text{ , if } \beta = -\alpha, \text{(summed over i)},$$

with $\alpha,\beta,\ldots$ denoting roots, and Σ the root system of the algebra
L. The indices $i,j = 1,2,\ldots,n$, with n the rank of the algebra.
The H_i are assumed to form an orthonormal basis for the Cartan
subalgebra,

$$(H_i , H_j) \equiv \text{trace}\,((\text{ad } H_i)\,(\text{ad } H_j)) = \delta_{ij} , \qquad\qquad (3.2)$$

where (ad H_i) is a matrix representing the element H_i in the
adjoint representation. A basis for the compact (simple) Lie
algebra associated with L is given by the elements

$$\sqrt{-1}\, H_i , \quad \sqrt{-1/2}\,(E_\alpha + E_{-\alpha}), \quad \sqrt{1/2}\,(E_\alpha - E_{-\alpha}) \qquad (3.3)$$

for all values i and all roots $\alpha > 0$. A (somewhat trivial)
example is given by the simple complex Lie algebra A_1. This
algebra has the two roots +1 and -1, with root system $\Sigma = \{1, -1\}$.
The Lie products are given by ($L_o \equiv H_i$, $L_{+1} \equiv L_+$, etc.)

$$[L_o , L_o] = 0$$

$$[L_o , L_\pm] = (\pm 1)\, L_\pm$$

$$[L_\pm , L_\pm] = 0$$

$$[L_+ , L_-] = (+1)\, L_o$$

The compact Lie algebra associated with A_1 has basis elements

$$L_3 = \sqrt{-1}\, L_o , \quad L_1 = \sqrt{1/2}\,(L_+ - L_-) , \quad L_2 = \sqrt{-1/2}\,(L_+ + L_-)$$

and the Lie products are given by

$$[L_i , L_j] = \varepsilon_{ijk} L_k ,$$

where ε_{ijk} represents the completely antisymmetric ε-tensor.

The Cartan Canonical Form for the algebra B_n , $n = 4\ell + 2$, of the preceding section is given by eqn (3.1), together with the root system

$$\Sigma = \{\pm (e_i - e_k), \ i \neq k; \ \pm (e_i + e_k), \ i \neq k; \ \pm e_i\}, \qquad (3.4)$$

$i, k = 1, 2 \ldots, 4\ell + 2$, $(e_i)_k = \delta_{ik}$ the i-th unit vector of a $4\ell + 2$ dimensional cartesian coordinate system.

The correspondence between the abstract algebra B_n , given by eqn (3.1) and eqn (3.4), and its realization in terms of creation and annihilation operators for fermions, eqn (2.3), is given by

$$
\begin{array}{ccc}
H_i & \longleftrightarrow & a^+_\eta a_\eta - 1/2 \\[6pt]
E_{\pm(e_i - e_j)} & \longleftrightarrow & a^+_\eta a_\rho \\[6pt]
E_{e_i} & \longleftrightarrow & a^+_\eta \\[6pt]
E_{-e_i} & \longleftrightarrow & a_\rho \\[6pt]
E_{e_i + e_j} & \longleftrightarrow & a^+_\eta a^+_\rho \\[6pt]
E_{-e_i - e_j} & \longleftrightarrow & a_\eta a_\rho
\end{array}
\qquad (3.5)
$$

The correspondence between the two kinds of labels can, for example, be chosen as $i = k + s (2\ell + 1)$ with $(m, m_s) = (\ell + 1 - k, 1/2 - s)$, $k = 1,2,\ldots, 2\ell + 1$ and $s = 0, 1$.

Given eqn (3.1) and eqn (3.4) for the algebra B_n , the structure constants $N_{\alpha,\beta}$ have still to be determined. This can be done in the abstract or, having the realization eqn (3.5) of B_n in terms of creation and annihilation operators, this can be done easily by using this particular realization.

The states of eqn (2.2) are eigenstates of the commuting elements H_i. It is easily seen that the eigenvalue of any H_i

acting on any of these states is $\pm 1/2$. The eigenvalue is $+ 1/2$, if the state contains the electron with index i, and it is $- 1/2$, if the state does not contain the electron with index i. Thus, the weights are of the form $(3d)$ (15)

$$(\pm 1/2, \pm 1/2, \ldots , \pm 1/2) \ , \ 4\ell + 2 \text{ components,} \qquad (3.6)$$

and thus the atomic shell transforms like the smallest spin representation of B_n $(SO(8\ell + 5))$. The highest weight of this representation is

$$(1/2, 1/2, \ldots , 1/2) \ , \ 4\ell + 2 \text{ components} \qquad (3.7)$$

i.e. the state which represents the closed shell. The <u>one particle</u> states correspond to the weight

$$(1/2, -1/2, -1/2, \ldots , -1/2) \ , \ 4\ell + 2 \text{ components,} \qquad (3.8)$$

and all weights obtained from it by permutation of components.

DYNKIN EMBEDDING THEORY

In the preceding section the correspondence between the creation-annihilation operator formulation of B_n and its abstract form was established. In the present section it will be indicated how Dynkin's theory for the embedding of semisimple Lie algebras in semisimple Lie algebras (16) (17) can be utilized to classify <u>all possible</u> symmetry chains that lead from the superalgebra B_n , $n = 4\ell + 2$, to the physically significant subalgebras of type A_1 $(SU_J(2)$, $SO_L(3) \times SU_S(2)$, $SO_L(3))$. No attempt is made to present here a complete review. It is rather attempted to give the reader some intuitive understanding and to elucidate the power of this approach. More complete information can be found in references (18) and (19).

In order to avoid unnecessary complications, the first step is to determine the embedding of the physically significant algebra(s) in the algebra B_n. The algebra B_n has for large values n many inequivalently embedded subalgebras of type A_1 or $A_1 + A_1$. Which one of these the relevant subalgebra is depends entirely on the physical problem considered. That is, the branching of the (large) representations of B_n into a direct sum of irreducible representations of A_1 (or $A_1 + A_1$) is distinct for different embeddings of A_1 in B_n. Physical arguments will decide which of the branching the relevant one is--and thus determine the embedding of A_1 in B_n. For the case in question it is known that the weights of the smallest spin $1/2$ representation of B_n which characterize <u>one particle states</u> in the atomic shell are $(1/2, - 1/2, - 1/2,$

. . . , $- 1/2$), $4\ell + 2$ components, and all weights obtained from
this weight by permutations of components. These $4\ell + 2$ weights
have to go over into $(4\ell + 2)A_1 + A_1$ $(SO_L(3) \times SU_S(2))$ weights
(m, m_s), with $m = \ell, \ell - 1, \ldots, -\ell$ and $m_s = \pm 1/2$, or into
$4\ell + 2\ A_1$ $(SU_J(2))$ weights $j = \ell + 1/2, \ell - 1/2, \ldots, -(\ell + 1/2)$,
$\ell - 1/2, \ell - 3/2, \ldots, -(\ell - 1/2)$. This then defines the
embedding of $A_1 + A_1$ and A_1 in B_n.

Having determined the physically significant subalgebra
$A_1 + A_1$ or A_1 in B_n , all subalgebras of B_n which do not in turn
contain the subalgebras $A_1 + A_1$ or A_1 can be eliminated from
consideration.

In the following a brief sketch of Dynkin's theory is presented. Let L denote a simple or semisimple Lie algebra with
elements H_κ , E_α , $\kappa = 1, 2, \ldots, n$ and $\alpha \in \Sigma$,

$$[H_\kappa , E_\alpha] = \alpha_\kappa E_\alpha$$

$$[E_\alpha , E_{-\alpha}] = \alpha_\kappa H_\kappa$$

(4.1)

and let L' denote another simple (or semisimple) Lie algebra
with elements H_i', E_α , $i = 1, 2, \ldots, m \le n$ and $\alpha' \in \Sigma'$,

$$[H_i', E_\alpha'] = \alpha_i' E_{\alpha'}$$

$$[E_{\alpha'} , E_{-\alpha'}] = \alpha_i' H_i'$$

(4.2)

Let $f(H_i')$ denote an element of L which represents the element H_i'
of L', and let $f(E_{\alpha'})$ denote an element of L which represents the
element $E_{\alpha'}$ of L'. Then the algebra L' can be embedded as a
subalgebra in L, if in L holds

$$[f(H_i'), f(E_{\alpha'})] = \alpha_i' \ f(E_{\alpha'})$$

$$[f(E_{\alpha'}), f(E_{-\alpha'})] = \alpha_i' \ f(H_i').$$

(4.3)

If eqn (4.3) holds, then the algebra L contains a subalgebra
which is isomorphic to the algebra L'. It then holds

$$f(H_i') = f_{i\kappa} H_\kappa$$

$$f(E_{\alpha'}) = \sum_{\alpha \in \Gamma_{\alpha'}} C_{\alpha'\alpha} E_\alpha$$

(4.4)

where the sum over the repeated root α is over all elements of
the set

$$\Gamma_{\alpha'} = \{\alpha \in \Sigma \ | f_{i\kappa} \alpha_\kappa = \alpha_i'\}$$

(4.5)

with $\alpha = (\alpha_1, \alpha_2, \ldots, \alpha_n) \in \Sigma$, the root system of the algebra L, and $\alpha' = (\alpha_1', \alpha_2', \ldots, \alpha_n') \in \Sigma'$, the root system of the algebra L'. The $f_{i\kappa}$ are real numbers, while the $C_{\alpha'\alpha}$ are, in general, complex. The $C_{\alpha'\alpha}$ are determined, up to phase factors, from the relation

$$\alpha_i' \, f_{i\kappa} = \sum_{\alpha \in \Gamma_{\alpha'}} |C_{\alpha'\alpha}|^2 \, \alpha_\kappa \qquad (4.6)$$

(sum over repeated indices). The relative phase factors for the coefficients $C_{\alpha'\alpha}$ can then be obtained from the conditions imposed by the Lie products

$$[\, f(E_{\alpha'}), \, f(E_{\beta'})] = N_{\alpha', \, \beta'} \, f(E_{\alpha' + \beta'}) \text{ if } \alpha' + \beta' \in \Sigma'$$

$$= 0, \text{ if } \alpha' \neq -\beta', \, \alpha' + \beta' \in \Sigma'. \qquad (4.7)$$

The sum

$$\sum_{\alpha \in \Gamma_{\alpha'}} |C_{\alpha'\alpha}|^2 = n_f \qquad (4.8)$$

defines an integer, called the index of embedding. This index of embedding n_f is used as a superscript to characterize the embedding of an algebra L' in an algebra L. This characterization is, however, not unique. That is, distinct embeddings of the same algebra L' in L may have the same index n_f.

The rectangular n x m matrix $f_{i\kappa}$ is called the embedding matrix. This matrix characterizes, up to conjugacy (inner automorphisms), and up to phase factors, uniquely the embedding of an algebra L' in any minimal regular subalgebra of L containing it, and thus in the algebra L itself. Two distinct embedding matrices for the embedding of an algebra L' in L correspond to conjugate subalgebras of L, if by means of a succession of Weyl reflections of the algebra L one of the two embedding matrices can be brought into the other. The action of the elements of the Weyl group of L is on a column as a unit. That is columns are to permuted as a whole, a change of sign is for all elements of a column simultaneously.

As an example the embedding of the semisimple algebra $A_1{}^a + A_2{}^b$, $a = (4/3) \, \ell(\ell + 1) \, (2\ell + 1)$, $b = (2\ell + 1) \, (SO_L(3) \times SU_S(2))$ in B_n, $n = 4\ell + 2$, is given in explicit form. The defining matrix is

$$f = \frac{(\ell, \ell - 1, \ell - 2, \ldots, -\ell, \ell, \ell - 1, \ldots, -\ell)}{1/2 \, (1, \, 1, \quad 1, \quad \ldots, \, 1, -1, \, -1, \quad \ldots, \, -1)} = \frac{f^a}{f^b}$$

where f^a is the defining matrix for A_1^a in B_n and f^b the defining matrix for A_1^b in B_n. First the embedding f^a is discussed. It holds

$$\Gamma_{+1} = \{ \alpha \in \Sigma \ (A_{2\ell} + A_{2\ell} \subset B_n)|\ f^a(\alpha) = + 1, \ f^b(\alpha) = 0\},$$

with $+ 1$ the positive (simple) root of A_1^a and $\Sigma \ (A_{2\ell} + A_{2\ell} \subset B_n)$

the root system of the minimal regular subalgebra of B_n which contains A_1^a as subalgebra. In fact Γ_{+1} can be obtained from the root system of any <u>minimal</u> subalgebra of B_n which contains A_1^a. The reason to choose a <u>regular</u> minimal subalgebra is due to the fact that its embedding in B_n is trivial. The reason that the root system Γ_{+1} is to be obtained from the root system of a <u>minimal</u> subalgebra of B_n which contains A_1^a , rather than from any larger subalgebra of B_n containing A_1^a or even from the root system of B_n itself, is a consequence of the fact that if additional roots were needed for the embedding then a larger subalgebra of B_n which contains also these additional roots would actually be the minimal subalgebra of B_n which contains A_1^a. The appropriately chosen root system of $A_{2\ell} + A_{2\ell}$ in B_n is

$$\Sigma = \{ \pm(e_i - e_j); \ \pm(e_{2\ell + 1 + i} - e_{2\ell + 1 + j}) \ ; \ i \neq j,$$

$$i, j = 1, 2, \ldots , 2\ell + 1\}$$

and thus

$$\Gamma_{+1} = \{ e_i - e_{i + 1} , \ e_{2\ell + 1 + i} - e_{2\ell + 2 + i} ,$$

$$i = 1, 2, \ldots , 2\ell\}.$$

The embedding of the operators L_o , $L_{\pm}$ which represents total orbital angular momentum in B_n is then given, using eqn (4.4) and eqn (4.6), by

$$f(L_o) = \sum_{i=1}^{2\ell + 1} (\ell + 1 - i) (H_i - H_{2\ell + 1 + i})$$

$$f(L_{\pm}) = \sum_{i=1}^{2\ell} \sqrt{i(\ell - (1/2)(i - 1))} \ (E_{\pm(e_i - e_{i + 1})}$$

$$+ E_{\pm(e_{2\ell + 1 + i} - e_{2\ell + 2 + i})})$$

Next the embedding f^b is discussed. It holds

$$\Gamma_{+1} = \{\alpha \in \Sigma \ (2\ell + 1) \ A_1 \subset B_n)|\ f^b(\alpha) = 1, \ f^a(\alpha) = 0\}$$

where $(2\ell + 1)\ A_1$ stands for the direct sum of $2\ell + 1$ algebras A_1 which are regularly embedded in B_n. Thus

$$\Sigma\ ((2\ell + 1)\ A_1) = \{\pm(e_i - e_{2\ell + 1 + i})\ ,\ i = 1,\ 2,\ \ldots\ ,\ 2\ell + 1\}$$

and

$$\Gamma_{+1} = \{(e_i - e_{2\ell + 1 + i})\ ,\ i = 1,\ 2,\ \ldots\ ,\ 2\ell + 1\}.$$

It follows, using eqn (4.4) and eqn (4.6), for the embedding of total spin in the algebra B_n ,

$$f(S_o) = 1/2\ \sum_{i=1}^{2\ell + 1}\ (H_i - H_{2\ell + 1 + i})$$

$$f(S_\pm) = 1/\sqrt{2}\ \sum_{i=1}^{2\ell + 1}\ E_{\pm(e_i - e_{2\ell + 1 + i})}.$$

Making use of eqn (3.5), these embeddings can be re-expressed in terms of the electron creation-annihilation operators.

The defining matrices have another important property. They serve as projection operators from the weight space of an algebra onto the weight space of the subalgebra whose embedding in the algebra they define. If $m = (m_1 ,\ m_2 ,\ \ldots\ ,\ m_n)$ is a weight of an algebra L, and $m' = (m_1',\ m_2',\ \ldots\ ,\ m_n')$, $m \leq n$, a weight of an algebra L' which has an embedding in L defined by the embedding matrix f_{iK} , then

$$f_{iK}\ m_K = m_i'\ , \tag{4.11}$$

defines the projection $f*(m) = m'$ of the weight m of L onto a weight m' of L'. Using the two defining matrices eqn (4.10) and eqn (4.11) it is easily seen that the one particle weights of B_n project onto the $A_1^a + A_1^b$ and A_1^{a+b} weights as described earlier in this section.

The details of the calculations leading to Figures one through four are to be published elsewhere.

SUMMARY

Imbedding theory provides a systematic method for the construction of networks of chains of subalgebras for any simple Lie group. Some of the networks relative to atomic shell models have been constructed and their physical significance discussed.

REFERENCES

1.   G. Racah, Phys. Rev., $\underline{76}$, 1352 (1949)

2a.   B. R. Judd, Operator Techniques in Atomic Spectroscopy
      McGraw-Hill, New York (1962)

2b.   B. R. Judd, Second Quantization and Atomic Spectroscopy
      Johns Hopkins Press, Baltimore (1963)

2c.   B. R. Judd, Group Theory and Applications
      (E. Loebl, Ed.) $\underline{1}$, 183 (1968)

2d.   B. R. Judd, Advances in Chemical Physics, $\underline{14}$, 91 (1969)

2e.   B. R. Judd, Topics in Atomic and Nuclear Physics
      U. of Canterbury, New Zealand (1970)

2f.   B. R. Judd, Advances in Atomic Physics, $\underline{7}$, 251 (1971)

2g.   B. R. Judd, Lecture Notes in Chemistry (Springer-Verlag)
      $\underline{12}$, 164 (1979)

3a.   B. G. Wybourne, Spectroscopic Properties of the Rare Earths
      Interscience, New York (1965)

3b.   B. G. Wybourne, Symmetry Principles in Atomic Spectroscopy
      Interscience, New York (1970)

3c.   B. G. Wybourne, New Directions in Atomic Physics, Yale, $\underline{1}$,
      63 (1972)

3d.   B. G. Wybourne, Classical Groups for Physicists, Interscience,
      New York (1974)

4.   C. Block and J. Horowitz, Nuclear Phys., $\underline{8}$, 91 (1958)

5.   P. G. H. Sandars, Advances in Chemical Phys., $\underline{14}$, 365 (1969)

6.   Y. Shadmi, Spectroscopic and Group Theoretical Methods in
      Physics, North-Holland, Amsterdam (1968)

7.   A. B. Midzal, Nuclear Theory: The Quasi-Particle Method
      Benjamin, New York (1968)

8.   H. Fukotome, Progress of Theoretical Physics $\underline{57}$, 1554 (1977)

9.   D. Tang and F. A. Matsen, Phys. Rev. submitted

10.   W. Harter, Phys. Rev., 8A, 2819 (1973)

11. F. A. Matsen, Int. J. Quantum Chem., $\underline{S8}$, 379 (1974)

12. N. Jacobson, "Lie Algebras", Interscience New York (1962)

13. P. A. Rowlatt, "Group Theory and Elementary Particles", American Elsevier Publishing Co., New York (1966)

14. R. Gilmore, "Lie Groups, Lie Algebras, and Some of Their Applications", John Wiley & Sons, New York (1974)

15. E. B. Dynkin, Mat. Sb. (N.S.), 30, 349 (1952), Am. Math. Soc. Transl. (2), 6, 111, (1965)

16. E. B. Dynkin, Am. Math. Soc. Transl. (2), 6, 245 (1965)

17. M. Lorente and B. Gruber, J. Math. Phys. $\underline{13}$, 1639 (1972)

18. B. Gruber and M. T. Samuel, "Semisimple Subalgebras of Semisimple Lie Algebras: The Algebra A_5 (SU(6)) as a Physically Significant Example", in "Group Theory and Its Applications", Vol. III, ed. E. M. Loebl, Academic Press, New York (1975)

19. B. Gruber and M. S. Thomas, "Symmetry Chains for the Atomic Shell Model: I. Classification of Symmetry Chains for Atomic Configurations", to be published

20. M. S. Thomas and B. Gruber, "Symmetry Chains for the Atomic Shell Model: II. Classification of Symmetry Chains for Atomic Shells", to be published

21. M. S. Thomas and B. Gruber, "Symmetry Chains for the Atomic Shell Model: III. Classification of Symmetry Chains Conserving Total Orbital Angular Momentum", to be published

22. M. S. Thomas and B. Gruber, "Symmetry Chains for the Atomic Shell Model: IV. Classification of Symmetry Chains for Mixed Configurations", to be published

** Supported by the Robert A. Welch Foundation of Houston, Texas

GROUP THEORY IN ATOMIC AND MOLECULAR PHYSICS

B.R. Judd

Physics Department
The Johns Hopkins University
Baltimore, MD   21218

In recent years, the role of group theory in atomic and molecular physics has stressed interpretation rather than calculation.
The trend has thus been to return to the style of the 1930's after
a period in the 1940's and 1950's when many problems (for example,
the calculation of the term energies of electronic states of the f
shell) were made accessible only by means of the technical break-
through achieved by Racah in his work on Lie groups and their applicability to atomic structure calculations. Modern computing facili-
ties are mainly responsible for this change in attitude. On the
other hand, the use of the laser has revealed fine structure in
atomic and molecular spectra that make new applications of group
theory possible. The detailed example of the vibration-rotation
fine structure of SF_6 is presented, and the recent work of Harter
in this area is reviewed. The usefulness of double tensors, in
which the two ranks refer to the frame of the molecule and the frame
of the laboratory, is illustrated by the derivation of selection
rules for several operators. The peculiar clustering of the fine-
structure components is described and related to the triple inter-
sections that Lea, Leask, and Wolf found for the crystal-field
splittings of the J levels of rare-earth ions such as Ho^{3+}, for
which the ground level possesses a value of 8 for J.

1. Historical Perspectives

    The use of group theory in atomic and molecular physics has
varied enormously in style and purpose over the last fifty years.
Wigner's classic work[1] of 1931 played a crucial role in showing how
the symmetry of a physical system can be exploited to evaluate quantities of physical interest. The groups brought into play here are
finite groups (such as the two-element group that leads to the

151

concept of parity) and R_3, the continuous group of rotations in ordinary three-dimensional space. By assigning group-theoretical labels to states and operators, we can relate certain matrix elements to one another and also determine selection rules. Results of this kind are embodied in what is now called the Wigner-Eckart theorem. Degeneracies receive a natural explanation when the corresponding wavefunctions are labelled by the irreducible representations of the symmetry group of the Hamiltonian.

A certain resistance to the concepts of group theory is apparent in Condon and Shortley's book on atomic spectra,[2] which appeared in 1935. The Slater determinant plays a central role in their approach. Given sufficient time, any problem in atomic spectroscopy can be solved without recourse to group theory. But the problems of handling electronic configurations such as d^2 or sp^2 by the methods of Condon and Shortley are insignificant compared to those that arise for f^5 or f^6 . No fewer that 72 determinants are required to represent the component of f^5 ^2P for which $M_S = 1/2$ and $M_L = 1$. What is worse is the fact that there are four ^2P terms in f^5, so some method has to be devised to separate and label them.

To cope with these difficulties Racah[3] introduced seniority and the coefficients of fractional parentage in 1943. A remarkable gestation period of some six years saw the reformulation and development of these concepts. The threads that led to Racah's article[4] of 1949 are most clearly visible in his Princeton lecture notes[5] of 1951. For the f shell, the groups R_7 (the rotation group in the seven-dimensional space spanned by the orbital components of an f electron) and G_2 (the first of Cartan's exceptional groups) are introduced to label the states and the operators so that the Wigner-Eckart theorem can be applied as in the case of R_3 . Unlike R_3, the groups R_7 and G_2 are noninvariance groups: their generators do not commute with the atomic Hamiltonian and the states that exist in nature are superpositions of the states labelled by the irreducible representations W and U of R_7 and G_2 . In spite of these defects, the mathematical simplifications that the use of these groups affords permitted Racah to derive the Coulomb energies of all the Russell-Saunders terms in the f shell.

The early 1950's saw the application of these ideas to the nuclear shell model. New groups were introduced by Elliott,[6] and it was becoming clear that there was scope for developments in atomic theory too. The techniques of second quantization make it possible to introduce groups in which the number of particles in a configuration is not conserved. This work culminated in the discovery[7] that an atomic shell can be subjected to a four-fold factorization. A summary of these methods has been provided by Wybourne[8].

At the same time, parallel developments in the use of noncompact groups had been pioneered by Barut and his collaborators.[9]

They uncovered an amazingly rich group-theoretical structure of the hydrogen atom. The historical origin of this work is Fock's discovery[12] of the R_4 symmetry of the non-relativistic hydrogen atom, a discovery that could have been made some six years earlier had Podolsky and Pauling[11] recognized that their wavefunctions of the hydrogen atom in momentum space were proportional to four-dimensional spherical harmonics. The extension of the hydrogen-atom analysis to more complex atoms is difficult: work in this direction has been carried out by Wulfman[12] and Herrick.[13]

As the techniques for applying group theory to atomic systems become more refined there is a tendency for the investigators to become more and more immersed in the mathematics. Rather than attempt a survey of current work in this direction, I shall describe a problem that is at present of considerable interest to molecular spectroscopists. Since molecules do not possess the R_3 symmetry of atoms, they are less susceptible to the kind of analyses outlined above. What follows exhibits the kind of group theory developed by Wigner; the methods are elementary though the physical system is not. My intention is to convey the flavor of a molecular problem so that some appreciation of what is involved can be gained.

2. Sulfur Hexafluoride

A good impression of the use of group theory in molecular physics can be gained by analyzing the far infra-red absorption spectrum of SF_6. The existence of a rich structure in the neighborhood of 10.6 microns - a wavelength corresponding to CO_2 laser radiation - has provoked a detailed study by means of tunable diode lasers. The transitions of SF_6 involve the simultaneous excitation of a vibrational quantum and a rotational quantum. The ground state of SF_6 corresponds to the fluorine ions occupying sites at the vertices of a regular octahedron, with the sulfur atom at its center. The general theory of spherical top molecules such as this has been given as long ago as 1960 by Hecht[14] and, more recently, with many pictorial and literary embellishments, by Harter, Patterson and Paixao.[15]

To properly understand the nature of the vibrational excitations that can occur, a normal mode analysis must be carried out. The seven atoms comprising SF_6 each provide three displacement coordinates. Of this set of 21, we can immediately discard the six linear combinations that correspond to translations and rotations of the molecule as a whole. The remaining 15 can be constructed to transform according to the irreducible representations of O_h, the octahedral group O augmented by inversion and reflections. The irreducible representations of O are traditionally labelled A_1, A_2, E, T_1, and T_2; their dimensions are 1,1,2,3, and 3 respectively. The subscripts g and u (for gerade and ungerade) are added for O_h.

The 15 linear combinations of atomic displacements are found to
correspond to

$$A_{1g} + E_g + 2T_{1u} + T_{2u} + T_{2g}.$$

The occurrence of two T_{1u} modes raises a familiar multiplicity
problem. The actual linear combinations of displacements cannot be
unambiguously determined (as they can for the modes A_{1g}, E_g, T_{2u}, and
T_{2g}) on group-theoretical grounds alone. The actual elastic character of the molecule enters into the calculation. The vibration
being excited in the vicinity of 10.6 microns is the T_{1u} mode
possessing the higher frequency of the two; according to Weinstock
and Goodman[16] the three normal coordinates correspond almost exactly
to axial displacements of just two fluorine ions and the central
sulfur ion, while the remaining four fluorines are undistrubed. The
three ways in which we can select pairs of oppositely situated
fluorine ions lead to the three normal coordinates.

The existence of these three coordinates makes the analysis
formally identical to that of a three-dimensional harmonic oscillator. Just as the latter can be solved in polar coordinates, with
the energy levels labelled by s, p, s + d, p + f, etc., so can the
former. The excitation of a single vibrational quantum corresponds
to a p state, for which the vibrational angular momentum L is unity.
The use of this kind of angular momentum is convenient when the
rotation of the entire molecule is included, since we can couple
angular momenta together in the usual way.

How this is done is a matter of some delicacy. One way to
avoid the problems of reversed angular momentum is to use double
tensors; the first rank refers to the frame of the laboratory, the
second to the frame of the molecule.[17] The wavefunctions for a
rigid rotator, which depend solely on the three Euler angles, are
rewritten in terms of the components of the double tensor $D^{(JJ)}$.
The two commuting angular momenta J and J', in terms of which the
ranks have a well-defined significance, are linear combinations of
partial derivatives with respect to the Euler angles. We can now
form vibronic wavefunctions by the coupling

$$(D^{(JJ)}p^{(01)})^{(JR)}. \tag{1}$$

The ranks (01) are assigned to p because the normal coordinates are
referred to the molecular frame. The coupling of J and 1 to R is
highly convenient because the wavefunction (1) is diagonal with
respect to the Coriolis term $AJ \cdot L$.

The usefulness of the double-tensor notation is illustrated
by the derivation of the selection rules for electric-dipole radia-

tion. The perturbation is E·r, where E is the electric vector of
the electromagnetic field. Since E is defined in the frame of the
laboratory, the assignment of rank to r yields $r^{(10)}$. Set between
states labelled by (J_1,R_1) and (J_2,R_2), the operator $r^{(10)}$ gives
vanishing matrix elements unless

$$\Delta R = 0, \quad \Delta J = 0,\pm 1.$$

The wavefunctions (1) possess a degeneracy $(2J + 1)(2R + 1)$.
This exceeds the degeneracy of $2J + 1$ coming from the overall R_3
symmetry, and cannot be justified on general grounds. In fact,
the simple model that we have used for SF_6 has neglected all anharmonic effects as well as higher-order vibration-rotation couplings
(which lead to R being only an approximately good quantum number).
But, however complicated the perturbations may be, we know that
their representation by effective operators must consist of terms
of the type $V^{(0k)}$, since they must be scalar with respect to rotations of the laboratory frame. Moreover, the octahedral symmetry
of SF_6 (or, more exactly, the octahedral symmetry of the Hamiltonian
of SF_6) implies that the only components of $V^{(0k)}$ required are those
of the type V_{0aA_1}. The multiplicity label a is not required for
small values of k. It is these that our attention is directed to,
since V is built up from coordinates and momenta that are of rank 1.
The first possibility, $V^{(00)}$, is discarded as it cannot split an R
level. The next is $V^{(04)}_{0A_1}$, and we would expect it to dominate the
splitting pattern. To describe the calculation in group-theoretical
terms, we break the irreducible representation of R_3 with dimension
$2R + 1$ into its irreducible representations P of O and then calculate the Clebsch-Gordan coefficients

$$(RaP, 4A_1|Ra'P').$$

However, the frequent necessity of introducing the multiplicity
labels a and a' makes it easier simply to diagonalize the linear
combination

$$Y_{40} + (5/14)^{\frac{1}{2}}(Y_{44} + Y_{4-4}) \qquad (2)$$

of spherical harmonics (this being the combination that is an octahedral scalar) within angular-momentum states of specified R.

3. Comparison with Experiment

The spectroscopist measures transition frequencies rather than
energies. Since $r^{(10)}$ is a scalar in the frame of the molecule,
all transitions are of the type $P \to P$ and $R \to R$. A set of transitions
(for various P) such as

$$J = R = 29 \to J - 1 = R = 29 \qquad (3)$$

give a pattern of absorption peaks that represent the difference
between the fine structures of the levels for which $(J,R) = (30,29)$
and $(29,29)$. The set (3) overlaps the CO_2 laser line and is called
R(29) by spectroscopists, the symbol R here denoting the branch.
The actual absorption spectrum[18] is considerably complicated by the
overlapping of R(29) with R(28), R(30), R(31), as well as by the
presence of many unidentified lines; however, a very convincing fit
has been made by McDowell et al.[18]

The striking feature of the data is not the quality of the fit,
but rather the peculiar clustering of the fine-structure lines.
Thus, for the case of R = 29, the lines (and hence the differences
in energy between corresponding fine-structure components of the
upper and lower levels) are given in Fig. 1. A qualitative explanation for the corresponding phenomenon in tetrahedral molecules
(such as CH_4) has been given by Dorney and Watson.[19] On a classical
picture, centrifugal distortions of a molecule such as SF_6 depend on
the orientation of the axis of rotation with respect to the octahedron. If this axis coincides with a four-fold axis of the octahedron, a distortion involves the stretching of four radial fluorine
bonds. If, on the other hand, the rotation axis coincides with a
three-fold octahedral axis, a distortion can occur with a mere
bending of the six bonds. As stressed by Harter and Patterson,[20]

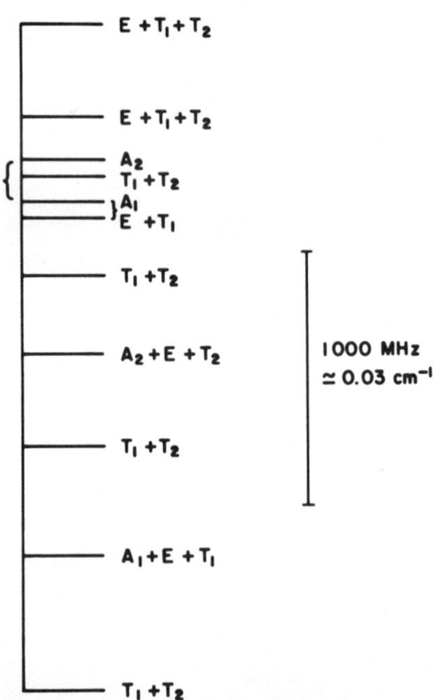

Fig. 1 Energy-level differences for SF_6

the former is a hard axis, the latter a soft one. For a given angular momentum, the moment of inertia will be increased more when rotations take place about the soft axis, thereby reducing the rotational energy by a greater amount. The upper and lower limbs of the fine structure of a given (J,R) level should thus correspond to rotations about the hard and soft axes respectively.

It is immediately apparent that there are three equivalent hard axes and four equivalent soft axes. Suppose we take a state for which R = 29 and choose a hard axis as the axis of rotation. In the classical regime of large R we know that the quantum mechanical state corresponds to the projection (or magnetic) quantum number K being either +29 or −29 (depending on the sense of the rotation). But, in addition to (R,K) = (29,29) and (29,−29), there are four other linear combinations of the states (R,K) corresponding to the two possible senses of rotation about the two remaining hard axes. These six states must transform according to a representation of O. To identify it, we work out the characters, this being a more straightforward method than that used by Harter and Patterson. We have only to recall that any operation of O that changes a direction of rotation contributes nothing to the character, whereas a rotation of θ about the axis of quantization multiplies the state (R,K) by $\exp(i\theta K)$. Under the five sample operations of O, one drawn from each class, the characters are

$$6,\ 0,\ 2\cos(29\pi),\ 0,\ 2\cos(29\pi/2) = 6,\ 0,\ -2,\ 0,\ 0. \qquad (4)$$

These can be formed from $T_1 + T_2$, which we therefore expect to label the upper cluster of the state for which R = 29. The fact that the cluster $T_1 + T_2$ appears to be the lowest in Fig. 1 indicates that the fine structure of the level for which (J,R) = (29,29) is greater than that for the upper level for which (J,R) = (30,29). Evidently the presence of a T_{1u} vibration has damped out the centrifugal distortions somewhat.

In the classical limit, the next state corresponds to (R,K) = (29,28). Replacing 29 by 28 in (4) yields the characters 6, 0, 2, 0, 2. These provide the representations $A_1 + E + T_1$, in accord with the second cluster from the bottom in Fig. 1. We can continue in this manner; however, if we start with a soft axis (with R=K=29) instead of a hard one, we arrive at the characters

$$8,\ 0,\ 0,\ 2\cos(29 \cdot 2\pi/3),\ 0,$$

which yield the uppermost cluster $E + T_1 + T_2$ of Fig. 1. Replacing 29 by 28 does not change the characters, so the next cluster from the top possesses the same labels. However, when 29 is replaced by 27 we get $A_1 + A_2 + T_1 + T_2$. This cluster occurs at an intermediate region where the eight-fold degenerate clusters coming down from the

top meet the six-fold degenerate clusters coming up from the bottom.
The component irreducible representations can be picked out, but the
clustering is not so pronounced.

4.   Crystal-Field Theory

The success of the explanation given above for the clustering
depends on K being a good quantum number. This is rigorously true
only in the classical limit of infinitely large K (and R). As R
decreases the clustering becomes progressively less apparent. However, even for R = 8 clusters can be picked out. This particular
angular momentum is of interest because the ground level of
Ho^{3+} is $^5I_8$. When the holmium ion is embedded in a cubic lattice
at a site of octahedral symmetry, the calculation of the crystal
splitting is very similar to the SF_6 analysis. The principal difference is that the perturbation $V_{A_1}^{(4)}$ has to be augmented by a

sixth-rank tensor $V_{A_1}^{(6)}$. Either part leads to a clustering phenom-

enon; the interesting feature is that a number of apparently exact
coalescences occur, as was noticed by Lea, Leask, and Wolf.[21]

It is not difficult to understand the underlying mechanism. If
the matrix of $V_{A_1}^{(4)} + V_{A_1}^{(6)}$ is set up using terms of the type (2), all
non-vanishing elements lie either on the diagonal or immediately
adjacent to it. By adjusting the relative strengths of $V_{A_1}^{(4)}$ and
$V_{A_1}^{(6)}$, any off-diagonal matrix element can be made to
vanish. For example, the (J,M) component (8,4) can be severed
from (8,8), and so (8,8) becomes an eigenstate. What would happen
only in the limit of large J if $V_{A_1}^{(4)}$ were considered now occurs

for J = 8. The arguments of Sec. 3 can now be repeated. There are
five companions to (8,8) that must form a degenerate set of eigenstates: the characters are 6, 0, 2, 0, 2, which determine the
coalescence of the three crystal-field levels A_1, E, and T_1. This
agrees with Fig. 1 of Lea, et al.[21] In a similar manner, we can
sever (8,7) or (8,5); but since the characters yield only T_1 and T_2,
we obtain a simple crossing of two levels. Of course, arguments
such as this assume that if (8,7) is an eigenstate, then the five
companion states obtained by rotating the axis of quantization to
the other five four-fold octahedral axes are linearly independent
of each other and of the original state (8,7). The smaller J and
M, the sooner this ceases to be valid.

A parallel analysis can be carried out for the ground level
$^4I_{15/2}$ of Er^{3+}. Because J is half integral, the double group of O

must be considered. There are three new irreducible representations
of dimensions 2,2, and 4. It turns out that the six-fold degenerate
representation obtained in the severing procedure always decomposes
into 2 + 4 and never into 2 + 2 + 2. Thus there are no triple
coincidences.

It is worth noting that no new coalescences can be obtained by
considering a three-fold axis because the energy matrix is no longer
tridiagonal, and the severing procedure breaks down.

As a final remark, we note that the breakdown of R as a good
quantum number in the SF_6 molecule is matched by the breakdown of J
in the crystal-field problem. The parallels can be extended: for
example, the $^6P_{7/2}$ level of Gd^{3+} shows a small splitting when the
gadolinium ion is at a site of cubic symmetry although a fourth-
rank tensor has vanishing matrix elements for a P state. Similarly,
anharmonic terms can contribute to the fine structure of the R(29)
line of SF_6, although $V_{A_1}^{(4)}$ cannot directly contribute to states for
which L = 0 or 1.

Partial support of the National Science Foundation of the work
described above is acknowledged.

References

1. E.P. Wigner, Gruppentheorie, Vieweg u. Sohn, Brunswick (1931).

2. E.U. Condon and G.H. Shortley, The Theory of Atomic Spectra,
    Cambridge University Press, New York (1935).

3. G. Racah, Phys. Rev. <u>63</u>, 367 (1943).

4. G. Racah, Phys. Rev. <u>76</u>, 1352 (1949).

5. G. Racah, Group Theory and Spectroscopy. Reprinted in Ergebnisse
    der Exakten Naturwissenschaften, Vol. 37, Springer, New York
    (1965).

6. J.P. Elliott, Proc. Roy. Soc. (London) <u>A218</u>, 345 (1953).

7. L. Armstrong, Jr. and B.R. Judd, Proc. Roy. Soc. (London) <u>A315</u>,
    27 and 39 (1970).

8. B.G. Wybourne, Symmetry Principles and Atomic Spectroscopy,
    Wiley-Interscience, New York (1970).

9. A.O. Barut, P. Budini, and C. Fronsdal, Proc. Roy. Soc. (London)
    <u>A291</u>, 106 (1966); C. Fronsdal, Phys. Rev. <u>156</u>, 1665 (1967);
    A.O. Barut and H. Kleinert, Phys. Rev. <u>156</u>, 1541; <u>157</u>, 1180;
    <u>160</u>, 1149 (1967).

10. V. Fock, Zeits. f. Phys. <u>98</u>, 145 (1935).

11. P. Podolsky and L. Pauling, Phys. Rev. <u>34</u>, 109 (1929).

12. C. E. Wulfman, Dynamical Groups in Atomic and Molecular Physics.
    In Group Theory and its Applications (Ed. E.M. Loebl) Vol. II,
    Academic Press, New York (1971); Chem. Phys. Lett. <u>23</u>, 370
    (1973).

13. D.R. Herrick, Phys. Rev. $\underline{A12}$, 413 (1975); $\underline{A17}$, 1 (1978); D.R.
     Herrick, and M.E. Kellman, Phys. Rev. $\underline{A18}$, 1770 (1978).
14. K.T. Hecht, J. Mol. Spectros. $\underline{5}$, 355 and 390 (1960).
15. W.G. Harter, C.W. Patterson, and F. J. da Paixao, Rev. Mod.
     Phys. $\underline{50}$, 37 (1978).
16. B. Weinstock and G.L. Goodman, Adv. Chem. Phys. $\underline{9}$, 169 (1965).
17. B.R. Judd, Angular Momentum Theory for Diatomic Molecules,
     Academic Press, New York (1975).
18. R.S. McDowell, H.W. Galbraith, G.J. Krohn, C.D. Cantrell, and
     E.D. Hinkley, Optics Comm. $\underline{17}$, 178 (1976).
19. A.J. Dorney and J.K.G. Watson, J. Mol. Spectros. $\underline{42}$, 135 (1972).
20. W.G. Harter and C.W. Patterson, Phys. Rev. Lett. $\underline{38}$, 224
     (1977); J. Chem. Phys. $\underline{66}$, 4872 and 4886 (1977).
21. K.R. Lea, M.J.M. Leask, and W.P. Wolf, J. Phys. Chem. Solids
     $\underline{23}$, 1381 (1962).

NEW APPROACH TO MATRIX ELEMENTS AND CLEBSCH–GORDAN COEFFICIENTS

FOR COMPACT AND NON–COMPACT GROUPS

A. U. Klimyk

Institute for Theoretical Physics
Kiev-I30
USSR

1. INTRODUCTION

Linear representations of Lie groups are useful in both physics
and mathematics. The most important applicable aspects of representations theory are matrix elements and Clebsch-Gordan (CG) coefficients of group representations. New results of representation
theory allow us to give a new approach to matrix elements and CG
coefficients. At first, matrix elements and CG coefficients of
unitary irreducible representations were studied separately for
compact and non-compact semisimple Lie groups and for each series
of representations (the principal unitary series, the supplementary
series, the discrete series and so on). These studies can now be
linked together. This link is realized by the principal nonunitary
series representations ("analytic continuation" in the continuous
parameters of the principal unitary series representations) of a
semisimple noncompact Lie group. The point is that every completely
irreducible representation of such a group is contained in some
representation of the principal nonunitary series. On the other
hand, matrix elements of the principal nonunitary series representations at fixed group element are entire analytic functions of
continuous parameters defining representations. Therefore, if matrix
elements of representations of one of the continuous series (for
example, of the principal unitary series) are known then those of
representations of other series can be obtained by an appropriate
analytic continuation. However, most of the unitary representations
which are contained in the principal nonunitary series representations are nonunitary. Therefore, it is necessary to evaluate the
matrices of unitarization of unitarizable representations. These
matrices bear a simple relationship with intertwining operators for
the principal nonunitary series representations. For this reason

an evaluation of matrices of intertwining operators is an important
task. Different intertwining operators are linked by CG coefficients
of the tensor product of the principal nonunitary series representations with finite dimensional representations of the group.

Since matrix elements of finite dimensional representations of
different real forms (compact and noncompact) of the same complex
Lie group are linked by an analytic continuation of group parameters,
then matrix elements of irreducible representations of compact real
form are related with those of the principal nonunitary series representations of noncompact real forms.

The complex approach, which we follow, is applicable to matrix
elements of representations of different unitary series of a semisimple noncompact Lie group and to those of irreducible representations of a compact group. The complex approach is called <u>a method
of the principal nonunitary series representations</u>.

This method can be used to study CG coefficients. In particular, it gives an immediate relation between CG coefficients of the
tensor product of finite dimensional representations and also of
the tensor product of finite and infinite dimensional representations. This method when applied to a complex semisimple Lie group
gives an interesting approach for the CG coefficients of its compact
form.

2. THE PRINCIPAL NONUNITARY SERIES REPRESENTATIONS

Let G be a linear connected noncompact semisimple Lie group with
Lie algebra $\hat{g}$, and K a maximal compact subgroup in G. The Lie subalgebra in $\hat{g}$ corresponding to K will be denoted by k. Let $\hat{g} = k + p$
be a Cartan decomposition of $\hat{g}$ corresponding to a Cartan involution
θ (see, for example, [1]). Let a be a maximal commutative subalgebra in p and A = exp a. Let G = ANK be an Iwasawa decomposition of
G (see [1]). If M is a centralizer of A in K then P = ANM is a
minimal parabolic subgroup in G.

Let δ be a unitary irreducible representation of M in the space
H_δ, and λ a complex linear form on a. Then the mapping h $\rightarrow$
exp(λ(log h)), for h $\in$ A, is a representation of A and the correspondence hnm $\rightarrow$ exp(λ(log h))δ(m), h $\in$ A, n $\in$ N, m $\in$ M, is an irreducible representation of P. Let $L_\delta^2(K, H_\delta)$ be a Hilbert space of
all measurable vector-functions from K into H_δ, such that

$$f(mk) = \delta(m)f(k) \quad \text{and} \quad \int \|f(k)\|_H^2 \, dk < \infty$$

where the integral is with respect to the invariant measure on K
(functions which coincide almost everywhere are not distinguished).
Then the mapping g $\rightarrow$ $\pi_{\delta,\lambda}(g)$ from G into a set of bounded operators

on $L^2_\delta(K,H_\delta)$, which are given as

$$\pi_{\delta,\lambda}(g)f(k) = \exp(\lambda(\log h))f(k_g))$$

where h and k_g are defined by the Iwasawa decomposition $kg = hnk_g$, $h \in A$, $n \in N$, $k_g \in K$, defines a representation of G in $L^2_\delta(K,H_\delta)$. The representations $\pi_{\delta,\lambda}$ are called the <u>principal nonunitary series representations</u> of G. Let us note that the space $L^2_\delta(K,H_\delta)$ of the representation $\pi_{\delta,\lambda}$ depends on δ but not on λ.

Let ρ be one-half of the sum of the restricted positive roots of the pair $(\hat{g},a)$. If the correspondence $h \to \exp((\lambda - \rho)(\log h))$ is a unitary character of the group A, then $\pi_{\delta,\lambda}$ is a unitary representation belonging to the principal unitary series.

Most of the principal nonunitary series representations are irreducible; however, some of them are reducible (but not necessarily completely reducible). Every reducible representation of the principal nonunitary series contains a finite number of completely irreducible representations, which are defined uniquely (up to an infinitesimal equivalence).

<u>Theorem 1</u> [3,4]. Every completely irreducible representation of G is infinitesimally equivalent to a completely irreducible constituent of some principal nonunitary series representation.

In particular, an irreducible unitary representation of G is infinitesimally equivalent either to some irreducible representation $\pi_{\delta,\lambda}$ or to a completely irreducible constituent of a reducible representation of the principal nonunitary series. However, often these representations are not unitary in the topology of the space $L^2_\delta(K,H_\delta)$. For example, the discrete series representations of G are infinitesimally equivalent to nonunitary (but unitarizable) completely irreducible constituents of the reducible representations $\pi_{\delta,\lambda}$.

According to Theorem 1 the finite dimensional representations of G are also contained in the representations $\pi_{\delta,\lambda}$.

The restriction of $\pi_{\delta,\lambda}$ to K acts on $L^2_\delta(K,H_\delta)$ by the formula $\pi_{\delta,\lambda}(k_o)f(k) = f(kk_o)$. It and the condition $f(mk) = \delta(m)f(k)$, $m \in M$, show that $L^2_\delta(K,H_\delta)$ is decomposed into an orthogonal sum of subspaces, in which the irreducible unitary representations μ of K are realized. The multiplicity of μ in $\pi_{\delta,\lambda}|_K$ is equal to a multiplicity of δ in μ. The complex linear hull of the subspaces of the irreducible representations of K in $L^2_\delta(K,H_\delta)$ will be denoted by $dL^2_\delta(K,H_\delta)$. It is invariant under the representation of $\hat{g}$ corresponding to the representation $\pi_{\delta,\lambda}$ of G. This representation of the Lie algebra $\hat{g}$ will be denoted by $d\pi_{\delta,\lambda}$.

The linear form λ which defines $\pi_{\delta,\lambda}$ can be characterized by ℓ complex numbers (ℓ is a real rank of g), the coordinates of λ with respect to some basis of a. We shall denote these numbers by c_1, c_2, ..., c_ℓ. Then the representation $\pi_{\delta,\lambda}$ can be denoted by $\pi_{\delta,c_1,c_2,\ldots,c_\ell}$.

Theorem 2 [5,6]. For fixed $g \in G$, λ and a fixed basis for the space $dL_\delta^2(K,H_\delta)$, the matrix elements of the operators $\pi_{\delta,\lambda}(g)$, $g \in G$, are the entire analytic functions of the parameters $c_1,c_2,\ldots,c_\ell$.

3. THE PRINCIPAL NONUNITARY SERIES REPRESENTATIONS AND THE FINITE
   DIMENSIONAL REPRESENTATIONS

According to Theorem 1 the finite dimensional irreducible representations of the group G are contained in its representations of the principal nonunitary series. The finite dimensional irreducible representations are interesting for us because of the relation between the finite dimensional representations of G and those of the compact form of the complexification [G] of G.

Let ω be a finite dimensional irreducible representation of G, Λ' will denote a weight of ω (with respect to a fixed Cartan subalgebra h of $\hat{g}$ containing the subalgebra a). Let $\Lambda'(\omega)$ be a restriction of the weight Λ' to a. Then $\Lambda'(\omega)$ will be called a restrictive weight. Let $\Lambda(\omega)$ be a lowest restrictive weight of ω with respect to some standard ordering. Let V denote a subspace of the space of the representation ω consisting of all weight vectors whose weights coincide on a with $\Lambda(\omega)$. In [7] it is shown that $\omega|_M$ realizes on V an irreducible representation of M, which we denote by δ. We shall say that the representation ω of G is an extension of the representation δ of M. A finite dimensional irreducible extension of the representation δ of M will be denoted by ω_δ.

Theorem 3 [7]. The principal nonunitary series representation $\pi_{\delta,\lambda}$ of G can contain at most one finite dimensional irreducible subrepresentation (i.e., a representation realized in an invariant subspace). In this case $\pi_{\delta,\lambda}$ does not contain other finite dimensional irreducible constituents. If $\pi_{\delta,\lambda}$ contains a finite dimensional subrepresentation then it coincides with ω_δ (i.e., with an extension of the representation δ of M). Moreover, $\pi_{\delta,\lambda}$ contains ω_δ as a subrepresentation if and only if $\Lambda(\omega_\delta) = \lambda$.

The last statement of Theorem 3 defines exactly the finite dimensional representation of G which is contained in $\pi_{\delta,\lambda}$ (if there is one). Really, the condition $\Lambda(\omega_\delta) = \lambda$ selects one extension of the representation δ of M.

Let us consider the set of all representations $\pi_{\delta,\lambda}$ of G with fixed δ. Let Ω_δ denote the set of all finite dimensional irreducible representations of G contained in these representations $\pi_{\delta,\lambda}$. Then Ω_δ coincides with the set of all extensions of the representations δ of M. The carrier spaces of the representations $\pi_{\delta,\lambda}$ with fixed δ coincide. Thus, every representation ω_δ of Ω_δ is realized in some subspace $V(\omega_\delta)$ of $L_\delta^2(K,H_\delta)$. Let V be the union of $V(\omega_\delta)$, for $\omega_\delta \in \Omega_\delta$. Due to Theorem 6.2 in [7] $V = dL_\delta^2(K,H_\delta)$. Let e_1 and e_2 be elements of an orthonormal basis of $dL_\delta^2(K,H_\delta)$. Due to the results of [7] there exists an infinite number of the representations ω_δ, $\omega_\delta \in \Omega_\delta$, such that $e_1 \in V(\omega_\delta)$ and $e_2 \in V(\omega_\delta)$. Therefore, if we consider the matrix element $<e_1|\pi_{\delta,\lambda}(g)|e_2>$, $g \in G$, then we know that there are an infinite number of $\omega_\delta \in \Omega_\delta$ for which $<e_1|\omega_\delta(g)|e_2>$, $g \in G$, has meaning.

We obtain the following corollary to Theorems 2 and 3.

Corollary 1. Matrix elements of every finite dimensional irreducible representation of G in the appropriate basis are matrix elements of the corresponding representation of the principal non-unitary series (this correspondence is defined by Theorem 3). Matrix elements of the principal nonunitary series representations $\pi_{\delta,\lambda}$ of G with fixed δ are analytic continuation of matrix elements of the finite dimensional representations $\omega_\delta \in \Omega_\delta$ in an appropriate basis.

This analytic continuation is not, however, unique. For uniqueness it is necessary to impose additional conditions on the procedure of analytic continuation. The conservation of the relation $(d/dt)\, g(t) = I\, g(t)$ for noncompact one-parameter subgroups $g(t)$ (I is an infinitesimal generator of $g(t)$) guarantees correct analytic continuation. For matrix elements this relation has the form

$$\frac{d}{dt} <n|g(t)|m> = \sum_{n'} <n|I|n'><n'|g(t)|m> .$$

If the basis of $L_\delta^2(K,H_\delta)$ consists of the bases of spaces of irreducible representations of K then the sum is finite.

According to Corollary 1, matrix elements of the principal nonunitary series representations (due to Theorem 2 they can be obtained by analytic continuation of matrix elements of the principal unitary series representations or of the representations of other continuous series) give matrix elements of the finite dimensional representations of G, which in turn lead to matrix elements of the irreducible representations of the compact form G_k of the complexification [G] of G. Thus, Theorems 2 and 3 and Corollary 1 connect matrix elements of the representations $\pi_{\delta,\lambda}$ of G with those of the irreducible representations of G_k.

4. MATRIX ELEMENTS AND INTERTWINING OPERATORS

Matrix elements of the principal nonunitary series representations in an orthonormal basis of $L^2_\delta(K,H_\delta)$ give matrix elements of
the principal unitary series representations in unitarized form. A
simpler statement is not correct for the representations of the complementary and the discrete series because they are not unitary in
the Hilbert space $L^2_\delta(K,H_\delta)$. In order to obtain unitarized matrix
elements it is necessary to introduce a new scalar product and to
choose an orthonormal basis in a new Hilbert space. The transition
to a new scalar product can be realized by self-adjoint operator A.
The operators A are related to intertwining operators of the principal nonunitary series representations (see Lemma 22 and Proposition
25 in [8]; see also [9]).

Intertwining operators define also the symmetry relations for
matrix elements of the unitary and the principal nonunitary series
representations as functions of the complex parameters c_1, c_2, ...,
c_ℓ of representations. This leads to relations for special functions
associated with these representations.

The explicit integral form for intertwining operators is introduced in [10,11]. For application of intertwining operators to
matrix elements of representations it is necessary to have them in
matrix form. Therefore we give the following definition of the
intertwining operators. The linear operator

$$\Pi \equiv \Pi^{\delta,\lambda}_{\delta',\lambda'}, \quad \text{from } dL^2_\delta(K,H_\delta) \text{ into } dL^2_{\delta'}(K,H_{\delta'}) \text{ for which}$$

$$\Pi d\pi_{\delta,\lambda} = d\pi_{\delta',\lambda'}\Pi$$

is called the intertwining operator for the representations $\pi_{\delta,\lambda}$ and
$\pi_{\delta',\lambda'}$. The representations $\pi_{\delta,\lambda}$ and $\pi_{\delta',\lambda'}$ can have a non-zero
intertwining operator if they have common irreducible representa-
tions. The latter condition holds if $\pi_{\delta,\lambda}$ and $\pi_{\delta',\lambda'}$ have the same
infinitesimal character (i.e., the same values of Casimir operators).
This is possible (see [12]) if the representation $\pi_{\delta',\lambda'}$ is related
to the representation $\pi_{\delta,\lambda}$ by some element of the Weyl group W_c of
the complexification $[G]$ of G. Intertwining operators which correspond to elements of the Weyl group W of the pair $(\hat{g},a)$ are analytic
functions of the complex parameters c_1, c_2, ..., c_ℓ [8,9]. An
explicit form of all intertwining operators is found for all the
principal nonunitary series representations of U(n,1) and $SO_o(n,1)$
in [13-15].

The operators A, which form the unitarization of the unitarizable representations, can be obtained from intertwining operators Π
by taking the square root of Π. See details in [6], Chapter 5.
For explicit form of the operators A for U(n,1) and $SO_o(n,1)$ see
in [5,6].

According to Corollary 1, matrix elements of the principal non-unitary series representations lead to matrix elements of the finite dimensional representations of G and, conversely, matrix elements of the finite dimensional representations continue analytically to matrix elements of the principal nonunitary series representations. For this reason sometimes it is necessary to have matrix elements of finite dimensional representations of G in an orthonormal basis of the space $L_\delta^2(K,H_\delta)$ if matrix elements of unitary finite dimensional representation of the compact form G_k of [G] are known and vice versa. The question is: What is the relation between them? It is known there is one-to-one correspondence between finite dimensional representations of G and G_k. This correspondence is realized in the following way. Matrix elements of the finite dimensional representations of G and G_k are real analytic functions of group parameters. Parameters of the group G_k are continued analytically to parameters of G. If this analytic continuation is successful for matrix elements of a finite dimensional representation of G_k as the functions of group parameters then we obtain corresponding matrix elements of a finite dimensional representation of G. This correspondence is invertible.

Suppose we have the matrix elements of a finite dimensional representation of G in an orthonormal basis of $L_\delta^2(K,H_\delta)$. We continue these matrix elements to matrix elements of a representation of G_k. The representation matrices so obtained are not unitary. A unitarization of these matrices is fulfilled by a transition to a new basis. This transition is given by the operator A which is related to the intertwining operator in the same manner as in the case of the operator A which unitarizes unitarizable representations of G.

Let us note that in the basis consisting of bases of the spaces of irreducible representations of the subgroup K (we shall call it a K-basis) the matrices Π and A have block form with finite dimensional blocks. Moreover, if the irreducible representations of K are contained in $\pi_{\delta,\lambda}$ with unit multiplicity then the matrices Π and A are diagonal (as in the cases of the groups U(n,1) and $SO_o(n,1)$).

5. MATRIX ELEMENTS OF IRREDUCIBLE REPRESENTATIONS OF COMPACT GROUPS IN DIFFERENT BASIS

We saw that using an analytic continuation and a unitarization by the operator A we may obtain matrix elements of the unitary irreducible representations of the compact group G_k from matrix elements of the principal nonunitary series representations of G. Here matrix elements are considered in a K-basis. In the same manner, the explicit form of infinitesimal operators of the unitary irreducible representations of G_k can be found from the explicit form of infinitesimal operators of the principal nonunitary series representations of G.

Let $\tilde{G}_k$ be a compact semisimple Lie group. Using this method we can find matrix elements and infinitesimal operators of unitary irreducible representations of $\tilde{G}_k$ in different bases. It is necessary to consider the complexification $[\tilde{G}_k]$ of $\tilde{G}_k$ and to take different real forms G^j of $[\tilde{G}_k]$ (j labels non-isomorphic real forms). The above method applied to different groups G^j leads to matrix elements and infinitesimal operators of the unitary representations of $\tilde{G}_k$ in K_j-bases, where K_j is a maximal compact subgroup of G^j. Since the groups G^j have different maximal compact subgroups then K_j-bases are different. Possibilities admitted by this method are shown in Table 1. It is seen that the above method allows us to obtain matrix elements of the unitary representations of SU(n) and its infinitesimal operators in bases corresponding to the subgroups Sp(n), if n is even, SO(n), $S(U_p \times U_q)$ and SU(p) $\times$ SU(q), p + q = n.

6. METHODS OF EVALUATION OF MATRIX ELEMENTS OF THE PRINCIPAL NON-UNITARY SERIES REPRESENTATIONS

Methods of evaluation of matrix elements of group representations by evaluation of the integral giving the scalar product in the carrier space [16] and by solving the system of differential equations (representing Casimir operators of a chain of subgroups) [17] are well known. Here we consider two methods which are not well known.

The first method uses the tensor product of the principal non-unitary series representations by the finite dimensional representations of G. To use this method it is necessary to know all matrix elements of those principal nonunitary series representations $\pi_{\delta,\lambda}$ for which δ is the identity representation of M, and CG coefficients for the tensor product of arbitrary principal nonunitary series representation by some simple finite dimensional representations of G. If D is a finite dimensional representation of G and C is a matrix realizing the decomposition of the tensor product $\pi_{\delta,\lambda} \otimes D$ into the

Table 1

Type of group	Noncompact real Lie group G	Corresponding compact Lie group G_k	Maximal compact subgroup in G	Remark
AI	SL(n,R)	SU(n)	SO(n)	n > 1
AII	SU(2n)	SU(2n)	Sp(n)	n > 1
AIII	SU(p,q)	SU(p+q)	S(U(p)×U(q))	
	U(p,q)	U(p+q)	U(p) × U(q)	
BI,DI	SO_o(p,q)	SO(p+q)	SO(p)×SO(q)	
DIII	SO*(2n)	SO(2n)	U(n)	n > 2
CI	Sp(n,R)	Sp(n)	U(n)	
CII	Sp(p,q)	Sp(p+q)	Sp(p)×Sp(q)	

principal nonunitary series representations (it is known that this tensor product can be decomposed into principal nonunitary series representations), i.e., C is a matrix consisting of CG coefficients, then

$$C(\pi_{\delta,\lambda} \otimes D)C^{-1} = \sum_{\delta',\lambda'} \pi_{\delta',\lambda'} .$$

If we write this relation with the help of matrix elements then we obtain the formula which connects matrix elements of $\pi_{\delta',\lambda'}$ with matrix elements of $\pi_{\delta,\lambda}$, D and corresponding CG coefficients. Beginning with matrix elements of the representations $\pi_{\delta,\lambda}$ with identical δ and iterating this procedure it is possible to find matrix elements of any fixed principal nonunitary series representations. For using this method we need CG series and CG coefficients for the tensor product $\pi_{\delta,\lambda} \otimes D$. These CG series are obtained in [6, 18]. CG coefficients are considered below.

Let us note that a combination of this method with other methods is useful. Other methods can be used for evaluation of matrix elements of $\pi_{\delta,\lambda}$ with the same δ.

The next method uses the explicit form of infinitesimal operators of the principal nonunitary series representations. If the explicit form of the infinitesimal operator I of the one-parameter subgroup g(t) is known then we can try to evaluate the matrix elements of exp I. For some groups (for example, for U(p,q), SU(p,q)) the infinitesimal operators I can be obtained from the matrix elements of the operators $(t^n/n!)I^n$ for some n. An action of the infinitesimal operators I on a K-basis element (K is a maximal compact subgroup of G) is expressed as finite linear combinations of K-basis elements. Therefore, matrix elements of the operator $(t^n/n!)I^n$ in K-basis are finite sums of products of matrix elements of I. But the infinitesimal operator I proves to be of such a simple form if I belong to the complexification g_c of $\hat{g}$. Hence exp tI is in the complexification [G] of G. The one-parameter subgroup g'(t) $\in$ G can be represented as a product of the subgroups exp tI, I $\in g_c$. This allows us to represent matrix elements of the representation operators corresponding to g'(t) in a form of a sum of products of matrix elements of the operators exp tI. In order to justify this representation of matrix elements it is necessary to use the results by R. Goodman [19] on analytic continuation of a representation of the real group G to a local representation of its complexification [G]. For details of this method, see [5, 6], where it is applied to find matrix elements of all principal nonunitary series representations of U(n,1).

In order to use this method it is necessary to know an explicit form of infinitesimal operators. For the principal nonunitary series representations of U(p,q) and $SO_o(p,q)$ they are found in [6, 20].

In [6] the method of evaluation of infinitesimal operators of the semisimple and the inhomogeneous Lie groups is given.

7. ANALYTIC PROPERTIES OF CG COEFFICIENTS. CG COEFFICIENTS FOR
    THE TENSOR PRODUCT OF A FINITE AND AN INFINITE DIMENSIONAL
    REPRESENTATIONS

Let G, K, $\hat{g}$, k be as before. There is the Cartan decomposition $\hat{g} = k + p$. Since $[k,p] \subset p$, the adjoint representation defines a linear representation of k in p. We denote it by d. Let p_i, i = 1,2, ..., dim p, be a basis of p.

Let us consider the representation $d\pi_{\delta,\lambda}$ of $\hat{g}$ in $dL_\delta^2(K,H_\delta)$. The operators corresponding to the elements $p_i \in p$ will be denoted by P_i. The K-basis elements of $dL_\delta^2(K,H_\delta)$ will be denoted by $|\delta,\lambda,\omega,n\rangle$ where ω denotes the unitary irreducible representations of K and n labels the basis elements of the space of ω. Then

$$P_i|\delta,\lambda,\omega,n\rangle = \sum_{\omega',n'} C_{\omega\omega'}(\delta,\lambda)\langle\omega,n;d,i|\omega',n'\rangle|\delta,\lambda,\omega',n'\rangle$$

where $C_{\omega\omega'}(\delta,\lambda)$ does not depend on n, n', and $\langle...;...|...\rangle$ is a CG coefficient. Here the sum is finite. Dependence on λ is completely contained in $C_{\omega\omega'}(\delta,\lambda)$.

Lemma 1. For the principal nonunitary series representations, $C_{\omega\omega'}(\delta,\lambda)$ depend linearly on the coordinates $c_1,c_2,...,c_\ell$ of the form λ.

According to Theorem 3 matrix elements of infinitesimal operators of the finite dimensional representations of G (and therefore of G_k) in an appropriate basis are linear functions of the representations parameters.

CG coefficients of the tensor product $\pi_{\delta',\lambda'} \otimes \pi_{\delta'',\lambda''}$ of the principal <u>unitary</u> series representations are introduced by the formula

$$|\delta,\lambda,\omega,n\rangle =$$

$$= \sum_{\substack{\omega',n' \\ \omega'',n''}} \langle\delta',\lambda',\omega',n';\delta'',\lambda'',\omega'',n''|\delta,\lambda,\omega,n\rangle$$

$$|\delta',\lambda',\omega',n'\rangle|\delta'',\lambda'',\omega'',n''\rangle. \qquad (1)$$

It leads to the following relations for any infinitesimal operator I:

$$\sum_{\omega_1',n_1'} \langle\delta',\lambda',\omega_1',n_1';\delta'',\lambda'',\omega'',n''|\delta,\lambda,\omega,n\rangle\ D^{\delta',\lambda'}_{(\omega',n')\,(\omega_1',n_1')}(I) \tag{2}$$

$$+\ \sum_{\omega_1'',n_1''} \langle\delta',\lambda',\omega',n';\delta'',\lambda'',\omega_1'',n_1''|\delta,\lambda,\omega,n\rangle\ D^{\delta'',\lambda''}_{(\omega'',n'')\,(\omega_1'',n_1'')}(I)$$

$$=\ \sum_{\omega_1,n_1} \langle\delta',\lambda',\omega',n';\delta'',\lambda'',\omega'',n''|\delta,\lambda,\omega_1,n_1\rangle\ D^{\delta,\lambda}_{(\omega,n)\,(\omega_1,n_1)}(I)\,.$$

In order to study analytic properties of CG coefficients we write
the relations (1) and (2) for any principal nonunitary series representations and call the coefficients of (1) CG coefficients too. If
the tensor product of the principal nonunitary series representations
$\pi_{\delta',\lambda'}$ and $\pi_{\delta'',\lambda''}$ does really contain in the decomposition the principal nonunitary series representation $\pi_{\delta,\lambda}$ then CG coefficients
of this decomposition satisfy the relations (1) and (2). Moreover,
if we consider in (1) and (2) only those basis vectors $|\delta',\lambda',\omega',n'\rangle$
and $|\delta'',\lambda'',\omega'',n''\rangle$ which belong to the spaces of the subrepresenta-
tions π' and π'' of $\pi_{\delta',\lambda'}$ and $\pi_{\delta'',\lambda''}$ and if the decomposition of
the tensor product $\pi'\otimes\pi''$ contains the representation $\pi_{\delta,\lambda}$ or its
subrepresentation, then the CG coefficients of this decomposition
satisfy the relations (1) and (2). Hence, studying the relations
(1) and (2) for all principal nonunitary series representations we
include in our investigation really existing CG coefficients.

The tensor product of the principal unitary series representa-
tions $\pi_{\delta',\lambda'}$ and $\pi_{\delta'',\lambda''}$ is decomposed into a direct integral of the
representations $\pi_{\delta,\lambda}$. This integral is taken over some measure.
This measure depends on δ and λ. The relations (1) and (2) do not
depend on this measure. Hence, the solution of the system (2) does
not depend on this measure. This means that the CG coefficients
obtained from (2) are not normalized. But the tensor product
$\pi_{\delta',\lambda'}\otimes\pi_{\delta'',\lambda''}$ does not uniquely define the decomposition measure.
Different decomposition measures have to be absolutely continuous
with respect to each other. To different measures there correspond
different normalizations of CG coefficients.

Theorem 4 [6]. Let the multiplicity of any representation
$\pi_{\delta,\lambda}$ of the principal unitary series of G in the tensor product of
the principal unitary series representations $\pi_{\delta',\lambda'}$ and $\pi_{\delta'',\lambda''}$ with
fixed δ' and δ'' be equal to 0 or m where m is a fixed integer or
infinity. Then for any principal nonunitary series representations

$$\langle\delta',\lambda',\omega',n';\delta'',\lambda'',\omega'',n''|\delta,\lambda,\omega,n\rangle$$

$$=\ \Phi(\delta',\lambda',\delta'',\lambda'',\delta,\lambda)\ \frac{P(\lambda',\lambda'',\lambda,\delta',\delta'',\delta,\omega',n',\omega'',n'',\omega,n)}{Q(\lambda',\lambda'',\lambda,\delta',\delta'',\delta,\omega',n',\omega'',n'',\omega,n)}$$

where Φ is a function (maybe a generalized one) which does not depend on ω', n', ω'', n'', ω, n and P, Q are polynomials of the coordinates $c_1',\ldots,c_\ell'$, $c_1'',\ldots,c_\ell''$, $c_1,\ldots,c_\ell$ of the forms λ', λ'', λ. The function Φ is not defined by the relations (2).

This theorem is proved with the help of Lemma 1 and has the following corollary.

<u>Corollary 2</u>. If the group G satisfies the condition of Theorem 4 then analytic continuation of CG coefficients of the tensor products of the finite dimensional representations of G in appropriate basis leads to CG coefficients for the principal nonunitary series representations of G.

Since the principal nonunitary series representations of G contain all irreducible representations of G we know all CG coefficients for irreducible representations (in particular for the principal unitary series representations, for the supplementary series representations, and for the discrete series representations) if we know all CG coefficients for the principal nonunitary series representations. But if we are interested in the tensor product of unitary representations of G then we have to take into account the following fact. If the unitarizable representation is not unitary in the topology of the space $L_\delta^2(K,H_\delta)$ then CG coefficients of the tensor product obtained by this method do not correspond to the orthonormal basis of the unitarized representation. In order to unitarize the representation we introduce new topology and with help of the operator A (see Section 4) we transform the basis $|\delta,\lambda,\omega,n\rangle$ into a new basis $|\delta,\lambda,\omega,n\rangle'$ which is orthonormal in new topology. We have to satisfy the corresponding transformation with CG coefficients, in other words, in (1), and we have to do a transition from the basis $|\delta,\lambda,\omega,n\rangle$ to the basis $|\delta,\lambda,\omega,n\rangle'$.

The procedure of analytic continuation allows us to obtain CG coefficients for the tensor product $D \otimes \pi_{\delta,\lambda}$ of a finite dimensional and a principal nonunitary series representation of G. We consider the tensor products $D \otimes \pi$ for $\pi \in \Omega_\delta$. Then the representations $D \otimes \pi$ are the subrepresentations of the representations $D \otimes \pi_{\delta,\lambda}$. Therefore, according to Corollary 2, CG coefficients of $D \otimes \pi_{\delta,\lambda}$ are analytic continuations in the representation parameters λ of CG coefficients of $D \otimes \pi$, $\pi \in \Omega_\delta$. It is clear that last CG coefficients must correspond to an orthonormal basis of the subspace of $L_\delta^2(K,H_\delta)$ in which π is realized. Usually CG coefficients of $D \otimes \pi$ are given in an orthonormal basis in which the representation π gives the unitary matrices of the corresponding representation of the compact form G_k of the complexification [G] of G. In that case it is necessary to do the transition to the new basis using the transition matrix A (see Section 4). Thus, the appropriate form of CG coefficients of the tensor products of the finite dimensional representations (which allows analytic continuation) and the transition matrix A lead us to CG coefficients of the tensor products $D \otimes \pi_{\delta,\lambda}$.

8. CG COEFFICIENTS AND INTERTWINING OPERATORS

The intertwining operators for the principal nonunitary series representations of G are connected with CG coefficients for the tensor products $\pi_{\delta,\lambda} \otimes D$ where D is a finite dimensional representation of G. Let us consider an intertwining operator for two principal nonunitary series representations which are linked by the element w of the Weyl group W of the pair $(\hat{g},a)$. We denote these representations by $\pi_{\delta,\lambda}$ and $w\pi_{\delta,\lambda}$. Let

$$\pi_{\delta,\lambda} \otimes D = \sum \pi_{\delta_i,\lambda_i} \quad , \quad (w\pi_{\delta,\lambda}) \otimes D = \sum \pi_{\delta_i',\lambda_i'} \, .$$

For convenience $dL_\delta^2(K,H_\delta)$ will be denoted by $dH(\delta,\lambda)$ where δ and λ correspond to $\pi_{\delta,\lambda}$. Let V be an operator from the space $dH(w(\delta,\lambda))$ of the representation $w(d\pi_{\delta,\lambda}) \otimes D$ into the space $\Sigma\, dH(\delta_i',\lambda_i')$ and let U be an operator from $\Sigma\, dH(\delta_i,\lambda_i)$ into $dH(\delta,\lambda) \otimes V$ for which

$$\{d\pi_{\delta,\lambda}(X) \otimes D(X)\}U = U\{\Sigma\, d\pi_{\delta_i,\lambda_i}(X)\} \, , \tag{3}$$

$$V\{w(d\pi_{\delta,\lambda})(X) \otimes D(X)\} = \{\Sigma\, d\pi_{\delta_i',\lambda_i'}(X)\}V \, , \tag{4}$$

where $X \in \hat{g}$. It is clear that matrix elements of the operators U and V are CG coefficients. Let $\Pi \equiv \Pi(\delta,\lambda,w)$ be an intertwining operator for the representations $\pi_{\delta,\lambda}$ and $w\pi_{\delta,\lambda}$:

$$\Pi d\pi_{\delta,\lambda} = w(d\pi_{\delta,\lambda})\Pi \, .$$

Then

$$(\Pi \otimes 1)(d\pi_{\delta,\lambda} \otimes D) = (w(d\pi_{\delta,\lambda}) \otimes D)(\Pi \otimes 1) \, .$$

Multiply both sides of this relation by V on the left hand side and by U on the right hand side. According to (3) and (4) we have

$$V(\Pi \otimes 1)U(\Sigma\, d\pi_{\delta_i,\lambda_i}) = (\Sigma\, d\pi_{\delta_i',\lambda_i'})V(\Pi \otimes 1)U \, .$$

Therefore, the operator $V(\Pi \otimes 1)U$ intertwines the representations $\Sigma\, d\pi_{\delta_i,\lambda_i}$ and $\Sigma\, d\pi_{\delta_i',\lambda_i'}$. It is possible to show that the operator $V(\Pi \otimes 1)U$ gives the intertwining operators for the principal nonunitary series representations from sums $\Sigma\, d\pi_{\delta_i,\lambda_i}$ and $\Sigma\, d\pi_{\delta_i',\lambda_i'}$.

This procedure allows us to evaluate the matrices of intertwining operators (and, consequently, of the operators A which unitarize unitarizable representations of $\pi_{\delta,\lambda}$) in K-basis if intertwining operators for the representations $\pi_{\delta',\lambda'}$ with some fixed δ' and CG coefficients are known. Let us note that it is enough to know CG coefficients of the tensor products $\pi_{\delta,\lambda} \otimes D$ for those D which generate any finite dimensional representation by tensoring. For example, in a case of real forms of SL(n,C) it is enough to have them for one case when D is the vector representation.

9. REPRESENTATIONS OF A COMPLEX SEMISIMPLE LIE GROUP AND CG COEFFICIENTS FOR ITS COMPACT FORM

Let G be a complex connected semisimple (or reductive) Lie group with maximal compact subgroup K. Then K is a compact real form of G. Let $\hat{g}$ be a Lie algebra of G. We consider G and $\hat{g}$ as real Lie group and algebra with double the number of parameters. Then any irreducible finite dimensional real-analytic representation of G is the tensor product of its complex-analytic and complex-anti-analytic irreducible representations of G, i.e., it has a form $g \to T_g^{\lambda} \otimes T_g^{\overline{\mu}}$, where $g \to T_g^{\lambda}$ and $g \to T_g^{\mu}$ are complex-analytic representations of G. Any complex-analytic irreducible representation of G under restriction into K leads to irreducible representation of K. Therefore, the representation $g \to T_g^{\lambda} \otimes T_g^{\overline{\mu}}$ of G after restriction into K gives the tensor product of the irreducible representations $k \to T_k^{\lambda}$ and $k \to T_k^{\overline{\mu}}$ of K (the last representation will be denoted by $k \to T_k^{\overline{\mu}}$). Hence, a basis of the tensor product of the representations $k \to T_k^{\lambda}$ and $k \to T_k^{\overline{\mu}}$ of K is a basis of the representation $g \to T_g^{\lambda} \otimes T_g^{\overline{\mu}}$ of G.

Choose the bases for the representations $k \to T_k^{\lambda}$ and $k \to T_k^{\overline{\mu}}$ of vectors for which the representation matrices are unitary. Denote the basis elements by $|\Pi_{\lambda}\rangle$ and $|\Pi_{\mu}\rangle$, correspondingly. If an action of infinitesimal operators of K onto the basis elements $|\Pi_{\lambda}\rangle$ and $|\Pi_{\mu}\rangle$ is known then an action of infinitesimal operators of G onto the basis elements $|\Pi_{\lambda}\rangle|\Pi_{\mu}\rangle$ is known. If I is an appropriate infinitesimal operator for G then

$$I|\Pi_{\lambda}\rangle|\Pi_{\mu}\rangle = \sum_{\Pi_{\lambda}',\Pi_{\mu}'} D(\Pi_{\lambda}',\Pi_{\lambda})D(\Pi_{\mu}',\Pi_{\mu})|\Pi_{\lambda}'\rangle|\Pi_{\mu}'\rangle . \tag{5}$$

In the space of the representation $g \to T_g^{\lambda} \otimes T_g^{\overline{\mu}}$ it is possible to choose other orthonormal bases. It is the basis consisting of orthonormal bases of the subspaces of irreducible representations of K. Denote its basis elements by $|\Omega/\Sigma\rangle$ where Σ labels irreducible representations of K and their basis elements, and Ω is a multiplicity index for the irreducible representations of K. If we consider restriction of the representation $g \to T_g^{\lambda} \otimes T_g^{\overline{\mu}}$ onto K then it is

clear that the transition from the first basis to the second one is
realized by CG coefficients of K.   Thus

$$|\Omega/\Sigma> = \sum_{\Pi_\lambda,\Pi_\mu} <\Pi_\lambda,\Pi_\mu|\Omega/\Sigma>|\Pi_\lambda>|\Pi_\mu> \,, \tag{6}$$

where $<\Pi_\lambda,\Pi_\mu|\Omega/\Sigma>$ are CG coefficients for the tensor product of the
representations $k \to T_k^\lambda$ and $k \to T_k^{\overline{\mu}}$ of K.   Suppose that we know the
formula of action of the infinitesimal operator I onto

$$I|\Omega/\Sigma> = \sum_{\Omega',\Sigma'} D(\Omega',\Sigma';\Omega,\Sigma)|\Omega',\Sigma'> \,. \tag{7}$$

From (5) and (6) we can obtain that

$$I|\Omega/\Sigma> = \sum_{\Pi_\lambda,\Pi_\mu} <\Pi_\lambda,\Pi_\mu|\Omega/\Sigma> \sum_{\Pi'_\lambda,\Pi'_\mu} D(\Pi'_\lambda,\Pi_\mu)$$

$$\times\ D(\Pi'_\mu,\Pi_\mu) \sum_{\Omega',\Sigma'} <\Omega'/\Sigma'|\Pi'_\lambda,\Pi'_\mu>|\Omega'/\Sigma'> \,.$$

Therefore

$$D(\Omega',\Sigma';\Omega,\Sigma) = \sum_{\Pi'_\lambda,\Pi'_\mu}\ \sum_{\Pi_\lambda,\Pi_\mu} <\Omega'/\Sigma'|\Pi'_\lambda,\Pi'_\mu>$$

$$\times\ D(\Pi'_\lambda,\Pi_\lambda)D(\Pi'_\mu,\Pi_\mu)<\Pi_\lambda,\Pi_\mu|\Omega/\Sigma> \,.$$

This relation is equivalent to the following one

$$\sum_{\Omega',\Sigma'} <\Pi'_\lambda,\Pi'_\mu|\Omega'/\Sigma'>D(\Omega',\Sigma';\Omega,\Sigma)$$

$$= \sum_{\Pi_\lambda,\Pi_\mu} D(\Pi'_\lambda,\Pi_\lambda)D(\Pi'_\lambda,\Pi_\mu)<\Pi_\lambda,\Pi_\mu|\Omega/\Sigma> \,. \tag{8}$$

This relation connects the CG coefficients for K with different
resulting representations in the tensor product. Really, the
infinitesimal operator I transfers the subspaces of the irreducible
representations of K (see (7)). The same irreducible representations of K are transfered on the left hand side of (8).

Since the representation $g \to T_g^\lambda \otimes T_g^{\overline{\mu}}$ of G is irreducible then
the relations (8) define uniquely (up to a constant which depends
only on λ and μ) all CG coefficients of the tensor product
$k \to T_k^\lambda \otimes T_k^{\overline{\mu}}$.

In order to use this method of evaluation of CG coefficients
we need an explicit form of infinitesimal operators of G in the

basis $|\Pi_\lambda\rangle|\Pi_\mu\rangle$ and in the basis $|\Omega/\Sigma\rangle$. In cases of the groups
GL(n,C) (or SL(n,C)) and SO(n,C) infinitesimal operators I in the
basis $|\Pi_\lambda\rangle|\Pi_\mu\rangle$ are known if the bases $|\Pi_\lambda\rangle$ and $|\Pi_\mu\rangle$ are the Gelfand-
Zetlin bases. Infinitesimal operators of G in the basis $|\Omega/\Sigma\rangle$ can
be found with the help of infinitesimal operators in a K-basis of
the principal nonunitary series representations of G. Under separating finite dimensional representations of TG from the principal nonunitary series representations we define the possible values of Ω,
i.e., we label multiple representations in the tensor product

$$k \to T_k^\lambda \otimes T_k^{\bar{\mu}}.$$

REFERENCES

1. Helgason, S., Differential geometry and symmetric spaces, Aca-
        demic Press, N. Y., 1962.
2. Warner, G., Harmonic analysis on semisimple Lie groups, Vol. I
        & II, Springer, Berlin & New York, 1972.
3. Harish-Chandra, Trans. Amer. Math. Soc., 76, 234 (1954).
4. Lepowsky, J., ibid, 176, 1 (1973).
5. Klimyk, A. U., Gavrilik, A. M., J. Math. Phys., 20, No. 9 (1979).
6. Klimyk, A. U., Matrix elements and Clebsch-Gordan coefficients
        of group representations, Naukova Dumka, Kiev, 1979 (in
        Russian).
7. Lepowsky, J., Wallach, N. R., Trans. Amer. Math. Soc., 184, 223
        (1973).
8. Knapp, A. W., Stein, E. M., Ann. Math., 93, 489 (1971).
9. Knapp, A. W., Auckerman, G., Lect. Notes Math., 587, 138 (1977).
10. Schiffmann, G., Bull. Soc. Math. France, 99, 3 (1971).
11. Kunze, R. A., Stein, E. M., Amer. J. Math., 89, 385 (1967).
12. Harish-Chandra, Trans. Amer. Math. Soc., 70, 28 (1951).
13. Gavrilik, A. M., Klimyk, A. U., Preprint ITP-75-63E, Kiev, 1975.
14. Gavrilik, A. M., Klimyk, A. U., Preprint ITP-75-18E, Kiev, 1975.
15. Klimyk, A. U., Gavrilik, A. M., Preprint ITP-76-39E, Kiev, 1976.
16. Vilenkin, N. Ya., Special functions and the theory of group
        representations, Amer. Math. Soc., R.I., 1968.
17. Barut, A. O., Raczka, R., Theory of group representations and
        applications, PWN, Warsaw, 1977.
18. Klimyk, A. U., Lett. Math. Phys., 1, 375 (1977).
19. Goodman, R., Trans. Amer. Math. Soc., 143, 55 (1969).
20. Klimyk, A. U., Gruber, B., J. Math. Phys., 20, (1979).

GROUP THEORY AND THE INTERACTION OF

COMPOSITE NUCLEON SYSTEMS

Peter Kramer

Institut für theoretische Physik
Universität Tübingen
Tübingen, West Germany

1. Survey

In the theory of nuclear structure and reactions, one often
splits the full system into composite systems and studies the dynamics of these composite systems. In this report, I shall describe
some methods of group theory which we have developed for dealing
with these systems. We shall describe three types of groups and
their application to composite particle theory. The symmetric group
will be associated with exchange, orbital partitions and the super-
multiplet scheme. The general linear group and its representations
will be applied to exchange decompositions. The inhomogeneous symplectic transformations of classical phase space and their representations will be used to describe the kinematics and dynamics of
composite particles.

2. The symmetric group $S(n)$

Consider the symmetric group $S(n)$ and pairs of subgroups
characterized by weights $\tilde{w}$ and w of n. By a weight w, we
mean a set of j integers $(w_1 w_2 .. w_j)$ such that

$$\sum_{\ell=1}^{j} w_1 = n$$

The group $S(w)$ of the weight w is defined as

$$S(w) = S(w_1) \times S(w_2) \times . . \times S(w_j)$$

2.1 <u>Proposition</u>: The double cosets of $S(n)$ with respect to the
subgroups $S(w)$, $S(w)$ are in one-to-one correspondence to the $j \times j$
matrices $k = \{k_{i\ell}\}$ with integer elements fulfilling

$$\sum_{i}^{j} k_{i\ell} = \tilde{w}_{\ell} \qquad\qquad \sum_{\ell}^{j} k_{i\ell} = w_{i}$$

Compare KR 1,2,3 for the proof.

The physical applications of this concept are illustrated by the following two examples.

<u>Example 2.1</u>: Occupation of non-orthogonal single-particle states consider j non-orthogonal single-particle states. Choose two weights $\tilde{w}$ and w and construct two n-body states $\psi_{\tilde{w}}$ and ψ_{w} by taking the components of the weights as occupation numbers of single-particle states. Clearly these states are stable under the action of the subgroups $S(\tilde{w})$ and $S(w)$, respectively. Hence we find for the matrix element of an arbitrary permutation p,

$$(\psi_{\tilde{w}} | U(p) | \psi_{w})$$

$$= (\psi_{\tilde{w}} | U(z_{k}) | \psi_{w})$$

$$= \prod_{i,\ell}^{j} (\varepsilon_{i\ell})^{k_{i\ell}} = \varepsilon^{k}$$

Here, z_{k} generates the double coset $k = \{k_{i\ell}\}$ and the matrix ε has the elements $\varepsilon_{i\ell} = (\phi_{i} | \phi_{\ell})$. We refer to the various matrix elements on different cosets as to the exchange integrals.

<u>Example 2.2</u>: The reaction ${}^{4}He + {}^{2}H \rightarrow {}^{3}He + {}^{3}H$

The occupation numbers of the two channels define two weights $\tilde{w} = (42)$ and $w = (33)$. The possible double cosets and hence, exchange contributions are characterized by the three matrices

$$k = \begin{Bmatrix} 3 & 1 \\ 0 & 2 \end{Bmatrix} , \begin{Bmatrix} 2 & 2 \\ 1 & 1 \end{Bmatrix} , \begin{Bmatrix} 1 & 3 \\ 2 & 0 \end{Bmatrix}$$

These three terms are often referred to as exchange of 0, 1 and 2 particles from ${}^{3}He$ to ${}^{2}H$. Fig. 1 illustrates the exchange in an elastic collision.

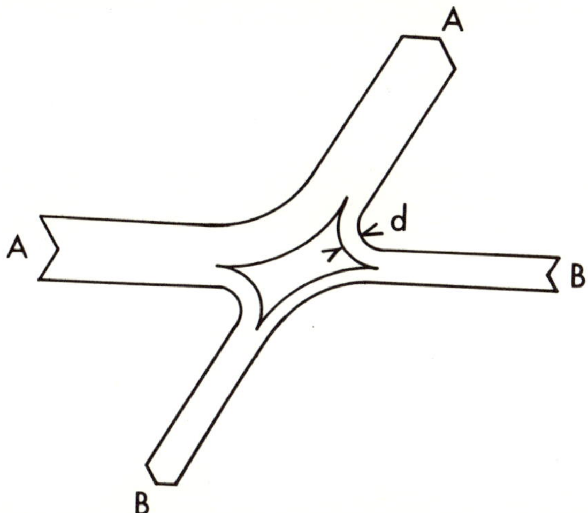

<u>Fig. 1</u>   Exchange in the elastic reaction $A + B \rightarrow A + B$. The
distance d is a measure for the number of nucleons
transferred from B to A and hence a measure of the
exchange type.

The representations of the symmetric group are characterized by a
partition f of n, we denote them by d^f. While in the standard
representation of Young and Yamanouchi we use the subgroups
S(n-1), S(n-2),.. to describe the basis, consideration of the subgroups $S(\tilde{w})$ and S(w) suggests the use of different chains of
subgroups which include these subgroups. A reasonable chain of subgroups is given by removing in each step the subgroup associated
with a single weight component. This procedure in general is not
free of multiplicity, but if the only representations of $S(\tilde{w})$ and
S(w) involved are the identity representations, the multiplicity
is fully specified by Gelfand patterns $\tilde{q}, q$ of intermediate partitions. In this case, the matrix elements of the representation d^f
become functions on double cosets which we denote as

$$d^{f}_{\tilde{q}q}(z_k)$$

We do not indicate the weights since they are implicit in the Gelfand patterns $\tilde{q}, q$. The properties and computation of these non-
standard representations of the symmetric group are discussed in
KR 3 . An application is given in

<u>Example 2.3</u>:   Inner products of n-particle states

By use of normalized Young operators   c(rfq)   from the group algebra
of   S(n), we may pass from the product states given in example 2.1
to bases characterized by a partition   f   of   n   and a standard
Young tableau   r.   We denote these new states as

$$\left|\psi_w \; qfr\right) = c(rfq)\left|\psi_w\right)$$

and state

<u>2.1 Proposition</u>:   The scalar product of two bases for different
weights and Gelfand patterns is given by

$$\left(\psi_{\underset{\sim}{w}}\left|c^+(rf\tilde{q})\circ c(rfq)\right|\psi_w\right)$$

$$= \left[\tilde{w}! \; w!\right]^{\frac{1}{2}} \sum_k d^f_{\tilde{q}q}(z_k) \; (k!)^{-1} \; \varepsilon^k$$

where

$$w! = \prod_i (w_i!) \qquad k! = \prod_{i,\ell} (k_{i\ell}!)$$

In composite particle theory, the representations d^f are used to
characterize the orbital states.   The spin-isospin partition is then
the partition associated to   f   and provides the labels for the
SU(4)   group of the supermultiplet scheme introduced by Wigner   WI 4.
The configurations of composite particles will later be characterized by orbital partitions   f   and it will be shown that this partition carries the basic properties of the underlying fermion system
into composite particle dynamics.   This connection between the super-
multiplet scheme and composite particle dynamics was already pointed
out by Wheeler   WH 5.

<u>Example 2.3</u>:   The system ^{4}He + ^{4}He

If the two nuclei ^{4}He   are taken as two composite particles, this
two-body configuration is still characterized by the orbital partition   f=[44]   of the underlying   8-nucleon system.

3.   The general linear group $GL(j,\mathbb{C})$

We recall the following properties of the representations of
the unitary group   U(j)   and the general linear group $GL(j,\mathbb{C})$:   The
finite irreducible representations of   U(j)   may be analytically
extended to irreducible representations of $GL(j,\mathbb{C})$.   These representations are characterized by partitions   f   of   n=0,1,2,..   .
The rows and columns of these representations may be characterized
by Gelfand patterns $\tilde{q}$ and   q, and thus, the matrix elements may
be taken as

$$D^f_{\tilde{q}q} = D^f_{\tilde{q}q}(g)$$

<u>3.1 Proposition</u>:   The matrix elements of the finite irreducible representations D^f of the group $GL(j,\mathbb{C})$ are given by

$$D^{f}_{\tilde{q}q}(g) = [\tilde{w}!w!]^{\frac{1}{2}} \sum_{k} d^{f}_{\tilde{q}q}(z_k) \ (k!)^{-1} \ g^k$$

where the coefficients d^f are the non-standard representations of the group $S(n)$ on double cosets. For the proof we refer to KR 3.

<u>Example 3.1</u>:   The orbital configuration ${}^4He + {}^4He$

For the weight $w = (44)$, the only admissible orbital partition for nucleons is $f = [44]$. The Gelfand pattern q reduces to the weight component $w_1 = 4$. From propositions 2.1 and 3.1 we get

$$(\psi_{(44)} | c^+(r[44]4) \ c(r[44]4) | \psi_{(44)})$$

$$= D^{[44]}_{4 \ 4}(\varepsilon) = (\varepsilon_{11}\varepsilon_{22} - \varepsilon_{12}\varepsilon_{21})^4$$

$$= \sum_{\gamma=0}^{4} (-1)^\gamma \binom{4}{\gamma} \varepsilon_{11}^{4-\gamma} \varepsilon_{22}^{4-\gamma} \varepsilon_{12}^{\gamma} \varepsilon_{21}^{\gamma}$$

The explicit form of the representation $D^{[44]}$ of $GL(2,\mathbb{C})$ is obtained from the highest weight polynomial given by Moshinsky, MO 6. The terms are clearly associated with exchange types characterized by the symbols

$$k = \left\{ \begin{matrix} 4-\gamma & \gamma \\ & 4-\gamma \end{matrix} \right\} , \gamma = 0 \ 1 \ 2 \ 3 \ 4 .$$

4.   The Weyl group and the symplectic group

We shall work in the Hilbert spaces of analytic functions introduced by Segal SE 7 and Bargmann BA 8 which may be briefly described as follows: For analytic functions

$$f(x) = f(x_1 x_2 \ \cdots \ x_m)$$

define the operators D_1, X_ℓ by

$$(D_i \ f)(x) = \frac{\partial}{\partial x_i} \ f(z)$$

$$i,\ell = 1 \ 2 \ .. \ m$$

$$(X_\ell \ f)(x) = x_\ell \ f(z)$$

and a scalar product

$$(f|g) = \int \overline{f(x)} \ g(x) \ d\mu(x)$$

such that D_i, X_ℓ obey the relations

$$[D_i, X_\ell] = \delta_{i\ell} \ , \ X_i^+ = D_i$$

The measure $d\mu$ is given by

$$d\mu(x) = \pi^{-m} \exp(-\bar{x}\cdot x) \ \prod_i dRe(x_i) \ d \ Im(x_i)$$

4.1 Definition: For

$$\alpha_c = (a_1 a_2 \cdot\cdot a_n \ \bar{a}_1 \bar{a}_2 \cdot\cdot \bar{a}_n) \ , \ a_\ell = \sqrt{\tfrac{1}{2}} \ (\alpha_\ell' + i\alpha_\ell'')$$

define the Weyl operator W_{α_c} by

$$(W_{\alpha_c} f) \ (x) = \exp\left[\bar{a} \cdot (x - \tfrac{1}{2} a)\right] \ f(x-a)$$

These operators are shown by Bargmann BA8 to fulfill the relations

$$W_{\alpha_c} \circ W_{\beta_c} = \exp\left[-\tfrac{1}{2} \ \{\alpha_c, \beta_c\}\right] \ W_{\alpha_c + \beta_c}$$

$$W_{\alpha_c} \circ \left\{\begin{matrix} X_\ell \\ D_\ell \end{matrix}\right\} = \left\{\begin{matrix} X_\ell - a_\ell I \\ D_\ell - \bar{a}_\ell I \end{matrix}\right\} \circ W_{\alpha_c}$$

The first one of these relations yields with $\{\alpha_c, \beta_c\} = a\cdot\bar{b} - \bar{a}\cdot b$
the multiplication law of the operators and the second one shows
that the Weyl operators represent translations of position and
momentum in classical phase space associated with the real and
imaginary parts α', α'' of the complex numbers a, respectively.

Example 4.1: Weyl translation of oscillator states

In Bargmann space, the oscillator states of a single particle with
angular momentum L are given by polynomials of the type

$$P_{LM}^N(x) = A_{NL} (x\cdot x)^{\frac{1}{2}(N-L)} \ \mathcal{Y}_{LM}(x)$$

Application of a Weyl operator yields the oscillator states displaced with respect to both position and momentum as

$$(W_{\alpha_c} P_{LM}^N) \ (x) = \exp\left[\bar{a}\cdot(x - \tfrac{1}{2} a)\right] P_{LM}^N(x-a)$$

In particular for NLM = 0 0 0 we obtain the normalized coherent
states

$$(W_{\alpha_c} P_{00}^0) \ (x) = \exp\left[\bar{a}\cdot(x - \tfrac{1}{2} a)\right]$$

In what follows we shall denote by $\hat{W}_{\alpha_c}$ the modified Weyl operator

$$\hat{W}_{\alpha_c} = W_{\alpha_c} \exp [\tfrac{1}{2} \bar{a}\cdot a]$$

The Weyl operators will be used in a later section 5 to describe composite particle configurations by displacement of configurations defined at a common center.

The symplectic group $Sp(2n,|R)$ is the group of linear canonical transformations in phase space. In the complex setting appropriate for Bargmann space, this group appears in a complex but equivalent form which may be characterized as

$$Sp(2n,\mathbb{C}) \cap U(n,n)$$

The representations of this group in Hilbert space we denote as S_{g_c} where g_c is the complex form of the real element $g \in Sp(2n,|R)$.

<u>4.2 Proposition</u>: The operators S_{g_c} and W_{α_c} are related by the equations

$$S_{g_c}^{-1} \circ W_{g_c \alpha_c} \circ S_{g_c} = W_{\alpha_c}$$

The operators S_{g_c} themselves form a projective representation of the symplectic group known as the metaplectic representation. In composite particle theory, the symplectic group appears in transformations between different sets of coordinates and in orbital permutations. Beyond these more kinematical transformations, a semigroup extension of the symplectic group allows one to describe a larger class of operators.

<u>4.3 Definition</u>: The semigroups $U^{>}(n,n)$ and $U^{\geqslant}(n,n)$ are defined by the conditions

$$U^{>}(n,n) = \{h_c \,|\, (h_c \xi \, h_c \xi)-(\xi\,|\,\xi)>0\}$$

$$U^{\geqslant}(n,n) = \{h_c \,|\, (h_c \xi\,|\,h_c \xi)-(\xi\,|\,\xi)\geq 0\}$$

where $(\,|\,)$ describes the inner product which is invariant with respect to the group $U(n,n)$.

<u>4.4 Proposition</u>: The unitary representation S_{g_c} may be modified into

(a)   a representation by Hilbert-Schmidt operators of the semigroup

$$Sp(2n,\mathbb{C}) \cap U^{>}(n,n),$$

(b)   a representation by bounded operators of the semigroup

$$Sp(2n,\mathbb{C}) \cap U^{\geq}(n,n).$$

For a detailed analysis we refer to BR 9,10.

<u>Example 4.2</u>: The Gaussian interaction

The Gaussian interaction may be written as an integral operator in Bargmann space. It is found that this integral operator is parametrized by an element of the semigroup $Sp(2n,\mathbb{C}) \cap U^{\geq}(n,n)$. Hence, Gaussian interaction may be treated by the extended representation theory of the symplectic group. The geometry and representation of the semigroups is discussed in BR 9,10,11.

5. Interaction of composite particles

To set up the dynamics of composite particles, one introduces internal ccordinates y and center coordinates z which, for two composite particles, are of the form

$$z_1 = [n_1]^{-\frac{1}{2}} \sum_{1}^{n_1} x_i$$

$$z_2 = [n_2]^{-\frac{1}{2}} \sum_{n_1+1}^{n} x_i$$

One fixes the internal state of the composite particles and, for a many-body Hamiltonian H applies the variational principle

$$(\delta\psi|c^+(r\ f\ q) \circ (H-E) \circ c(r\ f\ q)|\psi) = 0$$

with free variation of the state $u(z_1,z_2)$ for the motion of the two centers to obtain complex integral equations of the type

$$\int [\mathcal{H}(\tilde{z}_1\tilde{z}_2,z_1z_2) - E\,\mathcal{N}(\tilde{z}_1\tilde{z}_2,z_1z_2)]\,u(z_1z_2)\,d\mu(z_1z_2) = 0$$

The kernel $\mathcal{N}$ is called the normalization kernel, it reflects in composite particle theory the fermion property of the underlying system. The kernel $\mathcal{H}$, called the interaction kernel, stems from the underlying two-body interaction. The variational principle and its use are described in detail by Wildermuth and Tang WI 12.

We now indicate how the various groups mentioned earlier come into play. Denote by

$$|\psi) = |\alpha^{n_1}\alpha^{n_2})$$

a state of two harmonic oscillator shell configurations consisting of n_1 and n_2 occupied states, respectively. Introduce the Weyl operator $\hat{W}_{\tau_c}$ for the special choice

$$\tau_c = (\underbrace{t_1 \ldots t_1}_{n_1} \ \underbrace{t_2 \ldots t_2}_{n_2} \ \underbrace{\bar{t}_1 \ldots \bar{t}_1}_{n_1} \ \underbrace{\bar{t}_2 \ldots \bar{t}_2}_{n_2})$$

and put $\ t_1 = [n_1]^{-\frac{1}{2}} z_1 \ , \ t_2 = [n_2]^{-\frac{1}{2}} z_2 \ .$

5.1 <u>Proposition</u>: The normalization and interaction operator for two composite particles are given by the matrix elements

$$\left\{ \begin{matrix} \mathcal{N} \\ \mathcal{H} \end{matrix} \right\} (\tilde{\bar{z}}_1 \tilde{\bar{z}}_2, z_1 z_2)$$

$$= (\alpha^{n_1} \alpha^{n_2} | \hat{W}_{\tilde{\tau}_c}^+ \ \circ \ c^+(\tilde{r} \ \tilde{f} \ \tilde{q}) \ \circ \ \left\{ \begin{matrix} I \\ H \end{matrix} \right\} \ \circ \ c(r \ f \ q) \ \circ \ \hat{W}_{\tau_c} \ | \alpha^{n_1} \alpha^{n_2})$$

The proof of this equation is entirely based on the inhomogeneous symplectic group $I \, Sp(2m, |R)$, and use the commutation properties of the Weyl operators with the operators S_{g_c} which describe the transformation between different sets of coordinates, KR3.

Through this proposition, the computation of composite particle dynamics is reduced to a problem of non-orthogonal single-particle states. Full use can now be made of the groups $S(n)$ and $GL(j, \mathbb{C})$ to describe the fractional parentage methods needed in the analysis of interaction or transition properties. The symplectic group yields the analytic structure of the kernels as representations of finite inhomogeneous transformations in phase space. In particular, the Weyl operators serve to shift the single-particle states to their appropriate center positions. The semigroup extension of the symplectic group admits the treatment of Gaussian interactions.

We refer to KR 3 for the elaboration of the full technique of these groups in composite particle theory and concentrate here on a few aspects of the normalization kernel.

Example <u>5.1</u>: Configuration of unexcited oscillator states for two composite particles.

The computation of the normalization kernel proceeds by first com- puting the overlap between Weyl-shifted oscillator states $|N \ L \ M)$

where N L M = 0 0 0. This yields

$$\varepsilon_{i\ell} = (0\ 0\ 0\ |\ \hat{\tilde{W}}_\tau{}^+_c \circ \hat{W}_{\tau_c}\ |0\ 0\ 0)$$

$$= \exp\ [(n_1 n_\ell)^{-\frac{1}{2}}\tilde{z}_i \cdot \bar{z}\ ]$$

With these analytic overlaps, the methods of GL(j,ϕ) outlined in section 3 yields

$$\mathcal{N}\ (\tilde{z}_1 \tilde{z}_2, z_1 z_2) = D^f_{\tilde{q}q}(\varepsilon)$$

For the case of ^{4}He + ^{4}He, this expression becomes

$$\mathcal{N}\ (\tilde{z}_1 \tilde{z}_2, z_1 z_2) = \sum_\gamma (-1)^\gamma\ \binom{4}{\gamma}\ \exp[\ \frac{4-\gamma}{4}\ (\tilde{z}_1 \cdot \bar{z}_1 + \tilde{z}_2 \cdot \bar{z}_2) + \frac{\gamma}{4}(\tilde{z}_1 \cdot \bar{z}_2 + \tilde{z}_2 \cdot \bar{z}_1)]$$

Introducing the relative and c.m. coordinates

$$s_1 = \sqrt{\frac{1}{2}}\ (z_1 - z_2)$$

$$s_2 = \sqrt{\frac{1}{2}}\ (z_1 + z_2)\ ,$$

this kernel splits as

$$\mathcal{N}\ (\tilde{s}_1 \tilde{s}_2\ ,\ s_1 s_2)$$

$$= \exp\ [\tilde{s}_2 \cdot \bar{s}_2]\ \sum_\gamma (-1)^\gamma\ \binom{4}{\gamma}\ \exp\ [\ \frac{4-2\gamma}{4}\ \tilde{s}_1 \cdot \bar{s}_1]$$

into a first factor related to the c.m. motion and a second factor related to the relative motion. The first factor is just the reproducing kernel for the c.m. state. The second factor is a kernel whose eigenstates are easily found to be the harmonic oscillator states for the relative motion.

The interpretation of the kernel may be inferred from its eigenvalues η_N taken as functions of the relative oscillator excitation. Figure 2 gives these eigenvalues for the configurations ^{4}He + p, ^{4}He + ^{2}H , ^{4}He + ^{3}He and ^{4}He + ^{4}He. As mentioned earlier, the normalization kernel reflects the fermion nature of the nucleons. The normalization operator projects out the Pauli-forbidden oscillator excitation. It differs qualitatively from a projection

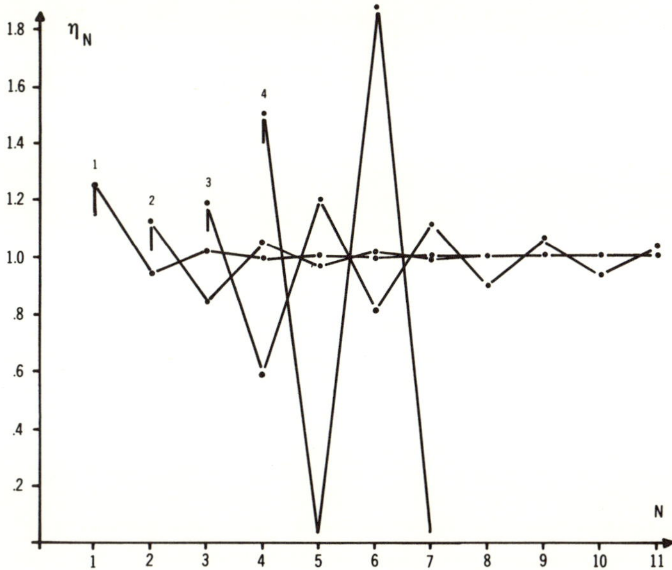

Fig. 2: Eigenvalues η_N of the normalization operator for the configuration $s^4 + s^{n_2}$ and $n_2 = 1,2,3,4$ as a function of the relative oscillator excitation N.

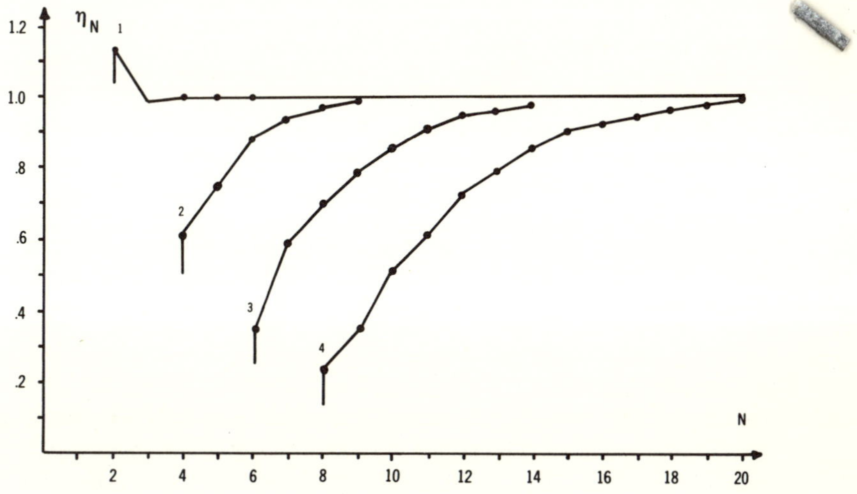

Fig. 3: Eigenvalues η_N of the normalization operator for the configuration $s^4 p^{12} + s^{n_2}$ and $n_2 = 1,2,3,4$ as a function of the relative oscillator excitation N.

operator on allowed states by oscillations between even and odd
excitations. For ^{4}He + ^{4}He, these oscillations increase dramatically
and lead to the birth of a pair of composite bosons from eight fermions. It can be seen that this boson pair is already preparing for
its appearance in the system ^{4}He + ^{3}He. Fig. 3 shows similar computations for a configuration of ^{16}O and n_2 = 1,2,3,4 nucleons in the
second composite particle. The onset of oscillations is expected
when approaching the system ^{16}O + ^{16}O, which again forms a boson
pair. These considerations may serve to illuminate some properties
of the normalization operators in composite particle dynamics.

For more complex composite particles, the normalization kernel becomes a more complicated function of the composite particle coordinates. Besides of the exponential terms, there appear polynomial
functions which reflect the complexity of the internal states. For
these configurations, it is essential to use a maximum of information from group representations to obtain closes analytic expressions,
and it is the interplay of the various groups that allows one to
write down closed analytic expressions for the dynamical equations.

References

1. Kramer, P. and T.H. Seligman, Nucl. Phys. A123 (1969) 161
2. Kramer, P. and T.H. Seligman, Nucl. Phys. A186 (1972) 49
3. Kramer, P., John, G., and Schenzle, D., Group Theory and the
    Interaction of Composite Nucleon Systems, Vieweg Wiesbaden 1979
4. Wigner, E.P., Phys. Rev. 51 (1937) 51
5. Wheeler, J.A., Phys. Rev. 52 (1937) 1083 and 1107
6. Moshinsky, M., J. Math. Phys. 4 (1963) 1128
7. Segal, I.E., Mathematical Problems of Relativistic Physics,
    Amer. Math. Soc., Providence, R.I., 1963
8. Bargmann, V., Group representations in Hilbert spaces of analytic
    functions, in: Analytic methods in mathematical physics, ed.
    by R.P. Gilbert and R.G. Newton, New York 1968
9. Brunet, M. and P. Kramer, Semigroups of length increasing trans-
    formations, to be published
10. Brunet, M. and P. Kramer, Complex extension of the representa-
    tion of the symplectic group associated with the canonical
    commutation relations, to be published
11. Brunet, M. and P. Kramer, Complex extension of the representa-
    tion of the symplectic group associated with the canonical
    commutation relations, in: Proc. IV. Int. Colloquium on Group
    Theoretical Methods in Physics, Nijmegen 1975, Springer Lecture
    Notes in Physics, vol. 50, ed. by A. Janner, T. Janssen, and M.
    Boon, Springer, Berlin Heidelberg, New York 1976
12. Wildermuth, K. and Y.C. Tang, A unified theory of the nucleus,
    Clustering Phenomena in Nuclei, vol. 1, ed. by K. Wildermuth
    and P. Kramer, Vieweg, Braunschweig 1977

TIME REVERSAL IN DISSIPATIVE SYSTEMS

M. Lax*

Physics Department
City College of the City University of New York
New York, New York 10031

and

Bell Laboratories
Murray Hill, New Jersey 07974

1. INTRODUCTION

An explicit formulation of the time reversal operator K with the property

$$K q K^{-1} = q; \quad K p K^{-1} = -p; \quad K \vec{\sigma} K^{-1} = -\vec{\sigma} \tag{1.1}$$

that it preserve the sign of the position operator q, and reverse the signs of the momentum operator p and the spin $\vec{\sigma}$ was made by Wigner (1932, 1959).[1] Wigner then developed theorems concerning additional degeneracy produced by time reversal when the Hamiltonian is even under time reversal

$$K H(q,p,\vec{\sigma})K^{-1} = H(q,-p,-\vec{\sigma}) = H(q,p,\vec{\sigma}). \tag{1.2}$$

[Here q, p and $\vec{\sigma}$, the position momentum and spin of a single particle, should be thought of as an abbreviated notation for the positions, momenta and spins of all the particles in the system.] In order to combine time reversal, an anti-linear, anti-unitary operator with the linear unitary operators of group theory, Wigner (1932, 1959) introduced the notion of corepresentation.

Wigner's ground breaking work did not address the problem of combining time-reversal with group operations to determine the effects of time reversal on selection rules in the evaluation of

matrix elements. This gap was filled by Lax (1962)[2] in a non-trivial
application to space groups. The method which avoided the use of
corepresentations was explained in more detail in a paper on subgroup
techniques, Lax (1965),[3] and a text, Lax (1974).[4]

Corepresentations were avoided by noting that although time-
reversal is anti-linear, so is Hermitian conjugation. The selection
rules turn out to be determined by the behavior of operators under
the combined operation F of Hermitian conjugation and time reversal.
In what follows, we shall assume that each operator a_i in the set $\underset{\sim}{a}$
is either even or odd under the linear "flip" operator F:

$$\overline{a}_i \equiv F\, a_i\, F^{-1} \equiv K\, a_i^{\dagger} = \varepsilon_i a_i \tag{1.3}$$

where $\varepsilon_i = \pm 1$ for even and odd operators respectively. The associated Heisenberg operator

$$a_i(t) = \exp(iHt)\, a_i\, \exp(-iHt) \tag{1.4}$$

(where $\hbar = h/2\pi$ has been set equal to unity) then obeys

$$\overline{a_i(t,H)} = \exp(-i\overline{H}t)\, \overline{a}_i(0)\exp(i\overline{H}t)$$

$$= \varepsilon_i \overline{a}_i(-t,\overline{H}) \tag{1.5}$$

where the Hamiltonian involved is displayed explicitly. In the
presence of a magnetic field $\vec{B}$, we would have

$$\overline{H(\vec{B})} = H(-\vec{B}). \tag{1.6}$$

In cases where H is even under time reversal, $\overline{H} = H$ and

$$\overline{a}_i(t) = \varepsilon_i a_i(-t). \tag{1.7}$$

The theorem, Lax (1974), for an arbitrary linear operator L,

$$(\Phi, L\Psi) = (K\Psi, \overline{L}K\Phi) \tag{1.8}$$

follows directly from the definition of the barring operation. If
$\{\Psi_n\}$ is any set of states which span a Hilbert space (or subspace)
σ, and $\overline{\sigma}$ is the corresponding subspace spanned by the states $\{K\,\Psi_n\}$,
then Eq. (1.8) permits us to assert an equality

$$\mathrm{Tr}_{\sigma} L = \mathrm{Tr}_{\overline{\sigma}} \overline{L} \tag{1.9}$$

between the trace of L over σ and that of $\overline{L}$ over $\overline{\sigma}$. Let us apply
this theorem to the case

$$L = a_i(t)\, a_j(0)\, \rho(\beta); \quad \rho(\beta) \equiv \exp(-\beta H)/\mathrm{Tr}[\exp(-\beta H)], \tag{1.10}$$

where $\rho(\beta)$ with $\beta = (k_B T)^{-1}$ describes the thermal equilibrium density operator,

$$\langle a_i(t,H) a_j(0,H)\rangle_\sigma = \varepsilon_i \varepsilon_j \langle a_j(0,\overline{H}) a_i(-t,\overline{H})\rangle_{\overline{\sigma}}$$

$$= \varepsilon_i \varepsilon_j \langle a_j(t,\overline{H}) a_i(0,\overline{H})\rangle_{\overline{\sigma}} . \tag{1.11}$$

Here $\langle\ \rangle$ implies a trace with $\rho(\beta)$ as a weight factor. The last step involves stationarity which follows if H does not depend explicitly on time. The physics of time reversal requires two assumptions

$$\overline{H} = H; \quad \overline{\sigma} = \sigma. \tag{1.12}$$

The assumption that the Hamiltonian is even under time reversal is easy to understand and verify in any particular application. If σ refers to some irreducible space, the statement that $\overline{\sigma} = \sigma$ is the statement that time reversal induces no additional degeneracy. This is the problem solved by Wigner (1932, 1959). In statistical mechanical applications, we wish to sum over all accessible states. If the space σ is made complete, it will contain all physical states $K\Psi_n$ as well as Ψ_n so that $\overline{\sigma} = \sigma$. Thus, whenever $\overline{H} = H$, we might expect that

$$\langle a_i(t) a_i(0)\rangle = \varepsilon_i \varepsilon_j \langle a_j(t) a_i(0)\rangle. \tag{1.13}$$

This seemingly trivial statement conceals an assumption, however, which we can explain using a ferromagnet as an example. In a ferromagnet, the Heisenberg spin Hamiltonian is necessarily time reversible. Nevertheless, even in an anisotropic crystal, the ground state Ψ_G will be degenerate: it will be a state with all the spins lined up, but it will be orthogonal to the state $K\Psi_G$ in which all spins are reversed. Although complete thermal equilibrium applies equal weight to the states Ψ_G and $K\Psi_G$ since they have equal energy, in fact there is a large energy barrier between these states. Thermal equilibrium can only be established if one waits a time long enough for a permanent magnet to demagnetize. For shorter times, it is appropriate to divide the complete set of staes into two subspaces σ and $\overline{\sigma} \neq \sigma$ such that equilibrium is only maintained in each subspace. For this case, in which $\overline{H} = H$, Eq. (1.11) then reduces to

$$\langle a_i(t) a_j(0)\rangle = \varepsilon_i \varepsilon_j \langle a_j(t) a_i(0)\rangle_{\overline{\sigma}} \tag{1.14}$$

with $\overline{\sigma} \neq \sigma$.

In the first application of time reversal to irreversible processes, Onsager (1931)[5] merely asserted the validity of Eq. (1.13) as a consequence of time reversibility, and then established the symmetry of transport coefficients now known as the Onsager relations.

Cognizant of all the ways in which time-reversal can be circumvented, the purpose of the present paper chapter is to discuss the consequences of time reversal for macroscopic, dissipative systems. We have not used the phrase time-reversal in irreversible systems, although that is indeed our subject, merely because that conjures up a larger paradox than in fact exists.

The microscopic picture of time reversal implies (in a classical description) that if one specifies all the dynamical variables at the time t' to take the values

$$\underset{\sim}{a}(t') = \underset{\sim}{a}' \qquad (1.15)$$

and finds that at a later time t they take the values

$$\underset{\sim}{a}(t) = \underset{\sim}{a} \qquad (1.16)$$

that by specifying a time reversed set of values $\overline{\underset{\sim}{a}}$ as an initial condition one would return to the state $\overline{\underset{\sim}{a}}'$ (the previous initial values, with reversed velocities), that is if $\underset{\sim}{a}(t') = \overline{\underset{\sim}{a}}$ then $\underset{\sim}{a}(t) = \overline{\underset{\sim}{a}}'$. In a dissipative system, or one in which the a's are not a complete set of microscopic variables, we no longer expect that a point in phase space will necessarily remain a point since diffusion takes place. The appropriate statement is a relationship

$$P(\underset{\sim}{a},t;\underset{\sim}{a}',t') = P(\overline{\underset{\sim}{a}}',t;\overline{\underset{\sim}{a}},t') \qquad (1.17)$$

between the probability of observing $\underset{\sim}{a}'$ at t', $\underset{\sim}{a}$ at t and the appropriate time-reversed probability.

Eq. (1.17) is indeed a consequence of our general time reversal arguments above. If we let

$$X(t) = \delta(\underset{\sim}{a}(t) - \underset{\sim}{b}) \qquad (1.18)$$

$$Y(t') = \delta(\underset{\sim}{a}(t') - \underset{\sim}{c}) \qquad (1.19)$$

then X and Y are neither even nor odd under time reversal, but Eq. (1.9) with the choice L = X(t) Y(t') $\rho(\beta)$ yields

$$\langle X(t)\, Y(t')\rangle = \langle \overline{Y}(t')\, \overline{X}(t)\rangle \qquad (1.20)$$

or

$$\langle\delta(\underset{\sim}{a}(t) - \underset{\sim}{b})\, \delta(\underset{\sim}{a}(t') - \underset{\sim}{c})\rangle = \langle\delta(\overline{\underset{\sim}{a}}(t') - \underset{\sim}{c})\, \delta(\overline{\underset{\sim}{a}}(t) - \underset{\sim}{b})\rangle$$

$$= \langle\delta(\underset{\sim}{a}(-t') - \overline{\underset{\sim}{c}})\, \delta(\underset{\sim}{a}(-t) - \overline{\underset{\sim}{b}})\rangle$$

$$= \langle\delta(\underset{\sim}{a}(t) - \overline{\underset{\sim}{c}})\, \delta(\underset{\sim}{a}(t') - \overline{\underset{\sim}{b}})\rangle \qquad (1.21)$$

where $\bar{a}_i(t) = \varepsilon_i a_i(-t)$ and $\bar{b}_i = \varepsilon_i b_i$ and stationarity is used in the last step. This last result is the statement

$$P(\underset{\sim}{b},t; \underset{\sim}{c},t') = P(\overline{\underset{\sim}{c}},t; \overline{\underset{\sim}{b}},t') \qquad (1.22)$$

of time reversibility we seek. Note, however, that its proof did not assume that the set of variables a constitute a complete set in a microscopic sense. Thus the conclusion is valid for macroscopic systems.

The heart of macroscopic thermodynamics and macroscopic statistical mechanics is the assumption that there exists a set of variables $\underset{\sim}{a} = [a_1, a_2, \ldots a_N]$ not complete in a microscopic sense but complete in a macroscopic sense: the future of $\underset{\sim}{a}$ is determined by the present of $\underset{\sim}{a}$ at least in a probabilistic sense. It is less restrictive to assume that the future of $\underset{\sim}{a}$ depends not only on its present values but also on its entire past history. One can, of course, in this history preserve an amount of information equivalent to the present values of all microscopic observables. The use of such a description with memory functions is therefore justified only in certain applications where it permits a drastic reduction in the number of variables needed. For simplicity, the ensuing discussion will specialize to the case of no memory, that is to the case in which the set of variables $\underset{\sim}{a}$ is sufficiently complete to permit a Markoffian description.

2. CONSEQUENCES OF TIME REVERSAL ON THE TIME EVOLUTION OPERATOR

In my first study of time-reversal and its relation to detailed balance in dissipative systems, [Lax (1966, 1968)],[6] my remarks were restricted to variables even under time reversal, and to systems describable by a Fokker-Planck equation. A generalization to include variables odd under time reversal was made by Graham and Haken (1971),[7] and by Risken (1972).[8] A proof appropriate to a master equation was given by Lax and Zwanziger (1973).[9] In this review we shall develop the consequences of time-reversibility in a manner valid for a mixture of even and odd variables, and for a time evolution operator of the differential (generalized Fokker-Planck) or integral (master equation) form.

When transition probabilities exist, i.e., when

$$\lim_{\Delta t \to 0} P(\underset{\sim}{a}, t + \Delta t \mid \underset{\sim}{a}'t)/\Delta t = w(\underset{\sim}{a},\underset{\sim}{a}') \qquad (2.1)$$

exists for $\underset{\sim}{a} \neq \underset{\sim}{a}'$, the Chapman-Kolmogoroff equation for a Markoff process can be written as a master equation

$$\partial P(\underset{\sim}{a},t)/\partial t = \int w(\underset{\sim}{a},\underset{\sim}{a}')d\underset{\sim}{a}' \, P(\underset{\sim}{a}',t) - \Gamma(\underset{\sim}{a})P(\underset{\sim}{a},t). \qquad (2.2)$$

The first term describes all transitions into state $\underset{\sim}{a}$ and with

$$\Gamma(\underset{\sim}{a}) \equiv \int w(\underset{\sim}{a}',\underset{\sim}{a})d\underset{\sim}{a}', \tag{2.3}$$

the second term describes all transitions out of state $\underset{\sim}{a}$. Eq. (2.2) can be written in the form

$$\frac{\partial P}{\partial t} = -LP = -\int \langle\underset{\sim}{a}|L|\underset{\sim}{a}'\rangle d\underset{\sim}{a}' \, P(\underset{\sim}{a}'t) \tag{2.4}$$

where

$$-\langle\underset{\sim}{a}|L|\underset{\sim}{a}'\rangle \equiv w(\underset{\sim}{a},\underset{\sim}{a}') - \Gamma(\underset{\sim}{a}) \, \delta(\underset{\sim}{a} - \underset{\sim}{a}')$$

$$= \lim_{\Delta t \to 0} \frac{P(\underset{\sim}{a},t + \Delta t|\underset{\sim}{a}',t) - \delta(\underset{\sim}{a} - \underset{\sim}{a}')}{\Delta t} \tag{2.5}$$

represents the time evolution operator in the Dirac notation customary in quantum mechanics. Alternatively, in III (5.14), we have shown that the time evolution operator can be written as a differential operator

$$-L = -L(i\partial/\partial\underset{\sim}{a},\underset{\sim}{a}) \tag{2.6}$$

where with $\underset{\sim}{y}_{op} = i \, \partial/\partial\underset{\sim}{a}$

$$-L(\underset{\sim}{y}_{op},a) \equiv \sum_{n=1}^{\infty} (i\underset{\sim}{y}_{op})^n : \underset{\sim}{D}_n(\underset{\sim}{a},t) \tag{2.7}$$

and the nth order diffusion coefficient D_n is defined by

$$n! \, \underset{\sim}{D}_n(\underset{\sim}{a}',t) = \lim_{\Delta t \to 0} \int (\underset{\sim}{a} - \underset{\sim}{a}')^n P(\underset{\sim}{a},t + \Delta t|\underset{\sim}{a}',t)d\underset{\sim}{a} \tag{2.8}$$

for $n = 1, 2, 3, \ldots$. A dyadic notation has been used to suppress the n indices in $(\underset{\sim}{y}_{op})^n$ and in $\underset{\sim}{D}_n$. The colon : implies contraction on all n indices. Note that it is consistent to truncate Eq. (2.7) at $n = 2$ but at no higher value of n except $n = \infty$ [Pawula (1967)].[10] In the latter case, it is appropriate to regard the infinite order differential operator as a particular representation of an integral operator. Even if one has a Fokker-Planck equation ($n \leq 2$), differential operators can be represented as integral operators in the manner used by Dirac to represent the momentum operator. Thus

$$i \frac{\partial}{\partial\underset{\sim}{a}} P(\underset{\sim}{a},t) = \int \langle\underset{\sim}{a}|\underset{\sim}{y}_{op}|\underset{\sim}{a}'\rangle \, d\underset{\sim}{a}' \, P(\underset{\sim}{a}',t) \tag{2.9}$$

where

$$\langle\underset{\sim}{a}|\underset{\sim}{y}_{op}|\underset{\sim}{a}'\rangle = -i\delta'(\underset{\sim}{a} - \underset{\sim}{a}'). \tag{2.10}$$

With this understanding, we can adopt the notation of Eq. (2.4) for the Fokker-Planck as well as the master equation case.

The time-reversal condition, Eq. (1.17), can be written in terms of conditional probabilities

$$P(\underset{\sim}{a}t|\underset{\sim}{a}'t')P_0(\underset{\sim}{a}') = P(\overline{\underset{\sim}{a}}'t|\overline{\underset{\sim}{a}}t')P_0(\underset{\sim}{a}) \tag{2.11}$$

where $P_0(a)$ is the stationary solution of the one time Eq. (2.4). By changing some dummy variables we obtain

$$P(\underset{\sim}{a},t + \Delta t|\underset{\sim}{a}'t')P_0(\underset{\sim}{a}') = P(\overline{\underset{\sim}{a}}',t + \Delta t|\overline{\underset{\sim}{a}}t)P_0(\overline{\underset{\sim}{a}}). \tag{2.12}$$

The initial condition

$$P(\underset{\sim}{a},t|\underset{\sim}{a}'t) = \delta(\underset{\sim}{a} - \underset{\sim}{a}') \tag{2.13}$$

leads to the requirement

$$\delta(\underset{\sim}{a} - \underset{\sim}{a}')P_0(\underset{\sim}{a}') - \delta(\overline{\underset{\sim}{a}}' - \overline{\underset{\sim}{a}})P_0(\overline{\underset{\sim}{a}})$$
$$= \delta(\underset{\sim}{a} - \underset{\sim}{a}')[P_0(\underset{\sim}{a}) - P_0(\overline{\underset{\sim}{a}})] = 0 \tag{2.14}$$

whose solution is

$$P_0(\underset{\sim}{a}) - P_0(\overline{\underset{\sim}{a}}) \equiv \text{a function which vanishes at } \underset{\sim}{a} = \underset{\sim}{a}'. \tag{2.15}$$

Since $\underset{\sim}{a}'$ is arbitrary, we have

$$P_0(\overline{\underset{\sim}{a}}) = P_0(\underset{\sim}{a}). \tag{2.16}$$

When transition probabilities exist, the term linear in Δt in Eq. (2.12) leads to the requirement

$$\langle\underset{\sim}{a}|L|\underset{\sim}{a}'\rangle P_0(\underset{\sim}{a}') = \langle\overline{\underset{\sim}{a}}'|L|\overline{\underset{\sim}{a}}\rangle P_0(\overline{\underset{\sim}{a}}). \tag{2.17}$$

The right hand side can be simplified by writing

$$\langle\overline{\underset{\sim}{a}}'|L|\overline{\underset{\sim}{a}}\rangle = \int \delta(\underset{\sim}{x} - \overline{\underset{\sim}{a}}')L(i\partial/\partial\underset{\sim}{x},\underset{\sim}{x})\delta(\underset{\sim}{x} - \overline{\underset{\sim}{a}})d\underset{\sim}{x} \tag{2.18}$$

changing the variable of integration to $\overline{\underset{\sim}{x}}$ and noting that $\delta(\overline{\underset{\sim}{x}} - \overline{\underset{\sim}{a}}) = \delta(\underset{\sim}{x} - \underset{\sim}{a})$ to obtain

$$\langle\overline{\underset{\sim}{a}}'|L|\overline{\underset{\sim}{a}}\rangle = \int \delta(\underset{\sim}{x} - \underset{\sim}{a}')L(i\partial/\partial\overline{\underset{\sim}{x}},\overline{\underset{\sim}{x}}) \delta(\underset{\sim}{x} - \underset{\sim}{a})d\underset{\sim}{x}$$
$$\equiv \langle\underset{\sim}{a}'| K_c L K_c^{-1} |\underset{\sim}{a}\rangle \tag{2.19}$$

where K_c is a classical time reversal operator defined by

$$K_c L(i\partial/\partial\underset{\sim}{x},\underset{\sim}{x})K_c^{-1} = L(i\partial/\partial\overline{\underset{\sim}{x}},\overline{\underset{\sim}{x}}). \tag{2.20}$$

That is, all variables x are replaced by their time reverses, but, in contrast to the quantum mechanical case, i is not replaced by

-i, i.e. K_c is linear rather than antilinear. No real contradiction
is implied because L is real so that KLK^{-1} and $K_c LK_c^{-1}$ are in fact
identical. With this understanding, we can interchange $\underset{\sim}{a}$ and $\underset{\sim}{a}'$ on
the right hand side of Eq. (2.19) by taking the Hermitian conjugate.
Eq. (2.17) then leads to the time reversal requirement

$$\langle \underset{\sim}{a}|L|\underset{\sim}{a}'\rangle \, P_0(\underset{\sim}{a}') = \langle \underset{\sim}{a}|\overline{L}|\underset{\sim}{a}'\rangle \, P_0(\underset{\sim}{a}) \tag{2.21}$$

where

$$\overline{L} = (K_c LK_c^{-1})^\dagger = (KLK^{-1})^\dagger . \tag{2.22}$$

If we define W to be an operator such that

$$W|\underset{\sim}{a}\rangle = P_0(\underset{\sim}{a})|\underset{\sim}{a}\rangle \tag{2.23}$$

i.e., such that W produces multiplication by P_0, Eq. (2.21) can be
written as

$$\langle \underset{\sim}{a}|\overline{L}|\underset{\sim}{a}'\rangle = \langle \underset{\sim}{a}|W^{-1}LW|\underset{\sim}{a}'\rangle \tag{2.24}$$

or more succinctly as

$$\overline{L} = W^{-1}LW. \tag{2.25}$$

If Eq. (2.24) is applied to the form, Eq. (2.5) of L it results in
the conditions

$$w(\overline{\underset{\sim}{a}}',\overline{\underset{\sim}{a}}) = (P_0(\underset{\sim}{a}))^{-1} \, w(\underset{\sim}{a},\underset{\sim}{a}') \, P_0(\underset{\sim}{a}') \tag{2.26}$$

$$\Gamma(\overline{\underset{\sim}{a}}) = \Gamma(\underset{\sim}{a}). \tag{2.27}$$

In the special case of all variables even, $\overline{\underset{\sim}{a}} = \underset{\sim}{a}$, $\overline{\underset{\sim}{a}}' = \underset{\sim}{a}'$, Eq. (2.26)
constitutes detailed balance but we do not restrict ourself to this
case. Even if the limit, Eq. (2.5) does not exist, e.g., in the
Fokker-Planck case, Eq. (2.4) possesses the solution

$$P(\underset{\sim}{a}t|\underset{\sim}{a}_0 t_0) = \int \langle \underset{\sim}{a}| \, \exp[-L(t - t_0)]|\underset{\sim}{a}'\rangle d\underset{\sim}{a}' \, \delta(\underset{\sim}{a}' - \underset{\sim}{a}_0)$$

$$= \langle \underset{\sim}{a}| \, \exp[-L(t - t_0)]|\underset{\sim}{a}_0\rangle \tag{2.28}$$

appropriate to the initial condition, Eq. (2.13). After making
the replacements $\underset{\sim}{a} \to \overline{\underset{\sim}{a}}_0$, $\underset{\sim}{a}_0 \to \overline{\underset{\sim}{a}}$ Eq. (2.28) can be rewritten as

$$P(\overline{\underset{\sim}{a}}_0 t \mid \overline{\underset{\sim}{a}} t_0) = \langle \overline{\underset{\sim}{a}}_0| \, \exp[-L(t - t_0)]|\overline{\underset{\sim}{a}}\rangle$$

$$= \langle \underset{\sim}{a}| \, \exp[-\overline{L}(t - t_0)]|\underset{\sim}{a}_0\rangle \tag{2.29}$$

with the help of Eq. (1.8). On the other hand, by the time reversal
condition, Eq. (2.11)

$$P(\overline{a}_0 t \,|\, \overline{a}t_0) = P_0(\overline{a})^{-1} \, P(at\,|\,a_0 t_0) \, P_0(a_0)$$

$$= P_0(a)^{-1} \, \langle a| \, \exp[-L(t - t_0)]\,|a_0\rangle \, P_0(a_0)$$

$$= \langle a| \, W^{-1} \, \exp[-L(t - t_0)] \, W\,|a_0\rangle . \qquad (2.30)$$

Comparison with Eq. (2.29) then yields

$$\exp(-\overline{L}t) = W^{-1} \, \exp(-Lt) \, W = \exp[-W^{-1}LWt] \qquad (2.31)$$

which confirms Eq. (2.25) even when differential transition probabilities do not exist.

A time symmetric form of the decay operator may be obtained by introducing

$$M \equiv (1/W)^{\frac{1}{2}} \, L \, W^{\frac{1}{2}} \qquad (2.32)$$

so that Eq. (2.25) reduces to

$$\overline{M} = M. \qquad (2.33)$$

If all the variables a are even under time reversal, then $\overline{M} = M^{\dagger}$ and, in this important special case

$$M^{\dagger} = M \quad \text{(all } a \text{ even)} \qquad (2.34)$$

so that the time decay operator L has been converted to a Hermitian operator M.

3.   EIGENFUNCTION EXPANSION OF THE CONDITIONAL PROBABILITY

The set of eigenfunctions of L

$$L \, P_n(a) = \Lambda_n \, P_n(a) \qquad (3.1)$$

and the set of eigenfunctions of $L^{\dagger}$

$$L^{\dagger} \, \Phi_n(a) = \Lambda_n^{\dagger} \, \Phi_n(a) \qquad (3.2)$$

are known to have conjugate eigenvalues and to form a biorthogonal set, [Morse and Feshbach (1953)].[11]

$$\int \Phi_n(a)^* \, P_n(a)\,da = \delta_{mn}. \qquad (3.3)$$

As a consequence, the expansion of a delta function can be written as

$$\delta(a - a') = \sum_n P_n(a)\Phi_n(a')^*. \qquad (3.4)$$

The conditional probability can therefore be written

$$P(\underset{\sim}{a}t|\underset{\sim}{a}'t') = \exp[-L(t - t')]\,\delta(\underset{\sim}{a} - \underset{\sim}{a}')$$

$$= \sum_n \exp[-\Lambda_n(t - t')]P_n(\underset{\sim}{a})\Phi_n(\underset{\sim}{a}')^* \qquad (3.5)$$

where we have used that L acts on a. The aim of this section will be to show that because of time reversal

$$\Phi_n(\underset{\sim}{a}')^* = P_n(\overline{\underset{\sim}{a}}')/P_0(\underset{\sim}{a}') \qquad (3.6)$$

so that

$$P(\underset{\sim}{a}t|\underset{\sim}{a}'t') = \sum_n \exp[-\Lambda_n(t - t')]P_n(\underset{\sim}{a})P_n(\overline{\underset{\sim}{a}}')/P_0(\underset{\sim}{a}') \qquad (3.7)$$

and the full two-time probability has the unusually symmetric form

$$P(\underset{\sim}{a}t;\underset{\sim}{a}'t') = \sum_n \exp[-\Lambda_n(t - t')]P_n(\underset{\sim}{a})P_n(\overline{\underset{\sim}{a}}') \qquad (3.8)$$

which automatically obeys the time reversal symmetry requirement, Eq. (1.17).

If we introduce the eigen-functions of M and $M^\dagger$

$$M\,Q_n(\underset{\sim}{a}) = \Lambda_n\,Q_n(\underset{\sim}{a}); \quad M^\dagger\hat{Q}_n(\underset{\sim}{a}) = \Lambda_n^*\,\hat{Q}_n(\underset{\sim}{a}) \qquad (3.9)$$

then the relation Eq. (2.32) leads to the relations

$$P_n(\underset{\sim}{a}) = Q_n(\underset{\sim}{a})[P_0(\underset{\sim}{a})]^{\frac{1}{2}}; \quad \Phi_n(\underset{\sim}{a}) = \hat{Q}_n(\underset{\sim}{a})/[P_0(\underset{\sim}{a})]^{\frac{1}{2}}. \qquad (3.10)$$

In the case in which all variables a are even, $M^\dagger = M$ so that

$$\hat{Q}_n(\underset{\sim}{a}) = Q_n(\underset{\sim}{a})^*; \quad \Phi_n(\underset{\sim}{a})^* = P_n(\underset{\sim}{a})/P_0(\underset{\sim}{a}) \quad \text{all } \underset{\sim}{a} \text{ even.} \qquad (3.11)$$

This is the result we had proven earlier, [Lax and Louisell (1967)],[12] with the help of detailed balance. When some variables are odd, we must establish Eq. (3.6) which in the present notation is

$$\hat{Q}_n(\underset{\sim}{a})^* = Q_n(\overline{\underset{\sim}{a}}) \qquad (3.12)$$

a result which reduces to Eq. (3.11) when $\overline{\underset{\sim}{a}} = \underset{\sim}{a}$.

To prove Eq. (3.12), we take Eq. (3.9b) and apply the anti-linear time-reversal operator K. Thus

$$K\,M^\dagger\,K^{-1}[K\hat{Q}_n(\underset{\sim}{a})] = \Lambda_n\,K\,\hat{Q}_n(\underset{\sim}{a}) \qquad (3.13)$$

or

$$\overline{M} \; \hat{Q}_n(\overline{\underset{\sim}{a}})* = \Lambda_n \; \hat{Q}_n(\overline{\underset{\sim}{a}})*. \tag{3.14}$$

But time reversal in Eq. (2.33) has established that $\overline{M} = M$. Thus $\hat{Q}_n(\overline{\underset{\sim}{a}})*$ must be an eigenfunction of M with eigenvalue Λ_n. Thus in the absence of degeneracy we must take:

$$\hat{Q}_n(\overline{\underset{\sim}{a}})* = Q_n(\underset{\sim}{a}). \tag{3.15}$$

In the degenerate case, $\hat{Q}_n(\overline{\underset{\sim}{a}})*$ need only be a linear combination of those $Q_m(\underset{\sim}{a})$ within the degeneracy set. However, if we require the original $Q_n(a)$ to obey the orthonormality condition

$$\int Q_m(\underset{\sim}{a}) \; Q_n(\overline{\underset{\sim}{a}}) \; d\underset{\sim}{a} = \delta_{mn} \tag{3.16}$$

which can always be achieved by diagonalizing a symmetric matrix [see problems 10.6.1-2 of Lax (1974)]) then the biorthogonality condition, Eq. (3.3), will require Eq. (3.15) to be obeyed. If we replace a in Eq. (3.15) by $\overline{a}$ we have established Eq. (3.12), hence also Eq. (3.6). This eigenfunction expansion, Eq. (3.8) and (3.7) of the two-time probability and the conditional probability were found to be extremely useful in calculating the correlation functions associated with the laser spectrum, and the laser intensity fluctuation spectrum, [Haken (1970),[13] Risken (1970,[14] Risken and Vollmer (1967),[15] Lax and Louisell (1967),[12] Hempstead and Lax (1967)].[16]

Note that in view of the relations Eq. (3.6), (3.12) between the eigenfunctions of $L^\dagger$ and $M^\dagger$ and the corresponding eigenfunctions of L and M, the biorthonormality condition, Eq. (3.3) takes the form

$$\int P_m(\overline{\underset{\sim}{a}}) \; P_n(\underset{\sim}{a})/P_0(\underset{\sim}{a}) d\underset{\sim}{a} = \delta_{mn} \tag{3.17}$$

or

$$\int Q_m(\overline{\underset{\sim}{a}}) \; Q_n(\underset{\sim}{a}) d\underset{\sim}{a} = \delta_{mn}. \tag{3.18}$$

4. TIME REVERSAL AND DETAILED BALANCE

The condition Eq. (2.26) on the transition probabilities

$$w(\overline{\underset{\sim}{a}}',\overline{\underset{\sim}{a}}) \; P_0(\underset{\sim}{a}) = w(\underset{\sim}{a},\underset{\sim}{a}') \; P_0(\underset{\sim}{a}') \tag{4.1}$$

reduces to the usual detailed balance condition when $\overline{a} = a$ and $\overline{a}' = a'$ for the case described by a master equation. The same generalized detailed balance requirement is therefore contained in the operation equation (2.25) and may be applied to the differential operator, Eq. (2.7) with the result

$$\sum_n D_n(\tilde{\underset{\sim}{a}}) : (\partial/\partial \tilde{\underset{\sim}{a}})^n = [1/P_0(\underset{\sim}{a})] \sum_n (-\partial/\partial \underset{\sim}{a})^n : D_n(\underset{\sim}{a}) \, P(\underset{\sim}{a}) \qquad (4.2)$$

where the colon : implies summation on the suppressed indices. Let this operator equation act on exp $(\underset{\sim}{z} \cdot \underset{\sim}{a})$ where $\underset{\sim}{z}$ is a constant vector, to obtain

$$\sum_n D_n(\tilde{\underset{\sim}{a}}) : (\underset{\sim}{z} \cdot \underset{\sim}{a})^n = [1/P_0(\underset{\sim}{a}] \sum_n (-1)^n [\partial/\partial \underset{\sim}{a} + \underset{\sim}{z}]^n : D_n(\underset{\sim}{a}) P_0(\underset{\sim}{a}). \qquad (4.3)$$

Comparing equal powers of z we obtain conditions on each diffusion tensor. If we define

$$\tilde{D}_n(\underset{\sim}{a}) \equiv \underset{\sim}{\varepsilon}^n : D_n(\underset{\sim}{a}). \qquad (4.4)$$

The condition on the drift vector is

$$\tilde{D}_i(\underset{\sim}{a}) \equiv \varepsilon_i D_i(\underset{\sim}{a}) = -D_i(\underset{\sim}{a})$$
$$+ \frac{2}{P_0(\underset{\sim}{a})} \frac{\partial}{\partial a_j} [D_{ij}(\underset{\sim}{a}) P_0(\underset{\sim}{a})]$$
$$- \frac{3}{P_0(\underset{\sim}{a})} \frac{\partial^2}{\partial a_j \partial a_k} [D_{ijk}(\underset{\sim}{a}) P_0(\underset{\sim}{a})] + \ldots \qquad (4.5)$$

where as usual ε_i is the signature (± 1) of a_i under time reversal. The second rank diffusion tensor obeys

$$\tilde{D}_{ij}(a) \equiv \varepsilon_i \varepsilon_j D_{ij}(\tilde{\underset{\sim}{a}}) = D_{ij}(\underset{\sim}{a})$$
$$- \frac{3}{P_0(\underset{\sim}{a})} \frac{\partial}{\partial a_k} [D_{ijk}(\underset{\sim}{a}) P_n(\underset{\sim}{a})]$$
$$+ \frac{6}{P_0(\underset{\sim}{a})} \frac{\partial^2}{\partial a_k \partial a_\ell} [D_{ijk\ell}(\underset{\sim}{a}) P_n(\underset{\sim}{a})] - \ldots \qquad (4.6)$$

and the third rank diffusion tensor obeys

$$\tilde{D}_{ijk}(\underset{\sim}{a}) \equiv \varepsilon_i \varepsilon_j \varepsilon_k D_{ijk}(\tilde{\underset{\sim}{a}}) = D_{ijk}(\underset{\sim}{a})$$
$$- \frac{4}{P_0(\underset{\sim}{a})} \frac{\partial}{\delta a_\ell} [D_{ijk\ell}(\underset{\sim}{a}) P_0(\underset{\sim}{a}) + \ldots \qquad (4.7)$$

etc. The numerical coefficients of the diffusion tensor of rank n on the right hand side is (aside from sign) the combinatorial coefficient $_nC_1$ in Eq. (4.5), $_nC_2$ in Eq. (4.6) and $_nC_3$ in Eq. (4.7).

In the Fokker-Planck limit, these conditions reduce to a drift vector condition and a diffusion condition. The drift vector condition is

$$D_i{}^{ir} = \frac{1}{P_0(\underset{\sim}{a})} \frac{\partial}{\partial a_j} [D_{ij}(\underset{\sim}{a})P_0(\underset{\sim}{a})]$$

$$= \frac{\partial D_{ij}}{\partial a_j} - D_{ij} \frac{\partial \phi}{\partial a_j} \tag{4.8}$$

where ϕ is the "potential" associated with P_0:

$$P_0(\underset{\sim}{a}) \equiv N \exp[-\phi(\underset{\sim}{a})] \tag{4.9}$$

with N a normalization constant and

$$D_i{}^{ir} \equiv \frac{1}{2}(D_i + \tilde{D}_i) = \frac{1}{2}[D_i(\underset{\sim}{a}) + \varepsilon_i D_i(\varepsilon\underset{\sim}{a})] \tag{4.10}$$

is the irreversible part of D_i. The diffusion condition is:

$$\tilde{D}_{ij}(\underset{\sim}{a}) \equiv \varepsilon_i \varepsilon_j D_{ij}(\underset{\sim}{\overline{a}}) = D_{ij}(\underset{\sim}{a}). \tag{4.11}$$

The drift vector Eq. (4.8) permits one to solve directly for the gradient of ϕ.

$$\frac{\partial \phi}{\partial a_i} \equiv U_i = (D^{-1})_{ik}\left(\frac{\partial D_{kj}}{\partial a_j} - D_k{}^{ir}\right) \tag{4.12}$$

and thus to express ϕ as a line integral

$$\phi = \int U_i \, da_i \tag{4.13}$$

this line integral is, of course, independent of the path only if curl $\underset{\sim}{U} = 0$, that is

$$\frac{\partial}{\partial a_j}\left[(D^{-1})_{ik}\left(\frac{\partial D_{k\ell}}{\partial a_\ell} - D_k{}^{ir}\right)\right] = \frac{\partial}{\partial a_i}\left[(D^{-1})_{jk}\left(\frac{\partial D_{k\ell}}{\partial a_\ell} - D_k{}^{ir}\right)\right]. \tag{4.14}$$

This result was obtained by Graham and Haken (1971)[7] and by Risken (1972).[8] In the case when all variables are even $D_k{}^{ir} = D_k$, and Eq. (4.14) reduces to the integrability conditions obtained earlier by Stratanovich (1963)[17] and Lax (1966).[6]

The Fokker-Planck equation can be rewritten as a conservation equation

$$\partial P(\underset{\sim}{a},t)/\partial t = -\partial J_i(\underset{\sim}{a},t)/\partial a_i \tag{4.15}$$

where the current density is defined by

$$J_i \equiv D_i P(\underset{\sim}{a},t) - \frac{\partial [D_{ij} P(\underset{\sim}{a},t)]}{\partial a_j}. \tag{4.16}$$

It is therefore appropriate to split the current density into a
reversible and an irreversible part

$$J_i = J_i^{rev} + J_i^{ir} \tag{4.17}$$

$$J_i^{rev} = D_i^{rev} P(\underset{\sim}{a},t) \tag{4.18}$$

$$J_i^{ir} = D_i^{ir} P(\underset{\sim}{a},t) - \frac{\partial}{\partial a_j} [D_{ij} P(\underset{\sim}{a},t)] \tag{4.19}$$

where

$$D_i^{rev}(\underset{\sim}{a}) = \frac{1}{2}[D_i(\underset{\sim}{a}) - \varepsilon_i D_i(\underset{\sim}{\varepsilon a})]. \tag{4.20}$$

The detailed balance condition, Eq. (4.8) is thus equivalent to

$$J_i^{ir}(\underset{\sim}{a},t) = 0 \tag{4.21}$$

whereas the reversible current obeys the weaker condition

$$\partial J_i^{rev}/\partial a_i = 0. \tag{4.22}$$

Thus in the presence of odd random variables, circulating currents are possible, and detailed balance in the traditional sense
is not valid.

5. THE ROTATING WAVE VAN DER POL OSCILLATOR

As an example of the application of time reversal, we use the
rotating wave van der Pol (RWVP) oscillator because we have shown
in VI Appendix B that it is a prototype for all well designed
oscillators in the vicinity of threshold. After an appropriate
scaling of amplitude and time, Hempstead and Lax (1967)[16] and Lax
(1968)[6] have reduced the RWVP to a Fokker-Planck description in
which the drift vectors for the variables x and y are

$$D_x = [p - (x^2 + y^2)](x - \alpha y) \tag{5.1}$$

$$D_y = [p - (x^2 + y^2)](y + \alpha x) \tag{5.2}$$

and the diffusion coefficients are

$$D_{xx} = D_{yy} = 1; \quad D_{xy} = 0. \tag{5.3}$$

Here p is a dimensionless pump parameter, positive above threshold,
negative below, and α is a (small) detuning parameter. The gradient
of the potential ϕ is given by Eq. (4.12):

$$\frac{\partial \phi}{\partial x} = U_x = -D_x^{\ ir} \ ; \quad \frac{\partial \phi}{\partial y} = U_y = -D_y^{\ ir} .$$ (5.4)

In an early treatment, Lax (1968), x and y were both treated as even variables. In that case $D^{ir} = D$ and the compatibility requirement that curl U = curl D = 0 fails. However, it is incorrect to treat y as even, as remarked by Risken (1972). The reason is that x − iy is a classical variable associated with the quantum destruction operator b [Lax and Louisell (1967)] which in turn is proportional to q + ip where q is an even position variable and p is an odd momentum variable. When y is treated as odd, we find that

$$D_x^{\ ir} = [p - (x^2 + y^2)]x$$ (5.5)

$$D_y^{\ ir} = [p - (x^2 + y^2)]y$$ (5.6)

so that

$$\partial \ D_x^{\ ir}/\partial y = \partial \ D_y^{\ ir}/\partial x$$ (5.7)

and the compatibility requirement, Eq. (4.14) is met. The potential ϕ may now be obtained immediately by integration of Eq. (5.4)

$$-\phi = \frac{1}{2} \ p\rho - \frac{1}{4} \ \rho^2 ; \quad \rho \equiv r^2 = x^2 + y^2 .$$ (5.8)

Thus $P_0(x,y)$ is a radial function only, and can be written

$$P_0(x,y)\,dxdy = \frac{d\phi}{2\pi} \ \frac{\exp(\frac{1}{2} \ pr^2 - \frac{1}{4} \ r^4)\,rdr}{\displaystyle\int_0^\infty \exp(\frac{1}{2} \ pr^2 - \frac{1}{4} \ r^4)\,rdr} .$$ (5.9)

The symmetry in ϕ of this result suggests that we transform the Fokker-Planck equation for the RWVP from the variables x, y to r, ϕ defined by

$$a \equiv x - iy = r \ \exp(-i\phi)$$ (5.10)

with the result

$$\frac{\partial P(r,\phi)}{\partial t} = \frac{\partial}{\partial r} \ [(pr - r^3 + \frac{1}{r})P] - \frac{\partial}{\partial \phi} \ [\alpha(p - r^2)P]$$

$$+ \frac{\partial^2 P}{\partial r^2} + \frac{1}{r^2} \ \frac{\partial^2 P}{\partial \phi^2} .$$ (5.11)

In these coordinates the stationary state solution of Eq. (5.11) is

$$P_0(r) = N \ r \ \exp(\frac{1}{2} \ pr^2 - \frac{1}{4} \ r^4); \quad \int_0^\infty P_0(r)\,dr = 1 .$$ (5.12)

Thus the r which appeared in Eq. (5.9) because of the volume element $r dr d\phi$ is now automatically included in P, that is, $P(r,\phi) dr d\phi$ is now the probability of finding r in dr and ϕ in $d\phi$.

The stationary state, Eq. (5.12) has a moment

$$\langle\rho\rangle = \langle r^2\rangle = p + [\exp(p^2/4) \int_{-p/2}^{\infty} \exp(-y^2)dy]^{-1}$$

$$= p + \frac{\sqrt{\pi}}{2} \exp(-p^2/4)[1 + \mathrm{erf}(p/2)]^{-1} \qquad (5.13)$$

that describes the mean number of photons in a laser and is expressible in closed form in terms of the error function. For large positive p, we see that

$$\langle\rho\rangle \approx p \qquad p \gg 0, \qquad\qquad (5.14)$$

whereas for $p < 0$, $\langle\rho\rangle$ is expressible as a continued fraction

$$\langle\rho\rangle = \frac{2}{|p|+} \frac{4}{|p|+} \frac{6}{|p|+} \frac{8}{|p|+} \cdots \qquad\qquad (5.15)$$

so that

$$\langle\rho\rangle \approx 2/|p| \qquad p \ll 0. \qquad\qquad (5.16)$$

See Figure 1 for a plot of $\langle\rho\rangle$ versus p. Higher moments may be computed directly from the distribution function or by means of the recursion formula

$$\langle\rho^n\rangle = p\langle\rho^{n-1}\rangle + 2(n-1)\langle\rho^{n-2}\rangle. \qquad\qquad (5.17)$$

In particular,

$$\langle\rho^2\rangle = p\langle\rho\rangle + 2; \qquad\qquad (5.18)$$

A plot of $\langle(\Delta\rho)^2\rangle/\langle\rho\rangle^2 = \langle\rho^2\rangle/\langle\rho\rangle^2 - 1$ against p is shown in Figure 1 together with experimental results of Gamo, Grace and Walter (1968),[18] Davidson and Mandel (1967)[19] Arecchi et al. (1967).[20]

The remaining eigenfunctions may be sought in the form

$$P_{\lambda n}(r,\phi) = R_n(r,\lambda) \frac{\exp(i\lambda\phi)}{2\pi} ; \qquad \lambda = 0, \pm 1, \pm 2, \ldots . \qquad (5.19)$$

The radial equation then reduces to [Lax (1966b)][21]

$$\frac{d^2}{dr^2} R_n + \frac{d}{dr} [(r^3 - pr - \frac{1}{r})R_n]$$

$$+ [i\alpha\lambda(r^2 - p) - \frac{\lambda^2}{r^2}]R_n = -\Lambda_{\lambda n}R_n . \qquad\qquad (5.20)$$

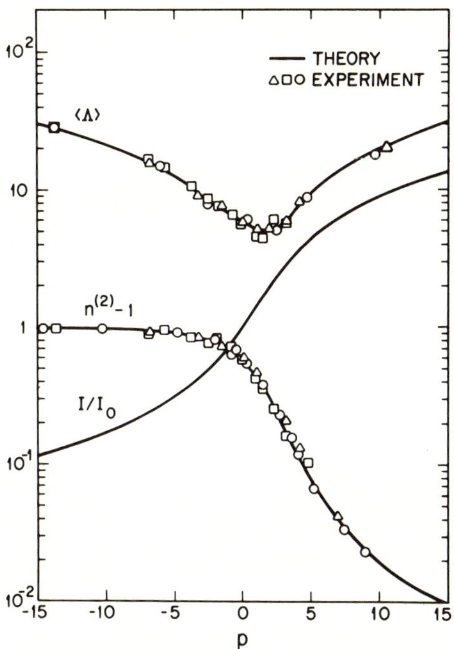

Fig. 1. Experimental verification of the Van der Pol model for
lasers operating near threshold. The theoretical curves
are from Hempstead and Lax (1967).[16] The experimental
points $\triangle$, $\square$, and $\bigcirc$ are from Gamo, Grace and Walter
(1968),[18] Davidson and Mandel (1967)[19] and Arecchi et al.
(1967)[20] respectively. Lower-case p is the pump parameter
of the model, ⟨Λ⟩ is the effective linewidth of intensity
fluctuations,

$$\langle\Lambda\rangle = \langle(\Delta\rho)^2\rangle / \int_0^\infty \langle\Delta\rho(t)\Delta\rho(0)\rangle \, dt,$$

and $n^{(2)}$ is the normalized second factorial moment
$\langle n(n-1)\rangle/\langle n\rangle^2$ of the photocount distribution $p(n,T)$ as
$T \to 0$. In this limit, $n^{(2)} - 1 = \langle(\Delta\rho)^2\rangle/\langle\rho\rangle^2$ and I/I_0
is the light intensity normalized to threshold value
$(p = 0$, i.e., $I/I_0 = \langle\rho\rangle/\langle\rho\rangle_{th} = \langle\rho\rangle/1.128$.

Note that detuning ($\alpha \neq 0$) has no effect on the $\lambda = 0$ eigenfunctions.
The two-time probability distribution, Eq. (2.7) can now be written

$$P(r\phi t; r_0\phi_0 0) =$$

$$\frac{1}{(2\pi)^2} \sum_{n,\lambda} \exp[-\Lambda_{\lambda n} t] R_n(r,\lambda) R_n(r_0,\lambda) \exp[i\lambda(\phi - \phi_0)] \qquad (5.21)$$

TABLE 1

Four Lowest Eigenvalues: Phase and Amplitude Noise

p	$\Lambda_{1,0}$	$\Lambda_{1,1}$	$\Lambda_{1,2}$	$\Lambda_{1,3}$
-10	10.376	32.858	57.133	82.994
-2	3.0840	12.889	25.219	39.593
-1	2.3316	11.008	22.166	35.381
0	1.668	9.400	19.468	31.591
1	1.1206	8.1331	17.186	28.284
2	0.7132	7.299	15.392	25.525
10	0.10212	19.237	27.687	35.064

TABLE 2

Fractional Contribution $p_{1,n}$ of nth Mode to Integrated

Spectrum for Amplitude and Phase Fluctuations

p	$p_{1,0}$	$p_{1,1}$	$p_{1,2}$	$p_{1,3}$
-10	1.0000	0.1347×10^{-3}	0.8812×10^{-6}	0.1185×10^{-9}
-2	0.9953	0.4584×10^{-2}	0.4790×10^{-4}	0.6263×10^{-6}
-1	0.9921	0.7801×10^{-2}	0.1223×10^{-3}	0.2222×10^{-5}
0	0.9867	0.1281×10^{-1}	0.3051×10^{-3}	0.7730×10^{-5}
1	0.9810	0.1923×10^{-1}	0.7066×10^{-3}	0.2518×10^{-4}
2	0.9739	0.2469×10^{-1}	0.1411×10^{-2}	0.7236×10^{-4}
10	0.9950	0.4829×10^{-2}	0.7023×10^{-6}	0.1337×10^{-3}

where we have used

$$\bar{r}_0 = r_0; \qquad \bar{\phi}_0 = -\phi_0 \tag{5.22}$$

to simplify the result.

The noise spectrum of phase and amplitude fluctuations in a self-sustained oscillator, aside from the scaling factor discussed earlier, can be defined by

$$G_p(\omega) = 4 \, \text{Re} \int_0^\infty e^{i\omega t} <a(0)^* \, a(t)> \, dt \tag{5.23}$$

where $a(t) \equiv x(t) - iy(t) \equiv r(t) \exp[-i\phi(t)]$. With the help of the two-time distribution Eq. (5.21) the spectrum is given by

$$G_p(\omega) = 4 \, \text{Re} \int_0^\infty e^{i\omega t} \, dt \int rr_0 e^{-i(\phi-\phi_0)} P(r\phi t; r_0\phi_0 0) \, dr dr_0 d\phi d\phi_0$$

$$= 4 \, \text{Re} \sum_n \frac{\left[\int_0^\infty r \, R_n(r,1) dr\right]^2}{\Lambda_{1n} - i\omega} \tag{5.24}$$

so that only the eigenfunctions and eigenvalues appropriate to $\lambda = 1$ contribute. Actually, there is also a contribution from $\lambda = -1$ that is concealed by taking the real part. The complex conjugate term involves $R_n(r,1)^* = R_n(r,-1)$. Only when $\alpha = 0$, that is, when there is no detuning, are the eigenfunctions $R_n(r,\lambda)$ and eigenvalues $\Lambda_{n\lambda}$ real.

It is understood, of course, that the eigenfunctions $R_n(r)$ are biorthogonal in the sense of Eq. (3.16). Since $\bar{r} = r$ this requires

$$\int_0^\infty R_m(r,\lambda) R_n(r,\lambda)/R_0(r,0) \, dr = \delta_{mn} \tag{5.25}$$

where $R_0(r,0) = P_0(r)$ is the steady state.

Moreover, Eq. (5.21) for $t = 0$ reduces to

$$[\delta(r - r_0) \, \delta(\phi - \phi_0)][P_0(r_0)/2\pi]$$

$$= \frac{1}{(2\pi)^2} \sum_{n,\lambda} R_n(r,\lambda) R_n(r_0,\lambda) e^{i\lambda(\phi-\phi_0)} \tag{5.26}$$

where the first factor on the left hand side is the conditional probability and the second factor is the steady state including a factor $1/2\pi$ for normalization of the angular dependence. Introduction of the Fourier series

$$\delta(\phi - \phi_0) = \frac{1}{2\pi} \sum_{\lambda=-\infty}^{\infty} e^{i\lambda(\phi-\phi_0)} \qquad (5.27)$$

and comparison of terms with the same angular dependence leads to the completeness relation

$$\sum_n R_n(r,\lambda) R_n(r_0,\lambda) = \delta(r - r_0) P_0(r_0). \qquad (5.28)$$

Multiplication of this last equation by rr_0 and integration over both these variables yields

$$\sum_n \left[\int_0^\infty r\, R_n(r,\lambda)dr \right]^2 = \int_0^\infty r^2 P_0(r)dr = \bar{\rho}. \qquad (5.29)$$

If we define

$$P_{\lambda,n} = \left[\int_0^\infty rR_n(r,\lambda)dr \right]^2 / \bar{\rho} \qquad (5.30)$$

we have

$$\sum_n P_{\lambda,n} = 1 \qquad (5.31)$$

so that the noise, Eq. (5.24) can be written as a weighted sum of Lorentzians:

$$G_p(\omega) = 4\bar{\rho} \operatorname{Re} \sum_n \frac{P_{1,n}}{\Lambda_{1n} - i\omega}. \qquad (5.32)$$

In the absence of detuning, $P_{1,n}$ is real so that

$$G_p(\omega) = 4\bar{\rho} \sum_n \frac{\Lambda_{1n}}{\Lambda_{1n}^2 + \omega^2} P_{1,n} \qquad (5.33)$$

centered at $\omega = 0$. This is so only because the rapid motion $\exp(-i\omega_0 t)$ was removed in constructing the model. Thus ω is really an abbreviation for $\omega - \omega_0$.

Numerical calculations of $P_{1,n}$ by Hempstead and Lax (1967) for the in tune case have shown that $P_{1,0}$ exceeds .97 at all operating levels. Thus the sums, Eq. (5.32), (5.33), are dominated by the first term and the spectrum is a simple Lorentzian whose line width $\Lambda_p = \Lambda_{1,0}$ is plotted versus p in Fig. 2.

The line width Λ_p is found to vary roughly as $1/\bar{\rho}$, i.e., 1/ the number of photons, over a broad range of power levels. However, $\bar{\rho}\Lambda_p$ varies smoothly from 2 below threshold to 1 above threshold. This variation eliminates an apparent discrepancy of a factor of

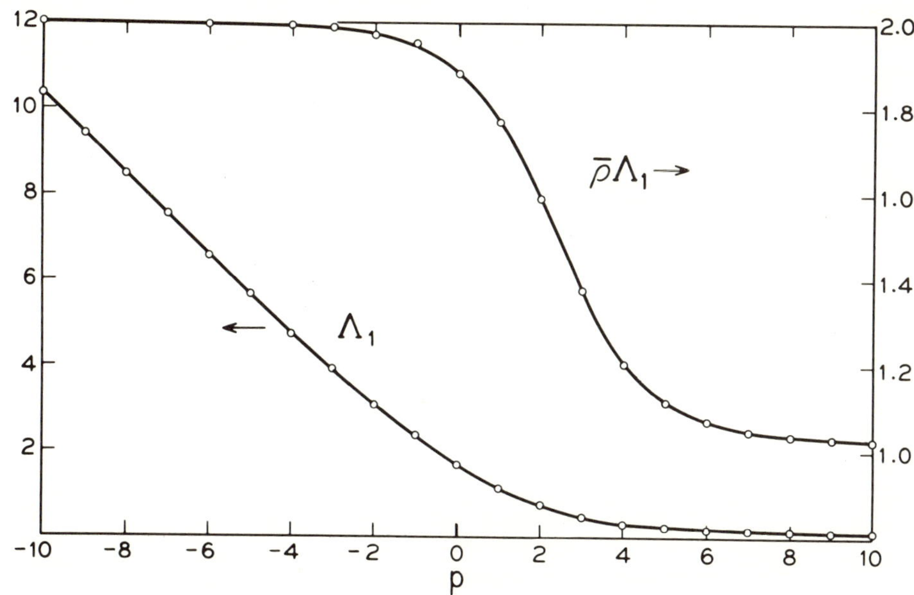

Fig. ·2.   The half width Λ^p for the combined amplitude and phase
fluctuations for a tuned homogeneously broadened laser is
plotted versus the net pump rate p. We also show $\langle\rho\rangle \Lambda_p$
which varies from 2 below threshold to 1 above threshold.

two between the mean-value approximation used by Pound (1957),[22]
Schawlow and Townes (1958),[23] Weber (1959)[24] and others, and the
quasilinear method employed by the Haken school, Haken (1966),[25]
Sauermann (1965),[26] Risken, Schmid and Weidlich (1966)[27] and by
Lax (1965b).[28]   Indeed, Lax (1965b) pointed out that the mean value
method is a good approximation well below threshold (p << 0) and
the quasilinear approximation is a good approximation well above
threshold (p >> 0).

Measurements of the laser line width were made by Gerhardt,
Welling and Güttner (1972)[29] using a Michelson interferometer technique with multiple-reflection delay paths as long as 2 km. The
observed line-widths, shown in Fig. 3, are in excellent agreement
with the calculations of Risken and Vollmer (1967) and Hempstead
and Lax (1967), described here. However, an attempt by Grossman
and Richter (1971)[30] to apply the Landau theory of phase transitions
leads to curves I and II which fit the experimental results much
worse than III, based on the theory presented here.

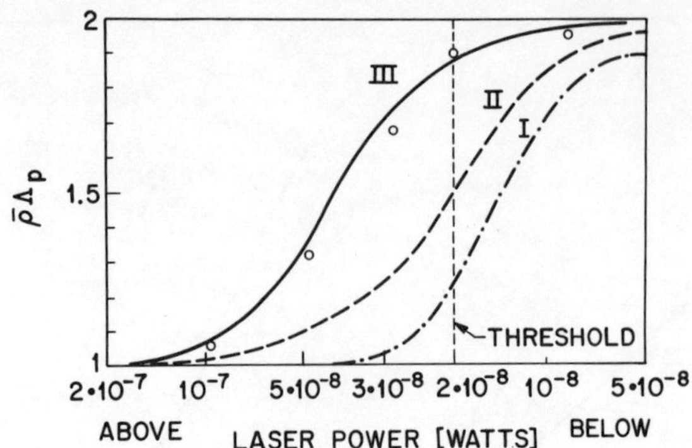

Fig. 3.   Spectrum for phase and amplitude fluctuations of a laser
is Lorentzian to a good approximation. The dimensionless
half-width Λ_p times the dimensionless mean power $\bar{p}$ are
plotted against laser power. The experimental results are
those of Gerhardt, Welling, and Güttner (Ref. 29). The
theoretical curves I and II are due to Grossman and Richter
(Ref. 30) and curve III is due to Risken and Vollmer (Ref.
15) and to Hempstead and Lax (Ref. 16).

Numerical Calculations

Actual numerical calculations were in fact performed, [Risken
and Vollmer (1967), Hempstead and Lax (1967)], with the help of the
transformation

$$R_n(r,\lambda) = [P_0(r)]^{\frac{1}{2}} Q_n(r,\lambda). \tag{5.34}$$

The Q eigen-functions obey

$$\left[\frac{d^2}{dr^2} - \frac{1}{4} r^6 + \frac{1}{2} pr^4 + (2 - \frac{1}{4} p^2)r^2 - p + \frac{1 - 4\lambda^2}{4r^2}\right.$$
$$\left. + i\alpha\lambda(r^2 - p)\right]Q_n(r,\lambda) = -\Lambda_{\lambda,n} Q_n(r,\lambda). \tag{5.35}$$

This is equivalent to a Schrodinger equation with energy eigenvalue

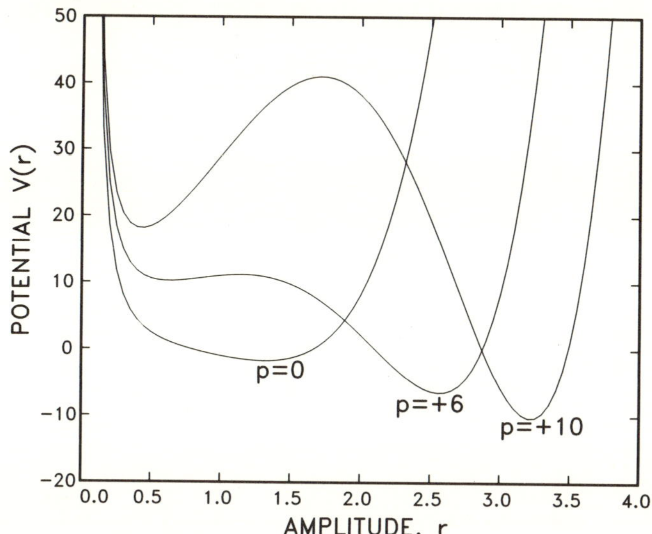

Fig. 4. The curves are the potentials for amplitude and phase
fluctuations [defined by Eq. (5.36) with $\lambda = 1$] for the
dimensionless pump parameter p equal to 0 (threshold of
oscillation), +6 and +10.

$E = \Lambda_{\lambda,n}$ and potential energy

$$V(r) = \frac{1}{4} r^6 - \frac{1}{2} pr^4 - (2 - \frac{1}{4} p^2)r^2 + p - \frac{1 - 4\lambda^2}{4r^2} - i\alpha\lambda(r^2 - p). \quad (5.36)$$

When there is no detuning, $\alpha = 0$, $V(r)$ is real and the operator M
is Hermitian. A plot of $V(r)$ for $\alpha = 0$, $\lambda = 1$ is given in Fig. 4.
For large positive p, $V(r)$ has a minimum at $r^2 \approx p$ in agreement
with Eq. (5.14).

Table 1 displays the three lowest eigenvalues for $\lambda = 1$, $\Lambda_{1,0}$,
$\Lambda_{1,1}$ and $\Lambda_{1,2}$ and Table 2 displays the coefficients $p_{\lambda,n}$ that
determine their relative importance. More complete tables are
available in VI. The dominance of the lowest eigenvalue guarantees
the Lorentzian shape of the laser line.

When $\alpha \neq 0$ complex eigenvalues result. The term $i\alpha(r^2 - p)$ in
first order produces a frequency shift and to second order in α produces an additional line broadening. Numerical calculations of
Risken and Seybold (1972)[31] shown in Fig. 5 show that the line width
is basically unchanged well below threshold ($p \ll 0$) and increases
quadratically with α well above threshold for α not too large.

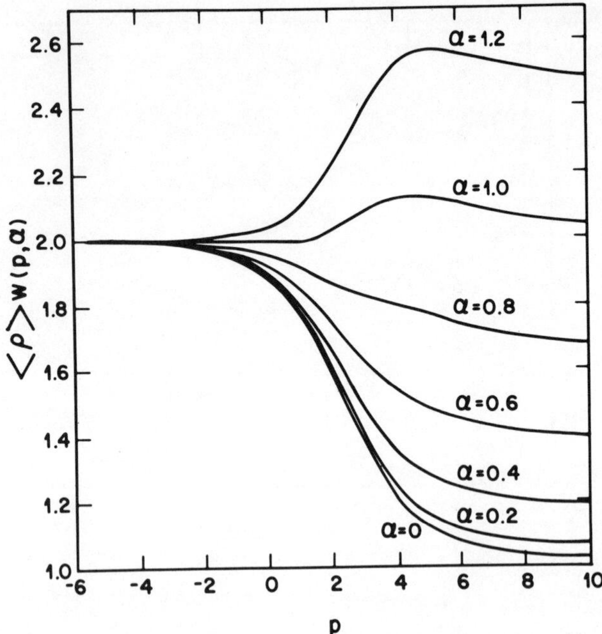

Fig. 5.   The line width factor $\langle\rho\rangle W(p,\alpha)$ as a function of the pump parameter p for various detuning parameters α.

These conclusions are in agreement with the quasilinear (QL) results V(6.6)[28] and QV(34) of Lax (1966c)[32] that well above threshold the line-width obeys

$$W_{QL}(p,\alpha) = W_{QL}(p,0)/\cos^2\beta = W_{QL}(p,0)(1 + \alpha^2) \qquad (5.37)$$

where $W = \text{Re}\,\Lambda$. Also, the mean-value approximation valid below threshold is found to obey

$$W_{MV}(p,\alpha) = W_{MV}(p,0) = 2\,W_{QL}(p,0). \qquad (5.38)$$

Indeed, Risken and Seybold (1972) claim that within 7% the simple interpolation formula

$$\langle\rho\rangle W(p,\alpha) = \langle\rho\rangle W(p,0) + [2 - \langle\rho\rangle W(p,0)]\alpha^2 \qquad (5.39)$$

fits at all operating levels. Here $W(p,0) = \Lambda_p = \Lambda_{1,0}$ is the line width in the absence of detuning. Note that as $p \to -\infty$, $\langle\rho\rangle W(p,0) \to 2$ and the second term disappears giving the result, Eq. (5.38).

REFERENCES

1. Wigner, E. P. (1932), "The Time Reversal Operation in Quantum
        Mechanics," Nachr. Ges. Wiss. Gottingen Math-Phys. Klasse
        (546-559).
    Wigner, E. P. (1959) Group Theory and Its Applications to Quantum
        Mechanics of Atomic Spectra, Academic Press.
2. Lax, M. (1962), "The Influence of Time Reversal On Selection
        Rules Connecting Different Points in the Brillouin Zone".
        Pp. 395-402 of Proceedings of International Conference on
        Physics of Semiconductors, Institute of Physics and Physical
        Society, London.
3. Lax, M. (1965), "Subgroup Techniques in Crystal and Molecular
        Physics", Phys. Rev. 138, A793-A802.
4. Lax, M. (1974), Symmetry Principles in Solid State and Molecular
        Physics, Wiley.
5. Onsager, Lars (1931), "Reciprocal Relations in Irreversible Pro-
        cesses", Phys. Rev. 37, (405-426); 38, (2265-2279).
6. Lax, M. (1966), Classical Noise III: "Nonlinear Markoff Pro-
        cesses", Revs. Modern Physics 38, 359-379 hereafter
        referred to as III.
    Lax, M. (1968), "Fluctuation and Coherence Phenomena" 1966
        Brandeis Summer Institute in Theoretical Physics, in Sta-
        tistical Physics, Phase Transitions and Superfluidity, Vol.
        2, edited by M. Chretien, E. P. Gross and S. Deser, Gordon
        and Breach, pp. 269-478.
7. Graham, R. and Haken, H. (1971), "Generalized Thermodynamic
        Potential for Markoff Systems in Detailed Balance and Far
        from Thermal Equilibrium", Z. Physik 243, 289-302.
8. Risken, H. (1972), "Solutions of the Fokker-Planck Equation in
        Detailed Balance", Z. Physik 251, 231-243.
9. Lax, M. and Zwanziger, M. (1973), "Exact Photocount Statistics:
        Lasers Near Threshold", Phys. Rev. A7, 750-771. See Appen-
        dix A.
10. Pawula, R. F. (1967), "Approximation of the Linear Boltzmann
        Equation by the Fokker-Planck Equation", Phys. Rev. 162,
        186-188.
11. Morse, P. M. and Feshbach, H. (1953), Methods of Theoretical
        Physics, McGraw-Hill.
12. Lax, M. and Louisell, W. H. (1967), "Quantum Noise IX: Quantum
        Fokker-Planck Solution For Laser Noise". IEEE J. Quant.
        Electronics QE-3, 47-58 Appendix A.
13. Haken, H. (1970), "Laser Theory" Encyclopedia of Physics vol.
        XXV/2C, Springer, Berlin.
14. Risken, H. (1970), "Statistical Properties of Laser Light" In
        Progress in Optics Vol. VIII, pp. 239-294, ed. E. Wolf,
        North Holland, Amsterdam.
15. Risken, H. and Vollmer, H. D. (1967), "Correlation Function of
        the Amplitude and of the Intensity Fluctuation Near
        Threshold", Z. Physik 191, 301-312.

16. Hempstead, R. D. and Lax, M. (1967), Classical Noise VI, Noise
         in Self-Sustained Oscillators Near Threshold", Phys. Rev.
         161, 350-366.
17. Stratanovich, R. L. (1963), "Topics in the theory of Random
         Noise", Vol. 1, pp. 78, 79, Gordon and Breach.
18. Gamo, H., Grace, R. E. and Walter, T. J. (1968), "Statistical
         Analysis of Intensity Fluctuations in Single Mode Laser
         Radiation near the Oscillation Threshold", IEEE J. Quantum
         Electron, QE-4, 344.
19. Davidson, F. and Mandel, L. (1967), "Correlation Measurements
         of Laser Beam Fluctuations Near Threshold", Phys. Lett.
         A25, 700-701.
20. Arecchi, F. T., Rodari, G. S., and Sona, A. (1967), "Statistics
         of The Laser Radiation At Threshold", Phys. Lett. A25,
         59-60.
     Arecchi, F. T., Giglio, M. and Sona, A. (1967), "Dynamics of the
         Laser Radiation At Threshold", Phys. Lett. A25, 341-342.
21. Lax, M. (1966b), Tokyo Summer Lectures in Theoretical Physics,
         Part I Dynamical Process in Solid State Optics, edited by
         R. Kubo and H. Kamimura. W. A. Benjamin Inc. Section 22.
22. Pound, R. V. (1957), "Spontaneous Emission and the Noise Figure
         of Maser Amplifiers", Ann. Phys. (N.Y.) 1, 24-32.
23. Schawlow, A. L. and Townes, C. H. (1958), "Infrared and Optical
         Masers", Phys. Rev. 112, 1940-1949.
24. Weber, J. (1959), "Masers", Revs. Modern Phys. 31, 681-710;
         erratum 32, 1033 (1960).
25. Haken, H. (1966), "Theory of Intensity and Phase Fluctuations
         of a Homogeneously Broadened Laser" Z. Physik 190, 327-356.
26. Sauermann, H. (1965), "Dissipation and Fluctuations in einem
         Zwei-Niveau-System", Z. Physik 188, 480-505.
27. Risken, H., Schmid, C. and Weidlich, W. (1966), "Fokker-Planck
         Equation for atoms and light mode in a laser model with
         quantum-mechanically determined dissipation and fluctuation
         coefficients," Phys. Letters 20, 489-491.
28. Lax, M. (1965b), Classical Noise V: "Noise in Self-Sustained
         Oscillators", Phys. Rev. 160, 290-307 referred to later
         as V.
29. Gerhardt, H., Welling, H. and Güttner, A. (1972), "Measurements
         of the laser line width due to quantum phase and quantum
         amplitude noise above and below threshold," Z. Physik 253,
         113-126.
30. Grossman, S. and Richter, P. H. (1971), "Laser Threshold and
         Nonlinear Landau Fluctuation Theory of Phase Transitions",
         Z. Physik 242, 458-475.
31. Risken, H. and Seybold, R. (1972), "Linewidth of a Detuned
         Single Mode Laser Near Threshold", Physics Letters 38A,
         63-64.
32. Lax, M. (1966c), "Quantum Noise V: Phase Noise in a Homogene-
         ously Broadened Maser," in Physics of Quantum Electronics,
         P. L. Kelley, B. Lax and P. E. Tanenwald, eds. (McGraw-Hill
         Book Co., Inc., N.Y.) pp. 735-747.

*Work at CCNY supported in part by the Army Research Office and by a grant from the City University of New York PSC-BHE Research Award Program.

LIE GROUPS, QUANTUM MECHANICS, MANY-BODY THEORY AND

ORGANIC CHEMISTRY

F. A. Matsen

Departments of Physics and Chemistry
The University of Texas
Austin, Texas   78712

1)   Introduction

Quantum mechanical perturbation theory provides a realization
of the classical Lie groups [1]:

i)   The Hamiltonian is a realization of an element in the
covering algebra of the group.

ii)   The zero-order Hamiltonian is an element in the subalgebra
spanned by the mutually commuting generators.

iii)   The perturbation Hamiltonian is a function of the "shift"
operators.

iv)   The invariant space of the Hamiltonian is a realization
of an irreducible subspace of the group space* and the quantum numbers are realizations of the label of the subspace.

v)   The zero-order states are realizations of that basis which
is an eigenvector basis to the algebra of the commuting generators.

A more explicit realization is provided by second quantized
many-body theory. The set of all linear and bilinear combinations
of the second quantized operators yield the Lie algebra of the
orthogonal group $SO(4\rho + 1)$ where ρ is the number (countable) of
relevant spin-free ordinary spin-free for electrons and isospin-free
for nucleons) one-body states. The subgroups $U(2\rho)$ and $U(\rho)$ are
the groups for particle number-conserved <u>fermions</u> and particle and
spin conserved <u>freeons</u> respectively. While traditional organic
chemistry might appear remote from Lie algebras $U(\rho)$ has been extensively applied to problems in quantum organic chemistry. We treat
butadiene as an example.

2)   Lie Groups and Quantum Mechanics

A Lie group [2] is a general mathematical structure widely used by nuclear and elementary particle physicists to express deep and subtle symmetries in nature.

Consider a Lie group

$$G = \{x,y,z \ldots \} \tag{1}$$

and its associated Lie algebra

$$LAG = \{X,Y,Z,\ldots \} \tag{2}$$

which is spanned by the infinitesimal generators of G,

$$LAG: \quad \{H_i, E_\alpha\ E_{\bar\alpha}; i,\alpha \text{ ranging}\} \tag{3}$$

with Lie products

$$[H_i, H_{i'}] = 0$$

$$[H_i, E_\alpha] = \alpha_i E_\alpha$$

$$[E_\alpha, E_\beta] = N_{\alpha,\beta} E_{\alpha,\beta}$$

$$[E_\alpha, E_{\bar\alpha}] = \sum_i \alpha\ H_i \tag{4}$$

G has a covering algebra

$$CAG: \{I, H_\alpha, E_\alpha, E_{\bar\alpha}, E_\alpha^{\,2}, \ldots \} \equiv \{W\} \tag{5}$$

The elements of G lie in CAG and are obtained from LAG by exponentiation

$$x = e^X$$

$$= \sum_n \frac{X^n}{n!} \tag{6}$$

Also in CAG are Casimir invariants $\{L_k\}$ such that

$$[I_k, X] = 0, \quad \forall I_k, X \tag{7}$$

Associated with each G is a group space

$$V_G = \sum_J \oplus\ V_G^{\,J} \tag{8}$$

where $J \equiv \{J_i\}$ identifies an invariant, irreducible space $V_G^{\ J}$ such that

$$X V_G^{\ J} = V_G^{\ J} \tag{9}$$

and where

$$V_G^{\ J}: \quad \{|J,M)\} \tag{10}$$

such that

$$I_K |J\ M) = J_k |JM)$$

and

$$H_i |JM) = M_i |JM) \tag{11}$$

where $M \equiv \{M_i\}$ is the weight of $|JM)$. Further $E_{\bar{\alpha}}$ and E_α are weight-lowering and weight raising operators respectively. From the highest weight state (denoted $|\ ))$ we obtain

$$|JM) = M(E_{\bar{\alpha}})|) \quad \text{(Generator States)}$$

where $M(\bar{\alpha})$ is a product of weight-lowering operators. The matrix elements over generator states can be evaluated algebraically.

For example

$$(JM|JM) = (|E_\alpha E_{\bar{\alpha}}|)$$

$$= (|E_{\bar{\alpha}} E_\alpha|) + \sum_i \alpha^i (|H_i|)$$

$$= \sum_i \alpha^i M_i \tag{12}$$

We now map objects of the group theory onto objects of many-body theory:

    i) $H \ \varepsilon \ CAG \rightarrow V_G^{\ J}$ invariant on $H \rightarrow J$ is a quantum number   (13)

    ii) $H^O = H^O(H_\alpha) \rightarrow |JM)$ is a zero order state where   (14)

           $E^O |JM) = H^O(M_O)|JM)$ (15)

    iii) $H' = H = H^O \rightarrow E_\alpha$ and $E_{\bar{\alpha}}$ are   (16)

perturbing generators and effect transition among the zero order states. The eigenvectors and eigenvalues are determined in the usual way:

$$\det\left[(JM|H|J'm') - \Lambda(JM|JM')\right] = 0$$

$$E(J,K), \ |JK\rangle = \sum_M |JM)\ (JM|K) \tag{17}$$

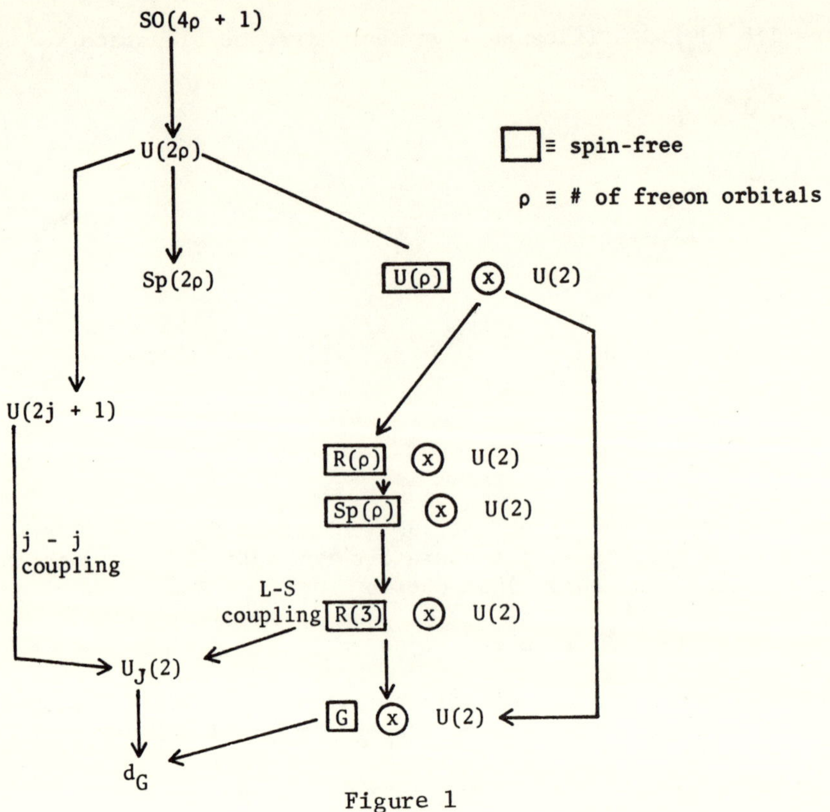

Figure 1

The unitary group is a Lie group which has been employed extensively
in spin-free quantum chemistry over the past several years. (See
Section 3)

$$\text{LAU}(\rho): \quad \{E_{rs}; \ r,s = 1 \text{ to } \rho\}$$

$$[E_{rs}, E_{tu}] = \delta(t,s)E_{ru} - \delta(r,u) E_{ts}$$

$$J = [\Lambda_1, \Lambda_2 \text{---} \Lambda] \to \text{YD}[\Lambda] \ \text{(Young diagram)}$$

$$V_G^{\ J} = V_\rho^{[\Lambda]}: \quad \dim = f_\rho^{[\Lambda]}, \ \text{Gel'fand or generator states} \tag{18}$$

Each Lie group is a member of one or more chains of subgroups
which descend from some supergroup.[3] We follow the construction of
Judd[4] and take for our supergroup the proper orthogonal (rotation)
group SO(4ρ + 1) associated with linear and bilinear combinations
of the creation and annihilation operators and where ρ is the number
of spin-free (freeon) orbitals. U(2ρ) is the unitary group for the
2ρ fermion orbitals for which fermion statistics require

$[\Lambda]_{2\rho} = [1^N]_{2\rho}$ which implies that $YD[\Lambda]_{2\rho}$ may contain only a single column. If the Hamiltonian is spin-free (ordinary spin-free for electrons or isospin free for nucleons) the relevant group is $U(\rho) \times U(2)$. Then

$$[1^N]_{2\rho} = [2^\rho, 1^{N-2b}]_\rho \times [N - p, p]_2 \quad 0 \leq p \leq N/2 \qquad (19)$$

That is $YD[\Lambda]_\rho$ may contain no more than two columns and

$$\left.\begin{array}{l} S(\text{ordinary spin for electrons}) \\[2ex] I(\text{isospin for nucleons}) \end{array}\right\} = N/2 - p \qquad (20)$$

or one-half the difference in the length of the two columns. In Figure 2 the reduction is given for $\rho = 4$.

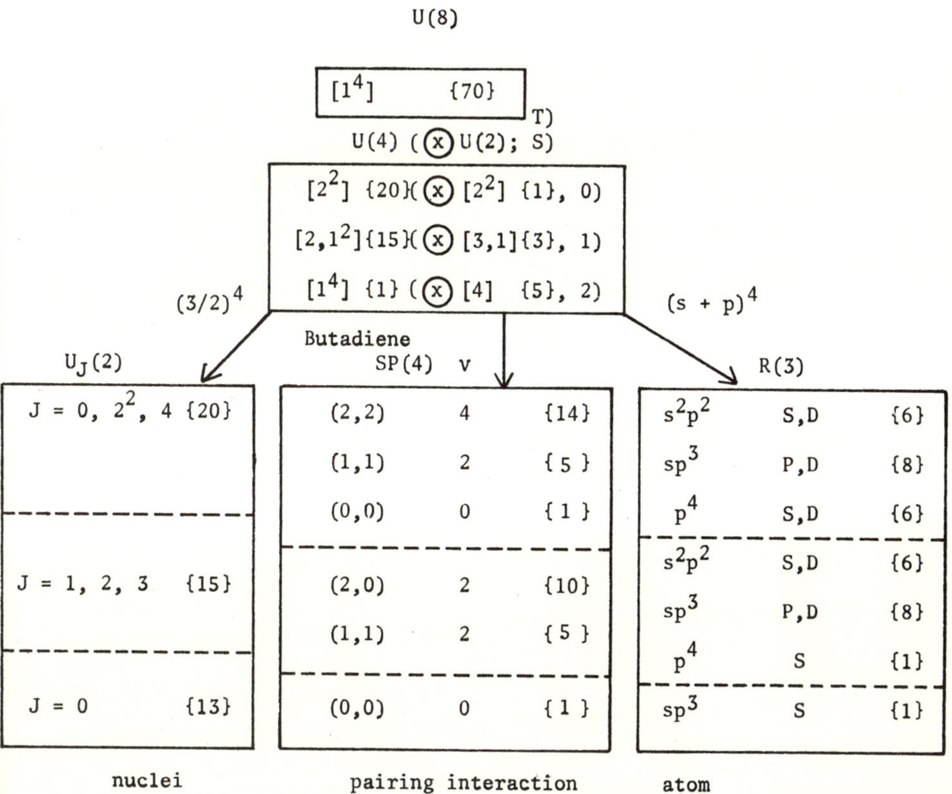

Figure 2. Decomposition of $U(\rho)$.

Since the unitary groups maps generally on many-body systems
we can expect to encounter interesting isomorphism between different kinds of matter. See Figure 3.

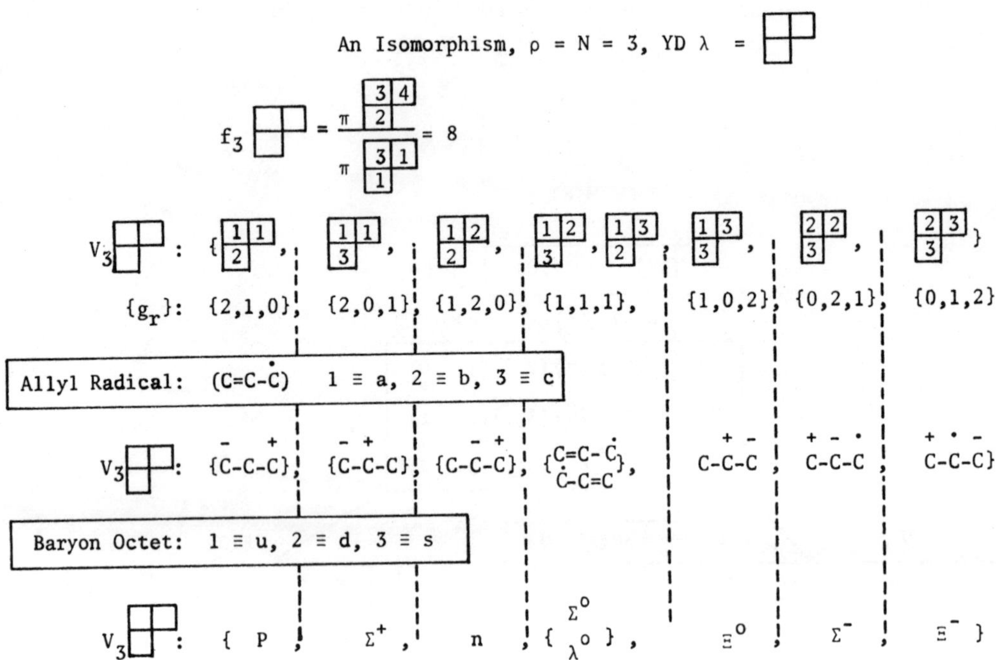

Figure 3. An isomorphism between organic chemistry and elementary
particle theory.

3)   The Unitary Group

The unitary group was introduced into many-body quantum mechanics from the very beginning by Weyl.[5] In 1935 Jordan[6] noted that the Hamiltonians of these systems could be expressed as a function of the infinitesimal generators of the unitary group. Further progress was delayed until the development of the representation theory for the classical groups in general by Cartan[7] and Racah[8] and for the unitary group by Gel'fand and Tsetlin,[9] Baird and Biedenharn[10] and Moshinsky.[11] In 1968 Moshinsky's book <u>Group Theory and the Many-Body Problem</u>[11] outlined a unitary group formulation of many-body theory and gave applications to nuclear physics. Recent contributors to the field are Harter and Patterson,[12] Paldus[13] and Shavitt.[14] Most of the above developments have in general used second quantization and the Bargman-Schwinger calculus. Matsen[15] on the other hand has employed only the infinitesimal generators of the unitary group and has imposed statistics at the end by restricting the irreducible representations of the unitary group, a procedure which is followed in the present article.

The unitary group formulation has added considerably to our understanding of many-body systems and has eased the computational burden. Conversely the application of the unitary group to many-body systems has added considerably to our understanding of the unitary group by providing both concrete realization and motivation. In the present development, we denote the spin-free unitary group by

$$U(\rho) = \{x, y, z...\} \tag{3.1}$$

with group elements obtained by exponentiation of its Lie algebra

$$LAU(\rho) = \{X, Y, Z,...\} \tag{3.2}$$

Thus

$$x \quad e^X$$

$$= \sum_n \frac{X^n}{n!} \tag{3.3}$$

Since for $U(\rho)$

$$x^\dagger = x^{-1}$$

$$X^\dagger = -X \tag{3.4}$$

The Lie algebra is spanned by the infinitesimal generators:

$$LAU(\rho): \quad \{E_{rs}; r,s = 1 \text{ to } \rho\} \tag{3.5}$$

with Lie products

$$[E_{rs}, E_{tu}] = \delta(t,s) E_{ru} - \delta(r,u) E_{ts} \tag{3.6}$$

and a general element

$$X = \sum_{rs} \sum X_{rs} E_{rs}, \quad X_{rs} \in \mathbb{C} \tag{3.7}$$

The elements of $U(\rho)$ lie in its covering algebra

$$CAU(\rho): \quad \{I, E_{rs}, E_{rs} E_{tu}, \ldots\} \equiv \{E(i,\tau)\} \tag{3.8}$$

where $E(i,\tau)$ denotes an ith degree generator product with $E(0,\tau) \equiv I$. A general element in $CAU(\rho)$ is

$$W = \sum_i \sum W(i,\tau) E(i,\tau), \quad W(i,\tau) \in \mathbb{C} \tag{3.9}$$

There exists in $CAU(\rho)$ a set of Casimir invariants $\{I_k, \ k = 1$ to $\rho\}$ homogeneous polynomials of degree k, such that

$$[I_k, W] = 0 \ \forall k, W \tag{3.10}$$

$U(\rho)$ has a group space denoted V which is the direct sum of irreducible invariant subspaces:

$$V_\rho = \sum_{[\Lambda]} \oplus \ V_\rho^{[\Lambda]} \tag{3.11}$$

where

$$[\Lambda] = [\Lambda_1, \Lambda_2, \ldots \Lambda_\rho], \quad \Lambda_1 \ \bar{>} \ \Lambda_2 \ \bar{>} \ \ldots \ \bar{>} \ \Lambda_\rho \ \bar{>} \ 0 \tag{3.12}$$

with

$$\sum_i \Lambda_i = k = 0, 1, 2, 3, \ldots \tag{3.13}$$

is a partition which is graphically realized by a Young diagram $YD[\Lambda]$, an array of k boxes with Λ_i boxes in the ith row. The dimension of $V_\rho[\Lambda]$, denoted $f_\rho[\Lambda]$ is given by a formula due to Robinson.

We denote the basis of $V_\rho[\Lambda]$ by

$$V_\rho[\Lambda]: \quad \{|\{m\}u)\} \tag{3.14}$$

where the basis elements are simultaneously eigenvectors of the mutually commuting set of diagonal generators, $\{E_{rr}\}$. Thus

$$E_{rr}|\{m\}u) = g_r|\{m\}u) \tag{3.15}$$

where

$$\{m\} \equiv \{g_r, \ r = 1 \text{ to } \rho\} \tag{3.16}$$

is called the <u>weight</u> of $|\{m\}u)$ and where $u = 1$ to $f\{m\}$ distinguishes among basis vectors with the same weight. Note that basis vectors of different weight are orthogonal since

$$(\{m\}u|E_{rr}|\{m'\}u') = g_r(\{m\}u|\{m'\}u') = g_r'(\{m\}u|\{m'\}u') \tag{3.17}$$

$|\{m\}u)$ is said to be of higher weight than $|\{m'\}u')$ if $g_1 > g_1'$, or if $g_1 = g_1'$ then $g_2 > g_2'$, etc. States with $g_i = 1$ are called <u>special</u> and their weights denoted $\{S\}$. $V_\rho[\Lambda]$ has a unique highest weight state with a weight

$$\{m*\} = [\Lambda] \tag{3.18}$$

and a unique lowest weight state with a weight denoted $\{m**\}$. Finally we note that the basis vectors of $V_\rho[\Lambda]$ are eigenvectors to the several Casimir invariants with eigenvalues:

$$I_k|\{m\}u) = I_k^{\Lambda}|\{m\}u) \tag{3.19}$$

We now proceed to the construction of these basis vectors.

The <u>generator</u> <u>basis</u> of $V_\rho^{[\Lambda]}$ is constructed from its highest weight state which we not denoted by $|$ $)$. Thus

$$V_\rho[\Lambda]: \quad \{|), |i,\tau)\} \tag{3.20}$$

where

$$| \ ) \equiv |\{m*\}) \tag{3.21}$$

and where

$$|i,\tau) \equiv E(i,\tau)|) \tag{3.21}$$

$$= E_{s's}E_{t't}\cdots 1)$$

$$\equiv \left| \begin{matrix} s't'\cdots \\ s \ t \end{matrix} \right) \tag{3.22}$$

$$\equiv |\{m\}u)$$

The weight $\{m\}$ of a generator state is computed as follows:

$$E_{rr} \Big|_{s\ t}^{s't'\cdots}\Big) = E_{rr}E_{s's}E_{t't}\cdots \Big|\ )$$

$$= E_{s's}(E_{rr} + \delta(r,s') - \delta(r,s))E_{t't}\cdots \ ) \qquad (3.23)$$

$$\text{etc.}$$

$$= g_r \Big|_{s\ t}^{s't'}\cdots\Big)$$

where

$$g_r \equiv g_r{}^* + \sum_{s'} \delta(r,s') - \sum_{s} \delta(r,s) \qquad (3.24)$$

Any set of $f_0[\Lambda] - 1$, $E(i,\tau)$ which generate a complete basis are called the <u>basis generators</u> for that basis and are subject to the following constraints:

i)   Each basis generator term-wise from the right may not raise weight of $|\ )$ and

ii) $i\tau)$ must have a weight lower than $\{m^*\}$ but no lower than $\{m^{**}\}$.

We distinguish among vectors of the same weight by weight-invariant operators (other than the Casimir invariants, I_k). An important example is

$$P_{rs} \equiv E_{rs}E_{sr} - E_{rr}$$

$$= E_{sr}E_{rs} - E_{ss}$$

$$\equiv P_{sr} \qquad (3.25)$$

Now

$$P_{rs}{}^2 = E_{rs}{}^2 E_{sr}{}^2 + E_{rr}{}^2 \qquad (3.27)$$

Consequently for a generator basis element with $g_r = g_s = 1$

$$P_{rs}{}^2 |\{m\}u) = |\{m\}u)$$

so P_{rs} is an orbital transposition. It follows that the space of special generator states

$$V\{S\}: \quad \{|\{s\}u)\ u = 1 \text{ to } f\{S\}\}$$

provide a representation of the symmetric group $S_N(N = \rho)$.

The unitary group form of the Hamiltonian is

$$H = \sum_{rs}\sum h_{rs}E_{rs} + \tfrac{1}{2}\sum_{rstu}\sum\sum\sum v_{rs,tu}(E_{rs}E_{tu} - \delta(s,t)E_{ru}) \tag{3.28}$$

where h_{rs} and v_{rstu} are matrix elements for one and two electron
operators respectively. The invariant $V_\rho[\Lambda]$ are invariant under
H so that $[\Lambda]$ is a good quantum number. The diagonalization of
H in $V_\rho[\Lambda]$ yields eigenvectors and eigenvalues.

4) Butadiene

The unitary group has been employed extensively in quantum
organic chemistry.[16] As an example we work out a theory of buta-
diene. Its classical structure is $\begin{smallmatrix}H\\ \end{smallmatrix}\!\!>\!C=C-C=C\!<\!\begin{smallmatrix}H\\H\end{smallmatrix}$ but we model it as
a four electron, four site $\oplus\,\oplus\,\oplus\,\oplus$ system with one atomic (or
$\quad\quad\quad\quad\quad\quad\quad\quad$ a b c d
Wannier) orbital attached to each site. For four electrons the

allowed spin-free Young diagrams are ⊞⊞ (single state)

⊞ (triplet state) and ▯ (quintet state). Since there are

four atomic orbitals, the group is U(4) so the invariant spaces
have dimensions given by a formula due to Robinson[16] as follows:

V_4 ⊞⊞ (dim = 20), V_4 ⊞ (dim = 15) and V_4 ▯ (dim = 1).

Each of these spaces are spanned by an orthonormal set of Gel'fand
states each of which is identified by a Gel'fand tableau constructed
by inserting the atomic orbitals into the several Young diagrams
in nondescending order along rows, in ascending order down columns
and with no orbital occurring more than once in a column. By
construction the number of electrons which can be assigned to a
given atomic orbital is 0, 1 and 2 giving rise to atomic structures
C^+, C and C^-, to which we assign energies 0, 0, and I respectively. In addition two orbitals, to each of which a single
electron is assigned, may be paired or <u>covalently</u> bonded. For
example, among the singlet states are

$$\boxed{\begin{array}{|c|c|}\hline a & a \\\hline b & b \\\hline\end{array}} \Rightarrow \overset{-}{C}-\overset{-}{C}-\overset{+}{C}-\overset{+}{C} \qquad E = 2I$$

$$\boxed{\begin{array}{|c|c|}\hline a & a \\\hline b & c \\\hline\end{array}} \Rightarrow \overset{-}{C}-C=C-\overset{+}{C} \qquad E = I$$

$$\left.\begin{array}{c}\boxed{\begin{array}{|c|c|}\hline a & b \\\hline c & d \\\hline\end{array}} \\[2em] \boxed{\begin{array}{|c|c|}\hline a & c \\\hline b & d \\\hline\end{array}}\end{array}\right\} \Rightarrow \left\{\begin{array}{c} C=C-C=C \\[2em] C-C=C-C\end{array}\right.$$

These are plotted on the right side of Figure 4.

An alternate description is in terms of Hueckel molecular orbitals which are linear combinations of the atomic orbitals,

$$|k> = \sum_{r=1}^{4} |r><r|k> \qquad k = 1, 2, 3, 4$$

with energies ε_k expressed in terms of the one-electron transfer integral T ($\equiv$ - Hueckel β). From these, construct molecular orbital Gel'fand states with energies given by $\sum_k \gamma_k \varepsilon_k$ where γ_k is the occupation number (0, 1, 2) of the molecular orbital Gel'fand state. The molecular orbital Gel'fand states and their energies are listed on the left side of Figure 4.

The molecular and atomic orbital Gel'fand states represent two extreme descriptions of butadiene. The "real" world lies in between. We can approximate the real world by means of the Hubbard Hamiltonian which in its atomic orbital unitary group form is

$$H = -T(E_{ab} + E_{ba} + E_{bc} + E_{cb} + E_{cd} + E_{dc})$$
$$+ I/2 \sum_{r=1}^{4} (E_{rr}^{2} - E_r)$$

This Hamiltonian is diagonalized in the space spanned by singlet (and triplet) atomic orbital Gel'fand states with results as shown in Figures 4 (and 5). These spectra are in strong qualitative agreement with experiment at $\frac{T}{I} = 0.5$.

The formulation outline above has been extensively used to study a large number of hydrocarbon molecules and their thermal and photochemical reactions.[16]

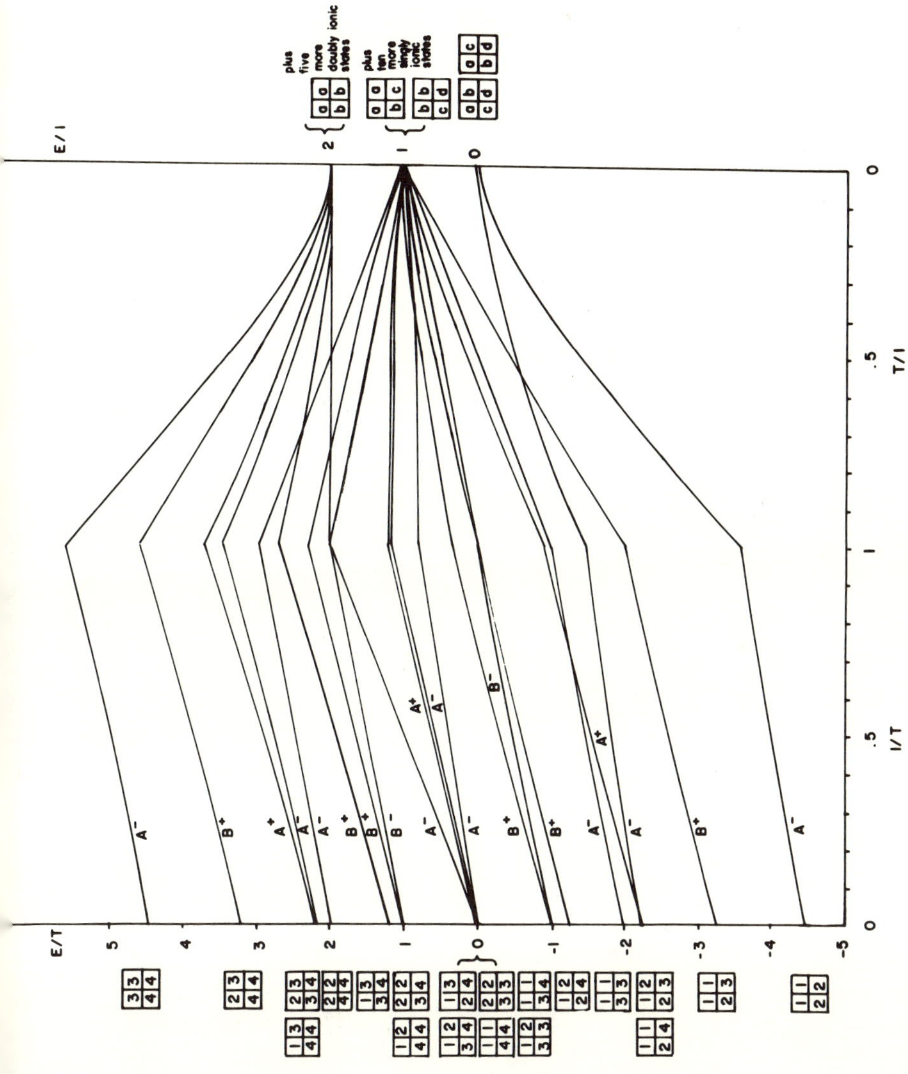

Figure 4. The singlet states of the Hubbard Hamiltonian for butadiene. The symbols {A,B} and {+,-} refer to point group and alternancy symmetry respectively.

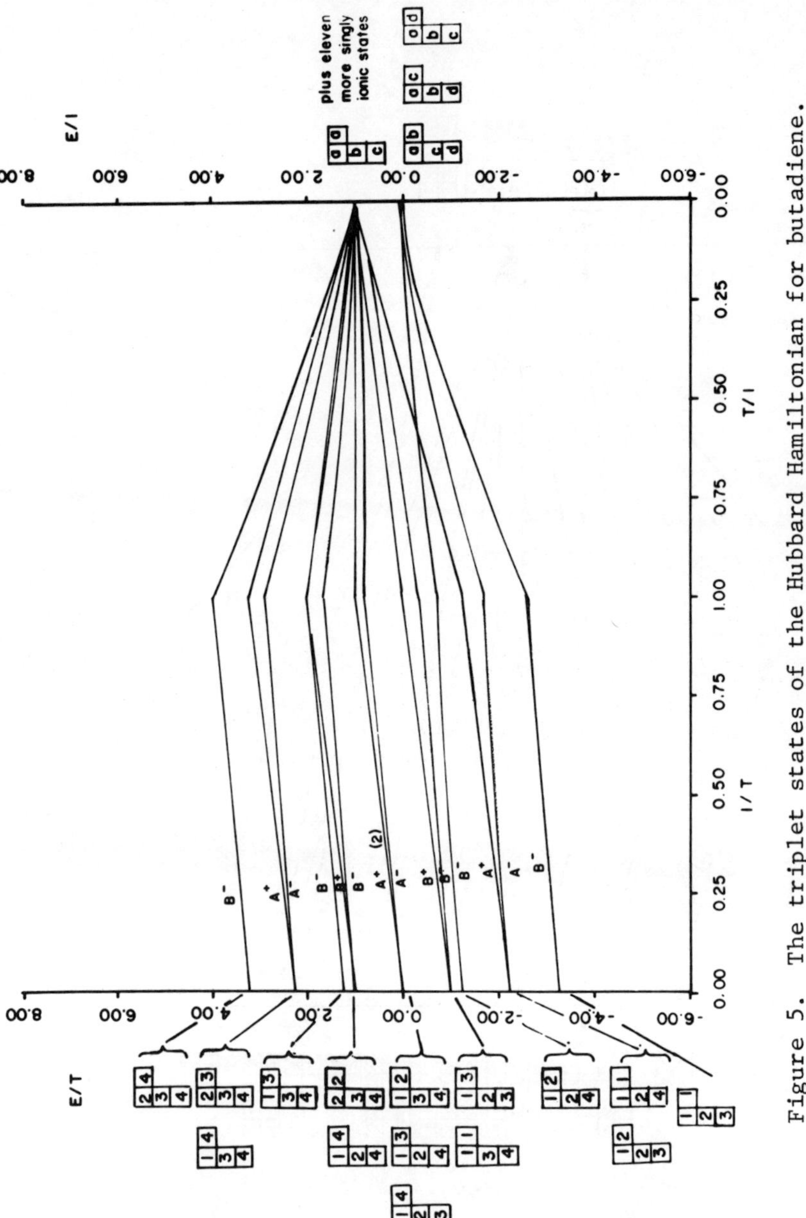

Figure 5. The triplet states of the Hubbard Hamiltonian for butadiene.

References

[1] R. A. Gilmore, Rev. Mexicana de Fisica (in press).

[2] See, for example, R. Gilmore, Lie Groups, Lie Algebras and Some of Their Applications, Wiley, N.Y. (1974).

[3] B. Gruber and M. T. Samuel, Group Theory and Applications, E. Loebl, ed., N. Y. Vol. III, 95 (1975).

[4] B. Judd, ibid. Vol. I, 183 (1968).

[5] H. Weyl, Theory of Groups and Quantum Mechanics, Dover, N.Y., (1928).

[6] P. Jordan, Z. Phys. $\underline{94}$, 531 (1935).

[7] Cartan, Sur la structure des groupes de transformation finis et continus, Thesis, Nony, Paris 1894.

[8] Racah, Group Theory and Spectroscopy, Ergeb. der exakten Naturwiss., Vol. 37, Springer, Berlin (1965).

[9] I. M. Gel'fand and M. E. Tsetlin, Dokl. Akad. Nauk SSSR, $\underline{71}$, 825, 1020 (1950).

[10] R. E. Baird and L. C. Biedenharn, O. Math. Phys., $\underline{4}$, 1449, $\underline{5}$, 1723 (1963).

[11] a) M. Moshinsky, J. Math. Phys., $\underline{4}$, 1128 (1963); $\underline{7}$, 691 (1966)
     b) Group Theory and the Many-Body Problem, Gordon and Breach, N. Y. (1968).

[12] a) W. G. Harter, Phys. Rev., $\underline{A8}$, 2819 (1973).
     b) W. G. Harter and C. W. Patterson, A Unitary Calculus for Electronic Orbitals, Springer-Verlag, Berlin, (1976).

[13] a) J. Paldus, J. Chem. Phys., $\underline{61}$, 5321 (1974).
     b) Theoretical Chemistry: Advances and Perspectives, Vol. 2, H. Eyring and D. J. Henderson, Eds. Academic Press, N. Y., (1976).
     c) Electrons in Finite and Infinite Structures, Plenum, N. Y. (1977).

[14] I. Shavitt, Intern. J. Quantum Chem. $\underline{11S}$, 131 (1977), $\underline{125}$ (in press).

[15] a) F. A. Matsen, Int. J. Quantum Chem. $\underline{S8}$, 379 (1974).
     b) F. A. Matsen, Int. J. Quantum Chem. $\underline{10}$, 511 (1976).
     c) F. A. Matsen, Int. J. Quantum Chem. $\underline{10}$, 525 (1976).
     d) F. A. Matsen, T. L. Welsher and B. Yurke, Int. J. Quantum Chem. $\underline{12}$, 985 (1977).
     e) F. A. Matsen and T. L. Welsher, Int. J. Quantum Chem. $\underline{12}$, 1001 (1977).
     f) F. A. Matsen and C. J. Nelin, Int. J. Quantum Chem. $\underline{15}$, 751 (1979).
     g) F. A. Matsen, Advances in Quantum Chem. $\underline{11}$, 223 (1978).
     h) C. J. Nelin, Int. J. Quantum Chem. $\underline{S13}$, 000 (1979).

[16] a) F. A. Matsen, Int. J. Quantum Chem. $\underline{S8}$, 379 (1974).
     b) F. A. Matsen, Int. J. Quantum Chem. $\underline{10}$, 511 (1976).
     c) E. A. Halevi, F. A. Matsen and T. L. Welsher, J. Am. Chem. Soc. $\underline{98}$, 7088 (1976).

d)   E. A. Halevi, J. Katriel, R. Pauncz, F. A. Matsen and
     T. L. Welsher, J. Am. Chem. Soc. 100, 359 (1978).
e)   T. L. Welsher, W. A. Seitz, B. Yurke, R. A. Gonzles and
     F. A. Matsen, J. Am. Chem. Soc. 99, 8389 (1977).
f)   W. A. Seitz, T. L. Welsher, B. Yurke and F. A. Matsen
     J. Am Chem. Soc. 100, 4679 (1978).
g)   F. A. Matsen, J. M. Picone and T. L. Welsher, Int. J.
     Quantum Chem. 9, 157 (1975).

*Group representation space

SYMMETRY AND VARIABLE SEPARATION FOR THE HELMHOLTZ, WAVE AND
HAMILTON-JACOBI EQUATIONS

Willard Miller, Jr.

School of Mathematics
University of Minnesota

1. INTRODUCTION

We are concerned with the relationship between the Laplace
equation

$$\Delta_4 \Psi = \Sigma_{i,j=1}^4 \; g^{-1/2} \partial_i (g^{1/2} g^{ij} \partial_j \Psi) = 0 \tag{L}$$

and the Hamilton-Jacobi equation (equation of the characteristics
of (L))

$$\Sigma_{i,j=1}^4 \; g^{ij} \partial_i W \partial_j W = 0 \; , \tag{H-J}$$

in a four dimensional Riemannian space, in particular with the interplay between the symmetries of these equations, the coordinate systems in which the equations admit solutions via separation of variables, and the properties of the separated solutions. Here $ds^2 =
\Sigma_{i,j=1}^4 \; g_{ij} dx^i dx^j$ is a complex Riemannian metric, $g = \det(g_{ij}) \neq 0$,
$\Sigma_{j=1}^4 \; g^{ij} g_{jk} = \delta_k^i$ (the Kronecker delta), $g_{ij} = g_{ji}$ and $\partial_j \Psi = \partial_{x^j} \Psi$.
(We allow the coefficients g_{ij} of the metric tensor and the coordi-
nates to be complex since all our results can be obtained for real
Laplace (or wave) equations by choosing appropriate real forms of
the complex metric tensor.) If the metric corresponds to flat space
then there exist coordinates $\{y^j\}$ such that (L) and (H-J) become

$$\Sigma_{j=1}^4 \; \partial_{y^j y^j} \Psi = 0, \quad \Sigma_{j=1}^4 (\partial_{y^j} W)^2 = 0 \; ,$$

respectively.

We begin by defining symmetry operators for (L). The operator $L = \Sigma \, \xi^j(x)\partial_j + \xi(x)$ is a (1st order) <u>symmetry operator</u> if L maps solutions of (L) into solutions: $\Delta_4(L\Psi) = 0$ whenever $\Delta_4\Psi = 0$. This is equivalent to the requirement that there exist a function $\rho(x)$ such that

$$[L,\Delta_4] \equiv L\Delta_4 - \Delta_4 L = \rho(x)\Delta_4 \; . \tag{1.1}$$

(In this paper all functions are assumed to be defined and analytic <u>locally</u>, i.e., in some coordinate patch.) The symmetry operators form a Lie algebra G with commutator bracket $[\cdot,\cdot]$. We note that if $L \in G$ then $\{\xi^j\}$ is a conformal Killing vector for the metric (g_{ij}), see [1].

A (<u>2nd order</u>) <u>symmetry operator</u>

$$S = \Sigma \, \eta^{jk}(x)\partial_{jk} + \Sigma \, \eta^\ell(x)\partial_\ell + \eta(x)$$

is a 2nd order differential operator that maps solutions of (1) into solutions. An equivalent requirement is

$$[S,\Delta_4] = K\Delta_4$$

for some 1st order differential operator K.

Example: Flat space

In this case $G \cong o(6,c) \oplus \{1\}$, where $L = 1$ is multiplication by the scalar 1 and $o(6,c)$ is the 15-dimensional conformal algebra. In cartesian coordinates $\{y^j\}$ a basis for $o(6,c)$ is:

$$P_j = \partial_j, \quad I_{k\ell} = y^k\partial_\ell - y^\ell\partial_k = -I_{\ell k}, \quad \ell \neq k$$

$$D = -(1 + \Sigma \, y^j\partial_j) \tag{1.3}$$

$$K_j = 2y^j + (2(y^j)^2 - \underset{\sim}{y}\cdot\underset{\sim}{y})\partial_j + 2y^j(\Sigma \, y^k\partial_k - y^j\partial_j) \; .$$

Except for "trivial" symmetries $f(y)\Delta_4$, all the 2nd order symmetries are 2nd order polynomials in the 1st order symmetries, a remarkable property of the flat space Laplace equation that is false for general Riemannian spaces.

Now we introduce the analogies of these definitions for (H–J). For this it is convenient to adopt a phase space formalism with coordinates (x^j, p_j), $j = 1,\ldots,4$, corresponding to coordinates $\{x^j\}$ on the original manifold. A change of coordinates $x^j = f^j(x)$ on the Riemannian manifold induces the coordinate change $\underset{\sim}{x}^j = f^j(x)$, $\underset{\sim}{p}_j = \Sigma_k \dfrac{\partial x^k}{\partial x^j} \, p_k$ in phase space. The <u>Poisson bracket</u> of two functions

$F(x,p)$, $G(x,p)$ on phase space is the function

$$\{F,G\}(x,p) = \Sigma((\partial_j G)(\partial_{p_j} F) - (\partial_{p_j} G)(\partial_j F)) \; . \qquad (1.4)$$

Let $H(x,p) = \Sigma \, g^{ij} p_i p_j$ be the <u>Hamiltonian</u>. We say that $L = \Sigma \, \xi^j(x)p_j$ is a (<u>1st order</u>) <u>symmetry</u> for (H-J) if $\{L,H\} = \rho(x)H$ for some function ρ. The 1st order symmetries form a Lie algebra G', dim $G' \leq 15$, under the Poisson bracket. (Note that $L \in G'$ if and only if $\{\xi^j\}$ is a conformal Killing vector. Furthermore, if $L = \Sigma \, \xi^j \partial_j + \xi \in G$ then $T(L) \equiv L = \Sigma \, \xi^j p_j \in G'$ and the map T is a Lie algebra homomorphism.) A (strictly) <u>2nd order symmetry</u> for (H-J) is a function $S = \Sigma \, \eta^{jk}(x)p_j p_k$ such that $\{S,H\} = \mu(x,p)H$ where μ is a linear function of p.

Example: Flat space.

Here $G' \cong o(6,c)$. In cartesian coordinates $\{y^j\}$ a basis for $o(6,c)$ is given by

$$P_j = p_j, \quad I_{k\ell} = y^k p_\ell - y^\ell p_k$$

$$D = -\Sigma \, y^j p_j \qquad\qquad (1.5)$$

$$K_j = (2(y^j)^2 - \underset{\sim}{y} \cdot \underset{\sim}{y})p_j + 2y^j(\underset{k}{\Sigma} \, y^k p_k - y^j p_j) \; .$$

Except for "trivial" symmetries $f(y)H$, all the strictly 2nd order symmetries are 2nd order polynomials in the 1st order symmetries.

2. SEPARATION OF VARIABLES FOR (H-J)

Next we discuss the meaning of (additive) variable separation for (H-J). Intuitively, one requires that there exists a coordinate system $\{y^k\}$ such that substitution of $W = \Sigma \, W^j(y^j)$ into (H-J) yields four separated ordinary differential equations for the $W^{(j)}$. Furthermore the separated equations should depend linearly on three independent parameters so that the solutions will yield a complete integral of (H-J). Rather than pursue this intuitive approach we shall adopt a very practical definition of variable separation which allows us to exhaustively classify the separable systems. Moreover, our definition turns out to be more general than the classical Stäckel definition, see [2], [3], and permits classification of non-orthogonal separable systems in addition to the usual orthogonal Stäckel systems.

Our classification is based on the number of ignorable and essential variables and starts with a listing of the separated

ordinary differential equations. A variable x^i in a separable system
is _ignorable_ if $L = p_i$ is a symmetry for (H–J); otherwise x^i is
essential. If the separated equation in the essential variable x^i
is first degree in $W^{(i)}$ then x^i is of _type 1_, if second degree then
x^i is of _type 2_.

To explain our method we treat one example in detail. We consider a separable system for (H–J) with two essential variables of
type 2 (x^1, x^2), one essential variable of type 1 (x^3), and one
ignorable variable (x^4). (This is called a type G equation.) With
$W = \Sigma_{j=1}^4 W^{(j)}(x^j)$, $W_j = \partial_j W$ we write the separated ordinary differential equations in the form

$$W_1^2 + f_1 W_4^2 + \lambda_1 a_1 + \lambda_2 b_1 \equiv \Phi_1 = 0 \qquad\qquad (2.1)$$

$$W_2^2 + f_2 W_4^2 + \lambda_1 a_2 + \lambda_2 b_2 \equiv \Phi_2 = 0$$

$$W_3 W_4 + \lambda_1 a_3 + \lambda_2 b_3 \equiv \Phi_3 = 0$$

$$W_4 = \lambda_3$$

where f_j, a_j, b_j are functions of x^j and λ_1, λ_2, λ_3 are the separation constants. Making the trivial change of variable $x^j = x^j(\tilde{x}^j)$
if necessary, we can assume without loss of generality that $a_1 =
b_2 = a_3 = 1$. To relate (H–J) with (2.1) we seek functions
$\Theta_j(x^1, \ldots, x^4)$ such that

$$\sum_{j=1}^3 \Theta_j \Phi_j \equiv \sum_{i,j=1}^4 g^{ij} W_i W_j \qquad\qquad (2.2)$$

identically in the separation constants, i.e., the coefficients of
λ_1, λ_2, λ_3 should vanish in (2.2). As is easily verified, this
condition determines the Θ_j up to an arbitrary multiple $Q(x^1, \ldots, x^4)$
and leads to the Hamilton–Jacobi equation

$$[G] \quad Q[(a_2 b_3 - 1)(W_1^2 + f_1 W_4^2) + (b_1 - b_3)(W_2^2 + f_2 W_4^2)$$

$$+ (1 - a_2 b_1) W_3 W_4] = 0. \qquad\qquad (2.3)$$

Similarly, by taking functional linear combinations of equations
(2.1) we can solve for the separation constants λ_1, λ_2, λ_3 in terms
of a_j, b_j, W_j and express the results in the form $L_i = \lambda_i$, $i = 1,
2, 3$. Here

$$L_1 = (a_2 b_1 - 1)^{-1}(p_1^2 + f_1 p_4^2 - b_1(p_2^2 + f_2 p_4^2))$$

$$L_2 = (a_2 b_1 - 1)^{-1}(p_2^2 + f_2 p_4^2 - a_2(p_1^2 + f_1 p_4^2)) \qquad\qquad (2.4)$$

$$L_3 = p_4 .$$

where $p_j = W_j$. The most general metric tensor yielding separation of this type can be read off from (2.3) and the separation is characterized by $L_i = \lambda_i$. One can check directly that the L_i are symmetries of (H-J) and that these symmetries are in involution, i.e., $\{L_i, L_k\} = 0$.

Taking all possible choices of ignorable and essential variables, we can obtain an exhaustive classification of separable coordinate systems for (H-J). In each case the separable solutions are characterized by a triplet of 1st or 2nd order symmetries L_1, L_2, L_3 which are in involution. The exact characterization is $L_i = \lambda_i$. (It is tedious but not difficult to provide a single general proof that the L_i satisfy the required commutation properties.) The (generic) Stäckel separation corresponds to the case of four essential variables of type 2, see [2], [3].

In addition to the type G separable equations above, the following Hamilton-Jacobi equations admit separation:

A. Four ignorable variables

$$[A] \quad Q \sum_{i=1}^{4} p_i^2 = 0$$

$$L_i = p_i^2, \quad i = 1,2,3$$

B. Three ignorable variables

$$[B] \quad Q \sum_{i,j=1}^{4} G^{ij}(x^4) p_i p_j = 0$$

$$L_i = p_i, \quad i = 1,2,3$$

C. Two ignorable variables with two essential variables of type 2

$$[C] \quad Q[p_1^2 + p_2^2 + (e_1+e_2)p_3^2 + 2(h_1+h_2)p_3 p_4 + (f_1+f_2)p_4^2] = 0$$

$$L_1 = p_3, \quad L_2 = p_4, \quad L_3 = p_1^2 + e_1 p_3^2 + 2h_1 p_3 p_4 + f_1 p_4^2$$

D. Two ignorable variables with one essential variable of each type

$$[D] \quad Q[p_1^2 + 2a_2 p_2 p_3 + 2b_2 p_2 p_4 + d_1 p_3^2 + 2(f_1+f_2)p_3 p_4 + e_1 p_4^2] = 0$$

$$L_1 = p_3, \quad L_2 = p_4, \quad L_3 = 2a_2 p_2 p_3 + 2b_2 p_2 p_4 + 2f_2 p_3 p_4$$

E. Two ignorable variables with two essential variables of
type 1

[E1] $Q(2a_1p_2p_3+2p_1p_4+2a_2p_2p_3+2p_2p_4+(c_1-c_2)p_3^2) = 0$

$$L_1 = p_3, \quad L_2 = p_4, \quad L_3 = 2a_2p_2p_3 + 2p_2p_4 + c_2p_3^2$$

[E2] $Q(2p_1p_4+2p_2p_3+2b_2p_2p_4+(d_1+d_2)p_3^2) = 0, \quad b_2 \neq 0$

$$L_1 = p_3, \quad L_2 = p_4, \quad L_3 = 2p_2p_3 + 2b_2p_2p_4 + c_2p_3^2$$

[E3] $Q(2p_1p_4+2p_2p_3+c_1p_3^2+d_2p_4^2) = 0$

$$L_1 = p_3, \quad L_2 = p_4, \quad L_3 = 2p_2p_3 + d_2p_4^2$$

F. One ignorable variable with three essential variables of
type 2

[F] $Q((q_2-q_3)p_1^2+(q_3-q_1)p_2^2+(q_1-q_2)p_3^2+[r_1(q_2-q_3)$

$$+ \; r_2(q_3-q_1)+r_3(q_1-q_2)]p_4^2) = 0$$

$$L_1 = p_4^2, \quad L_2 = Q[(q_3^2-q_2^2)P_1^2+(q_1^2-q_3^2)P_2^2+(q_2^2-q_1^2)P_3^2]$$

$$L_3 = Q[q_2q_3(q_2-q_3)P_1^2+q_1q_3(q_3-q_1)P_2^2+q_1q_2(q_1-q_2)P_3^2]$$

where

$$Q = [(q_1-q_2)(q_1-q_3)(q_2-q_3)]^{-1}, \quad P_i^2 = p_i^2+r_ip_4^2, \quad i=1,2,3.$$

H. No ignorable variables

[H] $Q\left(\sum_{j=1}^{4} M_{j,1}p_j^2\right) = 0$

$$L_i = \sum_{j=1}^{4} M_{j,i+1}\, p_j^2, \quad i = 1,2,3$$

where $M_{j\ell}$ is the (j,ℓ) minor of a 4×4 Stäckel matrix $(\Phi_{km}(x^k))$,
see [3].

3. R-SEPARATION OF VARIABLES FOR (L).

For (L) the appropriate concept of variable separation is
(multiplicative) R-separation. Intuitively, one requires that there

exists a coordinate system $\{y^k\}$ such that substitution of

$$\Psi = e^{R(y^k)} \, \Pi_{j=1}^{4} \, \psi^{(j)}(y^j) \tag{3.1}$$

into (L) yields four separated linear ordinary differential equations for the $\psi^{(j)}$, the equations depending linearly on three independent parameters. Here R is a fixed function which does not depend on the parameters. For $R \equiv 0$ we have <u>pure</u> separation. (Writing the separated solution (3.1) in the form $\Psi = e^R \Phi$ we see that $\Delta_4 \Psi = 0$ is equivalent to the equation

$$\Delta_4' \Phi = 0 \tag{L'}$$

where $\Delta_4' = e^{-R} \Delta_4 e^R$. Thus R-separation of (L) corresponds to pure separation of (L)'. Indeed

$$\Delta_4' = \Sigma \, b^{ij} \partial_{x^i x^j} \Phi + \Sigma b^i \partial_{x^i} \Phi + b^o \Phi = 0 \tag{3.1}$$

where

$$b^{ij} = g^{ij}, \quad b^i = \Sigma g^{ij} \partial_{x^j} \ell n[g^{1/2} g^{ij} M^2],$$
$$b^o = M^{-1}(\Delta_4 M), \quad M = e^R . \tag{3.2}$$

For the classification of R-separable systems we do not distinguish between purely separable and strictly R-separable systems for (L) because the conditions for pure separation can be obtained from those for R-separation by setting $M = 1$.)

The classification of R-separable types proceeds along the lines of the systems treated for the Hamilton-Jacobi equation. A variable x^i in a separable system is <u>ignorable</u> if for some analytic function ρ, $L = \partial_{x^i} + \rho(x)$ is a symmetry operator for (L); otherwise x^i is <u>essential</u>. If the separated equation in the essential variable x^i is first order then x^i is of <u>type 1</u>; if second order then x^i is of <u>type 2</u>. It is readily seen that for a given metric the separation of (H-J) is necessary for the R-separation of (L). Thus the only possible systems permitting R separation of (L) are those listed above. However, there are additional conditions that must be satisfied by the multiplier M in order for variables to R-separate.

To explain our method we treat one example, the analogy of the type G equation in detail. Here there are two essential variables of type 2 (x^1, x^2), one essential variable of type 1 (x^3), and one ignorable variable (x^4). With $\Psi = M \, \Pi_{j=1}^{4} \psi^{(j)}(x^j)$ we can write the separated ordinary differential equations as

$$\Psi_{11}^{(1)}+h_1\Psi_1+(f_1\lambda_3^2+\lambda_1 a_1+\lambda_2 b_1+K_1)\Psi^{(1)} \equiv \Phi_1\Psi^{(1)} = 0 \qquad (3.3)$$

$$\Psi_{22}^{(2)}+h_2\Psi_2^{(2)}+(f_2\lambda_3^2+\lambda_1 a_2+\lambda_2 b_2+K_2)\Psi^{(2)} \equiv \Phi_2\Psi^{(2)} = 0$$

$$\Psi_3^{(3)}\lambda_3+(\lambda_1 a_3+\lambda_2 b_3+K_3)\Psi^{(3)} \equiv \Phi_3\Psi^{(3)} = 0$$

$$\Psi_4^{(4)} = \lambda_3\Psi^{(4)}$$

where $\Psi_{jj}^{(j)} = \partial_{jj}\Psi^{(j)}$. To relate (L) with (3.3) one looks for functions $\Theta_j(x^1,\ldots,x^4)$ such that

$$\Phi \sum_{j=1}^{3} \Theta_j \Phi_j \equiv \sum_{i,j=1}^{4} b^{ij}\partial_{ij}\Phi + \sum_{i=1}^{4} b^i\partial_i\Phi + b^o\Phi = 0 \qquad (3.4)$$

where $\Phi = \Pi_{j=1}^{4}\Psi^{(j)}(x^j)$. Comparison of the coefficients of the
second derivative terms and the λ_i terms on both sides of (3.4)
leads to the same solutions for Θ_j and g^{ij} as found in (2.2). Comparison of the coefficients of the first derivative and constant
terms yields the R-separation conditions

$$\frac{M^2}{Q} = \left[\prod_{i=1}^{3} A_i(x^i)\right]\exp(\alpha x^4)(1-a_2 b_1)[(a_2 b_3-1)(b_1-b_3)]^{1/2}$$

$$\Delta_4 M = MQ[K_1(a_2 b_3 - 1) + K_2(b_1 - b_3) + K_3(1 - a_2 b_1)] \qquad (3.5)$$

$$\alpha \in \mathbb{C}.$$

By solving for λ_1, λ_2, λ_3 respectively, in (3.3) we can compute
partial differential operators L_i', $i = 1,2,3$, such that $L_i'\Phi = \lambda_i\Phi$.
(The highest derivative terms in these operators can be read off
from (2.4), replacing p_j with ∂_j.) The simplest of these is $L_3' = \partial_4$; the other two operators while straightforward to compute have
rather lengthy expressions which we will not bother to put down.
Finally, corresponding operators L_i for (L) such that $L_i\Psi = \lambda_i\Psi$
are given by $L_i = ML_i'M^{-1}$. The simplest of these is

$$L_3 = \partial_4 - \frac{1}{2}\frac{Q_4}{Q} - \frac{1}{2}\alpha. \qquad (3.6)$$

One can show directly that the L_i are symmetry operators of
(L) which are in involution. In fact, one can prove in general
that each R-separable system for (L) is characterized by a triplet
of commuting symmetry operators L_i in the sense that the R-separable
solutions Ψ are simultaneous eigenfunctions of the L_i with the separation constants λ_i as eigenvalues. A complete list of R-separable
systems for (L), corresponding to the list for (H-J) in section 2,
can be found in [4].

4. SEPARATION IN FLAT SPACE

For a general Riemannian space, multiplicative R-separation of
(L) in a given coordinate system implies additive separation of
(H-J) in the same coordinate system, but the converse is false.
However, in flat space these two equations separate in exactly the
same coordinate systems! (See [4] and [5] for the proof.) Moreover,
all separable systems can be characterized in terms of the symmetry
algebra $o(6,c)$. To be more specific we choose a basis $\{L_j\}$ for the
$o(6,c)$ symmetry algebra of (H-J), (1.5), and the corresponding basis
$\{L_j\}$ of the $o(6,c)$ symmetry algebra of (L), (1.3). Associated with
a given separable coordinate system $\{y^j\}$ for (H-J) there is a triplet of symmetries S_i in involution such that the separable solutions
are characterized by

$$S_i(x,p) = \lambda_i, \quad i = 1,2,3$$

$$S_i = 2\Sigma\ \alpha^{(i)}_{\ell m}\ L_\ell L_m, \quad \alpha^{(i)}_{\ell m} = \alpha^{(i)}_{m\ell}$$

(4.1)

where the $\alpha^{(i)}_{\ell m}$ are constants. The characterization of the corresponding R-separable solutions Ψ of (L) is in terms of a triplet
of commuting symmetry operators S_i. Indeed,

$$S_i\Psi = \lambda_i\Psi, \quad i = 1,\ 2,\ 3$$

$$S_i = \Sigma\ \alpha^{(i)}_{\ell m}(L_\ell L_m + L_m L_\ell).$$

(4.2)

Expressions (4.1) and (4.2) constitute a dictionary relating
separable solutions of (H-J) and (L).

Moreover, it is shown in [4] and [5] that for flat space there
are exactly 4 distinct possibilities for the operator Δ'_4, equation
(3.1): $\Delta'_4 = \hat{\Delta}_4 +$ const. where $\hat{\Delta}_4$ is the Laplace-Beltrami operator
for flat space, S^4 (the 4-dimensional complex sphere), $S^3 \times S^1$, or
$S^2 \times S^2$. The R-separable systems for the flat space Laplace equation correspond exactly to the purely separable systems (orthogonal
or not) for the Helmholtz equations $\hat{\Delta}_4\Phi = \lambda\Phi$ on each of these 4
manifolds.

We conclude with a number of explicit examples of the above
correspondence. These examples all refer to the wave equation and
the equation of its characteristics, real forms of (L) and (H-J):

$$\Psi_{tt} - \Psi_{xx} - \Psi_{yy} - \Psi_{zz} = 0 ,$$

(4.3)

$$W_t^2 - W_x^2 - W_y^2 - W_z^2 = 0 .$$

(4.4)

The real symmetry algebras of these equations are isomorphic to $o(4,2)$. Bases for the symmetry algebras can be obtained from (1.3) and (1.5) by setting $y^1 = t$, $y^2 = ix$, $y^3 = iy$, $y^4 = iz$, $i = \sqrt{-1}$ (and multiplying a basis element by i if necessary to obtain a _real_ element). For details and more examples see [6], [7], [8].

Let us consider the effect of diagonalizing the symmetry operator $P_o = \partial t$ for (4.3), i.e., let us look for solutions Ψ of (4.3) such that $P_o\Psi = i\omega\Psi$ where ω is a constant. We see immediately that $\Psi = \exp(i\omega t)\,\Phi(x,y,z)$ where Φ satisfies the reduced equation

$$(\partial_{xx} + \partial_{yy} + \partial_{zz} + \omega^2)\Phi = 0 , \qquad (4.5)$$

the Euclidean space Helmholtz equation.

For (4.4) we require, correspondingly, that $W_t \equiv P_o = i\omega$ and obtain $W = i\omega t + U(x,y,z)$ where U satisfies the reduced equation

$$p_x^2 + p_y^2 + p_z^2 + \omega^2 = 0 , \qquad (4.6)$$

the equation of geometrical optics. (Here, $p_x = U_x$, $p_y = U_y$, $p_z = U_z$.) The symmetry algebras of (4.5) and (4.6) are each isomorphic to $E(3)$, the 6-dimensional Lie algebra of the Euclidean group in 3-space. Each reduced equation separates in 11 coordinate systems, each coordinate system characterized by a commuting pair of 2nd order elements in the enveloping algebra of $E(3)$. (Note that the centralizer of P_o in $o(4,2)$ is $\{P_o\} \oplus E(3)$.) See [6] for the details. The correspondence between separable systems for (4.5) and (4.6) is, of course, perfect. Consider, for example, parabolic cylinder coordinates (ξ,η,z):

$$x = \frac{1}{2}(\xi^2 - \eta^2), \quad y = \xi\eta, \quad z = z . \qquad (4.7)$$

The separable solutions of (4.5) are characterized by

$$\frac{1}{2}(I_{xy}P_y + P_y I_{xy})\Phi = \lambda_1\Phi \qquad (4.8)$$

$$P_z\Phi = \lambda_2\Phi$$

and turn out to be products of parabolic cylinder functions in the variables ξ, η respectively, and $\exp(\lambda_2 z)$. The corresponding separable solutions of (4.6) are characterized by

$$I_{xy}P_y = \lambda_1, \quad P_z = \lambda_2 \qquad (4.9)$$

and they turn out to be a sum of elliptic functions in the variables ξ, η respectively, and $\lambda_2 z$.

If we diagonalize D in (4.3), $D\Psi = -i\nu\Psi$, then $\Psi = \rho^{i\nu-1}\Phi(s)$ where

$$t = \rho s_o, \quad x = \rho s_1, \quad y = \rho s_2, \quad z = \rho s_3 \tag{4.10}$$

$$s_o^2 - s_1^2 - s_2^2 - s_3^2 = 1 \tag{4.11}$$

and Φ satisfies the reduced equation

$$(I_{yx}^2 + I_{xz}^2 + I_{zy}^2 - I_{tx}^2 - I_{ty}^2 - I_{tz}^2)\Phi = (\nu^2 + 1)\Phi. \tag{4.12}$$

Here Φ is defined on the two-sheeted hyperboloid (4.11), the symmetry algebra of (4.12) is so(3,1), the Lie algebra of the homogeneous Lorentz group, and (4.12) is the Laplace-Beltrami eigenvalue equation on this hyperboloid. Similarly, setting $D = -i\nu$ in (4.4) we obtain the reduced equation

$$I_{yx}^2 + I_{xz}^2 + I_{zy}^2 - I_{tx}^2 - I_{ty}^2 - I_{tz}^2 = \nu^2 + 1 \ . \tag{4.13}$$

Equations (4.12) and (4.13) separate in 34 coordinate systems, each system corresponding to a pair of commuting second order elements in the enveloping algebra of so(3,1), [9].

The effect on (4.3) of diagonalizing the operator $\Gamma = \frac{1}{2}(P_t - K_t)$, $\Gamma\Psi = i\lambda\Psi$, is to yield

$$\Psi = (y_o - \cos\phi)\Phi(\underset{\sim}{y})\exp(-i\lambda\phi) \tag{4.14}$$

where

$$t = \frac{\sin\phi}{y_o - \cos\phi}, \quad x = \frac{y_1}{y_o - \cos\phi}, \quad y = \frac{y_2}{y_o - \cos\phi}, \quad z = \frac{y_3}{y_o - \cos\phi}$$

and Φ is defined on the 3-sphere

$$y_o^2 + y_1^2 + y_2^2 + y_3^2 = 1 \ . \tag{4.15}$$

The reduced equation is $\Lambda\Phi = (1 - \lambda^2)\Phi$ where Λ is the Laplace-Beltrami operator on the 3-sphere. The symmetry algebra is so(4) and the reduced equation separates in 6 coordinate systems, each corresponding to a commuting pair of second order elements in the enveloping algebra of so(4). We see from the factor $(y_o - \cos\phi)$ in (4.14) that the 6 separable systems for the reduced equation lead to 6 strictly R-separable systems for (4.3). We note that the equation $\Gamma\Psi = i\lambda\Psi$ is closely related to the quantum Kepler problem (more precisely, the pseudo-coulomb problem), see [7], [10]. Similarly the obvious Hamilton-Jacobi analogy of diagonalization for Γ leads to the classical Kepler problem.

For our final example we diagonalize $P_o + P_x$ in (4.3):
$(P_o + P_x)\Psi = i\beta\Psi$. Then

$$\Psi = \exp(is\beta)\Phi(z,y,\tau) \qquad\qquad\qquad (4.16)$$

$$2x = t + x, \quad 2\tau = x - t$$

and the reduced equation is the free-particle Schrödinger equation

$$(i\beta\partial_\tau + \partial_{yy} + \partial_{zz})\Phi = 0 . \qquad\qquad (4.17)$$

If in (4.4) we require $P_o + P_x = i\beta$ we similarly obtain the reduced equation

$$i\beta P_\tau + P_y^2 + P_z^2 = 0 , \qquad\qquad\qquad (4.18)$$

the Hamilton-Jacobi equation for a free particle. Both equations admit the 9-dimensional Schrödinger algebra as symmetry algebra and (R-)separate in 17 coordinate systems. These systems, via (4.16), lead to nonorthogonal separable systems for (4.3) and (4.4). The Schrödinger and Hamilton-Jacobi equations for the harmonic oscillator, repulsive oscillator and linear potential can also be obtained from (4.3) and (4.4) by this method.

See [6] for applications of this symmetry analysis of variable separation to obtain integral representations and expansion theorems for the separated (special function) solutions.

REFERENCES

1. Eisenhart, L. P., _Riemannian Geometry_, Princeton University
        Press, Princeton, N.J., 1949 (second printing), pp. 89-92.
2. Morse, P. and Feshbach, H., _Methods of Theoretical Physics_,
        Part I. McGraw-Hill, New York, 1953.
3. Moon, P. and Spencer, D. E., Theorems on separability in
        Riemannian n-space, Proc. Amer. Math. Soc. 3 (1952), 635-
        642.
4. Kalnins, E. G. and Miller, W. Jr., Nonorthogonal R-separable
        coordinates for four dimensional complex Riemannian spaces,
        SIAM J. Math. Anal. (to appear).
5. Kalnins, E. G. and Miller, W. Jr., R-separation of variables
        for the four-dimensional flat space Laplace and Hamilton-
        Jacobi equations, Trans. Amer. Math. Soc. 244 (1978), 241-
        261.
6. Miller, W. Jr., _Symmetry and Separation of Variables_, Addison-
        Wesley, Reading, Massachusetts, 1977.
7. Kalnins, E. G. and Miller, W. Jr., Lie theory and the wave equa-
        tion in space time. III: Semi-subgroup coordinates, J.
        Math. Phys. 18 (1977), 271-280.
8. Boyer, C. P., Kalnins, E. G., and Miller, W. Jr., Symmetry and
        separation of variables for the Hamilton-Jacobi equation
        $W_t^2 - W_x^2 - W_y^2 = 0$, J. Math. Phys. 19 (1978), 200-211.

9. Kalnins, E. G. and Miller, W. Jr., Lie theory and the wave
        equation in space-time, I: The Lorentz Group, J. Math.
        Phys. 18 (1977), 1-16.
10. Kalnins, E. G., Miller, W. Jr. and Winternitz, P., The group
        O(4), separation of variables and the hydrogen atom, SIAM
        J. Appl. Math. 30 (1976), 630-664.

REVIEW OF THE GROUP THEORY BEHIND THE INTERACTING BOSON MODEL OF

THE NUCLEUS

M. Moshinsky*

Instituto de Física, UNAM
Apdo. Postal 20-364
México 20, D.F.

1. INTRODUCTION

The present note will try to place into a wider context the
paper with a similar title by Castaños, Chacón, Frank and the present
author.[1] This paper was in turn prompted by the presentation of the
work on the interacting boson model of the nucleus by Arima in a
previous Oaxtepec Conference on nuclear physics.[2,3,4]

Why interacting boson models for nuclei composed of fermions?
Since long ago, from superconductivity[5] to nuclear matter,[6] it has
been useful to consider pairs of fermions, which thus would behave
as bosons, as fundamental structures that considerably simplify the
discussion of many body problems. In the nuclear case the fermion
states in a given configuration $n\ell j$ can be characterized by the
creation operator

$$A^+_{n\ell jm\tau} \tag{1.1}$$

where $m = -j, \ldots j$ is the projection of the total angular momentum
of the single particle and $\tau = \pm\frac{1}{2}$ is its isotopic spin. A boson
creation operator can then be defined as

$$b^+_{JM,t} =$$

$$\sum_{m_1,m_2,\tau_1,\tau_2} \langle jm_1,jm_2 | JM \rangle \langle \tfrac{1}{2}\tau_1, \tfrac{1}{2}\tau_2 | t\tau \rangle A^+_{n\ell jm_1\tau_1} A^+_{n\ell jm_2\tau_2} \tag{1.2}$$

where J, $M(t,\tau)$ are the total angular momentum (isospin) and its
projection. As the A^+ anticommute, we have from the symmetry properties of the Clebsch-Gordan coefficient,[7] that

$$b^+_{JM,t\tau} = -(-1)^{J+t}\, b^+_{JM,t\tau} \qquad\qquad (1.3)$$

and thus the boson operabor b^+ vanishes unless $J + t$ is odd. For
nucleons of the same type (i.e., protons <u>or</u> neutrons) $t = 1$ and thus
J can take only even values. For neutron-proton pairs the J is even
or odd depending on whether $t = 1$ or 0.

Considering identical nucleons we see that we can form bosons
of even angular momentum only and the $J = 0$ case has been extensively discussed in the literature.[5,6] On the other hand the first
paper of Arima and Iachello[2] deals only with $J = 2$ bosons, and one
can ask why. While they are not very explicit on this point it is
clear from their papers that they try to reproduce some of the
Bohr-Mottelson (BM) collective effects[8] and, in particular, those
appearing in vibrational nuclei. Thus they introduce a d-boson,
just as the one BM consider when dealing with the oscillator associated with the collective coordinate describing the quadrupole
deformation.[8]

Very soon though, Arima and Iachello added an s-boson to their
d, which they tried to justify from microscopic considerations providing a reasonable restriction to the large number of states available in a shell model calculation. I will take advantage of the
absence of Arima in this conference to give my own view of why they
introduced the s-boson. This view is colored by the fact that both
Arima and I are old hands in the structure game for nuclei in the
s-d shell.[9,10] In there we have 6 states in configuration space, 5
in a d and 1 in an s state. These introduce a U(6) group in the
picture which contains as a subgroup the six dimensional representation of SU(3). As Elliott[11] showed, this SU(3) provides us with
rotational bands.

Besides vibrational nuclei whose structure would already be
obtained with the help of a d boson,[2,3] Arima and Iachello wanted
rotational ones. The latter would appear naturally when dealing
with s and d bosons if we consider the SU(3) subgroup of U(6).

The s-d interacting boson model is the one that has been in
circulation for the past few years. As in the model game in general and the group theory game in particular (see for example the
extremes it has reached in the theory of elementary particles) the
justification stems mainly from its success in correlating a large
number of data. Trying to derive the model from microscopic considerations,[4] leads at best to a qualitative understanding of some
of its aspects. Thus in this note we shall only deal with the model
itself.[2,3] Furthermore, Arima gave last year an excellent presentation of some of its predictions and the correlation with experiment.
Many of those present here heard this presentation and the others
are likely to have seen the outlines of it in Physics Today[12] and
other journals.

Thus my purpose in this note will be restricted mainly to the conceptual development and, in particular, the group theory underlying it. I will not bore you though with extended calculations. Rather I will start with a very simple example, the asymmetric top, to indicate what are the essential ideas involved in calculations where we have different chains of subgroups of the main group. With this example in mind I will turn to the s-d boson model, discuss the appearance of $U(6)$ group and its different chains of subgroups[1] and, in particular, show that the most general two body interaction in this model can be expressed in terms of the Casimir operators of these chains of subgroups. The stage will then be set on how to derive group theoretically the spectra, states and transition probabilities in the model and what are the interesting extensions of it.

2. THE ASYMMETRIC TOP AS A GROUP THEORETICAL PROBLEM

The Hamiltonian of the asymmetric top can be written as

$$H = \frac{L_1^2}{2I_1} + \frac{L_2^2}{2I_2} + \frac{L_3^2}{2I_3} \qquad (2.1)$$

where L_1, L_2, L_3 are the components of the angular momentum in a system fixed in the body with its coordinates along the three principal axis and I_1, I_2, I_3 are the moments of inertia along those axis.

The commutation relations of the L_i taking $\hbar = 1$ and remembering that the L's are in a body-fixed system) is

$$[L_1, L_2] = -iL_3 \quad \text{and cyclically} \qquad (2.2)$$

and thus they are the generators of an $O(3)$ group. Clearly the Casimir operator L^2 of $O(3)$ commutes with the Hamiltonian.

If we consider the chain of groups $O(3) \supset O_3(2)$, (where the last $O_3(2)$ represents a two dimensional group of rotation around axis 3) then the states characterized by the irreducible representations L, K of this chain of groups are obviously eigenstates of the symmetric top for which $I_1 = I_2$ and where the eigenvalues are

$$E_{LK} = \frac{1}{2I_1} L(L + 1) + \frac{1}{2}\left(\frac{1}{I_3} - \frac{1}{I_1}\right)K^2 \; . \qquad (2.3)$$

If the top is symmetric around axis 1 or 2 rather than 3, the eigenvalues are also exact and have a similar form to (2.3), but we have to consider the chain of groups $O(3) \supset O_1(2)$ or $O(3) \supset O_2(2)$.

For the full asymmetric top (2.1) we have no choice but to take, for example, the states in $O(3) \supset O_3(2)$ chain i.e., the familiar[7]

$D^L_{MK}(\theta_i)$ functions of the Euler angles θ_i that are associated with the representation of the rotation group, and calculate the matrix elements of L^2_1, L^2_2 with respect to this basis, while L^2_3 is diagonal with value $K^2 \delta_{KK'} \delta_{LL'} \delta_{MM'}$. The matrix elements of L_1, L_2 with respect to the D's are a purely group theoretical problem involving the Clebsch-Gordan coefficients of a Racah tensor of order 1. From this, we could express L^2_1, L^2_2 in terms of a Racah tensor of order 2 and use the Wigner Eckart theorem, to determine them. Alternatively we note that, for example, L^2_1 is diagonal in the basis $O(3) \supset O_1(2)$ and we can pass from this to $O(3) \supset O_3(2)$ through the "transformation bracket" $D^L_{M'M''}(0, \pi/2, 0)$, thus being able to get the matrix element of L^2_1 in the basis associated with the latter chain of groups.

In this way we can obtain the finite i.e., $(2L + 1) \times (2L + 1)$ matrix associated with H of (2.1) for each given L, and through its diagonalization determine the energy levels of H and the corresponding eigenstates as combination of the $D^L_{MK}(\theta_i)$.

All of what we said here is very well known. The reason that we repeat it is that we wish to stress again the purely group theoretical character of the reasoning involved in all the steps of the calculations. Furthermore we also wish to stress the fact that the fundamental symmetry group $O(3)$ has three relevant subgroups $O_i(2)$ i = 1, 2, 3 and that, again with the help of group theory we can pass from the states of one chain to those of another, to reach finally a complete solution for the Hamiltonian (2.1).

3. GENERATORS FOR THE DIFFERENT CHAINS OF GROUPS IN THE s-d BOSON

MODEL AND THEIR CORRESPONDING CASIMIR OPERATORS

We shall start by introducing the coordinates associated with d and s states

$$\alpha_m \qquad m = 2, 1, 0, -1, -2, \qquad \alpha^m = (-1)^m \alpha_{-m}; \qquad \bar{\alpha} \qquad (3.1)$$

and the corresponding momenta

$$\pi_m = \frac{1}{i} \frac{\partial}{\partial \alpha^m}, \qquad \bar{\pi} = \frac{1}{i} \frac{\partial}{\partial \bar{\alpha}} \quad . \qquad (3.2)$$

When convenient we shall designate them also by

$$\alpha_{\ell m}, \qquad \pi_{\ell m}, \qquad \ell = 0, 2 \; . \qquad (3.3)$$

The creation and annihilation operators become then

$$\eta_{\ell m} = \frac{1}{\sqrt{2}} (\alpha_{\ell m} - i\pi_{\ell m}), \qquad \xi^{\ell m} = \frac{1}{\sqrt{2}} (\alpha^{\ell m} + i\pi^{\ell m}) \qquad (3.4)$$

which, as they are boson operators, satisfy the commutation rule

$$[\xi^{\ell'm'}, \eta_{\ell m}] = \delta_\ell^{\ell'} \delta_m^{m'} . \tag{3.5}$$

We can now write the generators of U(6) and its subgroups in terms of these creation and annihilation operators. To begin with

$$C_{\ell\ m}^{\ell'm'} \equiv \eta_{\ell m}\xi^{\ell'm'} \qquad \ell, \ \ell' = 0, \ 2 \tag{3.6}$$

are the 36 generators of U(6) as they satisfy the commutation relation

$$[C_{\ell\ m}^{\ell'm'}, C_{\ell''\ m''}^{\ell'''m'''}] = C_{\ell\ m}^{\ell'''m'''} \delta_{\ell''\ m''}^{\ell'\ m'} - C_{\ell''m''}^{\ell'm'}\delta_\ell^{\ell'''} \delta_m^{m'''} . \tag{3.7}$$

Obviously the generators of the U(5) subgroup of U(6) are given by (3.6) when we restrict ourselves to d states, i.e.,

$$C_{2m}^{2m'} = \eta_m\xi^{m'} . \tag{3.8}$$

As is well known,[9] for any number of dimensions, the generators of an orthogonal subgroup can be constructed from antisymmetric combinations of the generators of the corresponding unitary group. Thus

$$\Lambda_{\ell m, \ell'm'} \equiv \eta_{\ell m}\xi_{\ell'm'} - \eta_{\ell'm'}\xi_{\ell m} , \tag{3.9}$$

of which we have 15 independent ones, are the generators of O(6) and then of these

$$\Lambda_{2m,2m'} = \eta_m\xi_{m'} - \eta_{m'}\xi_m \tag{3.10}$$

are the generators of O(5).

In turn[9] the operators of angular momenta for d-bosons are

$$L_\sigma = \sum_{m,m'} \langle 2m'|L_\sigma|2m\rangle \eta_{m'}\xi^m =$$

$$\sqrt{\frac{5}{2}} \sum_{m,m'} \langle 2m,2m'|1\sigma\rangle (\eta_m\xi_{m'} - \eta_{m'}\xi_m) = \sqrt{10}\ [\underline{\eta} \times \underline{\xi}]_\sigma^1 \tag{3.11}$$

which shows that the L_σ $\sigma = 1,0,-1$ besides being generators of an O(3) group, form also a subalgebra of the algebra of generators (3.10) of the O(5) group.

We still require the determination of the generators of the SU(3) subgroup of U(6). The analysis developed in relation with the s-d shell picture[9] indicates that, besides the L_σ, the other generators Q_m of SU(3) are linear combinations of the $C_{\ell\ m}^{\ell'm'}$ with coefficients that are the matrix elements of $r^2Y_{2m}(\theta,\phi)$ with respect to quanta harmonic oscillator states $|2\ell m\rangle$, i.e.,

$$Q_m = -\sqrt{\frac{8}{15}} \sum_{\ell'm'}\sum_{\ell''m''} \langle 2\ell'm'|r^2Y_{2m}(\theta,\phi)|2\ell''m''\rangle \eta_{\ell'm'}\xi^{\ell''m''}$$

$$= \sqrt{\frac{7}{3}}\ [\underline{\eta} \times \underline{\xi}]_m^2 + \sqrt{\frac{4}{3}}\ (\overline{\eta}\xi_m + \eta_m\overline{\xi}) . \tag{3.12}$$

We easily check that the commutation of two Q's gives an L and the commutation of L and Q gives an O with precisely the coefficients that define a standard form[9] of the generators of SU(3).

Looking at the explicit generators of the different subgroups of U(6) we see that we can arrange them in the following chains where, underneath each group, we indicate its corresponding generators

$$U(6) \supset U(5) \supset O(5) \supset O(3) \supset O(2) \tag{3.13a}$$
$$C^{\ell'm'}_{\ell m} \quad C^{2m'}_{2m} \quad \Lambda_{2m',2m} \quad L_\sigma \quad L_0$$

$$U(6) \supset O(6) \supset O(5) \supset O(3) \supset O(2) \tag{3.13b}$$
$$C^{\ell'm'}_{\ell m} \quad \Lambda_{\ell'm',\ell m} \quad \Lambda_{2m',2m} \quad L_\sigma \quad L_0$$

$$U(6) \supset SU(3) \supset O(3) \supset O(2) \tag{3.13c}$$
$$C^{\ell'm'}_{\ell m} \quad (Q_m,L_\sigma) \quad L_\sigma \quad L_0$$

We now proceed to determine the Casimir operators for the groups in the different chains that commute with all their corresponding generators. To begin with the first order ones for U(6) and U(5) are

$$\hat{N} = \sum_{\ell m} \eta_{\ell m} \xi^{\ell m}, \quad \ell = 0,2, \quad \text{eigenvalue } N \tag{3.14}$$

$$\hat{n} = \sum_{m} \eta_m \xi^m, \quad \text{eigenvalue } \nu \tag{3.15}$$

and we only need them as we are dealing with symmetric i.e., single row representation of these groups. We indicate also the notation that we are going to use for the eigenvalues of the Casimir operators, where all symbols appearing in these eigenvalues represent integers.

The Casimir operator of the O(5) group, as discussed in reference 13, is given by

$$\Lambda^2 \equiv \frac{1}{2} \sum_{m,m'} \Lambda_{2m,2m'} \Lambda^{2m',2m}$$

$$= \hat{n}(\hat{n} + 3) - \left(\sum_m \eta_m \eta^m\right)\left(\sum_{m'} \xi_{m'}\xi^{m'}\right); \quad \text{eigenvalue } \lambda(\lambda + 3), \tag{3.16}$$

and with its help the Casimir operator of O(6) becomes

$$L^2 \equiv \frac{1}{2} \sum_{\ell m} \sum_{\ell' m'} \Lambda_{\ell m, \ell' m'} \Lambda^{\ell' m', \ell m}$$

$$= \hat{N}(\hat{N} + 4) - (\sum_m \eta_m \eta^m + \bar{\eta}^2)(\sum_m \xi_m \xi^m + \bar{\xi}^2)$$

$$= \Lambda^2 + K^2 , \quad \text{eigenvalue } \rho(\rho + 4) \tag{3.17}$$

where

$$K^2 = \hat{n}(\hat{N} - \hat{n} + 1) + (\hat{N} - \hat{n})(\hat{n} + 5)$$

$$- \sum_m (\eta_m \eta^m)\bar{\xi}^2 - \bar{\eta}^2 \sum_m (\xi_m \xi^m) . \tag{3.18}$$

The Casimir operator of O(3) is of course

$$L^2 = \sum_\sigma (-1)^\sigma L_\sigma L_{-\sigma}, \quad \text{eigenvalue } L(L + 1) \tag{3.19}$$

and with its help as well as with that of operator

$$Q^2 \equiv \sum_{m=-2}^{2} (-1)^m Q_m Q_{-m} \tag{3.20}$$

the second order Casimir operator[1,9] G of SU(3) becomes

$$G = Q^2 + \frac{1}{2} L^2, \quad \text{eigenvalue } \frac{2}{3}(p^2 + rp + r^2) + 2(r + p) \tag{3.21}$$

where (p,r) characterizes the irreducible representation in the fashion of Elliott[11] which uses the notation (λ,μ) that we avoid as these letters have other meanings in our series of papers on this subject.[1,13,14]

Having established our chains of groups and their Casimir operators, we proceed to show that the most general interaction in the s-d boson model can be given in terms of the latter.

4. THE GENERAL TWO BODY INTERACTION IN THE d–s BOSON MODEL

All operators appearing in the Hamiltonian for the d-s boson model must be of course invariant under rotation. Thus the one body interaction can only be a linear combination of the number operators $\hat{n}$, $\hat{N} - \hat{n}$ of the d and s bosons given in (3.14-15).

For the two body operators the situation is more interesting as the most general interaction would be of the form

$$V = \frac{1}{2} \sum_{\ell_1 m_1 \ell_2 m_2 \ell_1' m_1' \ell_2' m_2'} \{ \langle 2\ell_1 m_1, \ 2\ell_2 m_2 | V_{12} | 2\ell_1' m_1', \ 2\ell_2' m_2' \rangle$$

$$\eta_{\ell_1 m_1} \eta_{\ell_2 m_2} \ \xi^{\ell_1' m_1'} \ \xi^{\ell_2' m_2'} \}$$

$$(4.1)$$

$$= \frac{1}{2} \sum_{\ell_1 \ell_2} \sum_{\ell_1' \ell_2'} \sum_{L} \{ \langle 2\ell_1, \ 2\ell_2, \ L | V_{12} | 2\ell_1', \ 2\ell_2', \ L \rangle$$

$$(-1)^{\ell_1' + \ell_2'} (2L + 1)^{\frac{1}{2}} [[\eta_{\ell_1} \times \eta_{\ell_2}]^L \times [\xi_{\ell_1'} \times \xi_{\ell_2'}]^L]^0_0 \}$$

where $|2\ell m\rangle$ is an harmonic oscillator state of two quanta so that ℓ is restricted to the values $\ell = 0,2$ and in the last part of (6.1) we recoupled bra and ket to a total L, assuming of course that V_{12} is invariant under rotations.

If the matrix elements in (4.1) are real, as happens for example if $V_{12} = V(|\underline{r}_1 - \underline{r}_2|)$, then a brief examination indicates that we have only seven types of independent Hermitian operators[1]

$$A_L = [[\underset{\sim}{\eta} \times \underset{\sim}{\eta}]^L \times [\underset{\sim}{\xi} \times \underset{\sim}{\xi}]^L]^0_0 \qquad L = 0,2,4 \qquad (4.2a)$$

$$B = [\underset{\sim}{\eta} \times \underset{\sim}{\xi}]^0_0 \ \bar{\eta} \ \bar{\xi} \qquad (4.2b)$$

$$C = \bar{\eta}^2 \ \bar{\xi}^2 \qquad (4.2c)$$

$$D = [\underset{\sim}{\eta} \times \underset{\sim}{\eta}]^0_0 \ \bar{\xi}^2 + \bar{\eta}^2 [\underset{\sim}{\xi} \times \underset{\sim}{\xi}]^0_0 \qquad (4.2d)$$

$$E = [[\underset{\sim}{\eta} \times \underset{\sim}{\eta}]^2 \times \underset{\sim}{\xi}]^0_0 \ \bar{\xi} + \bar{\eta}[\underset{\sim}{\eta} \times [\underset{\sim}{\xi} \times \underset{\sim}{\xi}]^2]^0_0 \ . \qquad (4.2e)$$

The A_L involve only d bosons and already in the original work of Arima and Iachello[2] (and in fact much before in standard shell theory for d states[15]) it is developed in terms of $\hat{n}$, Λ^2 and L^2. The B, C concern only $\hat{N}$, $\hat{n}$ while D involves K^2 and thus L^2. The E requires a special analysis given in reference 1. We then arrive at the following expansion of our seven basic interactions in terms of the Casimir operators of the groups in the chains (3.13)

$$A_0 = \frac{1}{5} \hat{n}(\hat{n} + 3) - \frac{1}{5} \Lambda^2 \qquad (4.3a)$$

$$A_2 = \frac{1}{7\sqrt{5}} \{ -L^2 + 2\Lambda^2 + 2\hat{n}(\hat{n} - 2) \} \qquad (4.3b)$$

$$A_4 = \frac{1}{7} \{ \frac{1}{3} L^2 - \frac{1}{5} \Lambda^2 + \frac{6}{5} \hat{n}(\hat{n} - 2) \} \qquad (4.3c)$$

$$B = \frac{1}{\sqrt{5}} (\hat{N} - \hat{n})\hat{n} \tag{4.3d}$$

$$C = (\hat{N} - \hat{n})(\hat{N} - \hat{n} - 1) \tag{4.3e}$$

$$D = \frac{1}{\sqrt{5}} \{\Lambda^2 - L^2 + 2\hat{N}\hat{n} - 2\hat{n}^2 + 5\hat{N} - 4\hat{n}\} \tag{4.3f}$$

$$E = \frac{3}{4\sqrt{35}} \{Q^2 - \frac{2}{3}\Lambda^2 + \frac{4}{3}L^2 - \frac{1}{6}L^2 + \frac{14}{3}\hat{n}^2$$

$$+ \frac{22}{3}\hat{n} - \frac{8}{3}\hat{N}(2\hat{n} + 5\}. \tag{4.3g}$$

Thus the most general interaction in the d–s boson model is a linear combination of the Casimir operators of the groups in the chain (3.13).

This most general Hamiltonian is in a sense equivalent to the asymmetric top (2.1), i.e., there is no closed formula for the energy levels as the Hamiltonian is not associated with a <u>definite</u> chain of groups. A Hamiltonian corresponding to a symmetric top would be one related to one chain of groups, for example (3.13a). In that case its most general form would be

$$H = u_d\hat{n} + u_s(\hat{N} - \hat{n}) + v_0 A_0 + v_1 A_2 + v_2 A_4 + w_1 B + w_2 C \tag{4.4}$$

where the u, v, w are some constants. Clearly the eigenvalues of this Hamiltonian are given by

$$H = u_d\nu + u_s(N - \nu) + v_0[\frac{1}{5}\nu(\nu + 3) - \frac{1}{5}\lambda(\lambda + 3)]$$

$$+ v_1 \frac{1}{7\sqrt{5}} \{-L(L + 1) + 2\lambda(\lambda + 3) + 2\nu(\nu - 2)\}$$

$$+ v_2 \frac{1}{7} \{\frac{1}{3} L(L + 1) - \frac{1}{5}\lambda(\lambda + 3) + \frac{6}{5}\nu(\nu - 2)\}$$

$$+ w_1 \frac{1}{\sqrt{5}} (N - \nu)\nu + w_2(N - \nu)(N - \nu - 1) \tag{4.5}$$

where there are definite relations between the quantum number that appear[1] e.g., $\lambda = \nu, \nu - 2.....1$ or 0.

The more general Hamiltonian requires the explicit expression of the states for a given chain of groups and the matrix elements of the Casimir operators of the groups in the other chains with respect to these states. Again this has a counterpart in the asymmetric top model where we had to find the matrix elements of L_1^2 with respect to the $D_{MK}^L(\theta_i)$ functions.

Because of our original interest in the problem started with
the collective model of the nucleus, we used the chain of groups
(3.13a) as the standard one and obtained the explicit expression
of the states. In the next section we briefly describe these states
and the reduced 3j symbols for the $O(5) \supset O(3)$ chain of groups to
which they give rise.

5. THE EIGENSTATES IN THE STANDARD CHAIN OF GROUPS

To construct the eigenstates in association with the chain of
groups (3.13a), we start with the subchain $U(5) \supset O(5) \supset O(3) \supset O(2)$
appearing in the collective model[14] i.e., the eigenkets satisfying

$$\hat{n}|\nu\lambda sLM\rangle = \nu|\nu\lambda sLM\rangle \tag{5.1a}$$

$$\Lambda^2|\nu\lambda sLM\rangle = \lambda(\lambda + 3)|\nu\lambda sLM\rangle \tag{5.1b}$$

$$L^2|\nu\lambda sLM\rangle = L(L + 1)|\nu\lambda sLM\rangle \tag{5.1c}$$

$$L_3|\nu\lambda sLM\rangle = M|\nu\lambda sLM\rangle \, . \tag{5.1d}$$

There may be $d(\lambda,L)$ independent representations[13,14] L of $O(3)$
contained in a given representation λ of $O(5)$ and we introduce the
index

$$s = 1, 2, \ldots d(\lambda,L) \tag{5.2}$$

to distinguish them.

To obtain the explicit expressions of these eigenstates it is
convenient to go from the coordinates α_m to those of a system fixed
in the body that are related to them[8,13] through

$$\alpha_m = \sum_{m'} D^{2*}_{mm'}(\theta_i)a_{m'} \tag{5.3a}$$

$$a_2 = a_{-2} = \frac{1}{\sqrt{2}} \beta \sin \gamma, \; a_1 = a_{-1} = 0, \; a_0 = \beta \cos \gamma \tag{5.3b}$$

where $D^{\ell}_{mm'}(\theta_i)$ are the Wigner functions of the Euler angles, θ_i.
In terms of the coordinates β, γ, θ_i the eigenstates take the
form[8,13,14]

$$|\nu\lambda sLM\rangle = F^{\lambda}_j(\beta)\chi^{\lambda}_{sLM}(\gamma,\theta_i) \tag{5.4}$$

$$\chi^{\lambda}_{sLM}(\gamma,\theta_i) = \pi^{5/4}2^{-\lambda/2} \sum_K \phi^{\lambda sL}_K(\gamma)D^{L*}_{MK}(\theta_i) \, . \tag{5.5}$$

In (5.4) $j = (\nu - \lambda)/2$ and $F_j^\beta(\beta)$ is the radial eigenstate of a five dimensional oscillator and thus expressible in terms of Laguerre polynomials.[14] The $\chi_{sLM}^\lambda(\gamma,\theta_i)$ in (5.5) play for the $O(5) \supset O(3) \supset O(2)$ chain of subgroups the role that spherical harmonics play for the $O(3) \supset O(2)$ chain. The main new development achieved in reference 14 was the explicit determination of the $\phi_K^{\lambda sL}(\gamma)$ that permits the expansion of the $\chi_{sLM}^\lambda(\gamma,\theta_i)$ in terms of the $D_{MK}^{L*}(\theta_i)$ functions. Note that the χ's form a complete though <u>not orthonormal</u> set of functions.

When integrating the product of three χ over the angles γ, θ_i we obtain the Clebsch-Gordan coefficient for $O(5) \supset O(3) \supset O(2)$ chain of groups, in a similar way as the product of three $Y_{\ell m}(\theta,\phi)$ integrated over θ, ϕ gives the ordinary Clebsch-Gordan coefficients. Because the integrals of products of three $D_{MK}^L(\theta_i)$ over the θ_i is well known[7] we see that the important part of the Clebsch-Gordan coefficient is the expression[14]

$$\int_0^\pi \sum_{KK'K''} \begin{pmatrix} LL'L'' \\ KK'K'' \end{pmatrix} \phi_K^{\lambda sL}(\gamma)\phi_{K'}^{\lambda's'L'}(\gamma)\phi_{K''}^{\lambda''s''L''}(\gamma)\sin 3\gamma \, d\gamma$$

$$\equiv (\lambda sL, \lambda's'L', \lambda''s''L'') \tag{5.6}$$

to which we give the name of reduced 3j symbol in $O(5) \supset O(3)$ chain and denote it by the right hand symbol in (5.6). These reduced 3j-symbols have been explicitly determined[14] and also tabulated for[16]

$$(\lambda,s,L) = (1,1,2), \ (2,1,2) \text{ and } (\lambda,1,0) \tag{5.7}$$

where $\lambda \equiv 0 \bmod 3$

for arbitrary $\lambda's'L'$, $\lambda''s''L''$, and as we show later, they will play an essential role in determining the matrix elements of Q^2 for states associated with the standard chain of groups.

So far we have discussed only the states associated with the d-bosons but the extension to d-s case is trivial, as for the s part we have the eigenstates the one dimensional harmonic oscillator

$$\bar{\eta} \, \bar{\xi} \, \langle \bar{\alpha}|n\rangle = n\langle \bar{\alpha}|n\rangle \ . \tag{5.8}$$

Thus the eigenstates of the standard chain of groups (3.13a) take the form

$$\langle \alpha_m,\bar{\alpha}|N\nu\lambda sLM\rangle = \langle \alpha_m|\nu\lambda sLM\rangle\langle\bar{\alpha}|N - \nu\rangle \tag{5.9}$$

and we shall proceed to use them in the same way that the eigenstates of the symmetric top were used to determine the matrix of elements of the Hamiltonian of the asymmetric top.

6. TRANSFORMATION BRACKETS BETWEEN THE CHAINS OF GROUPS CONTAINING

U(5) and O(6)

In section 2 we indicated that one way to find the matrix elements of L_1^2 for states of the chain $O(3) \supset O_3(2)$ was to get the transformation brackets $D_{M'M''}^L(0, \pi/2, 0)$ between this chain and $O(3) \supset O_1(2)$. If we consider now the chains of groups (3.13a) and (3.13b) differing in the fact that the first contains the subgroup U(5) while the second has O(6), we can try to determine the transformation brackets that take us from one family of states to the other.

Actually the objective mentioned is easy to achieve in principle. The operator $\hat{N}$ gives a second order partial differential equation that is separable in the variables β and $\bar{\alpha}$, assuming for the eigenstate of Λ^2, L^2 and L_3 the $\chi_{SLM}^\lambda(\gamma,\theta_i)$ of (5.5). If we define b, δ by the expressions

$$\bar{\alpha} = b \cos \delta \qquad \beta = b \sin \delta \qquad\qquad (6.1)$$

the problem is still separable in these variables and it is easy to see[1] that we get in this way an eigenfunction of L^2 with eigenvalue $\rho(\rho + 4)$ which depends on δ, γ, θ_i. The dependence on δ is given in terms[1] of a Gegenbauer polynomial and that of b corresponds to a six dimensional oscillator and again can be expressed in terms of Laguerre polynomials. We have thus the eigenstates associated with the chain of groups (3.13b) and, using somewhat more sophisticated methods,[1] we can expand them in terms of the eigenstates of the standard chain of groups to obtain the transformation bracket

$$\langle N\nu\lambda | N\rho\lambda \rangle = (-1)^{(N-\rho+\nu-\lambda)/2} \left[\frac{2^{\rho-\lambda}(\rho+1)!(\rho+2)!}{2^{N-\rho}((N+\rho)/2+2)!(\rho-\lambda)!} \right.$$

$$\left. \frac{(\nu+\lambda+3)!(N-\nu)!((N-\rho)/2)!}{(\rho+\lambda+3)!((\nu+\lambda)/2+1)!((\nu-\lambda)/2)!} \right]^{\frac{1}{2}} \sum_s \frac{((N-\rho)/2+1)s}{s!}$$

$$\frac{((\lambda-\rho)/2)s((\lambda-\rho+1)/2)s}{((N-\rho-\nu+\lambda)/2+s)!(-\rho-1)s} \; ; \text{ with the notation}$$

$$(a)_n = a(a + 1) \dots (a + n - 1). \qquad\qquad (6.2)$$

where, as is well knwon from group theoretical arguments,[17] the scalar product is independent of the indices L, M characterizing the irreducible representations of $O(3) \supset O(2)$.

7.   MATRIX ELEMENTS OF THE CASIMIR OPERATOR OF O(6) IN THE STANDARD

     BASIS

We could use the transformation brackets (6.2) to determine the matrix elements of L^2 in the standard basis but a more direct method is also convenient. We note from (3.17) that $L^2 = \Lambda^2 + K^2$ and as Λ^2 is diagonal in the standard basis we need only to apply K^2 of (3.18) to it. But we note also from the definitions (3.4) of η_m, ξ_m in terms of α_m, π_m, that $\sum_m (\eta_m \eta^m)$, $\sum_m (\xi_m \xi^m)$ can be expressed in terms of the number operator $\hat{n} = \sum_m (\eta_m \xi^m)$ and some functions of β and of derivatives with respect to β. Thus we can write[1]

$$K^2 = \hat{n}(\overline{\eta}\,\overline{\xi} + 1) + (\hat{n} + 5)\overline{\eta}\,\overline{\xi}$$

$$-(\beta^2 - \beta\frac{\partial}{\partial\beta} - \hat{n} - 5)\overline{\xi}^2 - (\beta^2 + \beta\frac{\partial}{\partial\beta} - \hat{n})\overline{\eta}^2 \tag{7.1}$$

where

$$\beta^2 = \sum_m \alpha_m \alpha^m, \quad \beta\frac{\partial}{\partial\beta} = i\sum_m \alpha_m \pi^m . \tag{7.2}$$

As the application of $\overline{\eta}$, $\overline{\xi}$ to $\langle\overline{\alpha}|\eta\rangle$ and of β^2, $\beta\frac{\partial}{\partial\beta}$ to $F_j^\lambda(\beta)$ is elementary, we easily arrive at the relations

$$\langle N, 2j'+\lambda, \lambda sLM|L^2|N, 2j+\lambda, \lambda sLM\rangle = A^{N\lambda}_{jj'}, \tag{7.3}$$

where $A^{N\lambda}_{jj'}$ has the selection rule $j' = j \pm 1$, j and

$$A^{N\lambda}_{jj} = (2j+\lambda)(2N-2\lambda-4j+1)+5(N-\lambda-2j)+\lambda(\lambda+3) \tag{7.4a}$$

$$A^{N\lambda}_{jj+1} = 2[(N-\lambda-2j)(N-\lambda-2j-1)(j+1)(\lambda+j+\tfrac{5}{2})]^{\frac{1}{2}} \tag{7.4b}$$

$$A^{N\lambda}_{jj-1} = 2[(N-\lambda-2j+1)(N-\lambda-2j+2)j(\lambda+j+\tfrac{3}{2})]^{\frac{1}{2}} . \tag{7.4c}$$

Thus we obtained explicitly the matrix elements of the Casimir operator L^2 of O(6) in the standard basis.

8.   MATRIX ELEMENTS OF THE CASIMIR OPERATOR OF SU(3) IN THE STANDARD

     BASIS

In reference 1 an elaborate method was used to express the matrix elements of Q^2 of (3.20) with respect to states of the standard basis, making use of the reduced Wigner coefficients mentioned in (5.6), (5.7). Later a much more compact expression was obtained by Chacón which was added in proof in reference 1. He noted that rather than determine the matrix elements of Q^2 we could calculate those of the operator E of (4.2e) as the Λ^2, L^2, $\hat{n}$, $\hat{N}$ appearing in (4.3g) are diagonal in the standard basis while the matrix elements of L^2 in this basis are given by (7.3). From (4.2e) we see that we need to calculate the matrix elements of two operators that are the Hermitian conjugate of each other. Thus we can restrict ourselves to the calculation of one of them and, as the part connected with $\overline{\xi}$ is trivial we need only to determine the matrix element of

$$\langle \nu'\lambda's'LM \ [[\underset{\sim}{\eta} \times \underset{\sim}{\eta}]^2 \times \underset{\sim}{\xi}]^0_0 | \nu\lambda sLM \rangle \tag{8.1}$$

where we obviously have the selection rule $\nu' = \nu + 1$.

For this last task we see from the definition (3.4) and the properties of the Clebsch-Gordan coefficients of O(3) that

$$2\sqrt{2} \ [[\underset{\sim}{\alpha} \times \underset{\sim}{\alpha}]^2 \times \underset{\sim}{\alpha}]^0_0 =$$

$$[[\underset{\sim}{\eta} \times \underset{\sim}{\eta}]^2 \times \underset{\sim}{\eta}]^0_0 + 3[[\underset{\sim}{\eta} \times \underset{\sim}{\eta}]^2 \times \underset{\sim}{\xi}]^0_0 + [[\underset{\sim}{\xi} \times \underset{\sim}{\xi}]^2 \times \underset{\sim}{\xi}]^0_0$$

$$+ 3[[\underset{\sim}{\xi} \times \underset{\sim}{\xi}]^2 \times \underset{\sim}{\eta}]^0_0 . \tag{8.2}$$

Thus with the selection rule $\nu' = \nu + 1$ the matrix element (8.1) becomes proportional to the matrix element

$$\langle \nu + 1, \lambda's'LM | \beta^3 \cos 3\gamma | \nu\lambda sLM \rangle \tag{8.3}$$

where we used the relation

$$[[\underset{\sim}{\alpha} \times \underset{\sim}{\alpha}]^2 \times \underset{\sim}{\alpha}]^0_0 = -\sqrt{2/35} \ \beta^3 \cos 3\gamma. \tag{8.4}$$

The matrix element (8.3) factorizes into a product of that of β^3 with respect to the radial functions $F^\lambda_j(\beta)$ which is trivial to obtain, and the matrix element of $\cos 3\gamma$ with respect to the functions $\chi^\lambda_{sLM}(\gamma,\theta_i)$ which becomes the reduced 3j symbol

$$(310, \ \lambda's'L', \ \lambda''s''L'') \tag{8.5}$$

already tabulated.[16]

Thus we also have the matrix elements of E or equivalently, Q^2, with respect to the standard basis and we can obtain the matrix of the most general interaction in the s-d shell in this basis.

9.   EXAMPLES OF APPLICATIONS OF THE INTERACTING BOSON MODEL

We have established a systematic group theoretical procedure for deriving relevant predictions of the interacting d-s boson model.

These not only concern the spectra of Hamiltonians involving a general two body interaction, but also eigenstates derived as linear combinations of the states ascociated with the standard chain of groups that were described in section 5. Furthermore the probabilities for electromagnetic transitions between these states, as well as the branching ratios, can be derived from matrix elements (m.e.) of operators such as α_m or $[\underset{\sim}{\alpha} \times \underset{\sim}{\alpha}]^2_m$ with respect to the relevant states. As can be easily seen,[14] the determination of these m.e. involve the use of reduced 3j-symbols for the $O(5) \supset O(3)$ chain of groups of the type $(112, \lambda's'L', \lambda''s''L'')$ or $(212, \lambda's'L', \lambda''s''L'')$ which have already been tabulated[16] and which furthermore provide relevant selection rules.

The practical calculations have been done though by Arima, Iachello and their collaborators[2,3,4] using techniques less general than those outlined in the present note. Their starting point is first to consider the number of d, s bosons as given by the number of neutron and proton pairs outside closed shell. Thus for

$$^{156}\text{Gd} \qquad Z = 64 \qquad N = 92 ; \qquad Z - 50 = 14,$$

$$N - 82 = 10 \quad \left.\begin{array}{l} \text{Proton Pairs} = 7 \\[2ex] \text{Neutron Pairs} = 5 \end{array}\right\} \quad \text{Total number of bosons } N = 12$$

It turns out that this nucleus is an almost perfect example[4] of $SU(3)$ symmetry where the two body interaction is of the type Q^2 whose spectra, as indicated in (3.21), is given by

$$\frac{2}{3}(p^2 + rp + r^2) + 2(r + p) - \frac{1}{2}L(L + 1) \qquad\qquad (9.2)$$

where (p,r) is the representation of $SU(3)$. The number of $SU(3)$ quanta is double[9,11] the number of bosons and thus the representations of $SU(3)$ contained in the representation 12 of $U(6)$ are given by[9,11]

$$(p,r) = (24,0),\ (20,2),\ (16,4),\ (18,0) \qquad\qquad (9.3)$$

as indicated in a graph taken from an article appearing in Physics Today.[12]

Nuclei well described by purely d-boson interaction, i.e., by states in the standard chain of groups (3.13a), are for example ^{110}Cd and ^{188}Pt as indicated in reference 3, while some isotopes of Ba and Xe seem to correspond to states associated with the chain of groups (3.13b) that contains $O(6)$. The theoretical predictions fit experimental data both for the spectra and the transition probabilities.

The U(5), SU(3) and O(6) chain of groups correspond roughly to
the anharmonic vibrator, axial rotor and triaxial rotor with $\gamma = 30°$
of the collective model, thus providing an alternative way of deriving well known collective effects which may be more amenable to
correlations with the nuclear shell model.

10. GENERALIZATIONS AND CONCLUSIONS

Once one understands the structure of the d-s interacting boson
model many generalizations suggest themselves. For example, Arima,
Iachello, Talmi and their collaborators[18] considered the possibility
of distinguishing the s-d bosons coming from the neutron from those
coming from the proton pairs. In this way we have 12 bosons rather
than 6 and the fundamental group is U(12), but we can reduce it
immediately to the subgroup U(6) in which the representations will
have two rows rather than one. From there the different chains of
subgroups (3.13) continue to follow, only the representations are
more general in view of the two rowed representations of U(6). Much
can be done for this problem group theoretically as, in particular
Sharp and his collaborators[19] have looked into the states associated
with two rowed representations (λ_1, λ_2) of O(5) for the chain of
groups $O(5) \supset O(3)$. It is unlikely though that these types of representations will be required in their full generality.

An interesting generalization was proposed by Federman. As the
strongest interaction between nucleons is the isotopic spin $t = 0$
between protons and neutrons why not count the number of proton-
neutron pairs with $t = 0$ plus the number of the remaining neutron
pairs. As indicated in section 1, the proton-neutron pairs with
$t = 0$ correspond to J odd and the lowest value of J is $J = 1$. Thus
we would have p bosons added to d-s bosons to form a group U(9).
A simpler case of this problem appears when we consider just s
bosons for the remaining neutron pairs and thus have a p-s interacting boson picture that involves a U(4) group. Federman, Castaños
and Frank intend to explore these possibilities.

The essential step seems though to find a group theoretical
structure sufficiently rich to provide an interesting framework,
but not so rich that almost anything can be fitted into it. This
is not an easy matter and one must acknowledge that the picture
proposed by Arima and Iachello fits the above restrictions, thus
enhancing our understanding of many of the collective effects present
in nuclei.

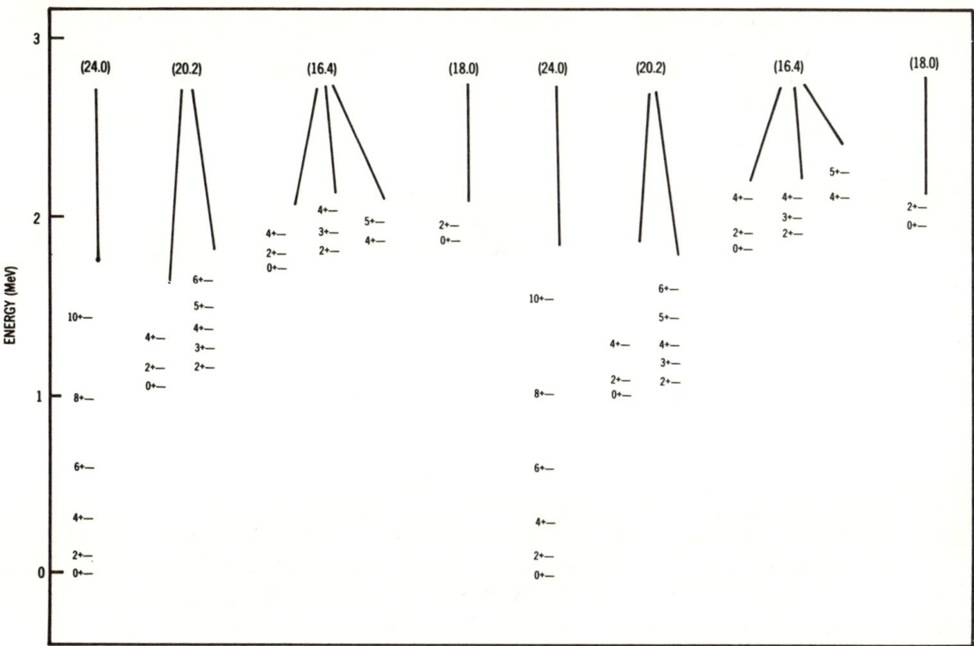

Fig. 1. Gd^{156} is an example of a spectrum with SU(3) symmetry, according to the interacting-boson model. At left are experimental data, at right the theoretical predictions. Note the repetition of the 0,2,2 pattern with the two 2^+ states almost degenerate in energy. Numbers in parentheses label representations of SU(3). (F. Iachello, to be publ. in Comments on Nuclear and Particle Physics, 7, 59, (1978).)

REFERENCES

1. O. Castaños, E. Chacón, A. Frank and M. Moshinsky, J. Math. Phys. 20, 35 (1979).
2. A. Arima and F. Iachello, Phys. Lett. B53, 309 (1974).
3. A. Arima and F. Iachello, Ann. Phys. (N.Y.) 99, 253 (1976), Ann. Phys. (N.Y.) 111, 201 (1978), Phys. Rev. Lett. 40, 385 (1978).
4. A. Arima, T. Ohtsuka, F. Iachello and I. Talmi, Phys. Lett. 66 B, 205 (1977).
5. J. R. Schrieffer, "Theory of Superconductivity" (W. A. Benjamin, Inc., New York 1964).
6. D. J. Thouless, "The Quantum Mechanics of Many Body Systems". (Academic Press, New York 1961).
7. M. E. Rose, "Elementary Theory of Angular Momentum" (Wiley, New York, 1957).

8.   A. Bohr and B. Mottelson, Kgl. Dan. Videnskab. Selks. Mat. Fys. Medd. $\underline{27}$, 16 (1953).

9.   M. Moshinsky, "Group Theory and the Many Body Problem" (Gordon and Breach, New York 1968).

10. E. Chacón, J. Flores, M. de Llano and P. A. Mello, Nuclear Physics $\underline{72}$, 352, 379 (1965).
     A. Arima, T. Inoue, T. Sebe, H. Hagiwara, Nuclear Physics $\underline{59}$, 1, (1964).

11. J. P. Elliott, Proc. Roy. Soc. London $\underline{A245}$, 562 (1958).

12. Unsigned article appearing in Physics Today - July 1978, p. 17.

13. E. Chacón, M. Moshinsky and R. T. Sharp, J. Math. Phys. $\underline{17}$, 668 (1976).

14. E. Chacón and M. Moshinsky, J. Math. Phys. $\underline{18}$, 870 (1977).

15. H. A. Jahn, Proc. Roy. Soc. $\underline{A201}$, 516 (1950); $\underline{A205}$, 192 (1951).

16. O. Castaños and A. Frank, "Computer programs for relevant reduced 3j-symbols of the $O(5) \supset O(3)$ chain of groups". To be published in forthcoming Ph.D. dissertations.

17. E. P. Wigner, "Group Theory" (Academic Press, New York, 1959).

18. A. Arima, T. Ohtsuka, F. Iachello, O. Scholten and I. Talmi, Proc. Int. Conference, Nuclear Structure, Tokyo 1977, J. Phys. Soc. Japan $\underline{44}$, 509 (1978).

19. R. Gaskell, A. Peccia and R. T. Sharp, J. Math. Phys. $\underline{19}$, 727 (1978).

*Miembro del Instituto Nacional de Energía Nuclear y de El Colegio Nacional.

REMARKS ON THE ALGEBRAIC STRUCTURE OF SPONTANEOUS SYMMETRY
BREAKING IN UNIFIED GAUGE THEORIES

L. O'Raifeartaigh,* S.-Y. Park,** and K.C. Wali**

*Dublin Institute for Advanced Studies
and
**Syracuse University, New York

1. Introduction

One of the most interesting recent developments in physics,
and indeed in many other branches of science too, has been the
gradual realization of the importance and universality of spontaneous symmetry breaking. For example, three out of four of this
mornings lectures dealt with the topic of spontaneous symmetry
breaking in three different areas--mathematics, physics and
biophysics. Within Physics, spontaneous symmetry breaking plays
an important role in many phenomena--magnetism, crystallography
and superconductivity for example are just macroscopic manifestations of spontaneous symmetry breaking for microscopic systems--and
there are even examples in classical physics such as the Jacobi
ellipsoids.[1] However, the subject of this talk will be the appearance
of spontaneous symmetry breaking in particle physics, in particular
in the context of unified gauge theories.[2] This context is, perhaps,
particularly appropriate for an Einstein symposium because unified
gauge theory represents a major step towards the realization of
Einstein's dream of unifying the fundamental interactions by means
of a single universal principle. It is also, perhaps, appropriate
to note that 1979 is not only the centenary of Einstein's birth but
is also the fiftieth anniversary of the publication of a paper by
Hermann Weyl which was the forerunner of unified gauge theory.[3] In
this paper Weyl showed that the Maxwell-Lorentz theory of electro-
magnetism was not only gauge-invariant but could be derived directly
and simply from the principles of gauge theory, using the gauge
group $U(1)$. Present-day unified gauge theories are essentially
the extension of Weyl's principles to arbitrary compact Lie gauge
groups--together with spontaneous symmetry breaking.

2. The Potential

Before placing the spontaneous breakdown in the context of gauge theory, we first consider it as a phenomenon in its own right. However, for simplicity, we consider it in the restricted form in which it is used in gauge theory, namely as the spontaneous breakdown of a potential $V(\phi)$, where ϕ denotes any finite number of Lorentz scalar fields $\phi(x)$. It is assumed that the fields ϕ transform vectorially with respect to some representation $U(g)$ of a compact Lie group G and that V is G-invariant, that is,

$$V(\phi) = V(U(g)\phi), \qquad\qquad (2.1)$$

$$g \in G.$$

Let $\overset{o}{\phi}$ be a constant vector (independent of x) for which V takes its absolute minimum value, $V(\overset{o}{\phi}) \leq V(\phi)$, all ϕ. The symmetry of V is said to be spontaneously broken if the vector $\overset{o}{\phi}$ is not G-invariant, that is, if

$$U(g)\,\overset{o}{\phi} \neq \overset{o}{\phi} \quad \text{for all} \ \ g \in G. \qquad\qquad (2.2)$$

The reason for the name is that the 'true' field i.e. the field which carries the energy, is then $\theta(x) = \phi(x) - \overset{o}{\phi}$ and $W(\theta) \equiv V(\phi)$ is not manifestly symmetric in θ, $W(U(g)\theta) \neq W(\theta)$, all $g \in G$. Of course, it may, and usually does, happen that

$$U(h)\,\overset{o}{\phi} = \overset{o}{\phi}, \qquad h \in H \quad G \qquad\qquad (2.3)$$

for <u>some</u> elements $h \in G$, and it is trivial to verify that these elements will form a subgroup H of G. This subgroup H is called variously the little group of $\overset{o}{\phi}$, the stability group of $\overset{o}{\phi}$, and the residual symmetry group of V, the latter since $W(U(h)\theta) = W(\theta)$, for all $h \in H$. Note that by definition, there is a spontaneous symmetry breakdown only if H is strictly less than G. More generally, the little group H and its place in the original group G determine the structure of the spontaneous breakdown.

An important concept in connection with spontaneous symmetry breaking is that of the <u>orbit</u> of $\overset{o}{\phi}$. The orbit is simply the set of all vectors $U(g)\overset{o}{\phi}$ as g runs through G. From (2.2) we see that if there is a spontaneous breakdown the orbit will contain at least one vector other than $\overset{o}{\phi}$, and hence will be non-trivial. Thus another, more geometrical way, to characterize a spontaneous breakdown of V is to say that the minimum of V falls on a non-trivial orbit. It is easy to verify that the little group H corresponding to points on the same orbit are the same (up to conjugation). Hence the little group H, and its place in the big group G, is determined by the orbit as a whole rather than any particular point on the orbit. The same is true of any quantity of physical or geometrical interest, and for this reason it is more effective to think in terms of the orbit as a whole rather than any particular point on the orbit.

Although all the above considerations are classical, they can be extended to the case of quantized fields, the most important new feature in that case being the identification of the vector $\overset{o}{\phi}$ with constant, c-number, vacuum expectation values $<|\phi(x)|>$ of the fields at the potential minimum.[4]

3. Unified Gauge Theory

As mentioned in the Introduction, it is natural to think of the Lagrangian for a unified gauge theory[2] as the generalization from the group $U(1)$ to an arbitrary compact Lie group of the well-known, renormalizable Lagrangian for the electromagnetic field interacting with any number of spin 0 and spin ½ fields, namely,

$$-\mathcal{L} = \tfrac{1}{4}(F_{\mu\nu})^2 + \tfrac{1}{2}(D_\mu \phi)^2 + V(\phi) \qquad\qquad 3.1$$

$$+ \psi(\not{D} + m)\psi + g\bar{\psi}\phi\psi \ ,$$

where $F_{\mu\nu}$ is the electromagnetic field, D_μ the usual covariant derivative $\partial_\mu + ieA_\mu$ and $V(\phi)$ a renormalizable potential for the scalar Higgs fields. The only difference, in fact, is that A_μ is no longer a single field but rather an expression $A_{n\mu}t_n$ where t_n are the group generators, $F_{\mu\nu}$ is no longer the simple curl of A_μ but rather $F_{\mu\nu} = \partial_\mu A_\nu - \partial_\nu A_\mu + [A_\mu, A_\nu]$ and instead of being assigned charges the matter fields are assigned to representations of G (with $V(\phi)$ and the Yukawa term assumed to be G-invariant). Note that the term $(F_{\mu\nu})^2$ contains a self-interaction of the gauge-field, which vanishes only if G is abelian. The great advantage of unified gauge theory, apart from the aesthetic appeal of the geometrical structure, is the almost unique determination of the interaction of the gauge (radiation) field A_μ with itself and with the matter fields, ϕ, ψ (through the covariant derivative). In fact, the only freedom available in the choice of interaction for the gauge field is (a) the choice of the group G (b) the choice of one coupling constant for each irreducible component (abelian or simple) of the Lie algebra G, and (c) the assignment of the matter-fields ψ and ϕ to representations of G. It is worth noting, because of the application to leptons and quarks, that any two fields which are assigned to the same representation of G are

constrained to interact with gauge-fields in exactly the same way
and with exactly the same strength.

It is clear from the above that the interaction of the gauge-
fields with matter is determined in a completely algebraic or
geometric manner. Unfortunately, as the theory now stands the same
cannot be said for the self-interaction of the matter fields, through
the potential V and the Yukawa term $g\bar{\psi}\phi\psi$. There is no principle
known by which these terms can be determined in advance (except in
supersymmetric theories) and usually they are chosen in an ad hoc
manner to provide the required mass-spectrum for the fermions after
spontaneous symmetry breaking.

4. Spontaneous Symmetry Breaking in Unified Gauge Theory

The algebraic structure of unified gauge theory discussed in
the previous section is formally the same whether or not the
potential V for the scalar fields is spontaneously broken. However,
if V is not spontaneously broken, the gauge fields in (3.1) will
have no mass-terms (and the masses of the matter fields will be
G-invariant). The role of the spontaneous symmetry breaking is to
supply masses for the gauge-fields, and mass-breaking for all fields,
without destroying the renormalizability of the theory. It is easy
to see that if V is spontaneously broken, so that $\phi(x) \to \theta(x) =$
$\phi(x) - \overset{o}{\phi}$, the Lagrangian (3.1) acquires the extra terms

$$-\Delta \mathcal{L} = \tfrac{1}{2} M_{nk} A_\mu^n A_\mu^k \; + \tfrac{1}{2}\mu_{ab} \theta_a \theta_b$$

$$+ \; \bar{\psi}\Delta m\psi + f^a_{nk} \, \theta_a A_\mu^n A_\mu^k + 0(\theta^3) , \tag{4.1}$$

where the mass-matricies indicated are given by

$$M_{nk} = (\overset{o}{\phi}, \, t_n t_k \overset{o}{\phi}), \quad \mu_{ab} = \frac{\partial^2 V}{\partial\phi_a \partial\phi_b} \Bigg|_{\phi = \overset{o}{\phi}} \tag{4.2}$$

$$\Delta m_{ij} = g^a_{ij} \, \overset{o}{\phi}_a ,$$

and the coupling constants for the cubic interaction shown are

$$f_{nk} = (t_n t_k \overset{o}{\phi})_a . \tag{4.3}$$

Thus the spontaneous breakdown supplies mass-terms for all three
kinds of fields and a trilinear interaction between two gauge fields
and a scalar field (together with a slight modification of the potential indicated by $0(\theta^3)$). The special form of the mass terms in Δ
and the occurence of the trilinear interaction (4.3) are what distinguish a spontaneously broken theory from one which is explicitly
broken by the ad hoc insertion of mass-terms. In particular, the
trilinear interaction (4.3) is crucial for the renormalizability.
We shall not be concerned with renormalizability here, but we shall

see another amusing role which is played by the trilinear term
later. Note that the additional part of the Lagrangian generated
by the spontaneous symmetry breaking is not G-invariant, but is
invariant with respect to the little group H.

The important point to note about the additional part
generated by the spontaneous symmetry breaking is that its form is
completely fixed, and apart from $\overset{\text{o}}{\phi}$ it contains no new parameters.
It follows that the terms in $\Delta\mathcal{L}$ and in particular the mass-spectra,
are determined completely by the spontaneous breakdown, that is by the
orbit of $\overset{\text{o}}{\phi}$. For this reason the study of the orbital possibilities
for spontaneous breakdown is of great importance. This study is
essentially algebraic, even when radiative corrections are included.

5. Physical Consequences of Spontaneous Symmetry Breaking

The fact that unified gauge theory has such a rigid structure
and has so few parameters even when spontaneously broken leads one
to expect it to predict a large number of correlations, and indeed
in principle, unified gauge theory makes very strong predictions.
Unfortunately, however, because of the largeness of the gauge-field
and Higgs masses only a few of these predictions are accessible to
present-day experiment. In fact the only fields seen at present
are the photon-field and some fermion fields (leptons and, indi-
rectly, quarks) and the only interaction terms of $\mathcal{L} + \Delta\mathcal{L}$ seen are
some of the fermion masses and the four-fermi (current-current)
interactions supposedly mediated by the massive gauge fields. The
massive gauge fields and scalar Higgs fields are not seen, much less
the form of their interactions with each other and with the fermions.
Thus most of the structure discussed in the previous two sections
concerns future experimental results, and indeed it is surprising
that any predictions can be made concerning present experiments.
However, the structure of unified gauge theory is so strong that
the unseen part of the structure places restrictions on the part
which can be seen at present. In this section we wish to illustrate
by means of two examples how such restrictions can arise by showing
how, in particular, the choice of representation for the Higgs
field can influence the current-current interaction.

The first example concerns the standard model of electro-weak
interactions with gauge group $G=U(2) \times U(1)$, little group $U(1)$,
and the electron and its neutrino asssigned to the singlet (e_R)
and the doublet (ν, e_L) representation of G. The gauge-field-fermion
part of this model (with <u>ad hoc</u> masses M_+ and M_0 for the charged
and neutral gauge-fields) was proposed by Salam and Ward[5] in 1964 and
in that form had <u>two</u> parameters, namely M_+ and M_0, in addition to
the standard Fermi and electromagnetic coupling constants G and e
(or equivalently the irreducible coupling constants g_1 and g_2 of

Thus when the Higgs fields are masslen we have the amusing result:
For ϕ in the adjoint representation (C=1) the electrostatic repulsion
of like (spin 1) charges is just balanced by the Higgs attraction.
For ϕ in any other representation (C > 1) the Higgs attraction is
stronger and like charges attract!

An interesting corollary to the above result concerns the Olive-
Mentonen hypothesis of duality between $W^{\pm}_{\mu}$'s and the 't Hooft-
Polyakov monopoles which they generate.[7] The hypothesis is based
partly on the fact that for C = 1 there is a cancellation between
the electrostatic repulsion and the Higgs attraction for the
monopoles[8] similar to that for the $W^{\pm}_{\mu}$. However, it turns out that
for C > 1 the similarity breaks down. That is, while for C > 1
the Higgs attraction overcomes the electrostatic repulsion also for
the monopoles, the dependence of the net attractive force for the
monopoles has a different C-dependence to that for the W's displayed
in (5.2). In fact, for the monopoles, although the net attraction
still increases with C, it is bounded above as C > ∞. Thus O-M
duality is limited to at most the case C = 1, when the Higgs field
is in the adjoint representation.

It is hoped that the above two examples will serve to illustrate
the physical consequences of the choice of representation of the
Higgs field, and so help to motivate the study of the structure
of spontaneous breakdown which will be sketched in the next three
sections.

6. Structural Properties of the Zero-Mass Fields

It is convenient to divide the study of the structure of the
spontaneous breakdown in unified gauge theory into two separate
parts. The first part concerns the study of the form of the terms
(4.2) and (4.3) in particular the form of the mass-spectrum, given
the group G, the representation R of ϕ and the orbit of the
spontaneous breakdown vector $\overset{o}{\phi}$. The second part concerns the study
of the possibilities for R and $\overset{o}{\phi}$ given G and the condition that the
potential be renormalizable. In this and the next section we con-
sider the structure of Δ^{o} , given G, R and $\overset{o}{\phi}$, and sketch some
general results as illustration.

The first major structural results vis-a-vis the mass-spectrum
concern the fields which may have zero-mass. Here the concept of
Goldstone directions in the group representation R, namely the
directions $t_n \overset{o}{\phi}$ where t_n are the group generators, play a fundamental
role. Since $t_n \overset{o}{\phi} = 0$ iff t_n is a generator of the little group H,
it is clear that the Goldstone directions span a (dim G/H)
-dimensional subspace of R. For the gauge-fields, then seen by, it is

inspection of (4.2) that a gauge-field will have mass-zero iff it
is associated with a generator of the little group H. Thus for the
gauge-fields, the result of the spontaneous symmetry breakdown is to
leave those fields which belong to the little group H massless (as one
might expect since H is the residual symmetry group) but to equip
all the remaining (dim G/H) gauge-fields with non-zero masses. For
the Higgs scalars the situation is described by the celebrated
Goldstone theorem,[9] which in this context states that, if the group
G is continuous and the potential $V(\phi)$ is group-invariant, the
scalar-field mass-matrix μ_{ab} will have (dim G/H) zeros, corresponding
to (dim G/H) massless scalar fields (Goldstone fields) in the Goldstone directions $t_n \overset{o}{\phi}$ of the group representation. In general the
remaining (dim R - dim G/H) Higgs field will be massive, but the
parameters can be such that some or all of them are fortuitously
massless.

The second major structural result is the Higgs mechanism
which states that there always exists a gauge (the unitary or physical gauge) in which the dim G/H massless Goldstone fields are
absorbed by the longitudinal component of the dim G/H massive gauge-
fields.[10] In fact in the classical limit, the gauge-transformation
from an arbitrary to the physical gauge can be written down explicitly.   The gauge is called physical because of the 5d fields
A_μ, ϕ where d = dim G/H, only the 3d space-like components $\vec{A}$ actually
appear, and these are the true physical fields.

Since the main requirement for the Higgs mechanism is that
there be enough massive gauge fields to absorb the Goldstone scalars,
it might be thought that, on occasion, the mechanism could work with
a subgroup $G' \subset G$ of gauge-fields, such that dim G'/H' = dim G/H.
However, it can be shown that such is never the case. The reason
is that each independent generator t_n corresponding to a Goldstone
field (i.e. such that $t_n \overset{o}{\phi} \neq 0$) must correspond to a gauge-field
for the Higgs mechanism to work, and the set of such t_n close under
commutation to form an invariant subgroup G' of G. Hence the gauge
group must be an invariant subgroup of G. But since a compact Lie
algebra allows can contain an invariant subalgebra only in direct
sum form, it follows that (up to discrete central elements) G must
then be of the form G = G' x G_u where G_u is totally unbroken (and
therefore irrelevant for the Higgs mechanism). This result means,
for example, that if the flavour group SU(n), say, of the strong
interactions is spontaneously broken, and the Goldstone fields
absorbed by the Higgs mechanism, then the whole of the flavour
group SU(n) must be a gauge-group. It also shows that if the
potential develops a larger symmetry group than the gauge-symmetry
on the minimal orbit of $\overset{o}{\phi}$, extra Goldstone fields will appear
which cannot be gauged away (pseudo-Goldstone phenomenon).

$SU(2)$ and $U(1)$). The Higgs field and spontaneous breakdown were added to the model by Weinberg[6] in 1967 and by assigning the Higgs field to the doublet or $I = \frac{1}{2}$ representation of G (some admixture of $I = \frac{1}{2}$ is a natural choice), he reduced the two parameters $M_{\pm}$ and M_0 to one. In fact, it follows directly from the spontaneous symmetry breaking mechanism that for $I = \frac{1}{2}$ the gauge-field masses are correlated by the formula $M_0^2/M_{\pm}^2 = (g^2 + f^2)/g^2$. The crucial point, however, is that the reduction in the number of parameters follows only if the Higgs field is pure $I = \frac{1}{2}$. Otherwise the formula is

$$\frac{M_0^2}{M_{\pm}^2} = 2 \, \frac{(g^2 + f^2)}{g^2} \, \frac{\Sigma I_3^2}{\Sigma(I^2 - I_3^2)} \, , \tag{5.1}$$

and then the second factor in (5.1) enters as a new parameter. Thus the choice of representation for the (unseen) Higgs field determines the ratio M_+/M_0 and hence determines the four-fermi interactions which, since they are mediated by the gauge-fields, have coupling constants of the form $G = g^2/M^2$.

The second example is less realistic since it is concerned with physical Higgs fields whose masses are fortuitously zero (and which therefore would be seen through their long-range effects if they really existed). The gauge group is SU(2) with little group U(1), so that there are two massive charged gauge fields $W_{\mu}^{\pm}$, a massless gauge-field photon W_{μ}^3 and one neutral Higgs field $\overset{o}{\phi}(x)$, which belongs to an integer spin representation of SU(2). Let us now consider the long-range, static, interaction of two massive gauge-fields of the same charge. There are two contributions to the interaction, namely the static limit of the Born graphs mediated by W_{μ}^3 and ϕ respectively. The vertices in the first graph come from the trilineary term $e \, \varepsilon_{\alpha\beta\gamma} A_{\mu}^{\alpha} A_{\nu}^{\beta} A_{\mu,\nu}^{\alpha}$ in $\frac{1}{4}(F_{\mu\nu})^2$ and hence have the static coupling constant eM, where M is the mass of the $W_{\mu}^{\pm}$. Since the gauge field is a vector the first graph therefore contributes a Coulomb repulsion of strength $(eM)^2/r^2$. The vertices in the second graph come from the trilinear term (4.3) in Δ induced by the spontaneous symmetry breaking and have the strength $e^2|\overset{o}{\phi}|C$, where C is the Casimir of the ϕ-representation. Since the

Higgs field is a scalar this graph contributes a Newtownian repulsion of strength $(e^2|\overset{o}{\phi}|C)^2/r^2$. But from the spontaneous symmetry breaking mechanism (4.3) we know that the mass of the $W_{\mu}^{\pm}$ is given by $M^2 = e^2|\overset{o}{\phi}|^2 C$. Hence the net force is an attractive one of strength

$$\frac{e^4|\overset{o}{\phi}|^2 \, C(C - 1)}{r^2} \tag{5.2}$$

7. Structural Properties of the Non-zero Masses

The results for the mass-zero fields are quite general and make no reference to the details of G, R, or to specific bases in the representation. The structure of the massive part of the mass-spectrum does depend on such details, but again there are some quite general results that can be obtained, and we present some as illustrations.

The first general result concerns the structure of the gauge-field mass-matrix $M_{nk} = e^2(\overset{o}{\phi}, t_n t_k \overset{o}{\phi})$ and may be formulated as follows: if $\overset{o}{\phi}$ is an eigenstate of flavour (i.e. of the Cartan generators H_i of G in the representation R) then M_{nk} is essentially diagonal in the Cartan basis. More precisely

$$M_{nk} = \delta_{\alpha\beta} M_\alpha + \lambda_i \lambda_j M, \quad \text{where } H_i \overset{o}{\phi} = \lambda_i \overset{o}{\phi}$$

and α are the $(r-1)/2$ root vectors of G. The proof is given in the appendix.

A second result concerns the mass of the physical Higgs fields, in particular the physical Higgs field $\overset{o}{\theta}(x) = (\overset{o}{\phi}, \overset{o}{\theta}(x)) \overset{o}{\phi}$ which lies in the direction of the spontaneous symmetry breaking. For short we shall call this Higgs field the <u>polar</u> Higgs field so that a Higgs multiplet consists of the Goldstone fields in the directions $t_n \overset{o}{\phi}$, the physical polar field in the direction $\overset{o}{\phi}$ (orthogonal to the $t_n \overset{o}{\phi}$), and the remaining physical Higgs fields in the directions (if any) orthogonal to $\overset{o}{\phi}$ and $t_n \overset{o}{\phi}$. Now suppose that (as is often the case, especially for renormalizable potentials) the potential takes the form

$$(\phi) = -\mu(\phi,\phi) + I_s(\phi), \tag{7.1}$$

where I_s is a monomial of degree $s > 2$ $s = 3$ or 4 if V is renormalizable). Then the extremal condition $\partial V | \partial \overset{o}{\phi} = 0$ implies that

$$\overset{o}{\phi} \, \frac{\partial V}{\partial \overset{o}{\phi}} = -2\mu(\overset{o}{\phi},\overset{o}{\phi}) + s I_s(\overset{o}{\phi}) = 0. \tag{7.2}$$

On the other hand, the expectation value $\mu/0$ of the mass of the polar Higgs field is given by

$$\mu_o(\overset{o}{\phi},\overset{o}{\phi}) = \overset{o}{\phi}_a \, \frac{\partial^2 V}{\partial \overset{o}{\phi}_a \partial \overset{o}{\phi}_b} \, \overset{o}{\phi}_b = -4\mu(\overset{o}{\phi},\overset{o}{\phi}) + s^2 I_s(\overset{o}{\phi}), \tag{7.3}$$

and hence from (7.2), for $(\overset{o}{\phi}, \overset{o}{\phi}) \neq 0$, we have

$$\mu_o = 2(s - 2)\mu. \tag{7.4}$$

Thus, independently of the group G, the representation R of ϕ or the form of $I_s(\phi)$, the expectation value for the mass of the polar Higgs field is given by the formula (7.4), which involves only the parameters μ and s of the potential (7.1), and hence can be obtained by inspection. In particular for $s = 3,4$ we have $\mu_o = 2\mu$ and 4μ respectively.

As a matter of fact, one can make the stronger statement that for the potential (7.1) μ_o is not merely the expectation value of the polar Higgs mass but the actual mass. This result follows as a corollary to the final structural theorem which we shall mention, namely:

For a general renormalizable potential

$$(\phi) = -\mu(\phi,\phi) + I_3(\phi) + I_4(\phi), \tag{7.5}$$

where $I_3(\phi)$ and $I_4(\phi)$ are monomials of degree three and four respectively, the extremal equation

$$\frac{\partial V}{\partial \overset{o}{\phi}} = -2\mu\overset{o}{\phi} + \frac{\partial I_3}{\partial \overset{o}{\phi}} + \frac{\partial I_4}{\partial \overset{o}{\phi}} = 0, \tag{7.6}$$

reduces to the two separate equations

$$\frac{\partial I_s(\overset{o}{\phi})}{\partial \overset{o}{\phi}} = \lambda(s)\overset{o}{\phi}, \quad \left(\lambda(s) = \frac{sI_s(\overset{o}{\phi})}{(\overset{o}{\phi},\overset{o}{\phi})}\right), \quad s = 3,4 \tag{7.7}$$

if, and only if, the polar Higgs field $\overset{o}{\theta}(x)$ is an eigenstate of the mass-operator. (Thus, in particular, if either I_3 or I_4 is zero, as in (7.1), $\overset{o}{\theta}(x)$ is automatically a mass-eigenstate). The proof follows by combining the extremal condition (7.6) with the mass-eigenstate condition

$$\mu_{ab}\overset{o}{\phi}_b = \mu_o\overset{o}{\phi}_a, \tag{7.8}$$

which may be written in the form

$$\frac{\partial^2 V}{\partial \overset{o}{\phi}_a \partial \overset{o}{\phi}_b}\overset{o}{\phi}_b = 3\frac{\partial I_4}{\partial \overset{o}{\phi}_a} + 2\frac{\partial I_3}{\partial \overset{o}{\phi}_a} - 2\mu\overset{o}{\phi}_a = \mu_o\overset{o}{\phi}_a \tag{7.9}$$

to eliminate either $\partial I_3 / \partial \overset{o}{\phi}_a$ or $\partial I_4 / \partial \overset{o}{\phi}_a$.

8. Orbital Structure of the Spontaneous Symmetry Breakdown

The second structural question mentioned above is that of
determining the possible orbits and little groups given the symmetry group G, the representation R of ϕ and the potential $V(\phi)$.
There are a number of ways one can tackle this problem. One way
is to take advantage of the renormalizability condition, which
limits the possible potentials to the general form (7.5) and hence
leads to trilinear extremal conditions of the form (7.6) which can
then be analyzed. The condition that the extrema be (absolute)
minima imposes further strong conditions. A second approach is to
analyze the orbital structure of given representations R of G, in
particular the lower dimensional representations. For example, it
is easily shown that the 3, 6 and 8 of SU(3) contain only orbits
with little groups {SU(2) }, {U(1), SU(2), O(3) } and {U(1) x U(1),
U(2) } respectively, and similar results for a wide class of groups
and representations have been found by Michel and Radicati and by
Li.[11] One can also combine these two approaches and then the extremal
condition (7.6) places further restrictions on the possible orbits.
For example, for the 8 of SU(3) there is no quadralinear invariant
I_4, but if   (ϕ) contains the trilinear invariant non-trivially, the
external condition (7.6) may be written as

$$\frac{\partial I_3}{\partial \phi_a} - 2\mu \, \phi_a = 3c \, d_{abc} \, \phi_b \phi_c - 2\mu\phi_a = 0, \qquad (8.1)$$

where c is a non-zero constant and d_{abc} are the Gell-Mann d-matrices.
Then the condition (8.1) forces the extremum to lie on the U(2) orbit,
thus excluding the little group U(1) x U(1) for this case.

A more general result in this direction which will be described
in detail in a later publication, is that for any representation R
for which the fields ϕ can be written as a matrix Φ which transforms
by conjugation i.e.

$$\Phi \rightarrow \Phi' = U(g) \, \Phi U^{-1}(g),$$

(e.g. the adjoint representations of SU(n), the symmetric tensor
representations of SO(n), or any reducible combination of these)
one has the results:
(a) For renormalizable potentials the extrema are matrices with
     at most <u>three</u> distinct eigenvalues
(b) the absolute maxima and minima are matrices with at most <u>two</u>
     distinctive eigenvalues.
It is clear that such matrices with only two distinct eigenvalues
have very large little groups e.g. $H = U(m_1) \times U(m_2)/U(1)$ and

$H = SO(m_1) \times SO(m_2)$ where $m_1 + m_2 = n$ for $SU(n)$ and $SO(n)$
respectively and that the mass-spectrum for the scalar fields, which
consists of H-multiplets will be highly degenerate. In fact they
consist only of the polar Higgs mass and one or two other masses.

   A third method of attacking the orbital problem is to con-
sider the extrema $\partial V(\phi)/\partial\phi = 0$ as critical points of the invariant
functions $V(\phi)$ on the manifold R and use results from the mathematical literature on critical points to obtain some general results
concerning the nature of the extrema. This approach has been adopted
particularly by Michel,[12] and some of his more salient results may be
described as follows: For any finite dimensional representation R
of a compact group G the little groups of the orbits can be partially ordered,

$$H_1 \subset \begin{matrix} H_2 \\ H_3 \end{matrix} \begin{matrix} H_4 \\ H_5 \\ H_6 \end{matrix} \begin{matrix} H_7 \\ H_8 \end{matrix} \tag{8.2}$$

where the order is defined by inclusion as a subgroup (up to con-
jugations) and then the minimal (but not necessarily the maximal)
little group is unique. For example, for the adjoint representation
of $SU(4)$ there are four possible orbits and the ordering of the
little group is

$$U(1) \times U(1) \times U(1) \subset U(2) \times U(1) \begin{matrix} U(2) \times U(2)/U(1) \\ U(3). \end{matrix}$$

It then turns out that, in general, the minimum of the potential
will fall on the minimal orbit H_1 only if the potential contains
no invariant except the second-order one $V(\phi) = V((\phi,\phi))$. If
V contains any higher order invariant $I_s(\phi)$, $s > 2$ then the
potential minimum tends to fall on one of the maximal orbits. For
example, as mentioned above, for the adjoint representations of
$SU(n)$ the minima of renormalizable potentials occur for matrices
with two distinct eigenvalues, and such matrices have maximal
little groups $U(m_1) \times U(m_2)/U(1)$ where $m_1 + m_2 = n$. The reason for
the abrupt change for $s > 2$ is that, unlike the second order invariant which merely fixes the norm of ϕ leaving the direction in
the representation unchanged, the higher order invariants are not
isotropic in the group representation and tend to pick out degenerate directions for their minima. One consequence of this tendency is that if one wishes to avoide large little groups (and
presumably one wishes to have only $U(1)$ (or $U(1) \times SU(3)$-colour) as
final little group) one is forced towards those representations
which simply do not have higher order invariants. These are

usually the lowest-dimensional representations, a result which may
explain why the lowest-dimensional representations, in particular
the fundamental representations, are used in most models, and seem
to work quite well.

Appendix

To show that the gauge-field mass-matrix is diagonal the
Cartan basis if $\overset{o}{\phi}$ is an eigenstate of the Cartan operators we
write

$$H_i \overset{o}{\phi} = \lambda_i \overset{o}{\phi} \,,$$

and then

$$M_{ij} = e^2(\overset{o}{\phi},H_iH_j\overset{o}{\phi}) = e^2(\overset{o}{\phi},\overset{o}{\phi})\, \lambda_i\lambda_j \,,$$

$$M_{\alpha j} = e^2(\overset{o}{\phi},\, E_\alpha H_j \overset{o}{\phi}) = e^2\lambda_i(\overset{o}{\phi},E_\alpha\overset{o}{\phi}) = 0,$$

and for $\alpha_i \neq \beta_i$

$$M_{\alpha\beta} = e^2(E_\alpha\overset{o}{\phi},E_\beta\overset{o}{\phi})$$

$$= e^2(\overset{o}{\phi},E_{-\alpha}E_\beta\,\overset{o}{\phi})$$

$$= \frac{e^2}{\beta_i - \alpha_i}\, (\overset{o}{\phi},\, [H_i,\, E_{-\alpha}E_\beta]\,\overset{o}{\phi}) = 0,$$

which establishes the result.

References

(1)   C. Jacobi, Acad. des Sciences, 1834, Gesammelte Werke, Vol.
      2 p. 17-72 (Berlin: Reimer 1897)
      S. Chandrasekhar, Ellipsoidial Figure of Equilibrium (Yale,
      University Press 1969).
(2)   J.C. Taylor, Gauge Theories of Weak Interactions (Cambridge
      University Press 1976)
      M. Beg and A. Sirlin, Ann. Rev. Nucl. Sci. 24, 379(1974)
      E. Abers and B. Lee, Phys. Rep. 9C, 1973.
(3)   H. Weyl, Zeitz. f. Phys. 56, 330 (1929).

(4)   R. Streater, Rep. Prog. Phys. $\underline{38}$, 771 (1975)
      L. Lopuszanski, Fort. Phys. $\underline{22}$, 295 (1974).
(5)   A. Salam and J. Ward, Phys. Lett. $\underline{13}$, 168 (1964).
(6)   S. Weinberg, Phys. Rev. Lett. $\underline{19}$, 1264 (1967).
(7)   P. Goddard and D. Olive, New Developments in the Theory of
      Magnetic Monopoles (Cern preprint TH 2445 1978) Rep. Prog.
      Phys. 1979.
(8)   L. O'Raifeartaigh, S-Y Park and K.C. Wali (Phys. Rev.).
(9)   J. Goldstone, Nuovo Cim. $\underline{19}$, 154 (1961).
(10) P. Higgs, Phys. Rev. Lett. $\underline{13}$, 508 (1964) Phys. Rev. $\underline{145}$, 1156
      (1966)
      T. Kibble, Phys. Rev. $\underline{155}$, 1554 (1967)
      S. Weinberg, Rev. Mod. Phys. $\underline{46}$, 255 (1974).
(11) L. Michel and L. Radicati, Symmetry Principles and High Energy
      (Benjamin New York 1968)) Evolution of Particle Physics (Aca-
      demic Press New York 1970)
      Ann. Inst. Poincare $\underline{18}$, 185 (1973)
      L-F. Li, Phys. Rev. $\underline{D9}$, 1723 (1974).
(12) L. Michel, Non-Linear Graph Action (Jerusalem, Israel Univ.
      Press 1972).

SYMMETRY BREAKING AND FAR-FROM-EQUILIBRIUM ORDER

Peter Ortoleva

Department of Chemistry
Indiana University
Bloomington, Indiana 47405

1. INTRODUCTION

If a drop of ink is placed in a glass of water the eventual out-
come is macroscopically homogeneous inky water. This randomization
is a result of the second law of thermodynamics. In contrast to this
is the familiar Bénard instability[1] wherein a temperature inversion
(i.e. heating a layer of liquid from below) causes the onset of an
organized flow pattern when the temperature difference between the
top and the bottom exceeds a critical value. This transition from
purely conductive to convective transport of energy is sustained
under the influence of a net driving force that maintains the system
sufficiently far from equilibrium. The self organization of the
molecular motion in the layer of liquid is not a violation of the
second law of thermodynamics. Indeed the onset and sustance of
pattern is at the expense of a net overall increase of entropy of
the universe as energy is transported from a hot to a cold reservoir.

A self organization may involve a breaking of symmetry either
in space or in time. In the case of the Bénard instability, symmetry
has been broken with respect to the direction parallel to the heated
layer. Examples of temporal symmetry breaking are perhaps most
common in biological systems.[2] A system fed with reactants (food)
and emptied of products may evolve in a temporally periodic fashion
as opposed to a simple steady state condition. Examples include
anerobic glycolytic oscillations in yeast and many other single
cellular systems and probably human (male and female) monthly and
other cycles.

An examination of the essential features needed to obtain these
phenomena of symmetry breaking reveals that the dynamics must be

both nonlinear and sufficiently far from equilibrium. The latter
requirement has been stressed by Prigogine and coworkers.[3] Indeed
that these conditions are closely related is clear when one recognizes that a measure of distance from equilibrium with respect to
inherent system properties (transport and reaction rate laws) can
only be cast in a nonlinear system. In the remainder of this lecture we shall examine pattern formation phenomena in physico-chemical
systems--demonstrating that these phenomena are not isolated but
ubiquitous.

2. BIO-PATTERNING: EARLY IUCUS DEVELOPMENT

The idea of time ordering in biology dates back at least as far
as Lotka and Volterra who, in the early part of this century, proposed that preditor-prey interactions could lead to oscillatory
populations.[4] Turing proposed that some of the spatial self organization phenomena in biology could be accounted for via the instability of reacting-diffusing systems to pattern forming perturbations.[5]

The best agreement between experiment and theory in the field
of nonequilibrium order in biology is the limit cycle oscillation of
glycolytic intermediates in yeast.[6] In this case workers have suc-
ceeded in reproducing oscillations in cell free stirred tank reactors. These oscillations have been described by a quantitative
chemical rate theory calculation. Also many examples of biological
oscillations have been found in heart and nerve cells, mitochondria,
intestinal muscle cells, slime molds, and many other systems.[2]

Turing's hypothesis has to date met with somewhat less success
in acceptance in the biological community. Let us consider in some
detail a very suggestive example--the early development of the sea
weed Fucus.[7]

At the one cell state the Fucus egg is a theorists delight.
It is spherically symmetric with nucleus at the center and no
apparent inherent asymmetry. However after the first division,
the two cell stage is very asymmetric; the rhizoid cell develops
into the roots and the thallus is committed to leaf-stem development. In a homogeneous culture medium the axis of the polarization
is random while in the presence of gradients the axis is fixed by
the externally imposed direction. For example in the presence of
unidirectional light the thallus arises at the pole facing the light
as one might expect (i.e. Nature would not like the roots to grow
to the sun!).

The process

$$(1 \text{ cell egg}) \rightarrow \text{Thallus} \cdot \text{Rhizoid}, \qquad\qquad (\text{II.1})$$

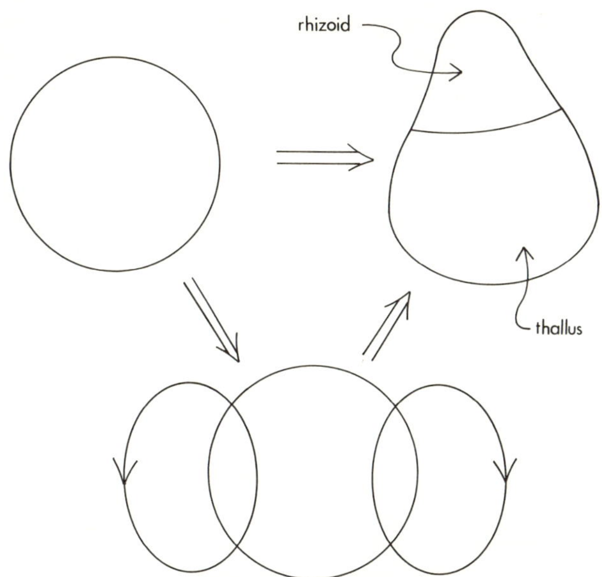

Fig. 1. Symmetry breaking in the egg of <u>Fucus</u> showing intermediate
electrically polarized state with attendant ionic currants
(directed lines).

shown in Fig. 1, is preceded by an apparently more fundamental--
although more subtle--process. Before cell division and even appre-
ciable protein synthesis, apparently ruling out genetic effects[8] an
electrical polarity with attendent ionic currents sets in as shown
also in Fig. 1.

The development of polarity in <u>Fucus</u> has been proposed as a
possible example of a Turing symmetry breaking phenomena.[9] In that
theory it is postulated that the phenomena could be described in
terms of the macroscopic continuity equations

$$\frac{\partial \underline{c}}{\partial t} = -\vec{\nabla} \cdot \underline{\vec{J}} + \underline{R} \qquad (II.2)$$

(c_i, $\vec{J}_i$ and R_i being concentration, flux and reaction rate for
species i) coupled to Poisson's equation

$$\nabla^2 V = \frac{-4\pi}{\varepsilon} \rho \qquad (II.3)$$

assuming constant dielectric constant ε. This equation for the
voltage is coupled to the concentrations through the charge density
ρ,

$$\rho = \sum_i z_i F c_i \qquad (II.4)$$

where z_i is the valence of species i and F is Faraday's constant.
The rates of reaction R_i are local processes and depend on $\underline{c}$. If
we assume Fick's law,

$$\vec{\underline{J}} = -(\underline{\underline{D}}\vec{\nabla}\underline{c} + \underline{\underline{M}}\underline{c}\vec{\nabla}V), \tag{II.5}$$

where $\underline{\underline{D}}$ and $\underline{\underline{M}}$ are matrices of diffusion and mobility coefficients,
then $(\overline{\overline{II.2,3}})$ comprises a closed set of equations. The theory is
completed by providing boundary conditions at the cell membrane that
relate the component of $\vec{J}$ normal to the membrane to the transmembrane
flux and possible membrane localized reaction. The membrane appears
in the theory as a boundary condition at the outer and inner sides
of the cell membrane. In particular let $\pm\hat{n}$ denote the unit vectors
to the cell membrane pointing into the outside world and inside the
cell. Then if J_i^m is the membrane flux (assumed perpendicular to
the membrane) and $G_i^{\pm}$ is the rate of surface localized reaction outside and inside the cell at the membrane, we obtain

$$\hat{n}\cdot\vec{J}_i^{\pm} = \mp J_i^m + G_i^{\pm}$$

where $\vec{J}_i^{\pm}$ is the flux of species i evaluated at the interfare between
the cell membrane and the outside world/cell interior.

In this treatment we focus on the onset of the electrical
polarization. In particular we seek solutions which depend on the
polar angle and not simply on the radial distance from the cell
center. Such solutions are of interest here if they are stable.
Because of rotational invariance we expect that polarity in this
model should actually be marginally stable to polar axis rotation.
This is indeed found to be the case for several hours after polarity
has been established. (At a later stage this lability of the direction of polarity is lost as another process seems to set in and
freeze the direction of polarity.) Because of rotational invariance
the axis of polarization is arbitrary--depending on initial fluctuations unless an imposed asymmetry is present or imposed.

Innumerable mechanisms can lead to self sustained asymmetry.
A necessary feature is nonlinearity in at least one of the phenom-
onological relations for R_i or the membrane transport or localized
reaction. Perhaps the simplest of these is a case where the pumping
in of some ionic species is promoted by having those ions at the
surface of the membrane inside the cell--an autocatalytic type feed-
back relationship. A detailed analysis of a model of this type has
been carried out and it is found that a state of self sustained currents through a cell could be obtained under physiological values
of cell parameters (cell size, ionic strength, transport coefficients).

A variety of qualitative features of biological self organization has been described by Turing like mechanisms--see the work of J. Cowan in this series of lectures for some interesting examples.

The group theoretical approach outlined by D. Sattinger in the previous lecture should be of great value in analyzing biological problems and in particular <u>Fucus</u>. This technique has the advantage that it can extract valuable information that transcends many details of the biological mechanism.

3. SOLITONS

Solitons are propagating disturbances of descriptive variables (such as concentration, pressure or temperature) which move with constant profile and velocity. They have been found in many non-linear field theories and reaction-diffusion continuum mechanics is no exception. However although no exception, the latter is probably unique because the nonlinearity--usually of a very narrowly defined type (as in hydrodynamics, elementary particle theory or magnetic systems) arises from chemistry (the term $\underline{R}$ in (II.2)) and can be of essentially any reasonable analytic form. This adds a unique richness to the theory of chemical waves. (See Ref. 10 for a review and citations on chemical waves.)

The simplest phenomena are plane waves. Let the wave propagate along the x axis. Then for a system of uncharged species the reaction-diffusion equations (II.2) with Fick's law (II.5) becomes

$$\frac{\partial \underline{c}}{\partial t} = \underline{\underline{D}} \frac{\partial^2 \underline{c}}{\partial x^2} + \underline{R} \ . \tag{III.1}$$

Waves may have constant or varying profiles or velocity. For stationary waves the profile is unchanged in the frame of reference moving with the wave, taken to have velocity v; it is convenient to make a Galelian transformation $\phi = x - vt$ to the wave fixed frame. In this case we have

$$\underline{\underline{D}} \frac{d^2 \underline{c}}{d\phi^2} + v \frac{d\underline{c}}{d\phi} + \underline{R} \ (\underline{c}) = \underline{0} \ . \tag{III.2}$$

An almost unbelievable number of phenomena have been found embedded in this equation. The solutions to (III.2) may be categorized as follows:
    (1) Pulses are of finite extent and leave the medium in the same state in their wake as it was in advance of their arrival (see Fig. 2a);
    (2) Fronts leave the system in a different state in the wake than it had in advance of their arrival (see Fig. 2b);
    (3) Infinite Periodic Wave Trains (see Fig. 2c).

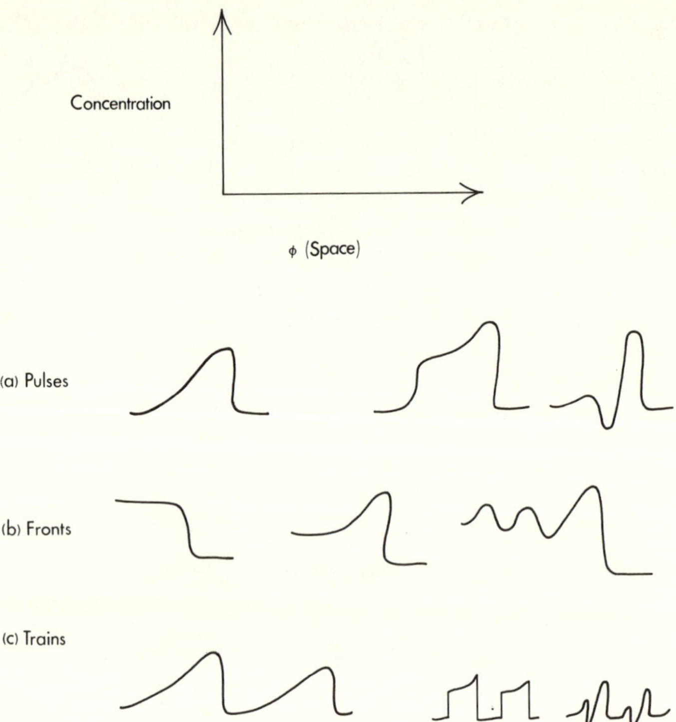

Fig. 2.   Examples of stationary wave profiles for (a) pulses, (b)
          fronts and (c) periodic wave trains as described in
          Sect. III.

Within each of these broad categories exists a variety of interesting phenomena. For example pulses can consist of a single maximum
(or minimum) or a finite train of undulations. Fronts can exist
with different transition profiles between the same asymptotic
($\phi \rightarrow \pm \infty$) states.[11] Hybrid situations may also exist as for a front
which is monotonic up to a transition region followed by a periodic
train in its wake.[12]

The richness does not stop here. In the transition from one
to two dimensions a wonderful new phenomena becomes possible--the
spiral wave of Winfree[13] shown in Fig. 3. Spiral waves are produced
in a thin layer of reactant. The two dimensional profile is stationary in a frame of reference rotating with an appropriate angular
frequency although the core of the spiral is sometimes chaotic.[13b]

Chemical waves are a symmetry breaking of the otherwise homogeneous medium. They may self organize if the medium is unstable
to either infinitesimal or finite amplitude pattern forming fluctuations. In the latter case thermal noise is typically very ineffi-
cient at creating sufficiently large fluctuations. An experimental
example of such an excitable medium is the system used by Winfree
to study waves[13] or the more familiar case of nerve impulse trans-
mission.[14]

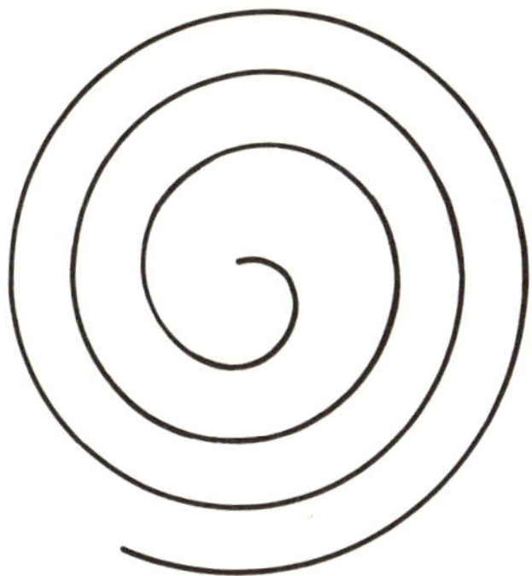

Fig. 3.   Schematic picture of spiral wave. The line indicates the
          concentration maximum of a species participating in the
          wave. The picture indicates a snapshot in time of a
          counter rotating wave.

A variety of successful approaches have been developed to study
nonlinear chemical waves. Many authors have used bifurcation theory
which is a small amplitude approach that balances off the tendency
toward exponential growth of the linearized equations with the lowest
order nontrivial nonlinearity.[15] A closely related scaling approach
results in Ginsbury-Landau type equations for the wave amplitude.[16]
Finite amplitude methods have also been successful. To one scheme
one seeks solutions which look locally like a stable temporal oscillation but which have weakly spatially varying phase.[17] The existence of homoclinic orbits in the homogeneous evolution can also be
used to generate wave solutions.[18]

A multiple time scale analysis has also led to strictly nonlinear techniques.[11,12,19] To illustrate the idea consider a two
species (X,Y) system with chemical rates given by

$$R_x = \frac{1}{\varepsilon} R(X,Y), \quad R_y = S(X,Y). \tag{III.3}$$

In these expressions R and S are independent of ε. It is common in
chemically reacting systems that processes evolve on vastly different
time scales. In the scheme (III.3) this would be indicated if
$\varepsilon << 1$. For waves in such systems we seek the solution of

$$D_x \frac{d^2X}{d\phi^2} + v \frac{dX}{d\phi} + \frac{1}{\varepsilon} R(X,Y) = 0 \tag{III.4}$$

$$D_y \frac{d^2Y}{d\phi^2} + v \frac{dY}{d\phi} + S(X,Y) = 0 . \tag{III.5}$$

Here we assume the matrix $\underline{\underline{D}}$ is diagonal. In the multiple time
scale limit, $\varepsilon \to 0$, we see that either X varies very rapidly with
ϕ or else X and Y lie near the "behavior surface" or "slow manifold"

$$R(X,Y) = 0 \quad . \tag{III.6}$$

Clearly the evolution of such a system depends crucially on the
topology of this surface. Catastrophe theory has been used to categorize the types of geometric features that may arise on these
behavior surfaces and with this categorize chemical waves in multiple
time scale systems.[12] This approach naturally led to the prediction
of a variety of new phenomena.

In some cases the presence of multiple scales can be used to
reduce the chemical wave equation to a Stefan or moving boundary
problem.[20] Moving boundary problems[21] in diffusing systems relate
the position of the boundary to the surface kinetics and medium
transport as in the determination of the shape of a growing body.
In a reduction of the chemical wave equations of the Field, Koros
and Noyes kinetics[22] for the Zhabotinsky chemical wave medium[23]
Murray[24] has cast the wave equations in the reduced form

$$\frac{d^2 u}{d\phi^2} + v\,\frac{du}{d\phi} + u(1 - u) - \frac{uw}{\varepsilon} = 0 \tag{III.7}$$

$$D\,\frac{d^2 w}{d\phi^2} + v\,\frac{dw}{d\phi} - \frac{buw}{\varepsilon} = 0 \tag{III.8}$$

where ε, b and D are constants. The boundary conditions are

$$u(-\infty) = 1, \quad u(+\infty) = 0 \tag{III.9}$$

$$w(-\infty) = 0, \quad w(+\infty) = 1 \quad . \tag{III.10}$$

As $\varepsilon \to 0$ we see that the product of the (scaled) concentrations
u, w must vanish or else u and w will vary rapidly with ϕ.
It is found[20] that the wave profile is schematically as in Fig. 4
in the limit $\varepsilon \to 0$. The velocity v of the boundary between the regions u = 0 and w = 0 can be found from asymptotic expansion methods.
(See Ref. 20 for details.)

Another finite amplitude method has been used to describe non-
planar waves such as those of circular or spiral geometry. The
method involves the use of Padé approximants to match solutions
valid near the wave center to those of essentially plane wave
character far from the center.[14]

The seemingly innocent act of applying an electrical field to
a propagating wave involving ionic species can have a profound
effect.[25,26] It was shown that waves could be stopped or reversed
and most interestingly that new types of waves, not present in the
field free medium, could be induced in experiments on a Zhabotinsky-
like medium.[26]

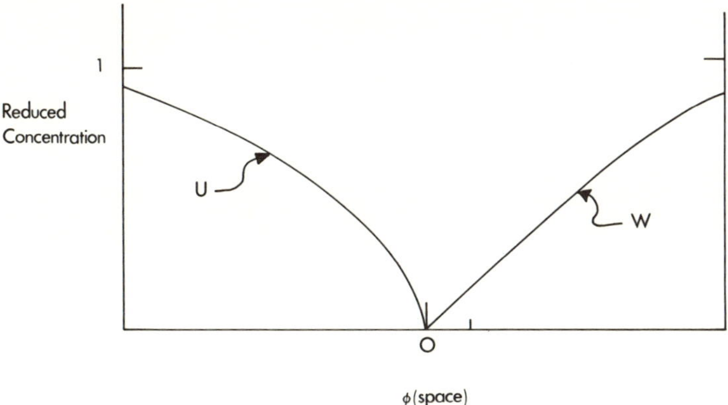

Fig. 4.   Schematic wave profile of reduced FKN waves showing
         Stefan like behavior.

Finally it should be mentioned that it is possible to find
solutions to the chemical wave equations which are unstable to
infinitesimal perturbations and hence are physically unobservable.
Results on this difficult mathematical question are infrequent
although some progress has been made. In many cases researchers
have resorted to numerical simulation of the partial differential
equations to test stability.

4. SELF ORGANIZATION DURING FIRST ORDER PHASE TRANSITIONS

Spatially periodic precipitation--the so called Liesegang bands,
was probably the earliest extensively studied example of self organization in a physico-chemical system. Indeed many papers on this
subject appeared in the latter part of the 19th century and Hedges
and Myers wrote a book entitled <u>Physico-Chemical Perrodicity</u> in the
1920's.[27]

In a typical experiment demonstrating this phenomena a solution
of potassium iodide is placed in a tube partially filled with a gel
solution containing lead nitrate. The result is regular (but not
periodic) bands of precipitation as the iodine diffuses into the
gel. An example is shown in Fig. 5a. If an electric field is
applied across the system the banding becomes essentially periodic
as in the example shown in Fig. 5b.

The traditional Liesengang experiment geometry is inherently
asymmetric in the direction of the patterning since the bands form

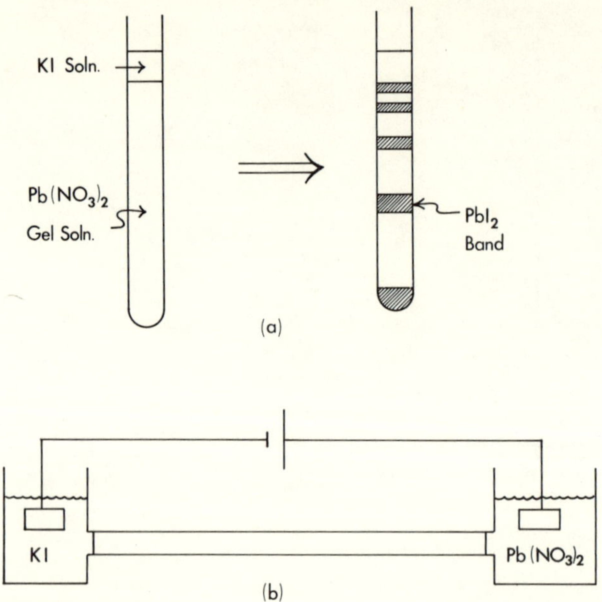

Fig. 5. Liesengang bands in a Pb(NO$_3$) – KI system (a) and in the
presence of an applied electric field[2] (b).

as a consequence of cross gradients of coprecipitates. However a
symmetry breaking does occur in the sense that the precipitation is
not monotonic down the tube as one might have expected.

More recently it was found that patterning could arise under
conditions of homogeneous precipitation.[28-30] In one set of experiments lead iodide, gel and water was heated and allowed to cool in
a thin layer in a petri dish. First the gel set. Then the system
became homogeneously yellow as the lead iodide precipitated out of
solution. After times ranging from hours to days (depending on
initial concentrations) the homogeneous yellow precipitate became
more and more inhomogeneous. Results are shown in Fig. 6. Typically the patterns are motted or mosaic as seen in Fib. 6a but spiral
(Fig. 6b) or halo (Fig. 6c) features occasionally appear.

The "Ostwald–Praeger" theory of Liesegang bands[31] depends on
the cross gradient geometry of the original Liesegang experiment.
This theory relies on sequential supersaturation, nucleation and
depleation resulting in a clear space separating precipitation bands.

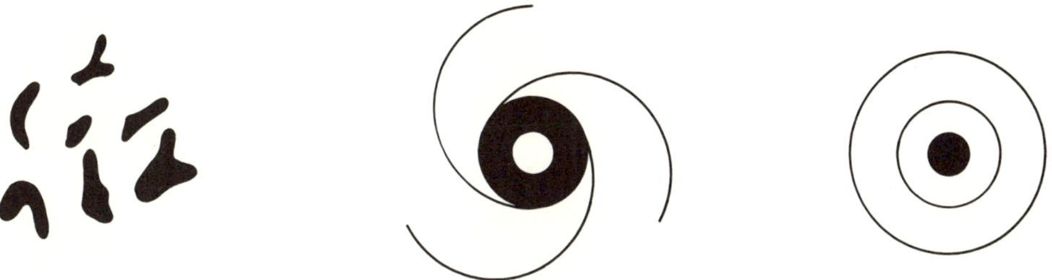

Fig. 6.   Spontaneous precipitation patterns arising from the aging
          of a PBI_2 sol (see Ref. 29).

This idea predicts that the band sparing should increase down the
tube as it typically--but not always--does. This theory may explain the old experiment but it certainly cannot explain the development of patterning from an initially homogeneous sol--the uniform
yellow state in the more recent experiments.

To explain these new experiments a theory based on the effect
of surface tension in the competion of particles of various sizes
to grow from Ph^{2+} and I^- has been introduced. Because of surface
tension particles of larger radii of curvature can exist at an
equilibrium concentration lower than that for smaller radii of
curvature. This idea was cast in mathematical form[29,30]. A
simple treatment will now be briefly considered to demonstrate
the idea.

Because of charge neutrality coupling in the PbI_2 system with
no background electrolytes the concentration of I^- is always twice
that of Ph^{+2} (in the absence of ion complexes). Thus we describe
the system locally with the salt concentration $c(\vec{r},t)$ at a point $\vec{r}$
and time t. Although the particles are distributed in size after
a very short time in any small local volume element, particles dis-
tinctly smaller than a typical size will dissolve and particles
appreciably greater than a typical size will not have had time to
grow. Thus for simplicity let us assume that we can describe the
system by the local typical particle radius $R(\vec{r},t)$. We shall assume
that nucleaton has occurred and hence R is fairly large at all points
in the system so that we can neglect diffusion of the precipitate.
This would be a particularly good approximation in a gel. Thus we
write

$$\frac{\partial R}{\partial t} = F(R,c) \ . \tag{IV.1}$$

$F(R,c)$ must vanish when $c = c^{eq}(R)$ where $c^{eq}(R)$ is the radius (R) of
curvature dependent equilibrium concentration. Thus we write

$$F(R,c) = \frac{1}{\varepsilon} \, q(R,c)\,[c - c^{eq}(R)] \tag{IV.2}$$

where q is some positive function of R and c which can be determined
explicitly if one makes specific assumptions as to whether the rate
of addition to the particle is diffusion (the usual assumption) or
surface kinetic limited. We introduce a factor $1/\varepsilon$ to suggest that
on the time scale of interest for the development of patterning
(i.e. hours or days) the evolution to the local equilibrium concentration, $c \to c^{eq}(R)$, is rapid.

The continuity equation for c may be easily written down from
conservation of mass. One finds, letting ρ be the molar density of
PbI_2 solid,

$$\frac{\partial c}{\partial t} = D\nabla^2 c - 4\pi n R^2 \rho \partial R/\partial t \ , \tag{IV.3}$$

where n is the number density of precipitate particles. For simplicity we focus on the post nucleaton phase so that n is to be taken
as a constant.

Since the time scale of interest is very long we have studied
the evolution of the system in the limit $\varepsilon \to 0$ using a multiple time
scale approach.[29] To lowest order in this scheme we find $c = c^{eq}(R)$
and to the next order a closed equation for c is obtained which involves nonlinear transport. For small derivations δR from homogeneity this equation is of the form

$$\frac{\partial \delta R}{\partial t} = -\overline{D}\nabla^2 \delta R$$

where $\overline{D}$ is positive. This is a diffusion equation with a negative
diffusion coefficient and hence patterns are amplified. The development of the patterns is given by a closed nonlinear equation for
$c(\vec{r},t)$ that results in the limit $\varepsilon \to 0$ from the scaling approach.[29]

Physically it is easy to understand the meaning of this theory.
Suppose in one region $R(\vec{r},0)$ is slightly larger than in its surroundings. Then the equilibrium concentration in the surroundings will
be larger and diffusion will drive salt into the region of initially
larger R. This makes c larger than the equilibrium value there and
hence these particles become even bigger. This leads to a runaway
phenomena creating a local maximum of precipitate surrounded by a
depleation ring. A more detailed discussion shows that there should
also be the induction of a ring of precipitation maximum around the
depleation ring and so forth. This is born out via numerical solution of the competitive particle growth equations (IV.1,3).[29]

It remains to show whether the competitive particle growth theory complements or superceeds the Ostwals–Praeger theory[31] in the cross gradient situation.

5. CRYSTAL GROWTH PATTERNS

Another example of self organization associated with a first order phase transition occurs at the level of the growth of an individual crystal. There are two basic phenomena. In crystalization very close to equilibrium conditions the crystal geometry is prismatic according to the equilibrium crystal habit, and, if the crystal is a solid solution, it is homogeneous in composition. Under far from equilibrium conditions one finds the two elementary phenomena of changes of crystal habit (shape) and compositional zoning (spatial variations).

A. Transitions to NonEquilibrium Crystal Habits

If conditions are sufficiently far from equilibrium, as from a highly supersaturated solution, there is the ever present tendency towards the growth of pertrusions into regions of higher concentration. This tendency is sometimes also enhanced by the ability of pertrusions to eliminate heat of crystalization more efficiently when the surface area is locally increased. These forces are counter-balanced by the tendency to minimize the local radius of curvature due to surface tension. When the system is sufficiently out of equilibrium the balance is shifted in favor of the growth of protrusions and, for example, the oftimes exotic formations of snow crystals are the result.[32] As we all know (a bit too well in the last few years) snowflakes may take on habits that are vastly different from the equilibrium hexagonal prismatic equilibrium ice form.

This phenomena was originally studied by Mullins and Sekerke[33] and more recently by Langer and Tarski[34] and by Chadam and the author.[35] This field promises much interesting new development in the near future.

B. Periodic Zoning

Under sufficiently far from equilibrium growth conditions, crystal zoning may be monotonic but in some instances it may be periodic. As indicated schematically in Fig. 7 where the mole fraction f of one of the two end members of the Plageoclase feldspar solid solution are plotted as a function of the distance from the crystal core. An interesting example of this is found in the case of periodic zoning in Plageoclase feldspars.[36] It was traditionally

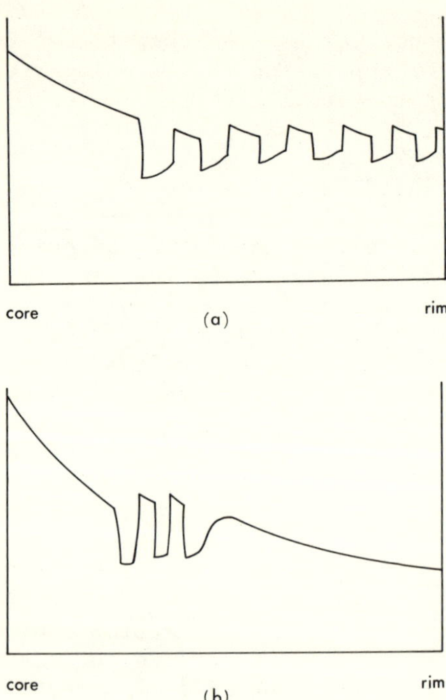

Fig. 7. Model predictions of periodic zoning in Plageclase feld-
        spars; f denotes mole fraction of Anorthite endmember.
        Shown (schematic) are examples of periodic (a) and tran-
        scent (b) banding.

believed in geological circles that this phenomena was due to
periodic variations of the conditions of crystalization in the
cooling magma. However, this seems somewhat unlikely since there
may be several hundred periodic variations.

    Recently a theory of periodic zoning has been formulated in
terms of equations of crystal growth kinetics and the diffusion of
ions in the magma that through surface attachment kinetics form
either end member crystal unit cell.[37] The theory proceeds by
coupling the continuity equation for the concentrations $c(\vec{r},t)$ of
species in the melt to the equation for the time evolution for the
surface $S(\vec{r},t) = 0$ denoting the crystal rim. In the simplest picture we have for one dimensional growth

$$\frac{\partial \underline{c}}{\partial t} = \underline{D}\, \frac{\partial^2 \underline{c}}{\partial x^2}\ , \qquad x > \overline{X}(t) \tag{V.1}$$

$$\frac{d\overline{X}}{dt} = V \tag{V.2}$$

where $\underline{D}$ is a matrix of diffusion coefficients, $\overline{X}(t)$ is the location
of the crystal rim and V is the velocity of advancement of the

crystal that depends on the concentrations $\underline{c}(\overline{\underline{X}}(t) + 0^+$, t) in the melt just at the crystal rim and on the composition of the solid in the crystal just at the rim, i.e., $\underline{X}(t) - 0^+$. The condition balancing the flux to the rim from the melt, the rate of sweeping out material by the advancing crystal and the surface attachment kinetics $\underline{G}$ (moles per area-time) is given by

$$\underline{\underline{D}} \left. \frac{\partial \underline{c}}{\partial x} \right|_{x = \underline{X}(t)} + V\underline{c} = \underline{G} \ , \qquad\qquad (V.3)$$

where $\underline{G}$ depends on the same quantities as V. The composition of the crystal at time t at the rim is determined either by the rate of deposition of the end members at a time $t - 0^+$ if the variation is smooth or via microscopic time scales if the surface composition is not compatible with that of the melt at $\underline{X}(t) + 0^+$. These considerations and surface chemical kinetics yield forms for the dependence of V and G on $\underline{c}$ and f and hence yield a complete theory (see Ref. 37 for details).

Numerical studies of the above set of crystal growth equations yield periodic zoning solutions as well as solutions which show a finite train of undulations followed by a transition to normal zoning, both types of which are found in the Plageoclase feldspar system, as seen in Fig. 7.

7. OVERVIEW OF MACROSCOPIC SELF ORGANIZATION

Macroscopic descriptions of many body systems are usually cast in terms of equations of balance for macrovariables such as local energy density, center of mass fluid velocity and the molar concentrations of various chemical species. These equations may be extended to include electrodynamic phenomena and nuclear or elementary particle reactions. The presence of nonlinearity of these equations makes it possible that sufficiently far from equilibrium multiple physical solutions may be possible. These solutions may be of essentially the same symmetry as the equilibrium state or may involve temporal or spatial symmetry breaking and the attendent self organization. Over the years it has been found that this symmetry breaking can occur in what appears to be an almost unending list of systems.

Self organization has been found or suggested in systems ranging from acoustic self amplification in the presence of reaction, membrane bound systems, chemical reactors, hydrodynamic flow, thermal diffusion coupled to reaction, metabolic pathways, ecology, plasmas, nuclear reactors, stars, the weather and socio-economic systems.

In conclusion it should be pointed out that self organization can also be analyzed in terms of a microscopic (statistical

mechanical) approach. The transitions between far from equilibrium
states have many features in common with equilibrium phase transitions. Hence critical behavior and first order phase transition
properties are being studied at this time. Typical "order parameters" are the amplitude of the spatial pattern or temporal oscillation. For example the phenomena of critical slowing down and the
divergence of correlation lengths is expected near a far from equilibrium critical point (see Ref. 38 for citations).

REFERENCES

1. See various articles in Synergetics, Proceedings of the Inter-
        national Workshop on Synergetics at Schloss Elmau, Bavaria,
        May 2-7, 1977, H. Haken, ed. (Springer-Verlag, Berlin,
        1977).
2. Periodicies in Chemistry and Biology, Vol. 4 of Theoretical
        Chemistry, H. Eyring and D. Henderson, eds. (Academic
        Press, N.Y., 1978).
3. G. Nicoles and I. Prigogine, Self-Organization in Nonequilibrium
        Systems (Wiley, N.Y., 1977).
4. See, for example, the discussion in P. Glansdorf and I. Prigo-
        gine, Thermodynamic Theory of Structure, Stability and
        Fluctuations (Wiley Interscience, N.Y., 1971).
5. A. M. Turing, Phil. Trans. Roy. Soc. (London) Ser. B, 237, 37
        (1952).
6. B. Hess and A. Boiteau, Ann. Rev. Biochem. 237, 40 (1971);
    B. Chance, E. K. Pye, A. K. Chosh and B. Hess, ed., Biological
        and Biochemical Oscillators (Academic Press, N.Y., 1973).
7. L. K. Jaffe, in Membrane Transduction Mechanisms, eds. R. A.
        Cone and J. E. Dowling (Raven Press, N.Y., 1979).
8. R. S. Quatrano, Develop. Biol. 30, 209 (1973).
9. R. Larter and P. Ortoleva, "A Theoretical Basis for Self-Elec-
        trophoreses", J. Theoret. Biol. (to appear 1980).
10. P. Hanusse, P. Ortoleva and J. Ross, Adv. in Chem. Phys.,
        XXXVIII, 317 (Wiley, N.Y., 1979).
11. P. Ortoleva and J. Ross, J. Chem. Phys. 63, 3398 (1975).
12. D. Feinn and P. Ortoleva, J. Chem. Phys. 67, 2119 (1978).
13. See article by A. T. Winfree in Ref. 2.
14. D. G. Aronson and H. F. Weinberger, Nonlinear Diffusion in Popu-
        lation Genetics, Combustion and Nerve Propagation, Lecture
        Notes in Math, 446 (Springer-Verlag, Heidelberg, 1975).
15. N. Kopell and L. Howard, Studies on Applied Math. 52, 291 (1973);
    J. A. Boa, Ph.D. Thesis, California Institute of Technology,
        Pasadena, California (1974);
    P. Ortoleva and J. Ross, J. Chem. Phys. 60, 5090 (1974).
16. Y. Kuramoto and T. Tsuzuki, Prog. Theor. Phys. 52, 1399 (1974);
        54, 687 (1975);
    A. Wunderlin and H. Haken, Z. Physik 393 (1975);

H. Haken, Z. Physik 329, 61 (1978) and references therein;
A. Nitzan, Phys. Rev. A, in press; and in Synergetics, A Workshop, H. Haken, ed., Springer 1977.

17. P. Ortoleva and J. Ross, J. Chem. Phys. 58, 5673 (1973);
P. Ortoleva, J. Chem. Phys. 64, 1395 (1976);
P. Ortoleva in Theoretical Chemistry, 4, 235 (ed. H. Eyring and
Henderson, Academic Press, N.Y., 1978).

18. N. Kopell and L. Howard, preprint.

19. P. Ortoleva and J. Ross, J. Chem. Phys. 63, 3398 (1975);
P. C. Fife, Rocky Mountain Journal of Math, 7, 309 (1977);
J. Stunshine and L. Howard, Studies in Applied Math. (1976);
J. Tyson, Lecture Notes in Biomathematics 10, (Springer-Verlag,
1976).

20. S. Schmidt and P. Ortoleva, "Asymptotic Solutions to the FKN
Chemical Wave Equations" (preprint).

21. J. R. Ockendon and W. R. Hodgkins, Moving Boundary Problems in
Heat Flow and Diffusion (Clarendon Press, Oxford, 1975).

22. R. Field, E. Koros and R. M. Noyes, J. Am. Chem. Soc. 94, 8649
(1972).

23. See in Tyson cited in Ref. 19.

24. J. Murray, J. Theoret. Biol. 56, 329 (1976).

25. S. Schmidt and P. Ortoleva, J. Chem. Phys. 67, 3771 (1977);
ibid, "Multiple Chemical Waves Induced by Applied Electric
Fields", J. Chem. Phys. (to appear) and ibid, "Electrical
Field Effects on FKN Chemical Waves" (preprint).

26. R. Feeney and P. Ortoleva, "Experimental Studies of Electrical
Field Effects on BZZW Chemical Waves" (in preparation).

27. E. S. Hedges and J. E. Myers, The Problem of Physico-Chemical
Periodicity (Longmans Green, N.Y., 1926).

28. M. R. Flicker and J. Ross, J. Chem. Phys. 60, 3458 (1974).

29. D. Feinn, P. Ortoleva, W. Scalf, S. Schmidt and M. Wolff, J.
Chem. Phys. 69, 27 (1978).

30. R. Louett, P. Ortoleva and J. Ross, J. Chem. Phys. 69, 949
(1978).

31. S. Prager, J. Chem. Phys. 25, 279 (1956).

32. G. W. Bryant, B. J. Mason, A. P. Vanden Heuvel, Phil. Mas. 8,
505 (1963).

33. W. W. Mullins and R. F. Sekerka, Crystal Growth (H. S. Peiser,
ed.) (Pergamont, Oxford) pg. 691.

34. J. S. Langer and L. A. Turski, "Studies in the Theory of Interfacial Stability I", (preprint);
J. S. Langer, "Studies in the Theory of Interfacial Stability",
(preprint).

35. J. Chadam and P. Ortoleva, "The Morphological Stability of
Growing Bodies".

36. Y. Bottinga, A. Kudo and D. Weill (1966) Amer. Mineral 51, 792.

37. J. Chadam, D. Feinn, S. Haase and P. Ortoleva, "A Theory of
Periodic Zoning in Plageoclase Feldspars", (preprint).

FINITE SUBGROUPS OF THE LORENTZ GROUP

AND THEIR GENERATING FUNCTIONS*

J. Patera and Y. Saint-Aubin

Centre de recherche de mathématiques appliquées
Université de Montréal
Montréal, Québec, Canada H3C 3J7

INTRODUCTION

Discretization of Euclidean spaces is a common feature in many
physical theories. Typically one is brought to that concept either
by the geometrical nature of the object under investigation such as
a crystal, or by considerations of technical character, for instance,
a desire to avoid occurrence of integrals diverging at small distances.

Possibilities of discretization of Minkowski space, where one
of the dimentions is interpreted as time, have not yet been thoroughly explored by physicists. Interesting in this direction might
be a field theory constructed on a lattice in Minkowski space.

Homogeneity and isotropy are the two basic properties of
Euclidean or non-Euclidean spaces with fundamental consequences
in physics. Thus there is a group G of congruent transformations
acting transitively on the points of the space. It contains a
subgroup G which acts transitively on the directions from any one
point of the space. In Minkowski space the Poincaré group plays
the role of G and the Lorentz group is G. It is natural to simulate
the basic properties of the continuous space by a lattice. For that
one chooses a discrete subgroup G of the Poincaré group acting
transitively on the lattice points and such that one of its finite
subgroups G acts transitively on the nearest neighbours of any given

point.   One thus arrives at the problem of describing finite subgroups of the Lorentz group.

Unlike the continuous subgroup of the Lorentz group $O(3,1)$
which are known for a long time[1], the finite subgroups have been
classified only recently. More precisely: An algorithm which
provides a list of representatives of $O(p,q)$-conjugacy classes of
finite subgroups of a pseudoorthogonal group $O(p,q)$, given the
finite subgroups of $O(p)$ and $O(q)$, has been developed recently[2].
This algorithm is easy to follow in the case of $O(3,1)$ group because of the particularly simple nature of the group $O(1)$, which
has only two subgroups: $\{(1)\}$ and $\{(1),(-1)\}$ of order one and two
respectively.

In this article we reproduce the list of finite subgroups of
$O(3,1)$, indicate the smallest of the five locally isomorphic
Lorentz groups which contains each finite subgroup
of generating function provide an efficient description of all
finite group invariants which are functions of space and time
coordinates.

I.   FINITE SUBGROUPS OF LORENTZ GROUPS

There are five locally isomorphic Lorentz groups:

$O(3,1)$:   the entire Lorentz group

$$\{g\,|\,g \in R^{4\times 4} \quad \& \quad g^T K_{3,1} g = K_{3,1}\},$$

where $K_{3,1} = I_3 \oplus (-I_1)$ and I_p is a $p\times p$ identity
matrix.

$SO(3)$:   the proper Lorentz group

$$\{g\,|\,g \in O(3,1) \quad \& \quad \det g = +1\}$$

$O_1(3,1)$:   the orthochronous Lorentz group

$$\{g\,|\,g \in O(3,1) \quad \& \quad \operatorname{sgn} g^t_{\ t} = +1\}$$

$O_2(3,1)$:   the orthochorous Lorentz group

$$\{g\,|\,g \in O(3,1) \quad \& \quad (\operatorname{sgn} g^t_{\ t})\times(\det g) = +1\}$$

$DO(3,1)$:   the restricted Lorentz group

$$\{g \mid g \in O(3,1) \ \& \ \text{sgn } g^t{}_t = \det g = +1\}.$$

Finite subgroups of $O(3,1)$, classified up to an $O(3,1)$ transformation, can be partitioned as follows:

(i)   the finite subgroups of $O(3,1)$ whose elements have the form $g \oplus 1$ where g is an element of $O(3)$ and 1 is the identity acting on the time coordinate

(ii)   the finite subgroups of $O(3,1)$ which are the direct product of a group of the class (i) with the group $\{I_3 \oplus (1), I_3 \oplus (-1)\}$. ($I_3$ is the identity of $O(3)$.)

(iii) the finite subgroups of $O(3,1)$ which are of the form: $((G\backslash N) \oplus (-1)) \cup (N \oplus (1))$ where G is a finite subgroup of $O(3)$ and N a representative of a $\text{Nor}_{O(3)}$ G-conjugacy class

of subgroups of G of index 2. (The subgroups are subdirect products of the group of the class (i) with the group $\{I_3 \oplus (1), I_3 \oplus (-1)\}$.)

In order to give a list of these classes, we define the following generators:

$$R_n = \begin{pmatrix} 1 & 0 & 0 & 0 \\ 0 & \cos\dfrac{2\pi}{n} & \sin\dfrac{2\pi}{n} & 0 \\ 0 & -\sin\dfrac{2\pi}{n} & \cos\dfrac{2\pi}{n} & 0 \\ 0 & 0 & 0 & 1 \end{pmatrix}, \qquad D = \begin{pmatrix} -1 & 0 & 0 & 0 \\ 0 & 1 & 0 & 0 \\ 0 & 0 & -1 & 0 \\ 0 & 0 & 0 & 1 \end{pmatrix},$$

$$R'_3 = \begin{pmatrix} 0 & 0 & 1 & 0 \\ 1 & 0 & 0 & 0 \\ 0 & 1 & 0 & 0 \\ 0 & 0 & 0 & 1 \end{pmatrix}, \qquad R'_2 = \begin{pmatrix} 0 & 0 & 1 & 0 \\ 0 & -1 & 0 & 0 \\ 1 & 0 & 0 & 0 \\ 0 & 0 & 0 & 1 \end{pmatrix},$$

$$K_{3,1} = \begin{pmatrix} 1 & 0 & 0 & 0 \\ 0 & 1 & 0 & 0 \\ 0 & 0 & 1 & 0 \\ 0 & 0 & 0 & -1 \end{pmatrix}, \quad -I_4 = \begin{pmatrix} -1 & 0 & 0 & 0 \\ 0 & -1 & 0 & 0 \\ 0 & 0 & -1 & 0 \\ 0 & 0 & 0 & -1 \end{pmatrix},$$

and

$$R''_3 = \begin{pmatrix} \alpha & \beta & 0 & 0 \\ -\beta & \alpha & 0 & 0 \\ 0 & 0 & 1 & 0 \\ 0 & 0 & 0 & 1 \end{pmatrix} \quad R'_3 \begin{pmatrix} \alpha & -\beta & 0 & 0 \\ \beta & \alpha & 0 & 0 \\ 0 & 0 & 1 & 0 \\ 0 & 0 & 0 & 1 \end{pmatrix}$$

with:

$$\alpha = \frac{1}{2}\left(\sqrt{2 + \sqrt{2\sqrt{5}-2}}\right) \quad \text{and} \quad \beta = -\sqrt{1-\alpha^2}.$$

The finite groups are described below by their generators. For each of them we also indicate its order N and the smallest of the Lorentz groups which it contains, that is the group generated by the finite group in question and by the restricted Lorentz group $DO(3,1)$. Inside each of the three classes we put together sub-groups under generic names: cyclic, dihedral, tetrahedral, octa-hedral, and icosahedral. A bracket $\langle ... \rangle$ stands for a group generated by the elements ... .

(i)   Cyclic groups:

$$\langle R_n \rangle \qquad\qquad n \geq 1, \qquad N = n \; , \quad DO(3,1)$$

$$\langle R_n \rangle \times \langle -K_{3,1} \rangle \qquad n \geq 1, \qquad N = 2n, \quad O_1(3,1)$$

$$\langle -R_n \rangle \qquad\qquad n \text{ even} \geq 2, \quad N = n \; , \quad O_1(3,1)$$

Dihedral groups:

$$\langle R_n, D \rangle \qquad\qquad n \geq 2, \qquad N = 2n, \quad DO(3,1)$$

$$\langle R_n, D \rangle \times \langle -K_{3,1} \rangle \quad n \geq 2, \qquad N = 4n, \quad O_1(3,1)$$

$$\langle R_n, -D \rangle \qquad\qquad n \geq 2, \qquad N = 2n, \quad O_1(3,1)$$

$$\langle -R_n, D \rangle \qquad\qquad n \text{ even} \geq 4, \quad N = 2n, \quad O_1(3,1)$$

Tetrahedral groups:

$$\langle R_2, D, R_3' \rangle \qquad\qquad N = 12 , \qquad DO(3,1)$$

$$\langle R_2, D, R_3' \rangle \times \langle -K_{3,1} \rangle \qquad N = 24 , \qquad O_1(3,1)$$

Octahedral groups:

$$\langle R_2', R_3' \rangle \qquad\qquad N = 24 , \qquad DO(3,1)$$

$$\langle R_2', R_3' \rangle \times \langle -K_{3,1} \rangle \qquad N = 48 , \qquad O_1(3,1)$$

$$\langle -R_2', R_3' \rangle \qquad\qquad N = 24 , \qquad O_1(3,1)$$

Icosahedral groups:

$$\langle R_2, R_3'' \rangle \qquad\qquad N = 60 , \qquad DO(3,1)$$

$$\langle R_2, R_3'' \rangle \times \langle -K_{3,1} \rangle \qquad N = 120, \qquad O_1(3,1)$$

(ii) There is a one-to-one correspondance between the groups of
the class (i) and of the class (ii). We obtain the latter
ones by doing the direct product of one of the former with
$\{I_3 \oplus (1), I_3 \oplus (-1)\}$. This operation multiplies the order by 2
and the minimal inclusion becomes $O_2(3,1)$ (resp. $O(3,1)$)
if the group of the class (i) minimally belongs to $DO(3,1)$
(resp. $O_1(3,1)$).

(iii) Cyclic groups:

$$\langle R_n K_{3,1} \rangle \qquad , \quad n \text{ even} \geq 2, \quad N = n, \quad O_2(3,1)$$

$$\langle R_n \rangle \times \langle -I_4 \rangle \qquad , \quad n \geq 1, \qquad N = 2n, \quad SO(3,1)$$

$$\langle R_n K_{3,1} \rangle \times \langle -K_{3,1} \rangle, \quad n \text{ even} \geq 2, \quad N = 2n, \quad O(3,1)$$

$$\langle R_n K_{3,1} \rangle \times \langle -I_4 \rangle \quad , \quad n \text{ even} \geq 2, \quad N = 2n, \quad O(3,1)$$

$$\langle -I_4 R_{2n} \rangle \qquad , \quad n \geq 1, \qquad N = 2n, \quad SO(3,1)$$

Dihedral groups:

$$\langle R_n, DK_{3,1} \rangle \qquad , \quad n \geq 2, \qquad N = 2n, \quad O_2(3,1)$$

$$\langle R_n K_{3,1}, D \rangle \qquad , \quad n \text{ even} \geq 4, \quad N = 2n, \quad O_2(3,1)$$

$$\langle R_n, D \rangle \times \langle -I_4 \rangle \quad , \quad n \geq 2, \qquad N = 4n, \quad SO(3,1)$$

$$\langle R_n, -I_4 D \rangle \times \langle -K_{3,1} \rangle, \quad n \geq 2, \qquad N = 4n, \quad O(3,1)$$

$$\langle R_n, DK_{3,1} \rangle \times \langle -I_4 \rangle \quad , \quad n \geq 2, \qquad N = 4n, \quad O(3,1)$$

$$\langle R_n K_{3,1}, D \rangle \times \langle -K_{3,1} \rangle, \quad n \text{ even} \geq 4, \quad N = 4n, \quad O(3,1)$$

$$\langle R_n K_{3,1}, D \rangle \times \langle -I_4 \rangle, \quad n \text{ even} \geq 4, \quad N = 4n, \quad O(3,1)$$

$$\langle R_n, -I_4 D \rangle, \quad n \geq 2, \quad N = 2n, \quad SO(3,1)$$

$$\langle R_n K_{3,1}, -DK_{3,1} \rangle, \quad n \text{ even} \geq 2, \quad N = 2n, \quad O(3,1)$$

$$\langle -R_{2n} K_{3,1}, DK_{3,1} \rangle, \quad n \geq 2, \quad N = 4n, \quad O(3,1)$$

$$\langle -I_4 R_{2n}, D \rangle, \quad n \geq 2, \quad N = 4n, \quad SO(3,1)$$

$$\langle -I_4 R_{2n}, DK_{3,1} \rangle, \quad n \geq 2, \quad N = 4n, \quad O(3,1)$$

Tetrahedral group:

$$\langle R_2, D, R_3' \rangle \times \langle -I_4 \rangle, \qquad\qquad N = 24, \quad SO(3,1)$$

Octahedral groups:

$$\langle R_2' K_{3,1}, R_3' \rangle, \qquad\qquad N = 24, \quad O_2(3,1)$$

$$\langle R_2', R_3' \rangle \times \langle -I_4 \rangle, \qquad\qquad N = 48, \quad SO(3,1)$$

$$\langle R_2' K_{3,1}, R_3' \rangle \times \langle -I_4 \rangle, \qquad\qquad N = 48, \quad O(3,1)$$

$$\langle R_2' K_{3,1}, R_3' \rangle \times \langle -K_{3,1} \rangle, \qquad\qquad N = 48, \quad O(3,1)$$

$$\langle -I_4 R_2', R_3' \rangle, \qquad\qquad N = 24, \quad SO(3,1)$$

Icosahedral group:

$$\langle R_2, R_3'' \rangle \times \langle -I_4 \rangle, \qquad\qquad N = 120, \quad SO(3,1)$$

II. GENERATING FUNCTIONS FOR INVARIANTS

An essential step toward description of invariants of a finite group acting in the Minkowski space is to find the degrees of invariant polynomials in the space coordinates x,y,z and the time coordinate t. This information is provided in a concise form by giving a corresponding generating function. More precisely, from the power series expansion

$$B(\lambda) = \sum_{n=0}^{\infty} c_n \lambda^n$$

of a generating function $B(\lambda)$ it follows that there are c_n linearly independent invariant polynomials of degree n.

When the representation acting in the Minkowski space is reducible, one uses as many distinct λ parameters as there are irreducible representations in the decomposition. Each λ_i indicates the degree in the invariant of the variables transforming themselves under the irreducible representation associated to λ_i.

The generating functions for invariants and other tensors of many finite groups of interest in physics have been recently calculated[3,4]. A recent paper[5] outlines a method for finding the generating functions in the case of $O(3,1)$ finite subgroups. For the finite groups of classes (i) and (ii) the generating functions are obtained directly from generating functions of corresponding finite subgroups of $O(3)$. Therefore here we restrict ourselves to the generating functions for the natural (reducible) representation of the groups of class (iii) only. The subscripts of λ refer to irreducible representations of finite groups as numbered in Ref.3; λ_t and λ_s refer to a representation that acts on the time and space coordinates respectively.

Cyclic Groups

$\langle R_n K_{3,1} \rangle$:

$$B(\lambda_1, \lambda_2, \lambda_n, \lambda_t) = (N_1 + \lambda_t N_2)/D_1$$

$\langle R_n \rangle \times \langle -I_4 \rangle$:

$$B(\lambda_1, \lambda_2, \lambda_n, \lambda_t) = \begin{cases} (1 + \lambda_1 \lambda_t) N_1 / D_1 & n \text{ even} \\ \\ [\; 1 + (\lambda_1 \lambda_2^n + \lambda_1 \lambda_n^n + \lambda_2^n \lambda_n^n) + \lambda_t (\lambda_1 + \lambda_2^n + \lambda_n^n) + \lambda_t \lambda_1 \lambda_2^n \lambda_n^n] N / D_2 \\ \hfill n \text{ odd} \end{cases}$$

$\langle R_n K_{3,1} \rangle \times \langle -K_{3,1} \rangle$:

$$B(\lambda_1, \lambda_2, \lambda_n, \lambda_t) = \begin{cases} (N_1 + \lambda_t N_2)/D_1 & n/2 \text{ even} \\ (N_1 + \lambda_1 \lambda_t N_2)/D_1 & n/2 \text{ odd} \end{cases}$$

$\langle R_n K_{3,1} \rangle \times \langle -I_4 \rangle$:

$$B(\lambda_1, \lambda_2, \lambda_n, \lambda_t) = \begin{cases} (N_1 + \lambda_1 \lambda_t N_2)/D_1 & n/2 \text{ even} \\ (N_1 + \lambda_t N_2)/D_1 & n/2 \text{ odd} \end{cases}$$

$\langle I_4 R_{2n} \rangle$:

$$B(\lambda_n, \lambda_{n+1}, \lambda_{n+2}, \lambda_t) = \begin{cases} [(N_4 + \lambda_{n+1} N_3) + \lambda_t (N_6 + \lambda_{n+1} N_5)]/E_2 & n \text{ even} \\[2ex] (N_6 + \lambda_t N_7)/E_1 & n \text{ odd} \end{cases}$$

Here

$$N_1 = \sum_{p=0}^{n-1} \lambda_2^p \lambda_n^p$$

$$N_2 = \sum_{p=0}^{n/2-1} \lambda_2^{n/2+p} \lambda_n^p + \sum_{p=n/2}^{n-1} \lambda_2^{p-n/2} \lambda_n^p$$

$$N_3 = \sum_{p=0}^{n-r} \lambda_n^p \lambda_{n+2}^{p+r+n-1} + \sum_{p=n-r+1}^{2n-1} \lambda_n^p \lambda_{n+2}^{p-n+r-1}$$

$$N_4 = \sum_{p=0}^{2n-r} \lambda_n^p \lambda_{n+2}^{p+r-1} + \sum_{p=2n-r+1}^{2n-1} \lambda_n^p \lambda_{n+2}^{p-2n+r-1}$$

$$N_5 = \sum_{p=0}^{3n-r} \lambda_n^p \lambda_{n+2}^{p+r-n-1} + \sum_{p=3n-r+1}^{2n-1} \lambda_n^p \lambda_{n+2}^{p-3n+r-1}$$

$$N_6 = \sum_{p=0}^{n-r} \lambda_n^p \lambda_{n+2}^{p+r-1} + \sum_{p=n-r+1}^{n-1} \lambda_n^p \lambda_{n+2}^{p-n+r-1}$$

$$N_7 = \sum_{p=0}^{2n-r} \lambda_n^p \lambda_{n+2}^{p+r-n-1} + \sum_{p=2n-r+1}^{n-1} \lambda_n^p \lambda_{n+2}^{p-2n+r-1}$$

$$E_i = (1-\lambda_{n+1}^2)(1-\lambda_n^{in})(1-\lambda_{n+2}^{in})(1-\lambda_t^2) \qquad i=1,2$$

$$D_i = (1-\lambda_1)(1-\lambda_2^{in})(1-\lambda_n^{in})(1-\lambda_t^2) \qquad i=1,2$$

Dihedral Groups

$\langle R_n, DK_{3,1} \rangle$:

$$B(\lambda_2, \lambda_5, \lambda_t) = (1 + \lambda_2 \lambda_5^n + \lambda_2 \lambda_t + \lambda_5^n \lambda_t)/D_5 \qquad n \text{ even}$$

$$B(\lambda_2, \lambda_3, \lambda_t) = (1 + \lambda_2 \lambda_3^n + \lambda_2 \lambda_t + \lambda_3^n \lambda_t)/D_3 \qquad n \text{ odd}$$

$\langle R_n K_{3,1}, D \rangle$:

$$B(\lambda_2, \lambda_5, \lambda_t) = (1 + \lambda_2 \lambda_5^n + \lambda_5^{n/2} \lambda_t + \lambda_2 \lambda_5^{n/2} \lambda_t)/D_5$$

$\langle R_n, D \rangle \times \langle -I_4 \rangle$:

$$B(\lambda_2, \lambda_5, \lambda_t) = (1 + \lambda_2 \lambda_5^n \lambda_t)/D_5 \qquad n \text{ even}$$

$$B(\lambda_2, \lambda_3, \lambda_t) = (1 + \lambda_2 \lambda_3^n + \lambda_3^n \lambda_t + \lambda_2 \lambda_3^{2n} \lambda_t)/E_3 \qquad n \text{ odd}$$

$\langle R_n, -I_4 D \rangle \times \langle -K_{3,1} \rangle$:

$$B(\lambda_2, \lambda_5, \lambda_t) = (1 + \lambda_5^n \lambda_t)/D_5 \qquad n \text{ even}$$

$$B(\lambda_2, \lambda_3, \lambda_t) = (1 + \lambda_2 \lambda_3^n + (\lambda_2 \lambda_3^n + \lambda_3^{2n}) \lambda_t)/E_3 \qquad n \text{ odd}$$

$\langle R_n, DK_{3,1} \rangle \times \langle -I_4 \rangle$:

$$B(\lambda_2, \lambda_5, \lambda_t) = (1 + \lambda_2 \lambda_t)/D_5 \qquad n \text{ even}$$

$$B(\lambda_2, \lambda_3, \lambda_t) = (1 + \lambda_2 \lambda_3^n + (\lambda_2 + \lambda_3^n) \lambda_t)/E_3 \qquad n \text{ odd}$$

$\langle R_n K_{3,1}, D \rangle \times \langle -K_{3,1} \rangle$:

$$B(\lambda_2, \lambda_5, \lambda_t) = (1 + \lambda_5^{n/2} \lambda_t)/D_5 \qquad n/2 \text{ even}$$

$$B(\lambda_2, \lambda_5, \lambda_t) = (1 + \lambda_2 \lambda_5^{n/2} \lambda_t)/D_5 \qquad n/2 \text{ odd}$$

$\langle R_n K_{3,1}, D \rangle \times \langle -I_4 \rangle$:

$$B(\lambda_2, \lambda_5, \lambda_t) = (1 + \lambda_2 \lambda_5^{n/2} \lambda_t)/D_5 \qquad\qquad n/2 \text{ even}$$

$$B(\lambda_2, \lambda_5, \lambda_t) = (1 + \lambda_5^{n/2} \lambda_t)/D_5 \qquad\qquad n/2 \text{ odd}$$

$\langle R_n, -I_4 D \rangle$:

$$B(\lambda_1, \lambda_5, \lambda_t) = (1 + \lambda_5^{n} \lambda_t)/F_5 \qquad\qquad n \text{ even}$$

$$B(\lambda_1, \lambda_3, \lambda_t) = (1 + \lambda_3^{n} \lambda_t)/F_3 \qquad\qquad n \text{ odd}$$

$\langle R_n K_{3,1}, -D K_{3,1} \rangle$:

$$B(\lambda_1, \lambda_5, \lambda_t) = (1 + \lambda_5^{n/2} \lambda_t)/F_5$$

$\langle -R_{2n} K_{3,1}, D K_{3,1} \rangle$:

$$B(\lambda_4, \lambda_{n+3}, \lambda_t) = (1 + \lambda_4 \lambda_{n+3}^{n} + (\lambda_{n+3}^{n} + \lambda_4 \lambda_{n+3}^{n})\lambda_t)/E_{n+3} \qquad\qquad n \text{ even}$$

$$B(\lambda_4, \lambda_{n+3}, \lambda_t) = (1 + \lambda_{n+3}^{n} \lambda_t)/D_{n+3} \qquad\qquad n \text{ odd}$$

$\langle -I_4 R_{2n}, D \rangle$:

$$B(\lambda_4, \lambda_{n+3}, \lambda_t) = (1 + \lambda_4 \lambda_{n+3}^{n} + (\lambda_{n+3}^{n} + \lambda_4 \lambda_{n+3}^{2n})\lambda_t)/E_{n+3} \qquad\qquad n \text{ even}$$

$$B(\lambda_4, \lambda_{n+3}, \lambda_t) = (1 + \lambda_{n+3}^{n} \lambda_t)/D_{n+3} \qquad\qquad n \text{ odd}$$

$\langle -I_4 R_{2n}, D K_{3,1} \rangle$:

$$B(\lambda_4, \lambda_{n+3}, \lambda_t) = (1 + \lambda_4 \lambda_{n+3}^{n} + (\lambda_4 + \lambda_{n+3}^{n})\lambda_t)/E_{n+3} \qquad\qquad n \text{ even}$$

$$B(\lambda_4, \lambda_{n+3}, \lambda_t) = (1 + \lambda_4 \lambda_t)/D_{n+3} \qquad\qquad n \text{ odd}$$

Here the following notations were used:

$$D_3 = (1-\lambda_2^2)(1-\lambda_3^2)(1-\lambda_3^n)(1-\lambda_t^2), \qquad E_3 = (1-\lambda_2^2)(1-\lambda_3^2)(1-\lambda_3^{2n})(1-\lambda_t^2);$$

$$D_5 = (1-\lambda_2^2)(1-\lambda_5^2)(1-\lambda_5^n)(1-\lambda_t^2);$$

$$D_{n+3} = (1-\lambda_4^2)(1-\lambda_{n+3}^2)(1-\lambda_{n+3}^n)(1-\lambda_t^2),$$

$$E_{n+3} = (1-\lambda_4^2)(1-\lambda_{n+3}^2)(1-\lambda_{n+3}^{2n})(1-\lambda_t^2);$$

$$F_3 = (1-\lambda_1)(1-\lambda_3^2)(1-\lambda_3^n)(1-\lambda_t^2), \qquad F_5 = (1-\lambda_1)(1-\lambda_5^2)(1-\lambda_5^n)(1-\lambda_t^2).$$

Tetrahedral Group

$$\langle R_2, D, R_3' \rangle \times \langle -I_4 \rangle:$$

$$B(\lambda_s, \lambda_t) = \frac{1+\lambda_s^6+\lambda_s^3\lambda_t+\lambda_s^9\lambda_t}{(1-\lambda_s^2)(1-\lambda_s^4)(1-\lambda_s^6)(1-\lambda_t^2)}.$$

Octahedral Groups

$$D = (1-\lambda_s^2)(1-\lambda_s^4)(1-\lambda_s^6)(1-\lambda_t^2)$$

$$\langle R_2'K_{3,1}, R_3' \rangle: \quad B(\lambda_s, \lambda_t) = (1+\lambda_s^9+\lambda_s^3\lambda_t+\lambda_s^6\lambda_t)/D$$

$$\langle R_2', R_3' \rangle \times \langle -I_4 \rangle: \quad B(\lambda_s, \lambda_t) = (1+\lambda_s^9\lambda_t)/D$$

$$\langle R_2'K_{3,1}, R_3' \rangle \times \langle -I_4 \rangle: \quad B(\lambda_s, \lambda_t) = (1+\lambda_s^3\lambda_t)/D$$

$$\langle R_2'K_{3,1}, R_3' \rangle \times \langle -K_{3,1} \rangle: \quad B(\lambda_s, \lambda_t) = (1+\lambda_s^6\lambda_t)/D$$

$$\langle -I_4 R_2', R_3' \rangle: \quad B(\lambda_s, \lambda_t) = \frac{1+\lambda_s^6\lambda_t}{(1-\lambda_s^2)(1-\lambda_s^3)(1-\lambda_s^4)(1-\lambda_t^2)}.$$

Icosahedral Group

$$\langle R_2, R_3'' \rangle \times \langle -I_4 \rangle: \quad B(\lambda_s, \lambda_t) = \frac{1+\lambda_s^{15}\lambda_t}{(1-\lambda_s^2)(1-\lambda_s^6)(1-\lambda_s^{10})(1-\lambda_t^2)}.$$

REFERENCES

1. P.Winternitz and I.Friš, Yad.Fiz.1:889 (1965) [Sov.J.Nucl. Phys.1:636 (1965)].

2. J.Patera, Y.Saint-Aubin and H.Zassenhaus, J.Math.Phys.21 (1980) (to be published); A.Janner and T.Janssen, Super-space Groups, Preprint of the Institute for Theoretical Physics, University of Nijmegen (1979).

3. J.Patera, R.T.Sharp and P.Winternitz, J.Math.Phys. 19:2362 (1978).

4. P.Desmier and R.T.Sharp, J.Math.Phys.20:74 (1979).

5. Y.Saint-Aubin, Can.J.Phys.58 (1980) (to be published).

ON DYNAMICAL SYMMETRIES IN RELATIVISTIC FIELD THEORIES*

R. Raczka**

Institute for Nuclear Research
00-681 Warsaw, Hoża 69
Poland

1. Introduction

One observes that fermions appear in fundamental strong electromagnetic and weak interactions in some towers

$$\Psi = \begin{pmatrix} \psi_1 \\ \vdots \\ \psi_n \end{pmatrix}$$

where ψ_k, $k = 1, \ldots, n$ are Dirac spinors. This is the case of the nucleon tower for Yukawa interactions, the electron-meson μ tower in the case of electromagnetic interactions, the electron-neutrino tower for Weinberg-Salam interactions and recently the u-d-s-c-b-t quark tower in the case of quark-gluon interactions.

The existence of tower structure of fundamental fermions suggests that there might be a higher dynamical symmetry group G for which a tower will provide a carrier space for an irreducible representation of G.

This problem is analyzed in the present paper. In Section 2 we show that the so called scalar Yukawa interaction has the local SO(4,2) conformal group as the dynamical higher symmetry group. This dynamical group realized linearly in space-time implies the existence of fermion doublet structure. The analysis of the obtained theory shows that a definite internal symmetry appears as a result of fermion dublet structure. This internal symmetry--

after natural identification of symmetry generators and scalar
fields--turns out to be a conventional isotopic spin symmetry.
This provides a natural dynamical explanation of isotopic spin
which otherwise is introduced by hands. It is interesting that
generators of isotopic spin Lie algebra are in six-dimensional
conformal space-time a representation of a group of reflections
in unphysical dimensions. This suggests that a conventional
introduction of isotopic spin rotation group might be physically
artificial.

In Section 3 we present a more general analysis of a theory
of fermion-boson system starting from the hypothesis that fermions
form a tower which transforms according to the irreducible representation of a certain Lie group G and that the total Lagrangian
is invariant under G. Starting from these natural assumptions
one derives the form of G, the number of fermions in the tower,
the explicit form of renormalizable trilinear interactions and
the specific tensor field content in considered Fermi-Bose system.
These results illustrate the usefulness of the concept of a dynamical group for a Fermi-Bose system.

2. Dynamical Symmetry in the Yukawa Theory

Consider first the so called scalar Yukawa model given by
the following total Lagrangian

$$L_T = L_0(\Phi) + L_0(\Psi) + g\overline{\Psi}\Gamma\Psi\Phi, \quad \Gamma = \pi \text{ or } \gamma_5, \tag{2.1}$$

where

$$L_0(\Phi) = \frac{1}{2}(\partial_\mu\Phi\partial^\mu\Phi - m^2\Phi^2) \text{ and } L_0(\Psi) = \overline{\Psi}(\frac{i}{2}\gamma_\mu\overleftrightarrow{\partial}^\mu - M)\Psi. \tag{2.2}$$

Here $\Phi(x)$, $x \in M^{3,1}$ is a classical scalar field and $\Psi(x)$ is the
Dirac field.

Proposition 2.1. The scalar Yukawa theory defined by the Lagrangian
(2.1) is locally $SO(4,2)$ covariant.

Proof. For dimensional reasons, the Lagrangian density (2.1) must
transform under a passive change of scale $x_\mu \to x'_\mu = \rho x_\mu$ as a dilation density of order -4, i.e.,

$$L_T(\Phi', \partial'_\mu\Phi', \Psi', \partial'_\mu\Psi', m', M')$$

$$= L_T(\rho^{n_\Phi}\Phi, \rho^{(n_\Phi-1)}\partial_\mu\Phi, \rho^{n_\Psi}\Psi, \rho^{(n_\Psi-1)}\partial_\mu\Phi, \rho^{-1}m, \rho^{-1}M)$$

$$= \rho^{-4}L_T(\Phi, \partial_\mu\Phi, \Psi, \partial_\mu\Psi, m, M). \tag{2.3}$$

Hence, from the form (2.2) of the free Lagrangian for Φ and Ψ it follows that $n_\Phi = -1$ and $n_\Psi = -3/2$; therefore g must be dimensionless.

It has now been proved by Flato, Simon and Sternheimer that the field equations derived from a density Lagrangian are local conformal covariant if and only if the following conditions are satisfied (cf. Ref. [1], Sec. 3):

1) L_T is a local Poincaré invariant Lagrangian depending on fields and their first derivatives, which is a dilation scalar density of order -4.

2) The vector field

$$R_\mu = -n_\Phi \frac{\delta L_T}{\delta(\partial_\nu \Phi)} g_{\mu\nu}\Phi + \frac{\delta L_T}{\delta(\partial_\nu \Psi)} (-n_\Psi g_{\mu\nu} + M_{\mu\nu}^\psi)\psi$$

$$+ \bar{\psi}(-n_{\bar\psi}g_{\mu\nu} + M_{\mu\nu}^{\bar\psi}) \frac{\delta L_T}{\delta(\partial_\nu \bar\Psi)} , \qquad (2.4)$$

where n_Φ, n_ψ, $n_{\bar\psi}$ are the conformal degrees of fields Φ, ψ, $\bar\psi$ and $M_{\mu\nu}$ is the spin part of generators of the Lorentz transformation, satisfies the condition

$$\delta \int R_\mu(x)d^4x = 0. \qquad (2.5)$$

Here δ means a variation due to the infinitesimal conformal transformations.

Condition 1) is satisfied in our case by the considerations given above. To verify condition 2) we note that if the coupling is nonderivative, the vector R_μ has contributions from the free Lagrangians only, i.e.,

$$R_\mu = \frac{1}{2} \partial_\mu \Phi^2 + R_\mu^\psi + R_\mu^{\bar\psi} .$$

It follows from (2.4) and (2.1) that $R_\mu^{\bar\psi} = -R_\mu^\psi$. In addition, in order that the energy operator P_0 in the Yukawa theory be finite, the field $\Phi(t,\vec{x})$ must go to zero for every $|x_k| \to \infty$, $k = 1,2,3$; moreover, in order that Φ has the asymptotic fields Φ_{in} and Φ_{out}, it must decay according to the law $\Phi(t,\vec{x}) \sim |t|^{-3/2}$ for large t. Hence,

$$\int R_\mu(x)d^4x = 0. \qquad (2.6)$$

Thus by the Flato, Simon and Sternheimer theorem, the assertion of Proposition 2.1 follows.

Proposition 2.1 shows that scalar Yukawa theory formulated on $M^{3,1}$ space-time is implicitly locally SO(4,2) covariant. However, our experience in Particle Physics teaches us that a good symmetry is the one which is realized as an explicit symmetry. To achieve this we extend the Lorentz spinors $\Psi(x)$ on $M^{3,1}$ to conformal spinors $\xi(\eta)$ on $M^{4,2}$. As it is well known, there are two conformal spinor representations D^+ and D^- characterized by the highest weights

$$m^+ = [\tfrac{1}{2}, \tfrac{1}{2}, \tfrac{1}{2}] \quad \text{and} \quad m^- = [\tfrac{1}{2}, \tfrac{1}{2}, -\tfrac{1}{2}] \tag{2.7}$$

respectively. Both D^+ and D^-, when restricted to SO(3,1), give the Dirac representation $D^{(\frac{1}{2},0)} \oplus D^{(0,\frac{1}{2})}$; however, under any reflection $D^+ \leftrightarrow D^-$. Consequently, in order to have a reflection invariant theory on $M^{4,2}$, one takes--similarly to the ordinary Dirac case--the combination $D = D^+ \oplus D^-$. This implies that we double the number of Dirac spinors and we pass to the conformal bispinor $\xi = \begin{pmatrix} \psi_1 \\ \psi_2 \end{pmatrix}$.

Thus the explicitly covariant total Lagrangian on $M^{4,2}$ takes the form:

$$L_T = L_0(S) + L_0(P) + \frac{1}{2} \bar{\xi} \, \Gamma_a \overset{\leftrightarrow}{\partial}{}^a \xi \; + M\bar{\xi}\xi$$
$$+ \; g_s \bar{\xi}\xi S + g_P \bar{\xi}\Gamma_7 \xi P \tag{2.8}$$

where

$$\bar{\xi}(\eta) = i\xi^+(\eta)\Gamma_0\Gamma_6, \tag{2.9}$$

and the Γ_a^{15} are 8×8 matrices realizing the Clifford algebra

$$\{\Gamma_a, \Gamma_b\} = 2g_{a,b}, \quad a, \; b = 0,1,2,3,5,6, \tag{2.10}$$

where

$$g_{a,b} = (1,-1,-1,-1,-1,1), \quad \Gamma_7 = -i\Gamma_0\Gamma_1\Gamma_2\Gamma_3\Gamma_5\Gamma_6. \tag{2.11}$$

Definition 1. The index symmetry Lie algebra is the Lie algebra generated by the matrices

$$\mathbb{1},\Gamma_a,\Gamma_7,\Gamma_7\Gamma_a, \; [\Gamma_a,\Gamma_b], \; \Gamma_7[\Gamma_a,\Gamma_b], \; \{\Gamma_a,\Gamma_b,\Gamma_c\}, \tag{2.12a}$$

$\{ \, , \, , \, \}$ - antisymmetric product or by matrices:

$$\mathbb{1} \otimes \sigma_\nu, \; i\gamma_5 \otimes \sigma_\nu, \; \gamma_\mu \otimes \sigma_\nu, \; i\gamma_5\gamma_\mu \otimes \sigma_\nu, \; \gamma_\mu\gamma_\rho \otimes \sigma_\nu, \tag{2.12b}$$

acting in the conformal bispinor space $\mathbb{C}^4 \otimes \mathbb{C}^2$.

Proposition 2.2. The index symmetry Lie algebra generated by matrices (2.12) is isomorphic to $u(4,4)$ Lie algebra. (The proof of this Proposition and the next ones in this Section are given in [2].)

Clearly, the most important transformations generated by elements (2.12) of index symmetry Lie algebra are those which commute with Poincaré transformations.

Definition 2.2. The kinematical internal symmetry Lie algebra is the subalgebra of index symmetry Lie algebra which commutes with Poincaré group.

In order to find the internal symmetry Lie algebra for Yukawa model in $M^{3,1}$, we have to find a method of passing from $M^{4,2}$ to $M^{3,1}$. This is carried out by the following coordinates and fields transformations [3]

$$\{\eta_a\} \to \{x_\mu, \eta^2, \kappa\} \tag{2.13}$$

where

$$x_\mu = \frac{\eta_\mu}{\eta_5 + \eta_6} \qquad \kappa = \eta_5 + \eta_6 \qquad \eta^2 = \eta_a \eta^a$$

$$\Phi(x) = \Phi(x_\mu, \eta^2 = 0, \kappa = 1),$$

$$\psi(x) = S^{-1}\xi(x_\mu, \eta^2 = 0, \kappa = 1) \tag{2.14}$$

$$S = e^{ix_\mu \Pi^\mu}, \qquad \Pi^\mu = -\frac{i}{2}\Gamma^\mu(\Gamma_5 + \Gamma_6).$$

These transformations imply:

$$\Box_6 \to \Box_4, \qquad S^{-1}\Gamma^a \partial_a S \to \Gamma^\mu \partial_\mu. \tag{2.15}$$

In the present case we have

Proposition 2.3. The kinematical symmetry Lie algebra of Yukawa theory is $su(2) \oplus su(2) \oplus u(1) \oplus u(1)$ which is generated by the following matrices

$$J_1 = -\frac{i}{2}\Gamma_6\Gamma_7, \quad J_2 = \frac{1}{2}\Gamma_5\Gamma_7, \quad J_3 = \frac{1}{2}\Gamma_6\Gamma_5$$

$$N_1 = -\frac{i}{2}\Gamma_5, \quad N_2 = \frac{1}{2}\Gamma_6, \quad N_3 = \frac{1}{2}\Gamma_7 \tag{2.16}$$

$$X_7 = \mathbb{1}, \quad X_8 = -\Gamma_5\Gamma_6\Gamma_7.$$

The interaction Lagrangian has the form:

$$L_I = g_J \tilde{\xi} \vec{J} \xi \vec{\phi}_J + i g_N \tilde{\xi} \vec{N} \xi \vec{\phi}_N + g_7 \tilde{\xi} X_7 \xi \phi_7 + g_8 \tilde{\xi} X_8 \xi \phi_8, \qquad (2.17)$$

where

$$\xi = \begin{pmatrix} \psi_1 \\ \psi_2 \end{pmatrix}, \quad \tilde{\xi} = (\psi_1^+ \gamma_0, \ \psi_2^+ \gamma_0) = \xi^+ \Gamma_0. \qquad \blacksquare$$

Notice that taking $\vec{N}$ generators in the so called Dirac basis $\vec{N} = i \gamma_5 \otimes \vec{\tau}/2$ and setting:

$$g_7 = g_8 = 0, \quad g_J = 0, \quad g_N = -2g, \quad \vec{\phi}_N \equiv \vec{\phi}$$

we obtain

$$L_I = g \tilde{\xi} \gamma_5 \vec{\tau} \xi \vec{\phi}. \qquad (2.18)$$

Clearly, the Lagrangian (2.18) coincides with the conventional one for Yukawa theory for proton-neutron-pi-mesons system. In that manner we have obtained that the isotopic spin invariance is a consequence of conformal covariance of Yukawa theory. This, we think, clarifies the dynamical origin of isotopic spin.

Let us examine in this example the mechanism of formation of the internal symmetry Lie algebra su(2) out of some generators of kinematical internal symmetry Lie algebra su(2) $\oplus$ su(2) $\oplus$ u(1) $\oplus$ u(1) given by (2.16). In the present case the isotopic spin su(2) Lie algebra was obtained out of $N_k = i/2 \ \gamma_5 \otimes \tau_k$, k =1,2,3 generators which do not even close to a Lie algebra: it follows from (2.17) and (2.18) that $1/2 \ \gamma_5$ factor in N_k gave a factor in L_I which represents the pseudo-scalar nature of $\vec{\phi}$ whereas τ_k factor of N_k, k = 1,2,3 acts in α_I on the proton-neutron tower $\begin{pmatrix} \psi_P \\ \psi_N \end{pmatrix}$ and represents generators of isotopic spin Lie algebra.

In order to understand better also the geometric meaning of isotopic symmetry, consider the free spinor equation in $M^{4,2}$, given by the formula:

$$(\Gamma_a \partial^a + M) \xi(\eta) = 0. \qquad (2.19)$$

One readily verifies that the generators N_k, k = 1,2,3 given by (2.16) imply the following coordinate transformations:

$$N_1 \leftrightarrow \begin{cases} \eta_5 \to \eta_5 \\ \eta_{a'} \to -\eta_{a'} \\ a'=0,1,2,3,6 \end{cases} , \quad N_2 \leftrightarrow \begin{cases} \eta_6 \to \eta_6 \\ \eta_{a'} \to -\eta_{a'} \\ a'=0,1,2,3,5 \end{cases} , \quad N_3 \leftrightarrow \begin{cases} \eta_a \to -\eta_a \end{cases} \qquad (2.20)$$

The generators τ_k of isotopic spin group are obtained from N_k by the formula:

$$\tau_k = 2i\gamma_5 N_k, \quad k = 1, 2, 3. \tag{2.21}$$

Since the action of the matrix γ_5 on equation (2.19) gives the total kinematical space-time reflection in $M^{3,1}$, $\eta_\mu \to -\eta_\mu$, $\mu = 0,1,2,3$, the formulas (2.20) and (2.21) imply that

$$\tau_1 \leftrightarrow \begin{cases} \eta_6 \to -\eta_6 \\ \eta_{a'} \to \eta_{a'} \\ a'=0,1,2,3,5 \end{cases}, \quad \tau_2 \leftrightarrow \begin{cases} \eta_5 \to -\eta_5 \\ \eta_{a'} \to \eta_{a'} \\ a'=0,1,2,3,6 \end{cases}$$

$$\tag{2.22}$$

$$\tau_3 \leftrightarrow \begin{cases} \eta_5 \to -\eta_5 \\ \eta_6 \to -\eta_6 \\ \eta_\mu \to \eta_\mu \\ \mu=0,1,2,3 \end{cases}$$

Consequently, the generators τ_k of isotopic spin Lie algebra represent in fact the reflection operators with respect to unphysical space-time directions X^5 and X^6 of conformal space $M^{4,2}$. Thus it is rather accidental that reflection generators close to su(2) Lie algebra. From this point of view the isotopic spin symmetry is closer to a discrete symmetry rather than to rotation symmetry in three-dimensional isotopic spin space.

It is interesting that originally Heisenberg also introduced isotopic invariance as a discrete symmetry [4].

3. Higher Dynamical Symmetries in Fermion-Boson Systems

In the previous section we proved, starting from a specific interaction Lagrangian, the existence of a higher dynamical symmetry in this fermion-boson system. This symmetry was realized as a local group of nonlinear transformation in space-time $M^{3,1}$. The existence of this group implied the tower fermion structure and the existence of a specific internal symmetry group realized in tower space.

Now we shall present a somewhat "inverse" approach to the problem of a dynamical symmetry of the fermion-boson system. Namely, we shall postulate that the fundamental fermions in our fermion-boson system transform according to an irreducible representation of a certain, a priori unknown, simple Lie group G.

Starting from this natural hypothesis, strongly motivated by experimental data for strong, electromagnetic and weak interactions, we determine the specific form of G, get a quantization rule for a number of fermions in the tower, find a specific form of trilinear interaction Lagrangian of fermions and bosons with a definite field content and obtain the specific internal symmetry group for physical system described by obtained interaction Lagrangian.

To begin with, we assume that our fermion-boson system satisfies the following conditions:

i) The fundamental fermions in the considered theory form a tower

$$\Psi = \begin{pmatrix} \psi_1 \\ \vdots \\ \psi_n \end{pmatrix} \tag{3.1}$$

which provides a carrier space for an irreducible representation of a certain simple Lie group G, containing the physical Lorentz group $O(3,1)$ as a subgroup.

ii) Similar to the Yukawa case, we assume that the physical space-time $M^{3,1}$ can be embedded in a certain space-time manifold M^s, $s > 4$, on which G is realized linearly in the following manner

$$U_g^F \Psi(\eta) = D^F(g)\Psi(g^{-1}\eta), \quad g \in G, \quad \eta \in M^s. \tag{3.2}$$

Here $g \to D^F(g)$ is a $4n \times 4n$-dimensional irreducible representation of G which, when restricted to $O(3,1)$, possesses the following decomposition:

$$D^F\Big|_{O(3,1)} = \bigoplus_1^n D^{(\frac{1}{2},0)} \oplus D^{(0,\frac{1}{2})}, \tag{3.3}$$

where $D^{(\frac{1}{2},0)} \oplus D^{(0,\frac{1}{2})}$ is the Dirac representation.

iii) The boson fields $B^{\rho_1\cdots\rho_\ell}$, $\ell = 0, 1, \ldots$ in the considered fermi-boson system extended to M^s also transform according to the representation $g \to U_g^B$ of G given by the formula

$$U_g^B B(\eta) = D^B(g)B(g^{-1}\eta), \quad g \in G, \quad \eta \in M^s$$

where $g \to D^B(g)$ is a finite dimensional representation of G, whose explicit form will be derived later. $g^{-1}\eta$ is linear in η.

We now describe the most important consequences of our basic hypotheses i), ii) and iii).

Proposition 3.1. The only simple Lie groups G with reflections which satisfy the assumptions i) and ii) are the pseudoorthogonal groups $O(p,q)$, $p + q = 2N$, or $p + q = 2\tilde{N} + 1$, $q \geq 1$ with

$$N = \ell g4n/\ell g2, \quad \tilde{N} = \ell g4n/\ell g2. \tag{3.4}$$

(The proofs of this and the following Propositions and Theorems are given in [5].)

Formula (3.4) implies that the number of fundamental fermions in the tower (3.1) is "quantized" and is given by the formula

$$n = 2^{\tilde{\tilde{N}}-2}, \quad \tilde{\tilde{N}} = N \text{ or } \tilde{N}, \quad \tilde{\tilde{N}} \geq 3, \tag{3.5}$$

i.e., $n = 2, 4, 8, 16, \ldots$.

The restriction that G is simple was only assumed for the sake of simplicity.

The general case where G is an arbitrary Lie group can also be treated by means of the Levi-Malcev decomposition. In what follows, we restrict our attention to the $O(p,q)$ dynamical symmetry group with $p + q = 2N$.

We now define the index symmetry and the internal symmetry Lie algebras. Let $\{\Gamma_A\}$, $A = 1, 2, \ldots, p + q = 2N$ be a Clifford algebra in $M^{p,q}$ space-time. The polyvectors

$$\Gamma_{C_1,\ldots,C_r} = (r!)^{-1}[\Gamma_{C_1}, [\Gamma_{C_2} [\cdots [\Gamma_{C_{r-1}}, \Gamma_{C_r}] \cdots]]] \tag{3.6}$$

$$r = 2, \ldots, N$$

and

$$\check{\Gamma}_{D_1,\ldots,D_s} = \Gamma_{D_1,\ldots,D_s} \Gamma_{2N+1}, \quad s = 2, \ldots, N - 1 \tag{3.7}$$

where

$$\Gamma_{2N+1} = (-i)^{p+N} \Gamma_0 \Gamma_1 \cdots \Gamma_{2N}$$

together with $\Gamma_{C_1,\ldots,C_r}\big|_{r=0} = I$ and $\Gamma_{C_1,\ldots,C_r}\big|_{r=1} = \Gamma_{C_1}$ form a basis in the space of all $2^N \times 2^N$-dimensional matrices. These polyvectors, similar to the case of weak interactions, will provide $O(p,q)$ bilinear covariants $\overline{\Psi}\Gamma_{C_1,\ldots,C_r}\Psi$ and $\overline{\Psi}\check{\Gamma}_{D_1,\ldots,D_s}\Psi$.

Proposition 3.2. The most general trilinear real spinor-boson interaction Lagrangian density, invariant under $O(p,q)$, which when restricted to $M^{3,1}$, gives a renormalizable theory, has the form:

$$L_I(B,\overset{\smallsmile}{B},\Psi) = -\left\{ \sum_{r=0}^{N} \frac{g_r}{r!} B^{[C_1,\ldots,C_r]} \, i^{r(q+\frac{r+1}{2})} \, \Gamma_{C_1,\ldots,C_r} \right.$$

$$\left. + \sum_{s=0}^{N-1} \frac{\overset{\smallsmile}{g}_s}{s!} \overset{\smallsmile}{B}^{[D_1,\ldots,D_s]} \, i^{(s-1)(q+\frac{s}{2})} \, \overset{\smallsmile}{\Gamma}_{D_1,\ldots,D_s} \right\}\Psi \qquad (3.8)$$

where

$$\overline{\Psi} = \Psi^{+}A \quad \text{with} \quad A = i^{r(r-1)/2} \, \Gamma_0 \Gamma_{p+1} \cdots \Gamma_{p+q} \qquad (3.9)$$

and

$$\overset{\smallsmile}{\Gamma}_{D_1,\ldots,D_r}\Big|_{r=0} = \Gamma_{2N+1}.$$

The class of dynamical higher symmetry groups is further restricted by the natural requirement that the restriction of free
Lagrangians from $M^{p,q}$ to $M^{3,1}$ gives the sum of Dirac free Lagrangians. Indeed we have:

Proposition 3.3. Density of free Lagrangian

$$L_0(\Psi)(\eta) = \overline{\Psi}(\eta)\left(\frac{iq}{2}\Gamma_A \partial^A - M\right)\Psi(\eta), \quad \eta \in M^{p,q}$$

when expressed in terms of fundamental Dirac fields ψ_k, $k = 1, \ldots,$
n reduces on $M^{3,1}$ to the sum of free Dirac Lagrangian densities
for ψ_k fields if and only if $O(p,q)$ groups are the Lorentz type
groups $O(p,1)$, $p > 3$.

Here and in the discussion of passing from the $M^{p,q}$ space to
the $M^{3,1}$ space which follows, the method of dimensional reduction
is employed [3].

It follows from (3.8) and (3.9) that effective matrix poly-
vectors which act in tower spinor space are

$$i^{r(q+\frac{r+1}{2})} \, \tilde{\Gamma}_0 A \Gamma_{C_1,\ldots,C_r}, \quad r = 0, 1, \ldots, N$$

and   (3.10)

$$i^{(s-1)(q+\frac{s}{2})} \, \tilde{\Gamma}_0 A \overset{\smallsmile}{\Gamma}_{D_1,\ldots,D_s}, \quad s = 0, 1, \ldots, N-1$$

where

$$\tilde{\Gamma}_0 = \gamma_\mu (\otimes \mathbb{1})^{N-2}.$$

Matrices (3.10) acting in tower fermion index space form a basis
in the space of all $2^N \times 2^N$ matrices and are closed therefore

under commutation relations. Consequently, they form a Lie algebra over the field of real numbers. This Lie algebra will be called the index symmetry Lie algebra.

Theorem 3.4. The index symmetry Lie algebra generated by the matrices (3.10) considered over the field of real numbers is isomorphic with $u(2^{N-1}, 2^{N-1})$ Lie algebra, for all pseudoorthogonal groups $O(p,q)$, $q > 1$.

Clearly, from the physical point of view, the most interesting subalgebra of the index symmetry Lie algebra is the one whose generators commute with the generators of the Poincaré group. We shall call this subalgebra the kinematical internal symmetry Lie algebra. This Lie algebra, whose properties are independent of any dynamics, should be distinguished from the internal symmetry Lie algebra with respect to which the total Lagrangian is invariant. In general, these Lie algebras are different. However we have:

Theorem 3.5. Consider a physical system of fermions and bosons described by the interaction Lagrangian (3.8). Then if we set $g_r = g_s = g$ for all r and s, the internal symmetry Lie algebra for the total Lagrangian $L_0 + L_I$ is isomorphic to the scalar part of kinematical internal symmetry Lie algebra and is equal to $u(2^{N-2}) = u(n)$. This Lie algebra is generated by matrices $\tau_{\rho_1} \otimes \ldots \otimes \tau_{\rho_{n/2}}$ where $\{\tau_\rho\} = \{I, \tau_k\}$ and τ_k are Pauli matrices.

The interaction Lagrangian (3.8) when restricted to $M^{3,1}$ takes the form

$$
L_I = g\tilde{\Psi}\{S^{\rho_1, \ldots, \rho_{N-2}} - P^{\rho_1, \ldots, \rho_{N=2}} \cdot \gamma_5
$$

$$
+ V_\mu^{\rho_1, \ldots, \rho_{N-2}}\gamma^\mu + A_\mu^{\rho_1, \ldots, \rho_{N-2}}i\gamma^\mu\gamma_5 \tag{3.11}
$$

$$
+ T_{[\mu,\nu]}^{\rho_1, \ldots, \rho_{N-2}} \frac{i}{4} [\gamma^\mu, \gamma^\nu]\} \otimes \tau_{\rho_1} \otimes \ldots \otimes \tau_{\rho_{N-2}} \}\Psi
$$

where

$$
\tilde{\Psi} = \Psi^+ \tilde{\Gamma}_0.
$$

Notice that if the coupling constants are set equal then the internal symmetry Lie algebra is maximal; if coupling constants are not equal, then the internal symmetry Lie algebra is a subalgebra of maximal internal symmetry Lie algebra (see [5] for examples).

It follows from Theorem 3.4 that the polyvector boson fields $B_{C_1,\ldots,C_r}$ and $B_{D_1,\ldots,D_s}$ in the interaction Lagrangian (3.8) when reduced to Minkowski space-time become, with respect to Poincaré group, one of scalar S, pseudoscalar P, vector V, pseudovector A and tensor T type. The remaining O(p,q) tensor indices of poly-vectors $B_{C_1,\ldots,C_r}$ or $\check{B}_{D_1,\ldots,D_s}$ after restriction to physical space-time $M^{3,1}$ produce indices $\rho_1, \ldots, \rho_{N-2}$ which describe the tensor transformation property of considered fields with respect to internal symmetry group.

Thus if our initial hypothesis reflects physical reality then in Nature bosons should also appear in tower

$$
B = \begin{pmatrix}
S^{\rho_1,\ldots,\rho_{N-2}} \\
P^{\rho_1,\ldots,\rho_{N-2}} \\
V^{\rho_1,\ldots,\rho_{N-2}} \\
A^{\rho_1,\ldots,\rho_{N-2}} \\
T^{\rho_1,\ldots,\rho_{N-2}}_{[\mu,\nu]}
\end{pmatrix}
\tag{3.12}
$$

each boson with the same tensor type with respect to $U(2^{N-2}) = U(n)$ internal symmetry group.

To illustrate the obtained results in more detail, consider the O(9,1) group as a dynamical symmetry group for the fermion tower (3.1). In this case N = 5 and by virtue of (3.5) we get that the number n of fermions in the tower equals 8. Theorem 3.4 implies that the index symmetry Lie algebra is u(16,16). In turn Theorem 3.5 implies that the kinematical internal symmetry Lie algebra is u(8). This Lie algebra contains

$$
su(3) \oplus su(2) \oplus su(2) \oplus u(1)
\tag{3.13}
$$

as a natural block diagonal subalgebra which is looked for in recent quark models.

The most general interaction Lagrangian connected with the O(9,1) dynamical group is given by formula (3.8) with N set to 5. The boson fields in this model S^{ρ_1,ρ_2,ρ_3}, P^{ρ_1,ρ_2,ρ_3}, etc. have three internal symmetry indices (ρ_1,ρ_2,ρ_3) which determine their covariance type w.r.t. internal symmetry group or its distinguished subgroups like (3.13).

Thus starting from the simple and natural assumptions i), ii) and iii) we have obtained a series of concrete and physically interesting predictions on a number of fermions in towers, field

content in fermion-boson system and the structure of internal
symmetry group. This illustrates that the exploration of a concept of higher dynamical symmetry groups of fermion-boson system
might provide a deeper insight into the nature of strongly interacting system, without imposing specific ad hoc dynamics from
the beginning.

Acknowledgement

The author thanks P. Budini and P. Furlan for inspiring
discussions and collaboration.

References

[1] M. Flato, J. Simon and D. Sternheimer, Ann. of Phys. $\underline{61}$, 78
         (1970).
[2] P. Budini, P. Furlan and R. Raczka, Hadronic Journal $\underline{1}$, 1364
         (1978).
[3] P. Budini, P. Furlan and R. Raczka, Nuovo Cim. $\underline{52A}$, 191 (1979).
[4] W. Heisenberg, Z. Physik $\underline{77}$, 1 (1932).
[5] P. Furlan and R. Raczka, "New Approach to Dynamical Symmetries
         in Fermion-Boson Systems", in preparation.

*Almost all results presented in this lecture were obtained in
         collaboration with P. Budini and P. Furlan.
**Partially supported by NSF Grant No. INT 73-20002 A01, formerly
         GF-41958.

ON THE GENERALISATION OF THE GELL-MANN-NISHIJIMA RELATION

Alladi Ramakrishnan

MATSCIENCE
The Institute of Mathematical Sciences
Madras 600 020, India

The Gell-Mann-Nishijima relation was formulated in 1954 to explain the existence of charge multiplets with different "charge spectra." It connects I_z, the z-component of the isotopic spin with the charge Q and the hypercharge Y, by the relation

$$Q = I_z + \frac{Y}{2} .$$

Experimental results since 1974 necessitated the inclusion of additional Quantum numbers like charm and the question therefore arises: Is there a logical scheme by which new Quantum numbers may be included?

Since 1967, at Madras, we have been working on the Clifford algebra and its generalisations which by a series of fortuitous circumstances led to a "generalisation" of the Gell-Mann Nishijima relation as follows:

$$I_{kl} = \frac{1}{2} (S_k - S_l)$$

where I_{kl} are the "z-components" of Isotopic spin like Quantum numbers and S_j are the scalar Quantum numbers (j = 1, 2, ..., n). All the n Quantum numbers S_j are of "equal status" i.e. they assume the same gamut of values and obey the condition

$$S_1 + S_2 + ... + S_n = 0.$$

The case n = 3 corresponds to the usual Gell-Mann-Nishijima relation with the identification

$$S_1 = Q, \quad S_2 = Y - Q, \quad S_3 = -Y$$

$$I_{12} = I_z, \quad I_{23} = U_z, \quad I_{31} = V_z$$

where I_z, U_z and V_z are the z-components of I-spin, U-spin and V-spin occurring in the formulation of SU(3) symmetry by Gell-Mann.

We can rewrite the generalised relation to look like the customary Gell-Mann-Nishijima relation as,

$$S_k = I_{k1} - \frac{1}{2} \sum_{j \neq k, 1} S_j \ .$$

This relation was suggested by recognising that the S_j can be written as linear combinations of the eigenvalues of matrices which are identified as the roots of the unit matrix,

$$S_1 = \frac{1}{n}[\eta_1 + \eta_2 + \ldots + \eta_{n-1}]$$

$$S_2 = \frac{1}{n}[\omega\eta_1 + \omega^2\eta_2 + \ldots + \omega^{n-1}\eta_{n-1}]$$

$$\cdot \ \cdot \ \cdot \ \cdot \ \cdot \ \cdot \ \cdot \ \cdot \ \cdot \ \cdot \ \cdot \ \cdot \ \cdot \ \cdot \ \cdot$$

$$S_n = \frac{1}{n}[\omega^{n-1}\eta_1 + (\omega^{n-1})^2\eta_2 + \ldots + (\omega^{n-1})^{n-1}\eta_{n-1}]$$

where $1, \omega, \omega^2, \ldots, \omega^{n-1}$ are the n roots of unity and $\eta_1, \eta_2, \ldots, \eta_n$ are the eigenvalues of the commuting matrices

$$A, A^2, \ldots, A^{n-1}$$

with $A^n = I$. Obviously η_j can assume the values $1, \omega, \ldots, \omega^{n-1}$.

An "elementary particle" is characterised by n Quantum numbers $\eta_1, \eta_2, \ldots, \eta_n$ and a "composite" particle is characterised by the sum of these quantum numbers,

$$[\eta_j] = \sum_i \eta_j^{(i)}$$

where the summation is over all the component particles.

In our formulation if we have n fundamental Quarks the Quantum numbers S_j can assume the values

$$-\frac{1}{n} \quad \text{or} \quad +\frac{n-1}{n} \ .$$

While at present, physicists are retaining the values $-1/3$ and $+2/3$ for the charge and "adjusting" the values of additional quantum numbers, according to our generalisation, charge,

hypercharge, charm and any additional quantum number should have
"equal status" and assume the same range of values.

We have to wait and see whether this principle of "equal
status" has any physical significance or justification.

REFERENCES

1. Alladi Ramakrishnan, P. S. Chandrasekaran, N. R. Ranganathan,
        T. S. Santhanam and R. Vasudevan, "The Generalised Clifford
        Algebra and the Unitary Group," J. Math. Analysis and
        Applications 27 (1969), 164.
2. Alladi Ramakrishnan, "L-Matrix Theory or The Grammar of Dirac
        Matrices," Tata McGraw-Hill (New Delhi) 1972.

RELATIVISTIC DYNAMICAL GROUPS IN QUANTUM THEORY AND SOME POSSIBLE

APPLICATIONS

P. Roman

SUNY at Plattsburgh
Plattsburgh, N.Y.   12901

There are many ways to measure the impact of a genius on the
future - yet all these assessments are ephemeral. For most physicists, the fame of Einstein rests on his deep understanding of the
fundamental coherence between space, time, and matter. Others
admire his breadth of command, encompassing, besides relativity, so
much of statistical physics and early quantum theory. But for many
of us, yet another aspect of Einstein's monumental work demands our
admiration and grateful respect: namely, his clear understanding
and insistent emphasis of <u>underlying symmetries that, in a very real
sense, govern the laws of Nature</u>. It is important to emphasize that
Einstein used the guiding principle of symmetry in its broadest
sense: not only as the manifestation of an invariance under some
permutation of elements, but also in the deeper connotation of an
<u>algebraic system</u> that lends unity to the parts and that permits the
"divination" of laws even when very little of the details pertaining
to the phenomena is known to start with.[1]

It is in this sense that we speak of <u>dynamical</u> (versus kinematical) <u>symmetries</u>. But the term "dynamical group" itself has been
used in a variety of different meanings. In our work, which was
done over the past nine years in cooperation with many colleagues
at Boston University and at the Max Planck Institut in Starnberg,
and about which we have now the honor to briefly report, we used the
term in a very specific and probably limited sense. We call a
groupstructure a <u>dynamical group</u> if in its Lie algebra it contains
an account of the geometric structure of the underlying event space
and if, in addition, the algebra gives a full account of the inertial
dynamics of an isolated system. If, moreover, the Lie algebra contains (as a subalgebra) the Heisenberg algebra, so that the laws of
quantization are a consistent part of the algebraic system, then we
talk of a <u>quantum dynamical group</u>.

The prototype of this approach to quantum dynamics is, of course, the central extension of the non-relativistic Galilei group, G_4. In the traditional introduction to this structure,[2] the underlying "geometrical substratum" is

$$E = E_3(\underset{\sim}{x}) \times E_1(t),$$

endowed with rotational, translational, and time-displacement symmetry. These are expressed by the Lie brackets

$$[\underset{\sim}{P},\underset{\sim}{P}] = [\underset{\sim}{Q},\underset{\sim}{Q}] = 0, \quad [\underset{\sim}{H},\underset{\sim}{J}] = 0 , \tag{1}$$

$$[\underset{\sim}{J},\underset{\sim}{J}] \sim \underset{\sim}{J}, \quad [\underset{\sim}{J},\underset{\sim}{P}] \sim \underset{\sim}{P}, \quad [\underset{\sim}{J},\underset{\sim}{Q}] \sim \underset{\sim}{Q} ,$$

where $\underset{\sim}{P}$, H, $\underset{\sim}{J}$ are the operators that correspond to spatial- and time-translations and to rotations, respectively, whereas $\underset{\sim}{Q}$ is assumed to represent a particle position. One then _assumes_ Galilean inertial symmetry, expressed by

$$[H,\underset{\sim}{P}] = 0, \quad [H,\underset{\sim}{Q}] = -i\underset{\sim}{P} . \tag{2}$$

These relations, together with

$$[\underset{\sim}{P},\underset{\sim}{Q}] = 0 \tag{3}$$

form the Lie algebra of the geometrical Galilei group G_4 with the transformation law

$$\underset{\sim}{x} \rightarrow R\underset{\sim}{x} + \underset{\sim}{a} + t\underset{\sim}{v} , \tag{4}$$

$$t \rightarrow t + \tau .$$

However, $\underset{\sim}{Q}$ cannot represent position in quantum theory: it is easy to see that, because of (3), we do not have localization. To remedy this situation, we pass on to the central extension G_4, in which (3) is replaced, by _fiat_, to become the Heisenberg relation

$$[P_k,Q_\ell] = -iM\delta_{k\ell} , \tag{5}$$

where M is a constant that (via the representation theory of $\tilde{G}_4$) can be identified with (Galilean) mass.[3]

The _ad hoc_ assumption of Galilean inertial symmetry and of the process of central extension was criticized by us[4] in 1974. We showed that a much deeper understanding can be obtained in the following way. We start, not with the space-time substratum, but only with the _event space_ , $E_3(\underset{\sim}{x})$ which is homogeneous and isotropic:

$$[\underset{\sim}{P},\underset{\sim}{P}] = 0, \quad [\underset{\sim}{J},\underset{\sim}{P}] \sim \underset{\sim}{P}, \quad [\underset{\sim}{J},\underset{\sim}{J}] \sim \underset{\sim}{J}.$$

Then we demand the <u>locality postulate</u> (gauge principle): To every transformation

$$\psi(\underset{\sim}{x}) \rightarrow e^{i\omega(\underset{\sim}{x})}\psi(\underset{\sim}{x}) \tag{6}$$

on the Hilbert space (that carries the familiar realization

$$P_k \sim -i\partial_k , \quad J_k \sim -i\varepsilon_{k\ell j}x_\ell \partial_j)$$

there corresponds in this Hilbert space a unitary operator U such that

$$(U\psi)(\underset{\sim}{x}) = e^{i\omega(\underset{\sim}{x})}\psi(\underset{\sim}{x}) . \tag{7}$$

If we then insist that the algebra of observables be large enough to ensure the existence of U (made up from observables only), we can easily show the <u>necessary presence</u> of operators Q_ℓ in the required algebra which obey

$$[\underset{\sim}{Q},\underset{\sim}{Q}] = 0, \quad [\underset{\sim}{J},\underset{\sim}{Q}] \sim \underset{\sim}{Q} ,$$

$$[P_k,Q_\ell] = -iM\delta_{k\ell} . \tag{8}$$

In fact, $U = \exp(iM^{-1}c_\ell Q_\ell)$ corresponds to linear phase transformations (with $\omega(\underset{\sim}{x}) = c_\ell x_\ell$). Thus, <u>Galilean boosts arise as particular local phase transformations</u>. In addition, one finds the realization $Q_k \sim Mx_k$ which completes the identification of $\underset{\sim}{Q}$ with position.

To incorporate dynamics, we then define <u>development transformation</u> of the isolated system as a kinematical symmetry

$$\underset{\sim}{P} \rightarrow \underset{\sim}{P}, \quad \underset{\sim}{J} \rightarrow \underset{\sim}{J}, \quad \underset{\sim}{Q} \rightarrow f(\underset{\sim}{Q},\underset{\sim}{P},\underset{\sim}{J}) .$$

This form merely expresses the requirement that the generator of intrinsic development be compatible with the geometry of the event space.[5] A few obvious requirements and the essential demand that the generator H of the development operator $U_\tau \equiv \exp(i\tau H)$ be already contained in the algebra of observables generated by $\underset{\sim}{P}$, $\underset{\sim}{Q}$, $\underset{\sim}{J}$ leads to the form

$$H = H(\underset{\sim}{P}^2,\underset{\sim}{TP},I) \tag{9}$$

where $\underset{\sim}{T} \equiv \underset{\sim}{J} - M^{-1}\underset{\sim}{Q} \times \underset{\sim}{P}$ is the spin. Next one observes that H and the generators of the kinematical group must form a <u>closed Lie algebra</u>.[6] This makes (9) <u>unique</u> and we find that

$$H = \underset{\sim}{P}^2/2M - C , \tag{10}$$

where C is an arbitrary constant. Now we trivially obtain

$$[H, \underset{\sim}{J}] = 0, \quad [H, \underset{\sim}{P}] = 0, \quad [H, \underset{\sim}{Q}] = -i\underset{\sim}{P},$$

which are precisely the <u>inertial motion equations</u>--this time not assumed, but deduced.

Thus, our local phase invariance postulate, together with a few almost obvious assumptions on dynamical development, led directly to the quantum mechanical dynamical Galilei group

$$\tilde{G}_4 = T_1^H \circledS \{SU(2)^J \circledS [T_3^P \circledS (T_3^Q \times T_1^M)]\} . \tag{11}$$

In a subsequent paper,[7] we showed that these ideas can be easily generalized so as to obtain the <u>relativistic enlargement of the Galilei group</u>. We start with the <u>Minkowski</u> event space $E_{3,1}(x)$ (rather than with the Euclidean E_3) and repeating the same line of arguments as in the nonrelativistic case, we arrive at the relativistic quantum dynamical group

$$\tilde{G}_5 = T_1^S \circledS \{SL(2,C)^J \circledS [T_4^P \circledS (T_4^Q \times T_1^{\ell^{-1}})]\} . \tag{12}$$

Writing this as $\tilde{G}_5 \equiv T_1^S \circledS K$, we see that the kinematical group K contains the Poincaré group (generated by P_μ and $J_{\mu\nu}$) and also the transformations generated by "relativistic Galilean boosts" Q_μ, again associated with linear local phase transformations. Apart from the familiar commutators of the Poincaré algebra, we also have the laws of quantization,[8]

$$[P_\mu, Q_\nu] = -i\ell^{-1} g_{\mu\nu} \tag{13}$$

and the dynamical development laws, with S being a <u>relativistic development operator</u>:

$$[S, J_{\mu\nu}] = 0, \quad [S, P_\mu] = 0, \quad [S, Q_\mu] = iP_\mu . \tag{14}$$

In analyzing the physical content of $\tilde{G}_5$, we found that Q_μ cannot be interpreted as particle-position operator: it is merely localizing <u>events</u>. However, we were able to show that the enveloping algebra of $\tilde{G}_5$ <u>does</u> contain in a natural way a configuration-space realization of the <u>Newton-Wigner position operator</u>. From the study of the representation theory of $\tilde{G}_5$ we also learned that <u>spin-towers</u> will contain the basis states of the system.

We note that, several years before the rigorous derivation and in-depth understanding of $\tilde{G}_5$, we suggested it as a dynamical group on purely heuristic grounds[9] and explored several aspects of it in subsequent papers.[10] Moreover, the Lie algebra of $\tilde{G}_5$ has been considered, along with several other related algebras, as a possible candidate for relativistic quantum dynamics in the subnuclear domain, by Castell,[11] several years before we became aware of this.

In the ultimate analysis, $\tilde{G}_5$ left us disappointed because it describes only <u>free motions</u>. We therefore started to search for a generalization which would account in a natural manner for the emergence of force-effects. A hint in this direction[12] came from the work of Bacry and Lévi-Leblond.[13] Starting with the deSitter group and using the method of group contractions, they obtained a nonrelativistic but "cosmologic" group $\tilde{H}_4$ which is now usually called the <u>Hooke group</u>. This algebraic system differs from $\tilde{G}_4$ only inasmuch as the inertial motion, given by the Lie brackets

$$[H,\underset{\sim}{Q}] = -i\underset{\sim}{P}, \quad [H,\underset{\sim}{P}] = i\nu^2\underset{\sim}{Q},$$

$$[H,\underset{\sim}{J}] = 0, \quad (\nu \equiv \text{const.})$$

(15)

is no longer a force-free motion but it rather corresponds to the behavior of a particle under the influence of a harmonic oscillator force. This is the manifestation of the long-range effect of curvature.

To understand the role of $\tilde{H}_4$ as a quantum dynamical group, we first showed[14] that $\tilde{H}_4$ can be derived in a natural manner (analogously to $\tilde{G}_4$) if one starts, not with flat Euclidean space but rather with a <u>uniformly curved 3-dimensional event space</u> and applies the locality principle as before.

Whereas this result told us that non-trivial quantum dynamics <u>can</u> arise from the geometry of a suitable event space, we had to face the new problem: where does the curvature of this space come from? We answered this question by demonstrating[15] that the required large curvature of space in a small region can arise, within the framework of a completely unified spontaneously broken gauge theory of the Yang-Mills-Einstein-Higgs type,[16] from the vacuum contribution to the hadronic energy-momentum tensor. Actually, a deSitter geometry emerges. We pursued this viewpoint of the relevance of the Hooke-dynamics inside nucleonic matter in some detail[15] and in fact established a relation between our nonrelativistic microuniverse endowed with $\tilde{H}_4$ dynamics on one hand and the SU(3) quark model with harmonic forces, on the other.

Whereas in these endeavours we confined ourselves to the low-speed approximation (which lead to $\tilde{H}_4$), we eventually had to ask the following question: If the completely unified and spontaneously broken gauge theory of nucleonic matter indeed leads to a deSitter microuniverse that corresponds to a hadron, what then is the exact, fully relativistic quantum dynamics in this space? From the philosophy of our previous pursuits, it is clear that the mathematical problem is the following: establish, via the locality principle, the algebraic structure that is based on the event space with the geometry of the homogeneous and isotropic (open) oscillating deSitter space. Following the well known pattern, there was no difficulty to derive[17] the "relativistic quantum dynamical Hooke group" $\tilde{H}_5$. The Lie algebra can be succinctly written in the following form:

$$[J_{\mu\nu}, J_{\zeta\sigma}] \sim iJ, \quad [J_{\mu\nu}, P_\zeta] \sim -iP, \quad [J_{\mu\nu}, Q_\zeta] \sim -iQ,$$

$$[P_\mu, P_\nu] = [Q_\mu, Q_\nu] = 0, \quad [P_\mu, Q_\nu] = -i\ell^{-1}g_{\mu\nu} \tag{16}$$

$$[S, J_{\mu\nu}] = 0, \quad [S, P_\mu] = i\nu^2 Q_\mu, \quad [S, Q_\mu] = -iP_\mu .$$

As expected, only the $[S, P_\mu]$ commutator differs from the corresponding $\tilde{G}_5$ relations, and we have a relativistic oscillator dynamics.

The Casimir invariants are

$$C_o = \ell^{-1}\mathbf{1}, \quad C_1 = 2\ell^{-1}S + P^2 + \nu^2\ell^2 Q^2 ,$$

$$C_2 = \frac{1}{2} T_{\mu\nu}T^{\mu\nu}, \quad C_3 = \frac{1}{4} \varepsilon_{\mu\nu\zeta\sigma}T^{\mu\nu}T^{\zeta\sigma} ; \tag{17}$$

$$(T_{\mu\nu} \equiv J_{\mu\nu} - \ell(P_\mu Q_\nu - P_\nu Q_\mu)).$$

The C_2 and C_3 describe the spin-towers (as in $\tilde{G}_5$) and only C_1 differs from the corresponding entity in $\tilde{G}_5$ (where the last term is not present). The important result is that the development operator ("relativistic Hamiltonian") S has now a <u>discrete spectrum</u>.

From the historical point of view, it is interesting to note that $\tilde{H}_5$ is not really "new": it is closely related (in the mathematical sense) to one of the structures studied by Castell[11] in his 1966 paper. As a matter of fact, $\tilde{H}_5$ is in close relation to several groups that arise from SO(p,q) with p + q = 6 by contraction.[17]

In ref. 17 we studied (by the Mackey method of induced representations) the irreducible unitary representations of $\tilde{H}_5$. We found the following classification scheme:

Class I: $u > 0, \quad p^2 > 0, \quad q^2 > 0$;

Class II: $u < 0, \quad p^2 < 0, \quad q^2 < 0$;

Class III: $u = 0, \quad p^2 = q^2 = 0, \quad p \neq 0, \quad q \neq 0$;

Class IV: $u = 0, \quad p^2 = q^2 = 0, \quad p = 0, \quad q = 0 .$

Here u is the eigenvalue of S. Clearly, only representations in Class I are relevant for hadrons.

In order to extract some possible physical results from the $\tilde{H}_5$ dynamics, we first must realize that in our picture the relativistic hadrons appear as <u>bilocal objects</u>: deSitter microuniverse "bubbles" embedded in flat external Minkowski space. As for all such models, the relation between the external and internal dynamics is not obvious or <u>a priori</u> given. However, it is well known that the algebraic approach of Nambu,[18] which we shall adopt, renders a plausible prescription. In this approach, the "internal group" is taken to contain a subgroup which characterizes the internal symmetry and the internal group must be large enough to contain within its generators some that form tensors under the Lorentz group. In view of the second requirement it is seen that the $\tilde{H}_5$ generators available for the internal symmetry are P_μ, Q_μ and S. To meet the first requirement, we note that setting

$$A_\mu^+ \equiv \frac{\nu}{\sqrt{2}} \, (\ell Q_\mu - i\nu^{-1}P_\mu), \qquad A_\mu \equiv \frac{\nu}{\sqrt{2}} \, (\ell Q_\mu + i\nu^{-1}P_\mu)$$

the elements $A_\mu^+ A_\nu$ in the enveloping algebra of $\tilde{H}_5$ span an U(3,1) algebra. If we fix the eigenvalue of

$$N \equiv A_\mu^+ A^\mu \, , \tag{18}$$

then we have the semisimple group SU(3,1) which is our internal symmetry. Actually, the eigenvalues of N are

$$n \equiv n_1 + n_2 + n_3 - n_0 \, , \tag{19}$$

with $n_\mu \geq 0$ integer. Fixing N implies (by (17)) that

$$S = -\ell\nu(N + 1) + \frac{1}{2} \, \ell C_1 \, . \tag{20}$$

The SU(3,1) generators $A_\mu^+ A_\nu$ form a Lorentz tensor, $A_\mu^+ + A_\mu$ is a (self-adjoint) vector, and S is a scalar. Thus, the most general Nambu-type equation is of the form

$$[A_\mu^+ A_\nu \, \mathbb{P}^\mu \mathbb{P}^\nu + \alpha(A_\mu^+ + A_\mu) \, \mathbb{P}^\mu + \beta \, \mathbb{P}^2 - \gamma S]\Psi^{n,b}(\mathbb{P}) = 0 \, . \tag{21}$$

Here $\mathbb{P}^\mu$ is the external fourmomentum; n labels a level of the (infinite component) wave function, and b is a degeneracy label.[19] Because of (20), the label n may be identified with the eigenvalue (19) of N.

Since A_μ^+ and A_μ are raising-lowering operators, they transform eigenvectors of S with a label n to those with a label n ± 1. Thus, (21) can be an eigenvalue equation for S only if $\alpha = 0$. Hence, the final form of the infinite component wave equation is

$$(A_\mu^+ A_\nu \, \mathbb{P}^\mu \mathbb{P}^\nu + \beta \, \mathbb{P}^2 - \gamma S)\Psi^{n,b}(\mathbb{P}) = 0 \, . \tag{22}$$

Going to the (external) rest frame with $\mathbb{P}_k = 0$, $\mathbb{P}_0 = m$ (= hadron mass), we have

$$(A_0^+ A_0 m^2 + \beta m^2 - \gamma S)\psi^{n,b}(\mathbb{P}_0) = 0 .$$

Using (20), (18), (19) (and taking[20] $C_1 = 0$) we obtain

$$[(n_0 + \beta)m^2 + \gamma \ell \nu (n_1 + n_2 + n_3 - n_0 + 1)]\psi^{n,b} = 0 ,$$

giving the <u>mass spectrum</u>

$$m^2 = -\gamma \ell \nu \frac{n_1 + n_2 + n_3 - n_0 + 1}{n_0 + \beta} . \tag{23}$$

Even though[21] $m^2 > 0$ and the spectrum is discrete, the presence of time-like oscillations leads to a feature which contradicts even qualitative evidence: if in (23) one takes the maximum possible values $n_0 = n_1 + n_2 + n_3$, the spectrum of m^2 has an accumulation point at $m^2 = 0$. To avoid this, we must <u>assume</u> that timelike oscillations are suppressed, $n_0 = 0$. Then (23) simplifies to an, at least qualitatively, acceptable spectrum

$$m^2 = \text{const. } (n + 1) , \tag{24}$$

$n \equiv n_1 + n_2 + n_3$; n_i = nonnegative integer.

Finally, we endeavored to establish a relation between our dynamical model and other <u>bilocal hadron models</u>. The wave function Ψ can be separated as

$$\psi^{n,b}(\mathbb{P}) \equiv \phi^n \psi_a(\mathbb{P})$$

where $\phi^n \equiv \phi(u;p)$ is the internal wave function, with $u = -\ell \nu(n_1 + n_2 + n_3 - n_0 + 1)$. The wave equation for the latter follows from the form of the Casimir invariant C_1 as given in (17), and we have[20]

$$(p^2 - \nu^2 \Box_p + 2\ell^{-1}u)\phi(u;p) = 0 . \tag{25}$$

This is of the same type as the one which is <u>assumed</u> in the relativistic harmonic oscillator model of Feynman et al.[22] and which was further studied in the context of the parton model by Kim and Noz.[23]

Suppressing again timelike oscillations (i.e., setting $A_0^+ A_0 \phi = A_0 A_0^+ \phi = 0$), Eq. (25) becomes

$$(\underset{\sim}{p}^2 - \nu^2 \nabla_{\underset{\sim}{p}}^2 + 2\ell^{-1}u)\phi(u;\underset{\sim}{p}) = 0 .$$

Going to configuration space, separating coordinates, and solving
the radial wave equation, one is eventually led to a family of
<u>Regge trajectories</u>,

$$j = Cm^2 - 2k - 2 \tag{26}$$

where C = const, $k = 0,1,2, \ldots$. Here j plays the role of the
"internal orbital momentum" which we interpret as the spin of the
physical hadron. Interesting as this may be, for reasons explained
at the end of ref. 17, formula (26) cannot yet be directly compared
with experiments. To extract more direct information, further work
is needed.

REFERENCES

1. Einstein's pioneering role in implanting the idea of symmetry
       as a crucial tool of research has been emphasized by E. P.
       Wigner, Proc. Amer. Phil. Soc. <u>93</u>, No. 7 (1949). See also
       Wigner's many further philosophical elaborations on the
       topic of symmetry, for example Proc. Internat. School of
       Phys. E. Fermi, <u>29</u>, 40 (1964), The Nobel Prize Lectures
       1964, Communic. Pure and Appl. Math. <u>13</u>, 1 (1960), etc.
2. A comprehensive review of the Galilei group was given by J.-M.
       Lévi-Leblond, in "Group Theory and its Applications", Vol.
       II. ed. by E. M. Loebl (Acad. Press, NY, 1971).
3. In conventional units, then, $\underset{\sim}{X} \equiv M^{-1}\underset{\sim}{Q}$ is the position operator.
4. P. Roman and J. P. Leveille, Jour. Math. Phys. <u>15</u>, 1760 (1974).
       See also P. Roman's contribution in "Quantum Theory & the
       Structures of Time & Space", p. 85. (Carl Hanser, München,
       1975).
5. I.e., that it be invariant under space translations and rota-
       tions.
6. This follows from the simplicity assumption that the development
       transformations form a <u>one-parameter</u> group.
7. P. Roman and J. P. Leveille, Jour. Math. Phys. <u>15</u>, 2053 (1974).
8. The constant ℓ has the dimension of length.
9. J. J. Aghassi, P. Roman, R. M. Santilli, Phys. Rev. <u>D1</u>, 2753
       (1970).
10. The more relevant papers are: J. J. Aghassi, P. Roman, R. M.
       Santilli, Jour. Math. Phys. <u>11</u>, 2297 (1970); J. J. Aghassi,
       P. Roman, R. M. Santilli, Nuovo Cimento <u>5A</u>, 551 (1971);
       R. M. Santilli, Particles and Nuclei <u>1</u>, 81 (1970); P. L.
       Huddleston, M. Lorente, P. Roman, Found. of Physics <u>5</u>,
       75 (1975).
11. L. Castell, Nuovo Cim. <u>46A</u>, 1 (1966) and ibid. <u>49A</u>, 285 (1967).
12. I am indebted to Prof. P. Winternitz (Montreal) who some years
       ago called my attention to ref. 13 which I once read but
       later forgot.

13. H. Bacry and J.-M. Lévy-Leblond, Jour. Math. Phys. $\underline{9}$, 1605 (1968).

14. P. Roman and J. Haavisto, Jour. Math. Phys. $\underline{17}$, 1664 (1976).

15. P. Roman and J. Haavisto, Internt'l Jour. of Theor. Phys. $\underline{16}$, 915 (1977).

16. Typical papers in this framework are: F. A. Bais and R. J. Russell, Phys. Rev. $\underline{D11}$, 2642 (1975); Y. M. Cho and P. G. O. Freund, Phys. Rev. $\underline{D12}$, 1588 (1975); E. Cremer and J. Sherk, Nucl. Phys. $\underline{B118}$, 61 (1977); and from another viewpoint: P. Nath and R. Arnowitt, Phys. Rev. $\underline{D15}$, 1033 (1977).

17. P. Roman and J. Haavisto, "A Relativistic Quantum Dynamical Group for Hadrons", Preprint BU-PNS-16; to be published. [See also J. Haavisto: "Quantum-Dynamical Symmetry Groups in Curved Spaces", Ph.D. Thesis, Boston University, 1978. (Caveat: This thesis contains several minor errors.)]

18. Y. Nambu, Progr. Theor. Phys. Supp. $\underline{1}$, (No's. 37-38), 368 (1966).

19. Possible external Poincaré labels are suppressed.

20. Because of ray-equivalence, setting $C_1 = 0$ is not a restriction of generality.

21. It is easy to see that, in Class I reps, m^2 will be indeed positive provided we take the constant $\beta \gtrless -n_0$ depending on whether the constant $\gamma \gtrless 0$. Note that, according to the results of ref. 7, $\ell = -h^{-1} < 0$.

22. R. P. Feynman, M. Kislinger and F. Ravndal, Phys. Rev. $\underline{D3}$, 2706 (1971).

23. Y. S. Kim and M. E. Noz, Phys. Rev. $\underline{D15}$, 335 (1977).

GENERALIZED COHERENT STATES

T. S. Santhanam*

Department of Theoretical Physics
Research School of Physical Sciences
The Australian National University
P. O. Box 4, Canberra, A.C.T. 2600, Australia

1. INTRODUCTION

The coherent states of the harmonic oscillator and hence the radiation field (which is considered as an assembly of oscillators) can be defined in many different, but essentially _equivalent_ ways.[1]

1. They are the quantum states which result in giving the minimum uncertainty in the simultaneous measurements of the canonical variables and therefore yield closest analogs of classical fields.[2]

2. They are obtained as displaced ground state of the oscillator, the unitary displacement operator representing an element of the Heisenberg-Weyl algebra.[3]

3. They are infinite superpositions of the eigenstates of the Number operator and therefore yield states with well-defined phase.

4. They are the eigenstates of the annihilation operator.[4]

5. They diagonalize the displaced or driven oscillator Hamiltonian with linear potential term.

Many attempts have been made to generalize the concept to more general situations. They can be broadly classified as follows:

1. They may be defined as eigenstates of the Ladder operators (raising and lowering) of Lie algebras. Barut and Girardello[5] have defined generalized coherent states in this way for the non-compact group $SU(1, 1) \sim SO(2, 1) \sim SL(2, R)$ whose unitary irreducible

representations are necessarily infinite dimensional. The method
cannot work for the finite dimensional irreducible representations
of compact groups, where the Ladder operators are the upper (or
lower) triangular matrices with diagonal entries being zero. Sharma,
Mehta and Sudarshan[6] have defined such states for para-bose oscillators, which mainly differ from the standard oscillators by the shift
of the ground state. These states do make the Schwartz inequality
an equality though they <u>do not</u>, in general, yield minimum uncertainty
states.

2. The generalized coherent states may be defined (Radcliffe,[7]
Perelemov[8]) as states displaced from the ground state (or a reference state). This definition works well both for compact and noncompact groups. The trouble is, however, that they give minimum
uncertainty only for the trivial states with maximum (or minimum)
projection (highest weight states) which are, in any case, the
eigenstates of the Ladder operators.

3. The generalized coherent states can be defined as the states
which make the Schwartz inequality an equality. These have been
called "intelligent" states (C. Aragone et al.).[9]

4. Recently Nieto and Simmons[10] have defined the generalized
coherent states as states which satisfy the classical equations of
motion. They have applied it to the case of Morse oscillator whose
ground state is also Gaussian. But, even in this case, some of the
equations are only <u>approximately</u> satisfied since these states are
superpositions of many energy (or number) eigenstates.

5. We have recently defined generalized coherent states for
angular momentum[11] as states with sharply defined phase (azimuthal
angle). They are obtained from the eigenstates of J_z (Zenithal
angle) by a finite Fourier transformation.

6. Generalized coherent states can be defined for Fermi systems as has been done by Martin[12] and by Ohnuki and Kashiwa[13] where
one employs Grassman variables in the place of complex numbers.

In section II we briefly review the many different ways of
defining the oscillator coherent states (OCS) and also some of their
interesting properties. In section III, we discuss the generalized
coherent states defined as eigenstates of Ladder operators. In section IV we discuss the Radcliffe-Perelemov generalized coherent
states. In section V , we discuss the "intelligent" states. In
section VI, we discuss our definition of generalized coherent states
as eigenstates of "phase." In section VII we discuss the generalized coherent states defined as states which obey the classical
equations of motion. In the last section VIII we briefly mention
the generalized coherent states for Fermi systems.

II. OSCILLATOR COHERENT STATES

a) Coherent states of the oscillator can be defined as the states which result in the minimum uncertainty of the measurements in the canonical variables, i.e.

$$\langle(\Delta q)^2\rangle\langle(\Delta p)^2\rangle = \frac{1}{4} , \tag{2.1}$$

where the self-adjoint position and momentum operators satisfy the Heisenberg relation

$$[q, p] = i1 , \tag{2.2}$$

and

$$\Delta X = X - \langle X\rangle , \tag{2.3}$$

and these states are the closest analogs of the classical states. An example of such a state is the ground state of the harmonic oscillator. In deriving Eq. (2.1) from Eq. (2.2) one makes use of Schwarz inequality which gives equality for the states ψ which satisfy

$$(\Delta q)\psi = -i\lambda(\Delta p)\psi$$

$$\lambda \text{ real} , \tag{2.4}$$

or

$$(q + i\lambda p)\psi = Z\psi , \tag{2.5}$$

with

$$Z = \langle q + i\lambda p\rangle . \tag{2.6}$$

b) The coherent states can be equivalently defined as

$$|Z\rangle = D(Z) |0\rangle , \tag{2.7}$$

$$Z = \text{complex number} ,$$

where $|0\rangle$ denotes the vacuum which satisfies

$$a | 0 \rangle = 0 , \tag{2.8}$$

$$N | 0 \rangle = 0 , \tag{2.9}$$

where the annihilation operator

$$a = \frac{q + ip}{\sqrt{2}} , \tag{2.10}$$

and the nubmer operator

$$N = a^\dagger a \,. \tag{2.11}$$

They satisfy the commutation relations

$$[a, a^\dagger] = 1 \tag{2.12}$$

$$[a, N] = a \tag{2.13}$$

$$[N, a^\dagger] = a^\dagger . \tag{2.14}$$

$D(Z)$ is the displacement operator. In order that $|Z\rangle$ are eigenstates of a, or

$$a \mid Z \rangle = Z \mid Z \rangle \,, \tag{2.15}$$

we have as a consequence of Eqs. (2.8) and (2.12) that

$$(\text{adj}\underset{\sim}{a})D \equiv [a, D] = \pm ZD \,, \tag{2.16}$$

i.e. D's furnish a basis for $(\text{adj}\underset{\sim}{a})$. Notice that $\{A, D\}_+ = ZD$ would also give Eq. (2.15). Since $a \sim \dfrac{\partial}{\partial a^\dagger}$ [in view of Eq. (2.12)], Eq. (2.16) gives the solution

$$D = \exp(Za^\dagger - Z{*}a) \tag{2.17}$$

where the second term has been added to make D unitary. Thus,

$$|Z\rangle = e^{-|Z|^2/2} \sum_{n=0}^{\infty} \frac{Z^n}{\sqrt{n!}} \, |n\rangle \,, \tag{2.18}$$

where

$$|n\rangle = \frac{(a^\dagger)^n}{\sqrt{n!}} \, |0\rangle \,. \tag{2.19}$$

It is important to realize that Eq. (2.18) is the direct consequence of the existence of $|0\rangle$ (lower bound for the spectrum of N or the Hamiltonian $H = N + \dfrac{1}{2}$ and the commutation relation (C.R.) Eq. (2.12) implying an infinite spectrum (no upper bound) of H or N.

It may be remarked that the statements Eqs. (2.7) and (2.15) are identical in content because of the commutation relation Eq. (2.12). The operators a, $a^\dagger$ and 1 form the Weyl group W whose general element is the unitary operator

$$D(t, Z) = e^{itl} \, e^{Za^\dagger - Z^*a}$$

$$= e^{itl} \, D(Z) \, , \qquad (2.20)$$

$$t = real \, .$$

It can be easily verified that

$$D(t_1, Z_1)D(t_2, Z_2)$$

$$= e^{i\phi} \, D(Z_1 + Z_2), \qquad (2.21)$$

where the phase factor

$$\phi = (t_1 + t_2 + Im \, Z_1 Z_2^*) \, . \qquad (2.22)$$

The operators $D \in W$ generate the coherent states $|Z>$ from the gound state $|0>$ which is also a coherent state. Since $D(Z)$ is unitary, all the states $|Z>$ result in minimum uncertainty in Eq. (2.1).

The displacement operators provide a complete and orthonormal basis for the adjoint group of the Weyl group with a scalar product given by

$$(D(Z), D(Z')) = Tr \, D(Z)D^\dagger(Z')$$

$$= \pi\delta(Z - Z') \, . \qquad (2.23)$$

Since the commutation and anticommutation operations are canonically conjugate, the solution of

$$\{a^\dagger, M\}_+ = mM$$

will be related to D by a Fourier transformation.

The coherent states $|Z>$ have many interesting properties and some of these are

a)   They are non-orthogonal since

$$<Z|Z'> = <0|D^\dagger(Z)D(Z')|0>$$

$$= e^{i\phi} \, <0|D(Z' - Z)|0>$$

$$\phi = Im \, (Z^*Z')$$

$$= 2\Delta(0, Z, Z') \, , \qquad (2.24)$$

where Δ is the area of the triangle with vertices at Z, Z' and 0. Also

$$|<Z|Z'>|^2 = \exp - |Z - Z'|^2 . \tag{2.25}$$

b) The resolution of identity reads as

$$1 = \int \frac{d^2Z}{\pi} |Z> <Z|$$

$$d^2Z = dZ_1 dZ_2 \tag{2.26}$$

$$Z = Z_1 + iZ_2 .$$

c) They give the minimum uncertainty in the product

$$<(\Delta q)^2> <(\Delta p)^2> = \frac{1}{4}$$

holding for all states $|Z>$.

d) They are overcomplete. A subset of $|Z>$ is already complete. In fact, coherent states on a lattice with

$$Z = \gamma(m + in) \equiv Z_{mn}$$

$$m,n = 0, \pm1, \ldots \tag{2.27}$$

are already complete for $\gamma = \sqrt{\pi}$ [Von-Neumann,[14] Bargmann et al.[15]] in the sense

$$<Z_{mn}|\psi> = 0 \quad \Rightarrow \quad |\psi> = 0 . \tag{2.28}$$

e) An arbitrary coherent state can be expanded as

$$|\psi> = \int \frac{d^2Z}{\pi} <Z|\psi> |Z> \tag{2.29}$$

$$<Z|\psi> = \exp - \frac{1}{2}|Z|^2 \cdot \sum_{n=0}^{\infty} \frac{<n|\psi>}{\sqrt{n!}} Z^n$$

$$= \exp\left(- \frac{1}{2}|Z|^2\right) F(Z) . \tag{2.30}$$

It follows that $F(Z)$ is an entire function.

The overcompleteness property enables that the operator A can be expanded as

$$A = \int \frac{d^2Z}{\pi} f(Z) |Z> <Z| \tag{2.31}$$

where $f(Z)$ may be a distribution, i.e. an operator can be expanded in terms of its diagonal matrix element only!

g) The object $|Z|^2$ is actually the average of the number operator N between coherent states i.e.

$$<Z|N|Z> = |Z|^2 . \qquad (2.32)$$

h) The coherent states remain coherent as time develops. Since the time evolution operator is the Hamiltonian ($H = N + \frac{1}{2}$ for the oscillator)

$$\begin{aligned}
|Z_t> &= e^{-iH\cdot t} |Z> \\
&= e^{-|Z|^2/2} \sum_{n=0}^{\infty} \frac{Z^n}{\sqrt{n!}} e^{-i(n+\frac{1}{2})t} |n> \\
&= e^{it/2} | Z e^{-it} > .
\end{aligned} \qquad (2.33)$$

Thus

$$\begin{aligned}
<Z_t|H|Z_t> &= <Z|H|Z> \\
&= |Z|^2 + \frac{1}{2} .
\end{aligned} \qquad (2.34)$$

Also we notice that

$$\begin{aligned}
<Z|q|Z> &= <Z| \frac{a + a^\dagger}{\sqrt{2}} |Z> \\
&= \sqrt{2}\ \text{Re}\ Z
\end{aligned}$$

$$<Z|p|Z> = \sqrt{2}\ \text{Im}\ Z \qquad (2.35)$$

$$\begin{aligned}
<Z_t|q|Z_t> &= <Z| e^{iHt} q\, e^{-iHt} |Z> \\
&= <Z| q \cos t + p \sin t |Z> \\
&= \sqrt{2}\ \{(\text{Re}\ Z) \cos t + (\text{Im}\ Z) \sin t\} \\
&= \sqrt{2}\ |Z|\ \sin (t + \phi) ,
\end{aligned} \qquad (2.36)$$

where we have denoted

$$Z = |Z|\ e^{i(\frac{\pi}{2} - \phi)} .$$

III. EIGENSTATES OF LADDER OPERATORS

We recall that the coherent states of the oscillator are eigenstates of the annihilation operator. One easily verifies that

$$a \, |Z> \; = \; a \left\{ e^{-\frac{1}{2}|Z|^2} \sum_{n=0}^{\infty} \frac{Z^n}{\sqrt{n!}} \, |n> \right\}$$

$$= \; e^{-\frac{1}{2}|Z|^2} \, Z \sum_{n=0}^{\infty} \frac{Z^{n-1}}{\sqrt{(n-1)!}} \, |n-1>$$

$$= \; Z \, |Z> \tag{3.1}$$

since

$$a \, |n> \; = \; n^{\frac{1}{2}} \, |n-1> \; . \tag{3.2}$$

Barut and Girardello[5] extended this idea from the Weyl group to the non-compact group $SU(1, 1)$.

The Lie algebra of this group is defined by the commutation relations in cartan's form

$$[L^+, \, L^-] \; = \; -2L_3 \tag{3.3}$$

$$[L^3, \, L^{\pm}] \; = \; \pm L^{\pm} \tag{3.4}$$

with $\quad L^{\pm} = L_1 \pm iL_2$.

The casimir operator is

$$Q \; = \; L^+ L^- - L_3(L_3 - 1)$$

$$= \; L^- L^+ - L_3(L_3 + 1) \; . \tag{3.5}$$

The group element can be parametrized in terms of the Euler angles as

$$W \; = \; e^{i\mu L_3} \, e^{i\xi L_{1,2}} \, e^{i\nu L_3} \tag{3.6}$$

$$= \; \begin{bmatrix} e^{i\mu/2} & 0 \\ 0 & e^{-\mu/2} \end{bmatrix} \begin{bmatrix} \cos(i\frac{\xi}{2}) & \sin(\frac{i\xi}{2}) \\ -\sin(\frac{i\xi}{2}) & \cos(\frac{i\xi}{2}) \end{bmatrix}$$

$$\begin{bmatrix} e^{i\nu/2} & 0 \\ 0 & e^{-i\nu/2} \end{bmatrix}$$

$$= \begin{pmatrix} \alpha & \beta \\ \overline{\beta} & \overline{\alpha} \end{pmatrix} \qquad (3.7)$$

where

$$\alpha = e^{i(\mu+\nu)/2} \cosh \frac{\xi}{2}$$

$$\beta = ie^{i(\mu-\nu)/2} \sinh \frac{\xi}{2}$$

$$|\alpha|^2 - |\beta|^2 = 1 \ . \qquad (3.8)$$

The group being non-compact, the unitary irreducible representations are infinite dimensional.

A basis $|\phi,m\rangle$ is chosen as**

$$\langle\phi,m|\phi,m'\rangle = \delta_{mm'} \qquad (3.9)$$

$$Q|\phi,m\rangle = \phi(\phi+1)|\phi,m\rangle \qquad (3.10)$$

$$L_3|\phi,m\rangle = m|\phi,m\rangle \ . \qquad (3.11)$$

The unitary irreducible representations are

1) Continuous principal series for which

$$\phi = -\frac{1}{2} + is \qquad 0 < s < \infty$$

with $m = 0, \pm1, \pm2, \ldots \qquad [c_\phi^0] \qquad (3.12)$

or $m = \pm\frac{1}{2}, \pm\frac{3}{2}, \ldots \qquad [c_\phi^1] \qquad (3.13)$

2) The supplementary series (E_ϕ) for which

$$-\frac{1}{2} < \phi < 0 \ , \quad m = 0, \pm1, \pm2, \ldots \qquad (3.14)$$

3. Discrete principal series for which

$$\phi = -\frac{1}{2}, -1, -\frac{3}{2}, \ldots$$

with $m = -\phi, -\phi + 1, \ldots \quad (D^+) \qquad (3.15)$

or $m = \phi, \quad \phi - 1, \ldots \quad (D^-) \qquad (3.16)$

We shall restrict here to the D^+ (or D^-) series while the same analysis has been carried out by Hongoh[16] for the continuous series. The action of the Ladder operator is

$$L^- \, |\phi,m\rangle = [(m + \phi)(m - \phi - 1)]^{\frac{1}{2}} |\phi, m - 1\rangle. \tag{3.17}$$

In analogy with the case of the harmonic oscillator, we can find the states

$$|Z\rangle =$$
$$= [\Gamma(-2\phi)]^{\frac{1}{2}} \sum_{n=0}^{\infty} \frac{Z^n}{n! \; \Gamma(n - 2\phi)} \, |\phi,n\rangle \tag{3.18}$$

where $n = m + \phi$ and is integral. It is easily verified using Eq. (3.17) that

$$L^- \, |Z\rangle = Z \, |Z\rangle \; \ldots \tag{3.19}$$

Also we can see that

$$\langle Z'|Z\rangle = \Gamma(-2\phi) \sum_{n=0}^{\infty} \frac{(Z'{*}Z)^n}{n!} \, \Gamma(-2\phi + n)$$

$$= \; {}_0F_1 \, (-2\phi, \, Z'Z) \; \ldots \tag{3.20}$$

Thus,

$$\| \, |Z\rangle \, \|^2 = {}_0F_1 \, (-2\phi, \, |Z|^2) \; . \tag{3.21}$$

These states are thus non-orthogonal and, in fact, over-complete states. A similar analysis can be carried out for D^-.

The operators L^+, L^- and L^3 of SU(1, 1) can be realized in the Hilbert space of entire functions as

$$L_3 = Z \frac{d}{dZ} - \phi \tag{3.22}$$

$$L_+ = Z \tag{3.23}$$

$$L_- = -2\phi \frac{d}{dZ} + Z \frac{d^2}{dZ^2} \tag{3.24}$$

The eigenfunctions of L_- are solutions of the equation

$$-2\phi \frac{df}{dZ} + Z \frac{d^2 f}{dZ^2} = \lambda f \tag{3.25}$$

and thus

$$f = {}_0F_1(-2\phi, \, \lambda Z) \; . \tag{3.26}$$

As is obvious this definition cannot work for compact groups whose irreducible representations are finite dimensional unitary ones. The Ladder operators cannot be diagonalized except for the states with highest weight which are annihilated by them.

Sharma et al.[6] have defined the coherent states of the parabose oscillators as the eigenstates of the annihilation operator. The parabose oscillators satisfy the equation of motion

$$[a, N] = a ,$$
(3.27)

where

$$N = \frac{1}{2} \{a, a^\dagger\}_+ - b_0 .$$
(3.28)

Here b_0 is the lowest eigenvalue of the Hamiltonian

$$H = \frac{1}{2} \{a, a^\dagger\}_+$$
(3.29)

The case $b_0 = \frac{1}{2}$ would then correspond to the standard oscillator. Using the fact that $a^{\dagger 2}$, a^2 and $\frac{1}{2}H$ close under $0(2, 1)$ and also H is positive definite, it has been possible to find the matrix elements of the annihilation operator [see ref. (17)] as (with a phase connection)

$$(a)_{2n, 2n+1} = [2(n + b_0)]^{\frac{1}{2}}$$
(3.30)

$$(a)_{2n-1, 2n} = [2n]^{\frac{1}{2}}$$
(3.31)

$$n = 0, 1, 2, \ldots$$

and also

$$\langle 2n| [a, a^\dagger] \ |2n\rangle = 2b_0$$
(3.32)

$$\langle 2n + 1| [a, a^\dagger] \ |2n + 1\rangle = 2(1 - b_0) .$$
(3.33)

The distinction between even and odd states will disappear for the normal oscillator case $(b_0 = \frac{1}{2})$. From these equations it is possible to deduce the states defined by

$$|\alpha\rangle_{b_0}$$

$$= \{f(|\alpha|^2)\}^{-\frac{1}{2}}$$

$$\sum_{n=0}^{\infty} \left\{ \frac{\Gamma(b_0)}{2^n \ \Gamma([\frac{n}{2}] + 1)\Gamma([\frac{n+1}{2}] + b_0)} \right\}^{\frac{1}{2}} \alpha^n |n\rangle_{b_0}$$
(3.34)

where $[\frac{n}{2}]$ and $[\frac{n+1}{2}]$ refer to their integral part and

$$|n\rangle_{b_0} = \left\{ \frac{\Gamma(b_0)}{2^n \Gamma([\frac{n}{2} + 1] + 1)\Gamma([\frac{n+1}{2}] + b_0)} \right\}^{\frac{1}{2}} (a^\dagger)^n \, |0\rangle_{b_0} \tag{3.35}$$

The factors inside the curly brackets replace the $(\sqrt{n!}\,)^{-\frac{1}{2}}$ factor which appears in the standard oscillator case. They in fact stand for $[(2n)!\, 2(n + b_0)!]^{-\frac{1}{2}}$, the two factors appearing for the even and odd cases respectively. The factor $\{f(|\alpha|^2)\}^{-\frac{1}{2}}$ is fixed by the normalization

$$_{b_0}\langle\alpha|\alpha\rangle_{b_0} = 1 \,\ldots \tag{3.36}$$

as

$$f(x) = \sum_{n=0}^{\infty} \frac{\Gamma(b_0)}{\Gamma([\frac{n}{2} + 1])\Gamma([\frac{n+1}{2}] + b_0)} (\tfrac{1}{2} x)^n \, . \tag{3.37}$$

In the case of the normal oscillator $(b_0 = \frac{1}{2})$,

$$f(x) = \exp(x) \, . \tag{3.38}$$

It has been demonstrated in ref. (6) that these states form an overcomplete non-orthogonal basis and the diagonal expansion can be extended to this situation as well. Since $[a, a^\dagger] \neq 1$, it follows from the Schwartz inequality that

$$\langle(\Delta q)\rangle^2 \langle(\Delta p)^2\rangle \geq \frac{1}{4} \left|\langle[q, p]\rangle\right|^2 \tag{3.39}$$

which for the coherent states $|\alpha\rangle_{b_0}$ becomes an equality. As has been repeatedly emphasized, the right hand side refers neither to a normalization nor to an uncertainty and depends on the "state" and thus the equality sign does not produce an absolute minimum. It may be remarked that the existence of a lower bound and an infinite spectrum for the Hamiltonian (or the Number operator) has made it possible to define the coherent states $|\alpha\rangle_{b_0}$.

IV. DISPLACEMENT OPERATORS

We have seen earlier that the oscillator coherent states can be defined as the displaced ground state. The displacement operators form the Weyl group. A generalization of this concept to an arbitrary group has been made by Radcliffe[7] (for the rotation group), by Perelemov[8] (for an arbitrary group) and in some sense by Atkins and Dobson[18] (for the rotation group). Let us now brief study the angular momentum coherent states.

Radcliffe made the formal analogy between the operators of the Heisenberg algebra and the angular momentum generators

$$|0> \rightarrow |j,j>$$

$$J^- \sim a^\dagger \tag{4.1}$$

to define the angular momentum coherent states as

$$|\mu> = D(\mu) |j,j> \tag{4.2}$$

where

$$D(\mu) = \exp (\mu J_- - \mu^* J_+) . \tag{4.3}$$

Since

$$(J^-)^p |j,j>$$

$$= \left\{ \frac{p! \, (2j)!}{(2j-p)!} \right\}^{\frac{1}{2}} |j,p> ,$$

$$0 \leq p = j - m \leq 2j$$

$$-j \leq m \leq j . \tag{4.4}$$

Eq. (4.2) can be rewritten as

$$|\mu> = \frac{1}{(1 + |\mu|^2)^j} \sum_{p=0}^{2j} \binom{2j}{p}^{\frac{1}{2}} \mu^p |p> \tag{4.5}$$

The normalization factor has been chosen so that

$$<\mu|\mu> = \frac{1}{(1 + |\mu|^2)^{2j}} \sum \binom{2j}{p} (|\mu|^2)^p$$

$$= 1 . \tag{4.6}$$

The bracket stands for the binomial coefficient. Actually the states $|\mu>$ are eigenstates of the operator

$$D J_z D^{-1} = J_z + \mu J_- \tag{4.7}$$

with eigenvalue j.

These coherent states become the oscillator coherent states in the Holstein-Primakoff[19] limit as $j \rightarrow \infty$

$$J^- \rightarrow (2j)^{\frac{1}{2}} a^\dagger$$

$$J_z \rightarrow 1$$

$$\mu \to \frac{z}{(2j)^{\frac{1}{2}}} \tag{4.8}$$

$$|\mu\rangle \xrightarrow{j\to\infty} \left(1 + \frac{|z|^2}{2j}\right)^{-j} e^{za^{\dagger}} |0\rangle$$

$$= e^{-\frac{1}{2}|z|^2} e^{za^{\dagger}} |0\rangle \tag{4.9}$$

$$= |z\rangle.$$

Actually Eq. (4.8) represents the contraction from the angular momentum algebra to the Heisenberg algebra.[35] These Radcliff states are non-orthogonal states since

$$\langle\lambda|\mu\rangle = \left\{\frac{1 + \lambda^*\mu}{[(1 + |\mu|^2)(1 + |\lambda|^2)]^{\frac{1}{2}}}\right\}^{2j} . \tag{4.10}$$

They form a complete set although it is necessary to include a weight function $M(|\mu|^2)$ in the integral. We have

$$\int d^2 \mu|\mu\rangle \langle\mu| \, M(|\mu|^2) = 1 \tag{4.11}$$

in which we choose

$$M(|\mu|^2) = \frac{(2j + 1)}{\pi} \frac{1}{(1 + |\mu|^2)^2} .$$

One can shoose an alternate parameterization in terms of the points (θ, ϕ) on a unit sphere as

$$\mu = \tan\frac{\theta}{2} e^{i\phi}$$

$$0 \le \theta \le \pi , \quad 0 \le \phi \le 2\pi . \tag{4.12}$$

This transformation $(\theta, \phi) \to \mu$ is nothing but a steographic projection of the unit sphere from the point $(0, 0, -1)$ into a complex plane μ followed by a reflection of the y axis. The coherent state $|\mu\rangle$ in these coordinates (θ, ϕ) reads as

$$|\mu\rangle = (\cos\frac{\theta}{2})^{2j} \exp\{\tan\frac{\theta}{2} e^{-i\phi}J_-\} |j,j\rangle . \tag{4.13}$$

The completeness relation becomes

$$\frac{2j + 1}{4\pi} \int d\,\Omega \, |\Omega\rangle \langle\Omega| = 1 . \tag{4.14}$$

In this representation the matrix elements of the angular momentum generators are given by

$$\langle\Omega|J_z|\Omega\rangle = j \cos \theta$$

$$\langle\Omega|J_+|\Omega\rangle = j \sin \theta \, e^{i\phi}$$

$$\langle\Omega|J_-|\Omega\rangle = j \sin \theta \, e^{-i\phi} \,. \tag{4.15}$$

Perelemov's construction[8] (which is essentially the same as the Radcliffe states for angular momentum) of a set of generalized coherent states is

$$|\Omega_g\rangle = T(g) \, |\psi_0\rangle \tag{4.16}$$

where $T(g)$ is a representation of the group G and $|\psi_0\rangle$ is a fixed vector in the vector space of the representation $T(g)$. If one chooses the reference state as the state with highest weight (diagonal in the cartan subalgebra) then Eq. (4.16) is equivalent to (apart from phase factors coming from the diagonal generators)

$$|\Omega_g\rangle \sim T(g) \, |\Omega_0\rangle \tag{4.17}$$

where g represents the Ladder operators. Thus, these are the same as Radcliffe states for angular momentum.

This definition yields the generalized coherent states for $SU(1, 1)$ as

$$|\zeta\rangle = \frac{1}{(1 - |\zeta|^2)^k} \, e^{\zeta L_-} \, |0\rangle \tag{4.18}$$

where $|0\rangle$ stands for the highest weight state i.e.

$$L_+ \, |0\rangle = 0 \,. \tag{4.19}$$

However, as has been demonstrated by Kolodziejczyk and Ryter[20] these states <u>do not</u> give the minimum uncertainty states except for $\mu = 0$ and these are the trivial states $|jj\rangle$ which are eigenstates of J_+ with eigenvalue zero. We will not expand the discussion (see ref. (21) for more details) except to make the following remarks.

The states $|\mu\rangle$ can also be written as

$$|\mu\rangle = \frac{1}{(1 + |\mu|^2)^j} \sum_p \binom{2j}{p}^{\frac{1}{2}} \mu^p \, v^{2j-p} \, |p\rangle$$
$$v = 1 \,. \tag{4.20}$$

Since $\mu^p \, v^{2j-p}$ stands for the homogeneous polynomial of deg $(2j)$, it has been noted[22] that $|\mu\rangle$ transforms as the $(2j + 1)$ dimensional vector representation for spin j.

If one writes

$$g^P(\mu, \nu) = \binom{2j}{p}^{\frac{1}{2}} \mu^P \nu^{(2j-p)} \tag{4.21}$$

then

$$\sum g^P(\mu', \nu')^* g^P(\mu, \nu)$$
$$= (\mu'^*\mu + \nu'^*\nu')^{2j} . \tag{4.22}$$

This is related[23] to a Vander Waerden invariant

$$W = (u_1 d_2 - u_2 d_1)^{2j} \tag{4.23}$$

where one makes the following identification

$$\begin{bmatrix} u_1 \\ d_1 \end{bmatrix} \rightarrow \begin{bmatrix} \mu \\ \nu \end{bmatrix}$$
$$\begin{bmatrix} d_2 \\ -u_2 \end{bmatrix} \rightarrow \begin{bmatrix} \mu'^* \\ \nu'^* \end{bmatrix} \tag{4.24}$$

The Clebsch-Gordon coefficients of the rotation group can be computed in the basis furnished by Radcliffe states.[24]

It should be also remarked that the coherent states for parabose oscillators introduced in the earlier section can be written in the Radcliffe-Perelemov form

$$|\alpha\rangle_{b_0} = [f(\alpha a^{\dagger}) \{f(|\alpha|^2)\}^{\frac{1}{2}}] |0\rangle_{b_0} . \tag{4.25}$$

The important point is that the operators $f(\alpha a^{\dagger})$ cannot be written in the form of a group generator except in the case of the normal oscillator $(b_0 = \frac{1}{2})$ when $f(\alpha a^{\dagger}) \rightarrow \exp(\alpha a^{\dagger})$.

Now, we shall discuss the method of Atkins and Dobson[18] to construct angular momentum coherent states. Here one makes use of Schwinger's method[25] of angular momentum and is based on two independent oscillators

$$\underline{a} = (a_+, a_-)$$
$$\underline{a}^{\dagger} = (a_+^{\dagger}, a_-^{\dagger})$$
$$[a_\zeta, z_{\zeta'}] = [a_\zeta^{\dagger}, a_{\zeta'}^{\dagger}] = 0$$
$$[a_\zeta, a_{\zeta'}^{\dagger}] = \delta_{\zeta\zeta'}$$
$$\delta = (+, -) . \tag{4.26}$$

The angular momentum operators can be expressed as

$$J_i = \underline{a}^{\dagger} \frac{1}{2} \sigma_i \underline{a} , \qquad i = x, y, z \tag{4.27}$$

where σ_i are the Pauli matrices. Also

$$J^2 = j(j + 1)1$$

$$j = a^{\dagger} (\tfrac{1}{2} 1) a \tag{4.28}$$

$$= \frac{1}{2} (n_+ + n_-)$$

$$n_{\pm} = a_{\pm}^{\dagger} a_{\pm} . \tag{4.28}$$

The angular momentum states $|j,m\rangle$ can be expressed as

$$|j,m\rangle = \frac{(a_+^{\dagger})^{j+m}}{\sqrt{(j+m)!}} \frac{(a_-^{\dagger})^{j-m}}{\sqrt{(j-m)!}} |0,0\rangle \tag{4.29}$$

where the vacuum states $|0,0\rangle$ satisfies

$$a_{\pm} |0,0\rangle = 0 . \tag{4.30}$$

The simultaneous eigenstates of a_+ and a_- are defined as angular momentum coherent states and these are given by

$$|\underline{f}\rangle = e^{-\frac{1}{2}(|k_+|^2 + |k_-|^2)}$$

$$\sum_{n_+,n_-} \frac{k_+^{n_+}}{\sqrt{n_+!}} \frac{k_-^{n_-}}{\sqrt{n_-!}} |n_+,n_-\rangle \tag{4.31}$$

where k_+, k_- are the two complex eigenvalues

$$a_s |\underline{f}\rangle = k_{\zeta}|\underline{f}\rangle \tag{4.32}$$

$$\underline{f} = (k_+, k_-)$$

$$\zeta = \pm .$$

As the state $|n_+,n_-\rangle$ refers to the angular momentum state $|j,m\rangle$, on identifying

$$n_{\pm} = j \pm m \tag{4.33}$$

we have

$$|f\rangle = e^{-\frac{1}{2}k^2} \sum_{j=0}^{\infty} \sum_{m=-j}^{j} \binom{2j}{j+m}^{\frac{1}{2}} (2j!)^{-\frac{1}{2}} k_+^{j+m} k_-^{j-m} |j,m\rangle \tag{4.34}$$

where

$$k^2 = |k_+|^2 + |k_-|^2 \ . \tag{4.35}$$

Eq. (4.34) can be reexpressed in a more compact form as

$$|\underline{f}> \ = e^{-\frac{1}{2}k^2} \sum_{j,m} \frac{(k_+ a_+^\dagger)^{j+m}}{(j+m)!} \ \frac{(k_- a_-^\dagger)^{j-m}}{(j-m)!} \ |0,0>$$

$$= \exp\{(\underline{f}[\underline{a}^\dagger - \tfrac{1}{2} \ f*])\}|0> \tag{4.36}$$

where the last bit is formal.

Eq. (4.36) tempts us to write such states for an arbitrary group G as

$$|\underline{g}> \ = \exp\{(g[\underline{a}^\dagger - \tfrac{1}{2} \ g*])\}|0> \tag{4.37}$$

where

$$a^\dagger = (a_1, \ \dots, \ a_r)$$

$$g \ = (k_1, \ \dots, \ k_r) \tag{4.38}$$

and $\underline{a}$ transforms as the natural representation of the group G. It has been shown by the author[26] that the angular momentum generators $(J_1, \ J_2, J_3)$ which satisfy Nambu brackets[27] and $\sqrt{J^2}$ can be rewritten in terms of two oscillator systems satisfying (in the limit $j \to \infty$)

$$[J_z, \ \phi] \ = i1$$

$$[\sqrt{J^2}, \ w] = i1 \tag{4.39}$$

and the simultaneous coherent states of these two statistically independent oscillators can be defined as generalized coherent states. Since the irreducible representations of the rotation group are finite dimensional, J_z is always bounded. In this case, the first of the relations in Eq. (4.39) has to be replaced by its finite analog. (See section VI).

V.   INTELLIGENT AND QUASI-INTELLIGENT STATES

From simple arguments based on the Schwartz inequality, it follows that since the angular momentum states satisfy the commutation relation

$$[J_x, \ J_y] = iJ_z \tag{5.1}$$

the eigenstates of the operator

$$A\psi \equiv \frac{J_x - i\lambda J_y}{\sqrt{1 + \lambda^2}} \, \psi = Z\psi$$

(5.2)

$$Z = \langle \frac{J_x - i\lambda J_y}{\sqrt{1 + \lambda^2}} \rangle \, , \quad \lambda \text{ real}$$

will give the equality (the sides proportional)

$$\langle(\Delta J_x)^2\rangle \, \langle(\Delta J_y)^2\rangle = \frac{1}{4} \, |\langle J_z\rangle|^2 \, .$$

(5.3)

When $\lambda = 1$, however, in (5.3) only eigenstate is $|j,j\rangle$ and the operator J_- is non-diagonalizable. If $\lambda \neq 1$, A is diagonalizable. These eigenstates are called intelligent states.[9] If the sides (ΔJ_x) and (ΔJ_y) are in addition equal it follows that

$$\langle(\Delta J_x)^2\rangle = \langle(\Delta J_y)^2\rangle = \frac{j}{2} \, .$$

(5.4)

If λ can also be complex, the states in Eq. (5.2) are referred to as quasi-intelligent states. Of course, as is clear, they will not give _even_ the equality in Eq. (5.3). As has been often emphasized, in either of these cases $\lambda =$ real or complex Eq. (5.3) does not, in general, give an absolute minimum uncertainty state. From the commutation relations of the angular momentum generators, it follows that

$$\Lambda = U \, J_z \, (U)^{-1} = (\cosh \lambda)J_x + i(\sinh \lambda)J_y$$

(5.5)

where

$$U = e^{\lambda J_z} \, e^{-i \frac{\pi}{2} J_y}$$

$$\lambda \text{ real} \, .$$

(5.6)

Therefore the states

$$|\lambda,j,m\rangle = U \, |j,m\rangle$$

(5.7)

are eigenstates of Λ with the same eigenvalue m as that of J_z.[28] Such states will certainly satisfy Eq. (5.3). If we choose

$$\cosh b = \frac{1}{(1 - \lambda^2)^{\frac{1}{2}}}$$

$$\sinh b = -\frac{\lambda}{(1 - \lambda^2)^{\frac{1}{2}}}$$

$$e^b = \left(\frac{1 - \lambda}{1 + \lambda}\right)^{\frac{1}{2}} \tag{5.8}$$

then the operator Λ will be the same as A defined in Eq. (5.2). From Eq. (5.7) we can see that

$$|b,j,m\rangle = \sum_{m'} e^{bJ_z} |jm'\rangle\langle jm'| e^{-i\frac{\pi}{2}J_y} |jm\rangle$$

$$= \sum_{m'} e^{bm'} d^j_{mm'}\left(\frac{\pi}{2}\right) |jm'\rangle \tag{5.9}$$

where $d^j\left(\frac{\pi}{2}\right)$ stands for Wigner reduced matrix elements. Bacry[29] has noted some geometrical properties of these states. The concept has been extended to the group SU(1, 1).[30] Rashid[31] has calculated the matrix elements of polynomials of the generators of the rotation group between quasi-intelligent states.

VI. EIGENSTATES OF PHASE AS ANGULAR MOMENTUM COHERENT STATES

In this section, we shall discuss our method[11] of defining the angular momentum coherent states as states with sharply defined phase.

In the spherical polar coordinate system in which

$$\underline{r} = (r \sin \theta \cos \phi, \ r \sin \theta \sin \phi, \ r \cos \theta) \tag{6.1}$$

the generators of angular momentum take the form

$$J_z = \frac{1}{i}\frac{\partial}{\partial\phi}$$
$$J^{\pm} = e^{\pm i\phi}\left(\pm\frac{\partial}{\partial\theta} + i \cot \theta \frac{\partial}{\partial\phi}\right)$$
$$J^{\pm} = J_x \pm i J_y \ . \tag{6.2}$$

Formally, therefore, J_z and ϕ are canonically conjugate as

$$[\phi, J_z] = i\, 1 \ . \tag{6.3}$$

Eq. (6.3) is not strictly correct, since, taking matrix elements with states $|j,m\rangle$ (eigenstates of J_z and J^2) this leads to a contradiction[32]

$$(m - m') \langle m'|\phi|m\rangle = i\, \delta_{mm'} \ . \tag{6.4}$$

This is because the domain of the operator J_z consists of functions periodic in ϕ.[33] On the other hand Weyl's commutation relation

$$UV = \varepsilon\, V\, U \tag{6.5}$$

$$U = \exp \frac{2\pi i\phi}{2j+1} = \begin{bmatrix} 0 & 1 & 0 & . & . & . & . & . \\ 0 & 0 & 1 & . & . & . & . & . \\ & & & & & & & \\ 1 & 0 & 0 & . & . & . & & 0 \end{bmatrix}$$

$$V = \exp\{(J_Z + jI)\} \tag{6.6}$$

$$\varepsilon = \exp \frac{2\pi i}{2j+1}$$

still holds.[36] Indeed, J_Z and ϕ satisfy the Quantum Mechanics in Finite Discrete space[37]

$$[\phi,\ J_Z]_{mm'} = \frac{(\log \varepsilon)^2}{2\pi} (m' - m) \left[\frac{1}{\varepsilon^{(m'-m)} - 1} \right]$$
$$m' \neq m$$
$$= 0 \quad m' = m \tag{6.7}$$

which as $j \to \infty$ tends to

$$[\phi,\ J_Z]_{mm'} \to i\delta(m - m') . \tag{6.8}$$

Eq. (6.7) can be reexpressed as

$$e^{-\frac{2\pi i\phi}{2j+1}} J_Z\, e^{2\pi i\phi/2j+1} = J_Z + 1 - L \tag{6.9}$$

$$L = (2j + 1)P_j \tag{6.10}$$

where P_j stands for the projection operator on the highest weight vector $|j,j\rangle$. The presence of L takes care of the <u>finite</u> dimensionality of the space. The angular momentum operators can be polar decomposed as

$$J_+ = J_T\, e^{i\phi} = e^{-i\phi}\, J_\perp$$
$$J_- = e^{-i\phi}\, J_T = J_\perp\, e^{i\phi} . \tag{6.11}$$

J_T and $J_\perp$ are singular hermitian operators, $e^{i\phi}$ is unitary. In the standard basis $e^{i\phi}$ is the same as the unitary operator U of Eq. (6.5) and thus will satisfy Eq. (6.9). Since $e^{i\phi}$ is a cyclic permutation matrix (a circulant), it is diagonalized by the Sylvester matrix (finite Fourier transform)

$$S = \frac{1}{\sqrt{2j+1}} \begin{bmatrix} 1 & 1 & 1 & \\ 1 & \varepsilon & \varepsilon^2 & \\ 1 & \varepsilon^2 & \varepsilon^4 & \\ \cdot & \cdot & \cdot & \cdots \\ 1 & \varepsilon^{n-1} & \varepsilon^{n-2} & \end{bmatrix}$$

$$SS^\dagger = S^\dagger S = 1$$

$$S^4 = 1 \tag{6.12}$$

and thus the eigenstates $|\phi\rangle$ of the hermitian "phase" operator ϕ, are given by

$$|\zeta\rangle = \frac{1}{\sqrt{2j+1}} \sum_{m=-j}^{j} \varepsilon^{\zeta m} |j,m\rangle . \tag{6.13}$$

These states $|\zeta\rangle$ are eigenstates of ϕ and are analogs of the oscillator coherent states in finite space. These form a complete set. It is then straightforward to calculate the matrix elements of the angular momentum operators in this basis which turn out to be

$$\langle J_x \rangle_\phi = \frac{1}{2j+1} [Tr(J_T)] \cos\left(\frac{2\pi i}{2j+1} \phi\right)$$

$$\langle J_y \rangle_\phi = \frac{1}{2j+1} [Tr(J_T)] \sin\left(\frac{2\pi i}{2j+1} \phi\right)$$

$$\langle J_z \rangle_\phi = 0$$

$$Tr(J_T) = \sum_m [(j-m)(j+m+1)]^{\frac{1}{2}}$$

$$= Tr(J_\perp) . \tag{6.14}$$

Also,

$$\langle (\Delta J_x)^2 \rangle_\phi = a \cos^2\left(\frac{2\pi i}{2j+1} \phi\right) + b$$

$$\langle (\Delta J_y)^2 \rangle_\phi = a \sin^2\left(\frac{2\pi i}{2j+1} \phi\right) + b \tag{6.15}$$

where

$$a = \frac{1}{2j+1} [Tr(J_T J_\perp)] - \frac{[Tr(J_\perp)]^2}{(2j+1)^2} \tag{6.16}$$

$$b = \frac{1}{3} j(j+1) - \frac{1}{(2j+1)} Tr(J_T J_\perp) .$$

and

$$\text{Tr}(J_T J_\perp) = \sum_m [(j-m)(j+m+1)(j+m)(j-m+1)]^{\frac{1}{2}} . \qquad (6.17)$$

Since $\langle J_z \rangle_\phi = 0$, we have

$$\langle (\Delta J_x)^2 \rangle_\phi \ \langle (\Delta J_y)^2 \rangle_\phi \geq 0 . \qquad (6.18)$$

The minimum on the left hand side is reached for

$$\phi = 0 \ \text{mod}\left(\frac{2j+1}{4}\right)$$

$$= \frac{k(2j+1)}{4} \qquad k = 0, 1, 2, \ldots . \qquad (6.19)$$

This, however, is different from zero. Also, only one component of angular momentum Viz J_z (identified with N) is picked. For general groups, the method relies on the reduction to the product of rotation groups. Each cartan generator with its phase provides an oscillator in the finite space.

VII.   COHERENT STATES FOR GENERALIZED POTENTIALS

We shall briefly discuss the approach of Nieto and Simmons[10] which emphasizes the equations of motion.

For the oscillator we have***

$$H = \frac{1}{2} p^2 + V(X)$$

$$V(X) = \frac{1}{2} X^2 m^2 w^2 . \qquad (7.1)$$

The Hamiltons equations which describe the time evolution of the variables in classical mechanics are

$$\dot{x} = \frac{\partial H}{\partial p} = \{x, H\}_{P.B}$$

$$-\dot{p} = \frac{\partial H}{\partial x} = \{H, p\}_{P.B} \qquad (7.2)$$

where the curly brackets stand for the Poisson bracket. For the potential in Eq. (7.1) they yield the solutions

$$X(t) = \left[\frac{2E}{mw^2}\right]^{\frac{1}{2}} \sin (wt + \phi)$$

$$p(t) = (2mE)^{\frac{1}{2}} \cos (wt + \phi) \qquad (7.3)$$

where the amplitudes are fixed by Eqs. (7.1) and (7.2). We had earlier seen from Eq. (2.36) that the average of X(t) between coherent states of the oscillator

$$\langle Z| \; X(t) \; |Z\rangle = [2\langle N\rangle]^{\frac{1}{2}} \sin (wt + \phi)$$

$$\langle N\rangle = |z|^2$$

which is exactly like Eq. (7.3), the solution of the classical
equations of motion. The only difference is that energy is replaced
by average number i.e. the zero point energy has been subtracted.

The method is to find variables

$$X_c(x) = A(E) \sin (w_c(E)t + \phi)$$

$$P_c(x) = p \; \frac{dX_c(x)}{dx} \qquad (7.4)$$

which obey the classical equations

$$m \; \dot{X}_c = P_c$$

$$\dot{P}_c = -m \; w_c^2 (E) \; X_c \qquad (7.5)$$

where

$$\frac{p^2}{2m} + V(x) = E \; . \qquad (7.6)$$

From Eq. (7.4) we see that

$$\frac{dX_c(x)}{dx} = \frac{P_c}{p}$$

$$= w_c \left[\frac{m(A^2 - X_c^2)}{2(E - V)} \right]^{\frac{1}{2}} \qquad (7.7)$$

and Eq. (7.6) now reads as

$$\frac{P_c^2}{2m} + \frac{1}{2} \; m \; w_c^2 \; X_c^2$$

$$= \frac{1}{2} \; m \; w_c^2 \; A^2 \; . \qquad (7.8)$$

Now we pass on to quantum operators

$$X = X_c$$

$$P = \frac{1}{2} \; (X_c' p + p \; X_c') \qquad (7.9)$$

with

$$p = \frac{\hbar}{i} \frac{d}{dx} \; . \tag{7.10}$$

These obey

$$[X, P] = i\hbar G \tag{7.11}$$

where

$$G = \left[\frac{dX(x)}{dx} \right]^2 \; . \tag{7.12}$$

The generalized uncertainty relation[34] is

$$\frac{<(\Delta X)^2> \, <(\Delta P)^2>}{\left| <G> \right|^2} \geq \frac{1}{4} \; . \tag{7.13}$$

Then one has to find those states which minimize this uncertainty
relation. What has been done is to transform a system with an
arbitrary potential into an _equivalent_ harmonic oscillator system

$$U^\dagger \left[\frac{P^2}{2m} + V(x) \right] U$$

$$= \frac{1}{2m} P_c^2 + \frac{1}{2} m \, w_c^2 \, X_c^2 \tag{7.14}$$

and then find out the coherent states of this harmonic oscillator
system by the well-known techniques, (finding generalized Ladder
operators, displacement operators, etc.). The method in this
respect is quite analogous to the Bogolubov transformation which
maps (_approximately_) a system with interactions into a free particle
system. Nieto and Simmons have applied it to Morse oscillator
(whose ground state is also Gaussian) with the potential

$$V(x) = U_o \, (1 - e^{-ax})^2$$

$$U_o = \frac{\lambda^2 \, \hbar^2 \, a^2}{2m} \; .$$

The crucial point is that these states (which are superpositions
of energy eigenstates) can obey some of the equations of motion
only _approximately_.

VIII.   COHERENT STATES FOR FERMI OPERATORS

    A description of a Fermi oscillator in classical dynamics
involves the introduction of anti-commuting Grassman coordinates.[12]
In fact, with simple complex numbers there does not exist a solution
to the equation

$$D(Z)\underline{b}_i D(Z)^{-1} = \underline{b}_i + Z \qquad (8.1)$$

where the Fermi operators b, $b^\dagger$ satisfy the standard anti-commutation relation

$$\{b_i, b_j^\dagger\}_+ = \delta_{ij} . \qquad (8.2)$$

However,

$$e^{\xi b_j^\dagger} b_i e^{-\xi b_j^\dagger} = b_i + \xi \delta_{ij} \qquad (8.3)$$

where the Grassman symbols ξ satisfy

$$\{\xi, b_i\}_+ = 0$$

$$\{\xi, b_i^\dagger\}_+ = 0$$

$$\{\xi_i, \xi_j\} = 0$$

$$\xi^2 = 0$$

$$[\xi \, |0>] = 0 . \qquad (8.4)$$

The coherent states can be constructed as[13]

$$|(\xi)_n> = \exp\left\{ \sum_{j=1}^{n} \xi_j \, b_j^\dagger \right\} |0> . \qquad (8.5)$$

These are eigenstates of the annihilation operators

$$b_j \, |(\xi)_n> = \xi_j \, |(\xi)_n> . \qquad (8.6)$$

The eigenvalues are now Grassman numbers. These states are complete.

ACKNOWLEDGMENTS

I am grateful to Professor K. J. Le Couteur for his gracious hospitality.

REFERENCES

1. A detailed discussion can be found in J. R. Klauder and E. C. G. Sudarshan, Fundamentals of Quantum optics 1968 (New York : Benjamin).

2. E. Schrödinger, Naturwissenschaftern <u>14</u>, 664 (1929).
        This has been discussed, for example, in L. I. Schiff,
        Quantum Mechanics (McGraw-Hill), N.Y., 1955), 2nd Ed.
        p. 67.
3. R. J. Glauber, Phys. Rev. <u>131</u>, 2766 (1963).
4. E. C. G. Sudarshan, Phys. Rev. Lett. <u>10</u>, 277 (1963).
5. A. O. Barut and L. Girardello, Commun. Math. Phys. <u>21</u>, 41 (1971).
6. J. K. Sharma, C. L. Mehta and E. C. G. Sudarshan, J. Math. Phys.
        <u>19</u>, 2089 (1978).
7. J. M. Radcliffe, J. Phys. <u>A4</u>, 313 (1971).
8. A. M. Perelemov, Commun. Math. Phys. <u>26</u>, 222 (1972).
9. C. Aragone, E. Chalband and S. Salamo, J. Math. Phys. <u>17</u>, 1963
        (1976).
    C. Aragone et al., J. Phys. <u>A7</u>, L149 (1974).
10. M. M. Nieto and L. M. Simmons, Phys. Rev. Lett. <u>41</u> 207 (1978)
        Los Alamos preprints 78-2137.
11. T. S. Santhanam, Found. Phys. <u>7</u>, 121 (1977), Phys. Lett. <u>56A</u>,
        345 (1976),
    See also, J. M. Levy-Leblond, Rev. Mexi. Fis. <u>22</u>, 15 (1973).
12. J. L. Martin, Proc. Roy. Soc. London. <u>A251</u>, 536 (1959).
13. Y. Ohnuki and T. Kashiwa, Nagoya Univ. preprint 12-78.
14. J. Von Neumann, Mathematical Foundations of Quantum Mechanics
        (Princeton, 1955).
15. V. Bargmann, P. Butera, L. Girardello and J. R. Klauder, Reports
        on Math. Phys. <u>2</u>, 221 (1971).
16. Hongoh, J. Math. Phys. <u>18</u>, 2081 (1977).
17. T. F. Jordan, N. Mukunda and S. V. Pepper, J. Math. Phys. <u>4</u>,
        1089 (1963).
    L. O' Raifeartaigh and C. Ryan, Proc. R. Irish, Acad. <u>A62</u>, 93
        (1963).
18. P. W. Atkins and J. C. Dobson, Proc. Roy. Soc. London <u>A321</u>, 321
        (1971).
19. F. Holstein and H. Primakoff, Phys. Rev., <u>58</u>, 1048 (1940).
20. L. Kolodziejczyk and A. Ryter, J. Phys. <u>A7</u>, 213 (1974).
21. A. M. Perelemov, Sov. Phys. Usp. <u>20(9)</u>, Sep 1977, Usp. Fiz.
        Nauk <u>123</u>, 23 (1977).
22. F. T. Hioe, J. Math. Phys. <u>15</u>, 11 74 (1974).
23. M. Hongoh, Rep. on Math. Phys. <u>13</u>, 305 (1978).
24. J. Bellissard and R. Holz, J. Math. Phys. <u>15</u>, 1275 (1974).
25. J. Schwinger, Quantum Theory of Angular Momentum, Eds., L. C.
        Biedenharn and H. Van Dam (Academic Press, N.Y.) 1965
        pp. 229-79.
26. T. S. Santhanam, Proc. International Conference on "Frontiers
        of Physics" held in Singapore, Singapore National Acad.
        of Sci. Eds. K. K. Phua et al., 1167-1196 (1978).
27. Y. Nambu, Phys. Rev. <u>7D</u>, 2405 (1973).
28. S. Ruschin and Y. Ben-Aryeh, Phys. Lett. <u>58A</u>, 207 (1976).
29. H. Bacry, J. Math. Phys. <u>19</u>, 1192 (1978)
        Phys. Rev. <u>18A</u>, 617 (1978).

30. G. Vanden Berghe and H. De Meyer, J. Phys. $\underline{A11}$, 1569 (1978).
31. M. A. Rashid, J. Math. Phys. $\underline{19}$, 1391 and 1397 (1978).
32. A detailed discussion of this appears in:
         P. Carruthers and M. M. Nieto, Rev. Mod. Phys. $\underline{40}$, 411
         (1968).
33. K. Kraus, Z. Phys. $\underline{188}$, 374 (1965), $\underline{201}$, 134 (1967).
34. R. Jackiw, J. Math. Phys. $\underline{9}$ 339 (1968).
35. H. Bacry, A. Grassman, and J. Zak, Proc. 4th Int. Colloq. on
         Group Theoretical Methods in Physics. Nijmegen, 1975.
         Springer.
36. For generalizations to many unitary operators see
         A. Ramakrishnan, these proceedings.
37. T. S. Santhanam and A. R. Tekumalla, Found. Phys. $\underline{6}$, 583 (1976).
     T. S. Santhanam, in "Uncertainty Principle and Foundations of
         Quantum Mechanics" Eds. W. Price and S. S. Chissick,
         John Wiley, 1977. pp. 227-243.
     T. S. Santhanam, Nuovo. Cim. Lett. $\underline{20}$, 13 (1977).
     T. S. Santhanam and K. B. Sinha, Aust. J. Phys. $\underline{31}$, 233 (1978).

*On leave of absence from MATSCIENCE, The Institute of Mathematical
         Sciences, Madras - 600020, India.
**ϕ has been used to denote many different things like phase, etc.
         Confusion cannot arise as the contexts are different.
***We retain the parameters of mass m and the frequency w without
         setting them unity to comply with the notation of the
         authors.

SPONTANEOUS SYMMETRY BREAKING IN BIFURCATION PROBLEMS

D. H. Sattinger

School of Mathematics
University of Minnesota
Minneapolis, Minnesota 55455

1. BIFURCATION AND LOSS OF STABILITY

In analyzing the dynamics of a physical system governed by non-linear equations the following questions occur: Are there equilibrium states of the system? How many are there? Are they stable or unstable? What happens as external parameters are varied? In particular, what happens when a known solution becomes unstable as some parameter passes through a critical value?

In nonlinear problems, when one solution becomes unstable, other solutions, either equilibrium solutions or time periodic solutions, may branch off. Thus, bifurcation is a phenomenon closely related to the loss of stability in a nonlinear physical system. Let us describe some simple types of bifurcation which typically occur in physical problems. For simplicity let

$$G(\lambda, u) = 0 \tag{1}$$

describe a nonlinear system of equations, for example ordinary or partial differential equations, where u is the state of the system and λ is some external parameter.

<u>Principle of Linearized Stability</u>: A solution u_0 of equation (1) is <u>stable</u> if the spectrum of the linearized operator $G'_u(\lambda, u_0)$ is contained in the left half plane; u_0 is unstable if $G'_u(\lambda, u_0)$ has spectrum in the right half plane.

This principle can be proved rigorously for a number of mathematical models of dissipative structures, for example systems of ordinary differential equations (Lyapounov's theorem), and a large class of partial differential equations (e.g. parabolic systems). Here G'_u denotes the <u>Frechet</u> derivative of the nonlinear mapping $G(\lambda,u)$.

<u>Bifurcation at a simple eigenvalue</u>: Let $u_0 = 0$ be a solution and suppose u_0 becomes unstable as λ crosses λ_c by virtue of a simple eigenvalue crossing the origin:

$$G'_u(\lambda,0)\phi(\lambda) = \sigma(\lambda)\phi(\lambda)$$

$$\sigma(\lambda_c) = 0 , \quad \sigma'(\lambda_c) > 0 .$$

Then the structure of the bifurcating solutions near $u_0 = 0$, $\lambda = \lambda_c$ is given by one of the three diagrams in Fig. 1. This result can be proved by topological degree theory--a kind of conservation of parity of solutions--or by a perturbation analysis of the spectrum of the linearized equations of the bifurcating solution.

A second type of transition phenomena which is very common is the bifurcation of time periodic solutions from an equilibrium solution when that solution loses stability by virtue of a complex conjugate pair of eigenvalues crossing the imaginary axis. This phenomenon is often referred to as Hopf bifurcation after E. Hopf, who gave a general proof of the result for ordinary differential equations with n degrees of freedom in 1942. Hopf's result has now been extended to general parabolic systems of partial differential equations, for example the Navier Stokes equations for a viscous fluid.

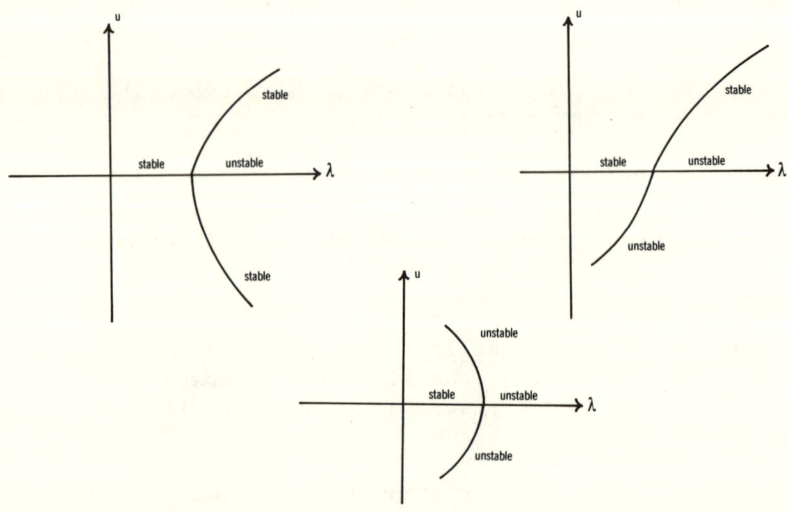

Fig. 1

Examples of Hopf bifurcation which are familiar in physics are the onset of oscillations in electrical networks or the firing of a laser when the pumping exceeds the critical threshold. (cf. Dicke, Haken, Lax, or Hepp and Lieb). In the Dicke-Haken-Lax model of the laser it is possible to describe the many body photon field by a mean field theory as N (the number of degrees of freedom) tends to infinity. Thus it is possible in this case to solve a non-linear quantum mechanical model, far from equilibrium, by reducing the problem to a system of ordinary differential equations for the expectation values of the extensive variables.

The two phenomena discussed above—bifurcation at a simple eigenvalue and the Hopf bifurcation theorem—comprise the simplest types of bifurcation phenomena which can occur in a dissipative system.

2. GROUP THEORY AND BIFURCATION THEORY

Let us now suppose that at criticality dim ker $L_0 > 1$. (By ker L_0 we mean the null space of the linear transformation L_0, and by L_0 we mean $G_u(\lambda_c, u_0)$.) This situation is commonly referred to as "bifurcation at a multiple eigenvalue." Such problems are considerably more complicated to treat, as one might expect. By the Lyapounov-Schmidt method the analysis of the solutions of (1) in a neighborhood of a bifurcation point can be reduced to an algebraic system of n equations in n unknowns,

$$F_i(\lambda, z_1, \ldots z_n) = 0 \qquad i = 1, \ldots n \qquad (2)$$

where $n = \dim \ker L_0$. This is nice in theory, but in practice the computation of even the lowest order terms of the bifurcation equations F_i is extremely complicated. Moreover, systems of n equations in n unknowns can display a bewildering variety of solution structures, and the algebraic problem can in general be quite complex.

In many problems of physical interest, however, the multiplicity of the branch point can be traced to an underlying symmetry of the problem. This phenomenon is well known in quantum mechanics, where the invariance of the Hamiltonian under a symmetry group leads to a degeneracy of the energy levels. Group representation theory is an important tool in analyzing the splitting of the energy levels of a Hamiltonian under symmetry destroying perturbations (e.g. the Stark effect); but these methods also apply in a natural and elegant way to the nonlinear problems of bifurcation theory. The applicability of group representation theory rests on the tensor character of the bifurcation equations on the one hand, and the theory of tensor products of group representations on the other.

Moreover, it is very often the case in physical applications, especially in the area of mechanics, that the systems of equations, even though nonlinear, are covariant with respect to a transformation group. For example, the Hamiltonian equations of celestial mechanics, or the partial differential equations governing the dynamics of a homogeneous continuum, are covariant with respect to the Euclidean group of rigid motions. Let us assume therefore, that the mapping G is <u>covariant</u> with respect to a transformation group $\mathcal{G}$. That is, let T_g be a linear representation of $\mathcal{G}$ on a Banach space $\mathcal{E}$ and assume that $T_g G(\lambda,u) = G(\lambda,T_g u)$. This is a natural assumption in physical theories and is a mathematical expression of the axiom that the equations of mathematical physics be independent of the observer.

From covariance it follows that $T_g G_u(\lambda,u) = G_u(\lambda,T_g u)T_g$; so if u_0 is an invariant solution of $\mathcal{G}$ then $T_g L_0 = L_0 T_g$. Therefore $N_0 = \ker L_0$ is invariant under T_g. If $\dim N_0 < +\infty$, as is often the case in applications $\Gamma \equiv T_g\big|_{N_0}$ is a finite dimensional representation of $\mathcal{G}$. Writing the bifurcation equations in the form $F(\lambda,v) = 0$, where $v \in N_0$ and $F: C \times N_0 \to N_0$, we have the following

<u>Theorem</u>. Let $G(\lambda,u)$ be an analytic operator from a complex Banach space $\mathcal{E}$ to $\mathcal{F}$, covariant with respect to a representation T_g of a compact group $\mathcal{G}$. Suppose that $G(\lambda_c,u_0) = 0$, $T_g u_0 = u_0$ for all $g \in \mathcal{G}$, and $G_u(\lambda_c,u_0)$ is a Fredholm operator of index zero with kernel N_0. Then N_0 is invariant under T_g and the bifurcation equations $F(\lambda,v)$ are covariant with respect to Γ, the restriction of T_g to N_0; that is $\Gamma_g F(\lambda,v) = F(\lambda,\Gamma_g v)$.

This theorem is easily proved by following through the details of the Lyapounov-Schmidt procedure.[22]

Let us now expand F in a power series in v:

$$F(\lambda,v) = A(\lambda)v + B_2(\lambda,v,v) + B_3(\lambda,v,v,v) + \ldots$$

Each term $B_k(\lambda,v,w,\ldots)$ is symmetric in its variables and covariant with respect to the representation Γ_g. In particular, $\Gamma_g A(\lambda) = A(\lambda)\Gamma_g$; so by Schur's lemma, if N_0 is irreducible $A(\lambda) = \sigma(\lambda)I$, where I is the identity transformation. The higher order terms B_2, B_3, ... can all be calculated, up to scalar factors, by the methods of group representation theory. For example, if $\mathcal{G} = 0(3)$ and N_0 transforms according to the irreducible representation D^ℓ, then the coefficients in the quadratic terms of the bifurcation equation are the ordinary Clebsch-Gordon coefficients for the rotation group.

Suppose for convenience $\lambda_0 = 0$ and $\sigma(\lambda) = C_1\lambda + C_2\lambda^2 + \ldots$. Then by various scaling arguments the bifurcation problem can be reduced to an analysis of the equations

$$\lambda w = B_k(w) \tag{3}$$

where B_k is the first nonvanishing term in F, homogeneous of degree
k. These equations are called the reduced bifurcation equations.
It can be shown ([23] Theorem 7.2) that the stability of the bifurcating solutions can be determined to lowest order from an analysis
of the Jacobian of (3) at a solution.

The group theoretic approach, then, is to compute the lowest
nonvanishing terms B_k, find all solutions of (3), and determine their
stability in the neighborhood of the branch point. This attack not
only allows us to bypass the numerical difficulties inherent in the
Lyapounov-Schmidt procedure; but it also provides us with a systematic approach to bifurcation at multiple eigenvalues and with a way
of classifying multiple eigenvalue bifurcation points.

In this way, transition phenomena can be classified according
to the geometry of the problem, rather than by the particular physical mechanisms involved. From this point of view, the onset of
convection in a spherical geometry, the buckling of a spherical
shell, or the onset of ionic currents in a developing spherical egg,
can all be given a unified mathematical treatment, even though the
physical mechanisms in each of the problems may be vastly different.
On the other hand, the specific physical mechanisms in the problem
make themselves apparent in the determination of the critical value
of λ and the transformation properties of the kernel N_0. In addition, the physics of the problem determine the values of the scalar
parameters multiplying the covariant terms in the bifurcation equations; these parameter values determine the direction of bifurcation
(supercritical or subcritical) as well as the stability of the various bifurcating solutions.

Let me return once more to equation (3). It turns out that,
depending on the representation Γ, there may be more than one covariant term of lowest degree k. In that case we would arrive at a
system of reduced equations of the form

$$\lambda w = a_1 B_k^{(1)}(w) + a_2 B_k^{(2)}(w) + \ldots + a_\ell B_k^{(\ell)}(w) \tag{4}$$

where the coefficients $a_1 \ldots a_\ell$ are parameters which depend on the
original parameters of the systems. The multiplicity ℓ of covariant
terms of degree k can be computed directly from a knowledge of the
representation Γ_0 and does not depend on the particular structure
of the equations at hand. (See Jarić and Birman [14] for a general
method of computing the multiplicities.) In fact, the multiplicity
ℓ is precisely the number of times the representation Γ is contained
in $(\Gamma^{\otimes k})$.

When there are multiple covariant tensors, as in (4') the possibility of selection mechanisms arises. The stability of the various
bifurcating solutions depends on the relative sizes of the parameters

$a_1, \ldots a_\ell$. In the Bénard problem, for example, $\ell = 2$ when $k = 3$, and there occurs a selection mechanism for the stability of rolls and hexagons [24].

The numerical values of the parameters $a_1, \ldots a_\ell$ are determined by the physics of the problem, but in general the computation of their exact dependence on the natural physical parameters of the problem is a difficult matter (equivalent to the direct calculation of the full set of bifurcation equations (2)).

Such calculations could be carried out numerically in specific cases, but it is not clear to me that such a direct numerical approach is necessary. In order to classify the types of transitions which can take place, it is sufficient to consider the parameters $a_1, \ldots a_\ell$ as free parameters, more or less like the <u>control parameters</u> in Thom's classification of the elementary singularities. In that way we can classify the spontaneous symmetry breaking transitions which a physical system may undergo purely on algebraic and geometric grounds.

I want to make one final remark concerning the reduced bifurcation equations (3). These are identical in structure to the equations arrived at by Michel and Radicati in their study of symmetry breaking in the physics of the hadrons. The relevant transformation groups there are SU(3) or SU(3) × SU(3). They write their equations for symmetry breaking in the form x ∨ x = λx, where ∨ is a symmetrical non-associative product. In the case where one is interested, say, in the bifurcation from rotationally invariant states when the kernel of L_0 transforms line D^ℓ, the reduced bifurcation equations take the form

$$\lambda z_m = (-1)^m \sum \begin{pmatrix} \ell & \ell & \ell \\ m_1 & m_2 & -m \end{pmatrix} z_{m_1} z_{m_2}$$

where $\begin{pmatrix} \ell & \ell & \ell \\ m_1 & m_2 & -m \end{pmatrix}$ are the Wigner 3-j symbols. The quadratic coupling represented by these symbols corresponds to the ∨ product in Michel and Radicati's notation. O'Raifeartaigh, in these proceedings, also discusses the equations for spontaneous symmetry breaking in elementary particle theory, and arrives at a similar set of equations. He poses the problem as one of minimizing an invariant functional. The corresponding Euler-Lagrange equations, which are the bifurcation equations, are then covariant with respect to that representation. Thus we see that there is a further unification between problems in classical and modern physics, and group theoretic methods should play the same role in bifurcation phenomena in classical physics that they do in quantum mechanics and elementary particle theory.

3. SPONTANEOUS SYMMETRY BREAKING IN PHYSICAL PROBLEMS

At the transition point the bifurcating solutions generally have
less symmetry than the original solutions, which have become unstable. This is so despite the fact that the symmetry group of the
equations remains unchanged as the parameter λ crosses through its
critical value λ_c. In such a case we say that symmetry is broken
spontaneously.

We see spontaneously broken symmetry already in the two simple
examples of bifurcation discussed in §I. The equations governing
Benjamin's machine are covariant under the reflection $u \to -u$ (where
u denotes the lateral displacement). The vertical solutions (null
solutions--no displacement) are invariant under this reflection, but
the branched solutions are not. In fact there are two branches u_1
and u_2, and the reflection symmetry interchanges u_1 and u_2; so in
fact at the bifurcation point an _orbit_ of solutions branches off.

The same is true in the case of Hopf bifurcation. The equations
take the form $\dot{x}_i = f_i(x,\lambda)$ and these are _autonomous_: they are covariant under time translations $t \to t + \gamma$. An equilibrium solution
x_0 is invariant under the entire group of time translations; but any
time periodic solutions which branch from x_0 as λ crosses λ_c are invariant only under the discrete subgroup of translations $t \to t + nT$,
where n is an integer and T is a period. Again it is actually an
orbit of solutions which branches off, for the time translation
$t \to t + \delta$ merely shifts the phase of the oscillations, and we have
a one-parameter family of solutions.

Another well known physical example of the breaking of symmetry
is the Benard problem. If a layer of fluid is heated from below,
convective instabilities set in when the temperature drop across the
layer exceeds a certain critical value; and the convective motions
which arise often display a striking cellular pattern (see
Koschmieder). The appearance of convection cells constitutes an
excellent example of a symmetry breaking instability. In the infinite plane layer model, the solution prior to the onset of instability is invariant under the entire group of rigid motions; whereas
after convection sets in the solutions are invariant only under a
crystallographic subgroup. Busse has also discussed the onset of
convection in spherical geometries and applications to geophysical
problems.

Other physical problems where symmetry breaking plays a role
are buckling problems in elasticity, pattern formation in reaction-
diffusion processes (see Auchmuty and Nicolis, Fife, and Turing);
neurobiological problems (Cowan and Ermentrout); and problems in
physical chemistry (Ortoleva, these proceedings). The importance
of group theoretic methods in the analysis of symmetry breaking
phenomena in these areas is just now beginning to be recognized.

Cowan and Ermentrout have argued, from an examination of experimental
data in psychobiological experiments, that patterns observed in
hallucinatory phenomena are similar to the patterns of instability
which would be observed in any problem where Euclidean invariance
is broken. Thus, from the geometrical point of view the analysis
of pattern formation in the Bénard problem is also applicable to
hallucinatory phenomena.

Erneux and Hershkowitz--Kaufmann have analyzed the onset of wave
motion in a circular geometry. Their analysis shows two distinct
types of wave phenomena--standing waves and rotating waves. Moreover, a stability analysis reveals that these two mode phenomena
cannot be simultaneously stable. Cowan and Ermentrout have discovered a similar phenomenon in their analysis of wave propagation
on linear neural networks. Their model exhibits both standing and
traveling waves; and mode selection mechanisms between the two are
predicted by a group theoretical analysis of the problem: there are
two distinct covariant mappings of third degree in the reduced bifurcation equations. Cowan and Ermentrout believe the two modal
character of this model is representative of clonic and tonic patterns of activity in epileptic seizures.

The idea that morphogenesis might be modeled mathematically by
reaction-diffusion equations is due to Turing. Turing formulated
the problem virtually as a bifurcation problem in reaction-diffusion
equations. These equations are generally a system of parabolic equations which are supposed to describe the electrical, chemical and
diffusion processes in a bio-chemical system. Turing's approach
has been discussed by many authors, for example Auchmuty and Nicolis,
and Fife.

The breaking of Euclidean symmetry also occurs in bifurcation
models for phase transitions, as discussed by Raveche and Stuart,
and Kozak, Rice and Weeks. Bifurcation models are generally valid
in statistical physics only when the system may be described by a
mean field theory. Mean field theories, on the other hand, are
valid only when the fluctuations remain small; and at a phase transition large fluctuations usually play a dominant role. In that
case, mean field theory cannot be expected to give an exact description of the problem. The usual consequence is that bifurcation
models for phase transitions yield the classical critical exponents
for the problem, which are at variance with the experimentally observed values. Nevertheless, mean field theory has until recently
been the primary approach in the analysis of phase transitions.
Two well-known mean field theories are the Van der Waals and the
Curie-Weiss models. These simple models are basically a "catastroph
theory" analysis of critical phenomena. A more sophisticated
approach is that of Raveche and Stuart, and Kozak, Rice and Weeks.
These researchers derive an integral equation for the single particle density function. The bifurcation points of this integral

equation then represent the critical points. Green, Luks, and Kozak
have obtained non-classical critical exponents in such a bifurcation
model.

One other theory that should be mentioned here is the Landau
theory of second order phase transitions. Landau's theory is a
phenomenological one. Furthermore, since it is again a mean field
theory, it does not give the correct critical exponents. But it
does give correct symmetry predictions. (Birman)

4. BIFURCATION IN THE PRESENCE OF O(3).

Let us discuss some of the results obtained, open problems, and
applications when one considers bifurcation in the presence of the
symmetry group O(3). The irreducible representations of SO(3) are
denoted by D^ℓ, $\ell = 0,1,\ldots$ and are of dimension $2\ell + 1$. They arise
when one considers the transformation properties of the spherical
harmonics $Y_m^\ell(\theta,\phi) = P_{\ell,m}(\cos\theta)e^{im\phi}$, $-\ell \leq m \leq \ell$. The representations
of O(3) are denoted by D_+^ℓ and D_-^ℓ; $D_+^\ell \equiv D^\ell$ and D_-^ℓ has the property
that $D_-^\ell(Ig) = -D^\ell(g)$, where I is the inversion and $g \in SO(3)$. We
consider the bifurcation problem obtained when the kernel N_0 transforms as D_+^ℓ.

Let the infinitesimal generators of SO(3) be L_1, L_2, and L_3;
these satisfy the commutation relations

$$[L_i, L_j] = \varepsilon_{ijk} L_k$$

where ε_{ijk} is the completely antisymmetric tensor. Putting $J_+ = \pm L_2 + iL_1$, $J_3 = -iL_3$ we obtain for J_1, J_2, and J_3 the commutation
relations

$$[J_+, J_-] = 2J_3 \qquad [J_3, J_\pm] = J_\pm$$

<u>Lemma</u>. Let V be a real vector space which transforms irreduci-
bly under the rotation group according to the representation D^ℓ.
Then there exists a basis $\{f_m\}$ for the complexification of V such
that

$$J_3 f_m = m f_m \tag{5a}$$

$$J_\pm f_m = \beta_{\pm m} f_{m\pm 1} \tag{5b}$$

$-\ell \leq m \leq \ell$ and $\beta_m = \sqrt{(\ell - m)(\ell + m + 1)}$. In addition, the f_m can be
normalized so that

$$\overline{f}_m = (-1)^m f_{-m}. \tag{5c}$$

The relations (5a,b) can be derived entirely from the commutation relations, as is well known.

Let N be the kernel of the linearized operator $L_0 = G_u(\lambda_c, 0)$. Identify N with linear polynomials in the variables $z_{-\ell}, \ldots, z_\ell$ which transform under the Lie algebra according to (4.2). The algebra $K[z_{-\ell}, \ldots, z_\ell]$ of polynomials in the variables $z_{-\ell}, \ldots, z_\ell$ is isomorphic to the algebra of symmetric tensors over N. Extend the operators $J_3, J_\pm$ to be __derivations__ on the algebra K:

$$J(\alpha f + \beta y) = \alpha J f + \beta J g$$

$$J(fg) = (Jf)g + f(Jg)$$

where α, β are scalars and f, g are polynomials in K. Let the bifurcation equations be

$$F_m(\lambda, z_{-\ell}, \ldots, z_\ell) = 0 \ .$$

These will be covariant with respect to D^ℓ provided the F_m transform as the z_m: that is,

$$J_3 F_m = m F_m, \quad J_\pm F_m = \beta_{\pm m} F_{m\pm 1}$$

where J_3, J_+, and J_- act as derivations of F_m. For example, the quadratic polynomials F_m are obtained as follows. The action of J_3 on $z_j z_k$ is

$$J_3(z_j z_k) = (J_3 z_j)z_k + z_j(J_3 z_k) = (j + k)z_j z_k$$

so $J_3 z_j z_k = m z_j z_k$ if and only if $j + k = m$. Therefore

$$F_m = \sum_{m_1 + m_2 = m} a_{m_1 m_2} z_{m_1} z_{m_2} \ .$$

In particular, when ℓ is even,

$$F_\ell = a_0 z_\ell z_0 + a_1 z_{\ell-1} z_1 + \ldots + a_{\ell/2}(z_{\ell/2})^2 \ .$$

Furthermore, $J_+ F_\ell = \beta_\ell F_\ell = 0$ and this condition gives us a set of linear equations for the coefficients $a_0, \ldots, a_{\ell/2}$. In the case $\ell = 2$

$$F_2 = a z_2 z_0 + b z_1^2$$

$$J_+ F_2 = \alpha \beta_0 z_2 z_1 + 2b \beta_1 z_1 z_2$$

$$= (a\beta_0 + 2b\beta_1)z_1 z_2 = 0 \ ,$$

so

$$a\beta_0 + 2b\beta_1 = 0 \ .$$

The last equation determines the coefficients a and b, hence F_2, up to a scalar multiple. Once F_ℓ is known we get $F_{\ell-1}$ from

$$J_-F_\ell = \beta_{-\ell}F_{\ell-1}$$

and so forth. In this way we construct all the F_m's.

This procedure extends immediately to higher order terms. For example, to get third-order terms we write

$$F_\ell = \sum_{i+j+k=\ell} a_{ijk}z_i z_j z_k$$

and apply $J_+F_\ell = 0$ to get a linear system of equations for the a_{ijk}. For $\ell = 1$ there is only one solution, but for $\ell = 3$ there are two independent solutions. In fact, the condition $J_+F_3 = 0$ in that case leads to five equations in seven unknowns. First

$$F_3 = az_3^2 z_{-3} + bz_3 z_2 z_{-2} + cz_3 z_1 z_{-1} + dz_3 z_0^2$$

$$+ \ ez_2 z_1 z_0 + fz_2^2 z_{-1} + gz_1^3 \ .$$

The condition $J_+F_3 = 0$ then leads to the system of equations

$$\begin{bmatrix} \beta_{-3} & \beta_2 & 0 & 0 & 0 & 0 & 0 \\ 0 & \beta_{-2} & \beta_1 & 0 & 0 & 2\beta_2 & 0 \\ 0 & 0 & \beta_{-1} & 2\beta_1 & \beta_2 & 0 & 0 \\ 0 & 0 & 0 & 0 & \beta_1 & \beta_{-1} & 0 \\ 0 & 0 & 0 & 0 & \beta_0 & 0 & 3\beta_1 \end{bmatrix} \begin{bmatrix} a \\ b \\ c \\ d \\ e \\ f \\ g \end{bmatrix} = 0 \ .$$

One solution is obtained by setting g = 0 and d = 1; then e = f = 0 and we get

$$F_3 = z_3(z_0^2 - 2z_1 z_{-1} + 2z_2 z_{-2} - 2z_3 z_{-3}) \ .$$

The quantity in parentheses is the second-order invariant, and so is annihilated by the application of any of the J operators. Therefore one mapping is

$$F_m = z_m(z_0^2 - 2z_1 z_{-1} + 2z_2 z_{-2} - 2z_3 z_{-3}) \ .$$

A second choice is $g \neq 0$, $d = 0$. The choice $g = \sqrt{7}$ leads to

$$G_3 = 9\sqrt{\frac{60}{7}}\, z_3^2 z_{-3} - 9\sqrt{\frac{60}{7}}\, z_3 z_2 z_{-2} + 3\sqrt{\frac{60}{7}}\, z_3 z_1 z_{-1}$$

$$- 3\sqrt{10}\, z_2 z_1 z_0 + \frac{30}{\sqrt{7}}\, z_2^2 z_{-1} + \sqrt{7}\, z_1^3 \ .$$

The lower weight polynomials are obtained by successively applying the lowering operator J_-. The general reduced bifurcation equations in this case take the form

$$\lambda z_m = A F_m + B G_m$$

where the parameters A and B depend on the external physical parameters of the problem. Such a situation occurs in the Benard problem, and gives rise to mechanisms for pattern selection.

For even ℓ the quadratic terms of the covariant mapping are given by

$$F_m = \sum_{m_1 + m_2 = m} (-1)^m \begin{pmatrix} \ell & \ell & \ell \\ m_1 & m_2 & -m \end{pmatrix} z_{m_1} z_{m_2}$$

where $\begin{pmatrix} j_1 & j_2 & j_3 \\ m_1 & m_2 & m_3 \end{pmatrix}$ are the Wigner 3j symbols for the rotation group. For odd ℓ the covariant quadratic mappings turn out to be anti-symmetric, so for odd ℓ one must go to third-order terms in the bifurcation equations. For even ℓ the quadratic terms possess a gradient structure, as follows. Consider the homogeneous polynomial of degree 3

$$p(z_{-\ell}, \ldots, z_\ell) = \frac{1}{3} \sum_{-\ell}^{\ell} F_m \bar{z}_m$$

restricted to the real subspace of N for which $\bar{z}_m = (-1)^m z_{-m}$. There we have

$$p(z_{-\ell}, \ldots, z_\ell) = \frac{1}{3} \sum_{-\ell}^{\ell} (-1)^m F_m z_{-m}$$

$$= \frac{1}{3} \sum_{m_1, m_2, m_3 = -\ell} \begin{pmatrix} \ell & \ell & \ell \\ m_1 & m_2 & m_3 \end{pmatrix} z_{m_1} z_{m_2} z_{m_3} \ .$$

For even ℓ the 3j symbols are completely symmetric in m_1, m_2, m_3 and therefore

$$\frac{\partial p}{\partial z_m} = F_m(z_{-\ell}, \ldots, z_\ell) \ .$$

Therefore the reduced bifurcation equations (linear plus quadratic terms) take the form

$$\sigma z_m + \frac{\partial p}{\partial z_m} = 0 \ ,$$

which are the Euler-Lagrange equations for the variational problem

$$\begin{array}{c} \text{Min} \\ |z|=1 \end{array} p$$

where $|z|^2 = \sum\limits_{-\ell}^{\ell} (-1)^m z_m z_{-m}$. The function p is the third order invariant for D^ℓ; that is, $p(D^\ell z) = p(z)$ and the Euclidean norm $|z|^2$ is the second-order invariant.

Following Michel and Radicati this variational problem may be cast in a slightly different form. For even ℓ we have the Clebsch-Gordon series

$$D^{\ell/2} \oplus D^{\ell/2} = D^\ell \oplus D^{\ell-1} \oplus \ldots \oplus D^0 \tag{6}$$

and the associated representation

$$U_g A = D^{\ell/2}(g) A D^{\ell/2}(g^{-1})$$

on $(\ell + 1) \times (\ell + 1)$ matrices A. This representation is unitary relative to the inner product

$$<A,B> = \frac{1}{2} \, \text{tr} \, AB* \tag{7}$$

where B* is the Hermitian conjugate of B. The third-order invariant (there is only one since $D^\ell \otimes D^\ell \otimes D^\ell$ contains D^0 only once) is

$$p(A) = \frac{1}{3} \, \text{tr} \, A^2 A* \ .$$

The highest weight space, the one that transforms like D^ℓ in (6), consists of Hermitian symmetric matrices, so we may rephrase our variational problem as

$$\text{Min} \, \frac{1}{3} \, \text{tr} \, A^3$$

subject to

$$\frac{1}{2} \, \text{tr} \, A^2 = 1 \quad \text{and} \quad \text{tr} \, AB_j = 0$$

where the B_j are symmetric matrices which lie in the lower weight invariant subspaces. In particular, $\text{tr} \, AI = \text{tr} \, A = 0$. For $\ell = 2$ the Clebsch-Gordan series in $D^1 \otimes D^1 = D^2 \oplus D^1 \oplus D^0$, but the matrices transforming as D^1 are anti-symmetric; so we have only the constraints $\text{tr} \, A = 0$, $\text{tr} \, A^2 = 2$, and the Euler-Lagrange equations are

$$A^2 = \lambda A + \gamma I \tag{8}$$

where A and I are 3×3 matrices. (The gradient of the functional $\frac{1}{3} \operatorname{tr} A^3$ is the mapping $A \to A^2$.) This equation can be completely solved as follows.

Taking the trace of (8) we get $\gamma = \frac{2}{3}$. When $\ell = 2$, A is a 3×3 symmetric matrix; and we can choose a rotation g such that $D^1(g)AD^1(G^{-1})$ is diagonal, since $D^1(g)$ ranges over all orthogonal matrices as g ranges over $O(3)$. So, assuming A is diagonal we can write (8) as

$$\mu_i^2 = \lambda \mu_i + \frac{2}{3}$$

where μ_1, μ_2, μ_3 are the eigenvalues of A. The constraints are

$$\mu_1^2 + \mu_2^2 + \mu_3^2 = 2 \ , \qquad \mu_1 + \mu_2 + \mu_3 = 0 \ .$$

There are two sets of solutions to these equations

$$\begin{pmatrix} \dfrac{1}{\sqrt{3}} & 0 & 0 \\[2mm] 0 & \dfrac{1}{\sqrt{3}} & 0 \\[2mm] 0 & 0 & \dfrac{-2}{\sqrt{3}} \end{pmatrix} \qquad \lambda = -\dfrac{1}{\sqrt{3}}$$

and

$$\begin{pmatrix} \dfrac{1}{\sqrt{3}} & 0 & 0 \\[2mm] 0 & \dfrac{-1}{\sqrt{3}} & 0 \\[2mm] 0 & 0 & \dfrac{2}{\sqrt{3}} \end{pmatrix} \qquad \lambda = \dfrac{1}{\sqrt{3}}$$

Any permutation of the diagonal elements produces a point on the same orbit; for any such permutation is accomplished by the operation PAP^{-1}, where P is a permutation matrix, and such a P is an element of $O(3)$. The two orbits described above give the maximum and minimum values of the functional $\frac{1}{3} \operatorname{tr} A^3$ on the sphere $\frac{1}{2} \operatorname{tr} A^2 = 1$. The isotropy subgroup in each case is $O(2)$ (rotations which leave $\begin{pmatrix} 0 \\ 0 \\ 1 \end{pmatrix}$ fixed, so each external is axisymmetric. The case

$\ell = 2$ was first treated by Busse by another method. The above method, while quite straightforward in the case $\ell = 2$, becomes extremely complicated already in the case $\ell = 4$ and so does not seem to be a practical approach to the resolution of the bifurcation equations in the general case. It is interesting, nevertheless, to compare this apporach with that of L. Michel and Radicati in their work on symmetry breaking problems in physics.

They study the action of SU(n) on the vector space Q of Hermitian traceless matrices A with the inner product (4.4). It can be proved that there are two linearly independent trilinear invariants of this action, <u>viz.</u>

$$\{A,B,C\} = \frac{\sqrt{n}}{2} \, tr(AB + BA)C$$

$$\{A,B,C\} = -\frac{i}{2} \, tr[A,B]C$$

with $\{ \; , \; , \; \}$ completely symmetric and $[\; , \; , \;]$ completely antisymmetric. The bilinear form (7) is the only second order invariant. From this it can be concluded that there are only two linearly independent algebras with SU(n) as automorphism group. One is the Lie algebra whose multiplication law is

$$x \wedge y = -\frac{i}{2}[x,y]$$

and the other is that with multiplication law

$$x \vee y = \frac{\sqrt{n}}{2} \, (xy + yx) - \frac{1}{\sqrt{n}} \, tr \; xy \; .$$

Michel and Radicati are led to study the equation

$$q \vee q + \eta(q)q = 0 \tag{9}$$

where $\eta(q)$ is a real number. This equation is precisely equivalent to (8).

In the case $\ell = 4$, the quadratic terms are

$$F_4 = \frac{1}{\sqrt{5}} \, z_4 z_0 - \frac{1}{\sqrt{2}} \, z_3 z_1 + \frac{3}{2\sqrt{14}} \, z_2{}^2$$

$$F_3 = \frac{1}{\sqrt{2}} \, z_4 z_{-1} - \frac{3}{2\sqrt{5}} \, z_3 z_0 + \frac{1}{\sqrt{14}} \, z_2 z_1$$

$$F_2 = \frac{3}{\sqrt{14}} \, z_4 z_{-2} - \frac{1}{\sqrt{14}} \, z_3 z_{-1} - \frac{11}{14\sqrt{5}} \, z_2 z_0 + \frac{3}{7\sqrt{2}} \, z_1{}^2$$

$$F_1 = \frac{1}{\sqrt{2}} \, z_4 z_{-3} + \frac{1}{\sqrt{14}} \, z_3 z_{-2} - \frac{6}{7\sqrt{2}} \, z_2 z_{-1} + \frac{9}{7\sqrt{20}} \, z_1 z_0$$

$$F_0 = \frac{1}{\sqrt{5}}\, z_4 z_{-4} + \frac{3}{2\sqrt{5}}\, z_3 z_{-3} - \frac{11}{14\sqrt{5}}\, z_2 z_{-2} - \frac{9}{14\sqrt{5}}\, z_1 z_{-1} + \frac{9}{14\sqrt{5}}\, z_0^{\,2}\,.$$

The remaining polynomials are found from those above by the relation-ship $F_{-m}(z_{-4},\dots,z_{+4}) = (-1)^m F_m(z_{-4},\dots,z_4) = (-1)^m F_m(\dots(-1)^m z_{-m}\dots)$. There are many possible solutions to the bifurcation equations in this case. Busse has found two special solutions:

(1)   Axisymmetric solutions: $z_{\pm 1} = \dots = z_{\pm 4} = 0,\ z_0 \neq 0$;

(2)   Octahedral solutions: $z_4 = z_{-4} = 5/\sqrt{14},\ z_0 = \sqrt{5},\ z_{\pm 1} = z_{\pm 2} = z_{\pm 3} = 0$.

Busse conjectures, on the basis of numerical work, that the second solution is the one which maximizes the third-order invariant. An analysis of the Jacobian shows that the axisymmetric solution is a saddle point and that the octahedral solution is a candidate for the maximum. The complete set of eigenvalues for the octahedral solution is [25]:

$$\left\{0,0,0,-\frac{20}{7}, -\frac{20}{7}, -\frac{20}{7}, -\frac{5}{7}, -\frac{5}{7}, 1\right\}.$$

Since only one eigenvalue is positive this octrahedral solution is a possible candidate for the maximum of the extremal problem. The axisymmetric solution above, however, is definitely a saddle point of the variational problem; the eigenvalues of the Jacobian are

$$\left\{0,0,\frac{20}{9},\frac{20}{9},\frac{10}{3},\frac{10}{3}, -\frac{5}{9}, -\frac{5}{9}, -1\right\}.$$

Busse's article also contains a discussion of the situation for higher values of ℓ, and special solutions are given for $\ell = 6,8$. His special solutions belong to one of two classes (besides the axisymmetric solutions)

$$z_0 \neq 0, \qquad z_n,\, z_{2n} \neq 0 \qquad \tfrac{1}{3}\,\ell < n \le \tfrac{1}{2}\,\ell$$

$$z_m = 0 \quad \text{otherwise}$$

$$z_0 \neq 0, \qquad z_n \neq 0 \quad \text{for a single } n > \ell/2$$

$$z_m = 0 \quad \text{otherwise}\,.$$

The axisymmetric solutions never give a maximum except in the case $\ell = 2$. The solution of the variational problem for $\ell = 4,6,8,\dots$ is at present unknown, and appears to be a difficult problem.

6.   APPLICATIONS

   Convection problems in spherical geometries arise naturally in geophysical problems and have been discussed by many authors.

Convective phenomena in fluid media are generally modeled by the
Boussinesq equations, which, in dimensionless variables, take the
form

$$\frac{\underline{u}}{\partial t} + \underline{u} \cdot \nabla \underline{u} = \Delta \underline{u} - \nabla p + \lambda g_1(r)\theta \underline{r} + \varepsilon \omega \underline{u} \times \hat{k}$$

$$\frac{\partial \theta}{\partial t} = \frac{1}{Pr}(\Delta\theta + \lambda\beta_1(r)\underline{u} \cdot r) - u \cdot \nabla\theta \quad \text{div } \underline{u} = 0$$

where $\underline{u}$ is the fluid velocity field, θ is the temperature perturbation, $\underline{p}$ is the hydrodynamic pressure, and $\underline{r} = x\hat{i} + y\hat{j} + z\hat{k}$ is the
position vector. Pr is the Prandtl number and λ is the Rayleigh
number; $g_1(r)$ is the gravitational field and $\beta_1(r)$ is the steady
state temperature gradient. The term $\omega\underline{u} \times \hat{k}$ is the coriolis term
due to rotation of the fluid. The operator $\underline{u} \times \hat{k}$ breaks O(3) symmetry, as it is only invariant under rotations about the k-axis.

In geophysical applications these equations are considered on
a spherical shell $\eta < r < 1$ with appropriate boundary conditions.
When both surfaces are free the kernel of the linearized equations
contains the space V^1 (which transforms as D^1). When both surfaces
are rigid and η is in the vicinity of 0.3 the kernel of L_λ for the
critical value of λ_c transforms as D^2; but as $\eta \to 1$ the kernel of

L_{λ_c} transforms as D^ℓ for higher values of ℓ. Chossat's thesis*
contains an extensive discussion of the linearized eigenvalue problem for the Boussinesq equations in a spherical shell, and also dis-
cusses the effect of the symmetry breaking term $\omega\underline{u} \times \hat{k}$ on the bifurcation point. Depending on the sign of ω, one gets either a bifurcation of stationary solutions or time periodic solutions.

The buckling of perfectly spherical shells has also been the
subject of much investigation. Many of the investigations have been
limited to axisymmetric buckling, as in Bauer, Keller, and Reiss.
This restriction is certainly justified if Ker $G_u(0,0)$ transforms
as D^ℓ for $\ell = 1,2$; but already in the case $\ell = 4$ Busse's result
shows that the axisymmetric solutions are generally not the relevant
ones.

One of the outstanding problems in geophysics is the

<u>Dynamo Problem</u>: How is the earth's magnetic field maintained? The
current consensus is that the magnetic field is sustained by electric
currents flowing in the electrically conducting earth's core. It is
known that purely axisymmetric fluid motions cannot sustain a magnetic field; but if convective motions accounted for asymmetric fluid
motions, these might sustain such an electromagnetic field. The
equations of dynamo theory are the convective equations plus an
equation governing the magnetic field:

$$\frac{\partial \vec{u}}{\partial t} + (\vec{u} \cdot \nabla)\vec{u} + 2\vec{\Omega} \times \vec{u} = -\nabla p - \beta \vec{g}\theta + \nu \Delta \vec{u} + \frac{1}{\mu}(\nabla \times \vec{B}) \times \vec{B} \qquad (10a)$$

$$\text{div } \vec{u} = 0 \qquad (10b)$$

$$\frac{\partial \theta}{\partial t} + (\vec{u} \cdot \nabla)\theta = -u \cdot \nabla T_0 + \eta \Delta \theta \qquad (10c)$$

$$\frac{\partial \vec{B}}{\partial t} = \text{curl}(\vec{u} \times \vec{B}) + \eta \Delta \vec{B} \qquad (10d)$$

where T_0 is the base temperature profile in the absence of convection and $2\vec{\Omega} \times \vec{u}$ is the Coriolis term.

When no magnetic field is present the quadratic term $(\nabla \times \vec{B}) \times \vec{B}$ in the first equation vanishes, and we have a pure convection problem. When the convective velocity $\vec{u}$ arises to a magnitude and configuration which can sustain a growing magnetic field in (1.2d) then a bifurcation may take place and a non-trivial magnetic field be sustained. Equation (10d) is linear so for a given $\vec{u}$ one should expect an exponentially growing magnetic field $\vec{B}$; but the nonlinear coupling between $\vec{B}$ and $\vec{u}$ via (10a) may be expected to prevent unlimited growth, so that a stable equilibrium is attained. Since (10d) is homogeneous in $\vec{B}$ some external magnetic field (for example, the sun's) is required to "seed" the dynamo process. For further discussion and a survey of the literature, see the article by Roberts [20].

REFERENCES

1. Auchmuty, J. F. G. and Nicolis, G., "Dissipative Structures, Catastrophes, and Pattern Formation: A Bifurcation Analysis," Proc. Nat. Acad. Sci. USA, 71 (1974), 2748-2751.

2. Birman, J. L., "Symmetry change in continuous phase transitions in crystals," Second International Colloquium on Group Theory in Physics, (Nijmegan, 1972) ed. by A. Janner.

3. Busse, F., "Patterns of Convection in Spherical Shells," J. Fluid Mech. (1975), 72, 67-85.

4. Cowan, J. D. and Ermentrout, G. B., "Secondary Bifurcation in Neural Nets," SIAM Journal Applied Mathematics, to appear.

5. Cowan, J. D. and Ermentrout, G. B., "A mathematical theory of visual hallucination patterns," Biological Cybernetics, to appear.

6. Dicke, R. H., Phys. Rev. 93 (1954), p. 99.

7. Erneux, T. and Herschkowitz-Kaufman, "Rotating waves as asymptotic solutions of a model chemical reaction," Jour. Chem. Phys. 66 (1977), 248-250.

8. Erneux, T. and Herschkowitz-Kaufman, "The Bifurcation Diagram of Model Chemical Reactions," Annals of the New York Academy of Sciences, 316.

9.   Fife, P., "Pattern formation in reacting and diffusing systems,"
        Jour. Chem. Phys. 64 (1976), 554-564.
10. Green, Luks, and Kozak, "A precise determination of the critical
        exponent γ for the YBG square-well fluid," to appear.
11. Haken, H. "Laser Theory," in Handbuch der Physik, vol. XXV/2C
        Springer-Verlag, Berlin, 1970.
12. Hopf, E., "Abzweigung linear periodischer Lösung eines Differen-
        tial Systems," Berichte der Math.-Phys. Klasse der Sächsi-
        schen Akademie der Wissenschaften zu Leipzig XCIV (1942)
        1-22.
13. Hepp, K. and Lieb, E. H., "Phase transitions in reservoir-driven
        open systems with applications to lasers and superconduc-
        tors," Helvetica Physica Acta, 46 (1973), 574-603.
14. Jaric, M. V. and Birman, J. L., "New algorithms for the Molien
        functions," Jour. Math. Phys. 18 (1977), 1456-1458.
15. Kozak, Rice, and Weeks, "Analytic approach to the theory of
        phase transitions," J. Chem. Phys. 52 (1970), 2416.
16. Koschmieder, E. L., "Benard Convection," Advances in Chemical
        Physics 26 (1974), 177-212.
17. Lax, M., "Phase transitions and superfluidity," Brandeis Lec-
        tures, 1956, (Gordon and Breach, N.Y. (1968)).
18. Michel, L. and Radicati, L. A., "The geometry of the octet,"
        Ann. Inst. Henri Poincaré Sect. A. Physique, Vol. 18 (1973),
        185-214.
19. Michel, L. and Radicati, L. A., "Properties of the breaking of
        hadronic internal symmetry," Annals of Physics 66 (1971),
        758-783.
20. Raveche, H. J., and Stuart, C., "Towards a molecular theory of
        freezing," J. Chem. Phys. 63 (1975), 136-152.
21. Roberts, P. H., "Dynamo Theory," in Mathematical Problems in
        the Geophysical Sciences, Lectures in Applied Mathematics,
        XIV, Amer. Math. Soc., Providence, 1971.
22. Sattinger, D. H., "Group Theoretic Methods in Bifurcation
        Theory," Lecture Notes in Mathematics, #762, Springer-
        Verlag, 1979.
23. Sattinger, D. H., "Group representation theory, bifurcation
        theory, and pattern formation," Jour. Functional Analysis
        28 (1978).
24. Sattinger, D. H., "Selection rules for pattern formation,"
        Arch. Rat. Mech. Anal. Rat. Mech. Anal. 66 (1977), 31-42.
25. Sattinger, D. H., "Bifurcation from rotationally invariant
        states," Jour. Math. Phys. 19 (1978), 1720-1732.
26. Turing, A. M., "The chemical basis of morphogenesis," Philo.
        Trans. Roy. Soc. London B237 (1952), 37-72.

*Chossat, P., "Etude, par la theorie des bifurcations, de la convec-
        tion dans un domaine spherique en rotation," Thèse, Univer-
        sité de Nice, 1977.

TIME, ENERGY, RELATIVITY, AND COSMOLOGY

I. E. Segal

Mathematics Department
Massachusetts Institute of Technology
Cambridge, Massachusetts 02139

1.   _Introduction_. One of Einstein's major messages is that
nothing is necessarily _a priori_; even the most unlikely subjects
can be discussed and analyzed, sometimes with revolutionary
conclusions. Most notably, he showed that space, time, and of
course gravitation, were not at all _a priori_. Mathematical progress -- at a sophisticated ideational level, rather than at an
elementary or classical problem-solving level -- provided the tools
for a cogent physical analysis having striking observational consequences, explaining existing anomalies and making predictions
which have been precisely confirmed.

What I should like to do here is to discuss and analyze, in
a similar spirit (albeit very briefly), the concepts of time-and-
energy (which are even more closely connected than space-and-time),
as well as some physical implications of the discussion in one
area of observation -- cosmology. What makes the mathematical
parameter t which we use in so many ways the _time_; or certain
expressions E , the _energy_? In short, apart from tradition,
authority, and crude pragmaticism, what makes the time the time,
or the energy the energy, as they are used in special and general
relativity, in classical and quantum mechanics?

What are the intrinsic characteristics of time and energy; and
to what extent are these characteristics uniquely definitive? Are
there other forms of time and energy that are physically conceivable,
besides the conventional ones?--just as, as been realized in the
past century or two, there are other forms of space-time, besides
the Newtonian, or even the special (or even general) relativistic?
The answers will be both explicit and physically applicable.

2. <u>The Einstein-Minkowski cosmos</u>. The Einstein-Minkowski
time was a much more sophisticated concept than the Newtonian time,
and disposed in a convincingly simple manner of the Michelson-
Morley anomaly. But with the advent of quantum mechanics, the
question arose of why time was not (or did not appear to be) an
operator, like other observables. Further, what was the relation
between time as a coordinate of space-time events, and time as a
group parameter, as in the temporal evolution group $U(t) = e^{itH}$, H
being the quantum hamiltonian? What, in these terms, is the observed time, in laboratories?

To work towards answers to these questions I begin with the
central discovery of Einstein-Minkowski, which in modern terms
is stated simply as: space-time is a manifold; the physical time
is defined only relative to a particular coordinate system, which
is arbitrary within a stated group of transformations, which
physically represents the possible relations between the totality
of observers. Among the various coordinates, the time was distinguished by its connection with causality, as noted by Minkowski
and emphasized in the work of Robb. The point of view here goes
back at least as far as Maxwell, who regarded our perception of
the serial order of events as the most primitive manifestation of
time.

3. <u>The Minkowski-Robb concept of causality</u>. Let us try to
formulate the causal structure of the world of events in general
but succinct terms, following the philosophical ideas of Einstein,
Minkowski, and Robb. Starting from the concept of the physical
cosmos -- i.e. 'space-time' -- as a manifold, say M , the following
axioms seem about as primitive and unexceptionable as, say, the
axioms of Peano for the integers.

<u>Axiom 1</u>. <u>The cosmos is a 4-dimensional manifold</u>.

<u>Axiom 2</u>. <u>At each point of the cosmos there exists an infi-
nitesimal notion of causality: specifically, there is given in the
tangent space T_p at each point p , a non-trivial closed convex cone
C_p -- representing physically the totality of 'future' directions
as perceived by an observer at</u> p .

In Axiom 2, 'non-trivial' means that C_p and $-C_p$ have only 0 in
common, a formulation that excludes Newtonian causality and subtly
insinuates Einstein's requirement that there is a limiting finite
velocity to all phyical processes (that of light); but a slightly
different formulation would admit the Newtonian concept, in which
the cones C_p are half-spaces. This matter will not be discussed
here, nor will conventional points of mathematical grammar-- the
degree of smoothness of the manifold, of the 'cone-field' C_p, etc.
interesting and non-trivial as they are in certain connections,
they are largely orthogonal to my present considerations. Let me

remark also that space-time limitations make it impossible for me
to pause to justify physically the convexity and closure of the
cones C_p, and similar points in the future, but that such justification exists.

4. <u>Leray's causal treatment of partial differential equations.</u>
The concept of 'causal cosmos' defined by Axioms 1 and 2 is a very
general 'soft' one, but it is cogent for at least one non-trivial
scientific purpose -- the study of causality features of dynamical
partial differential equations embodying finite propagation
velocity. In fact, in terms of the work of Leray, 'finite propagation velocity' means essentially a dynamical development which is
consistent with a certain causal structure, which defines domains
of dependence and regions of influence comparable to those considered in the classical theory of hyperbolic partial differential
equations. The very requirement that the dynamical development be
defined by such equations appears in fact as not merely an academic
or technological restriction, but to be implied by finiteness of
the propagation velocity--as shown in a simple but representative
case by S. Berman.

Leray begins with a given partial differential equation whose
characteristics define a cone-field of the foregoing type, and
treats its fundamental solutions. In order to have global advanced
and retarded elementary solutions, and not merely local ones, some
form of global causality is evidently required. (Evidently, 'if
time winds back on itself,' there is no distinction between the
advanced and retarded solutions.) Leray's concept of global hyper-
bolicity suffices; it could be adjoined as an additional axiom
formulating global causality, but in the presence of symmetry conditions to be developed later, the following weaker assumption is
adequate:

<u>Axiom 3. The cosmos is globally causal in the sense that it
admits no closed time-like loops.</u>

The point of view here now reverses the roles of the basic
partial differentiation equations and the causal structure of the
cosmos, relative to the approach of Leray. The causal structure
in the cosmos is here assumed given or determinable in some definite
manner; the dynamics of objects in the cosmos is then constrained
to be causal with respect to the given structure. On the other
hand, the Einstein equations join the points of view by making the
causal structure itself (apart from the scalar field defined by the
scale) the object whose dynamics is sought. But empty, or reference,
space-time, in terms of which gravitational effects are observed
and operationally described, has a more a priori causal structure,
related to its symmetries, to which I now turn.

5. <u>Causal symmetries.</u> In order to correlate theory with

physical observation in a well-defined manner, it seems necessary
to treat causal symmetries, which must exist, at least the symmetries which define temporal evolution, if physics is to be based
on measurements at all similar to those employed in laboratories
today. For example, there is no clear theoretical notion of
temporal duration with which the usual measurements of duration--
one of the most precise and fundamental of all measurements--may be
identified, in a generic causal manifold.

Intuitively, duration has to do with the flowing of time, much
as Newton conceived it. In quantum mechanics, the paramter t which
labels the temporal evolution operator $U(t) = e^{itH}$, H being the
Hamiltonian, may be correlated with laboratory time, as measured
e.g. by an atomic clock, quite readily; but the opposite is the
case for the general time-like coordinate on a causal manifold.
What then is the relation between these two conceptions of time?

To make a long and not necessarily conclusive discussion short,
I shall simply describe a concept of time which I believe does the
essential job of explicating the connection between time as a
coordinate and time as a group parameter. In a very simple way it
imposes a theoretically cogent limitation on time--_physical_ time,
that is, which has been presumed unique, within some definite class
of equivalences, as opposed to an infinitude of subjective concepts
of time, as treated by Bergson and his successors--which appears to
correspond to the intrinsic usage in both micro- and macrophysics.

First, the general concept of a causal symmetry: this is
defined as a transformation on a causal manifold which preseves
causality. That is to say, a one-to-one transformation T on the
causal manifold M is defined as _causal_ if it carries a future tangent vector λ at a point p (i.e. a vector in the future cone C_p)
into a future tangent vector $dT(\lambda)$ at the point Tp. It would be
the same to say that T preserves at least local temporal precedence--
where, locally, x 'precedes' y in the manifold M (both being in a
sufficiently small neighborhood of a given point) in case there
exists a time-like arc from x to y--or in other words, that for
any given point p, there is a neighborhood N of p such that if x
and y are in N and x << y (<< means 'precedes'), then Tx << Ty. On
Minkowski space, any orthochronous Lorentz transformation has this
property, and it is a theorem (Alexandrov-Ovchinnikova-Zeeman) that
no others, apart from scale transformations and their composites
with Lorentz transformations, do so _globally_. Presumably every
truly physically possible transformation on the cosmos--as opposed
to coordinate transformations, which merely relabel the points--is
a causal symmetry. I note in passing, however, and this will be
important later, that the local transformations (in a well-defined
mathematical sense) which preserve causality form a properly larger
local group, which is in fact the 15-dimensional conformal group,
4 dimensions greater than the globally causal group of Minkowski

space.

Next the concept of a temporal (forward/backward) displacement:
this is defined as a causal symmetry T such that for all points p,
either p << Tp (forward displacement) or Tp << p (backward displacement). For example, on Minkowski space every vector translation by
a vector in the interior of the future light cone is a forward
temporal displacement, and conversely, these are the only such. In
the case of a general causal manifold, there is a convex cone not
in the manifold itself, but in the infinitesimal group of causal
symmetries which corresponds to the forward displacements. More
specifically, if G denotes the group of all causal symmetries of
the cosmos M, the totality F of forward displacements of M forms an
'invariant semigroup,' having the properties:

$$F^2 \subset F, \quad aFa^{-1} \subset F \quad (a \in G), \quad F \cap F^{-1} = \{e\}$$

There is then a closed convex cone $\underline{F}$ in the infinitesimal group $\underline{G}$
which generates F.

6. <u>Temporal duration and invariant clocks</u>. With this sketch
of mathematical background, I define the notion of a <u>temporally
invariant clock</u> on the given causal manifold M--physically, one for
which there is an invariant notion of duration, independent of the
instants of measurement, as the following couple:

1) a function τ , say, which assigns to each point p of M a
real number $\tau(p)$ (its 'time' coordinate, physically speaking);

2) a one-parameter temporal evolution group T_t: i.e. each T_t
is a forward displacement for t > 0, and $T_{t+t'} = T_t T_{t'}$ for all real
values of t and t'; which are connected by the following basic
constraint:

3) for all points p and all real values t,

$$\tau(T_t p) = \tau(p) + t.$$

The last condition in effect states that the clock gives an invariant
notion of duration, and severely limits the possible transformations
of the time coordinate; for example, an arbitrary smooth monotone
transformation $t \to f(t)$ is no longer permitted.

In the case of Minkowski space, for example, the following
definitions

$$\tau(x_0, x_1, x_2, x_3) = x_0, \quad T_t(x_0, x_1, x_2 \cdot x_3) = (x_0 + t, x_1, x_2, x_3)$$

satisfy the foregoing conditions. Conversely, every invariant
clock on Minkowski space takes this form relative to a suitable

Lorentz frame and scale in the space of the parameter t. Together,
these two mathematical facts provide some validation for the
philosophically reasonable but physically tentative notion of clock
just proposed.

A general class of examples of invariant clocks is obtainable
in a similar fashion by the replacement of the flat euclidean space
by an arbitrary 3-dimensional Riemannian manifold, say S, with the
causal structure on the cosmos $R^1 \times S$ defined from the analogous
Lorentzian structure $dx_0^2 - ds^2$, where ds denotes the element of
distance on S. The special case in which $S = S^3$, the 3-sphere
(surface of the unit sphere in 4-dimensional euclidean space) will
be seen to have some remarkable properties, but more clearly to
the point at the moment is the fact that direct observation of the
cosmos is necessarily local; that the notion of invariant clock
extends directly to a correspondingly local notion, mathematically
entirely well-defined; and that there exist invariant local clocks
on Minkowski space that are essentially different from any global
clock, i.e. the presently standard one. "Local" here is on a scale
far exceeding even the solar system, so the question must arise of
whether the standard clock is entirely correct.

In fact, an alternative local clock exists which cannot be
distinguished from the standard clock by direct laboratory experiments of the present order of precision, and yet would have natural
large-scale implications that are quite different from those of the
standard clock, which implications nevertheless are not in disagreement with observation. This means that it is a material physical
issue, whether the standard clock or an alternative clock, is correct on a large scale. To describe explicitly this alternative
clock, consider the case of $R^1 \times SU(2)$ with the unique invariant
metric on SU(2) (under both left and right group translations),
and the causal structure indicated in the preceding paragraph.
Locally, say in the vicinity of the point (0,I), where I is the
unit matrix in SU(2), this cosmos is identical, as a causal manifold with Minkowski space; specifically, the mapping which carries
the point (t,U) of $R^1 \times SU(2)$ into the inverse Cayley transform
of $e^{it}U$ yields a 2×2 hermitian matrix H, which corresponds to a
unique point (x_0, x_1, x_2, x_3) of Minkowski space via the representa-

tion $H = \begin{pmatrix} x_0 + x_3 & x_1 + ix_2 \\ x_1 - ix_2 & x_0 - x_3 \end{pmatrix}$; and this mapping is causal.

It follows that the clock on $R^1 \times SU(2)$ earlier indicated--
according to which $\tau(t,U) = t$ and $T_s(t,U) = (t+s,U)$--is carried
locally into a local clock on Minkowski space. This local clock
is not only distinct from any standard clock, it is not even
transformable into a standard clock by any local causal transformation. It is simply physically essentially different; but this
does not necessarily mean that it is physically incorrect. An

analysis and computations which can only be summarized here indicate that in order to distinguish the non-standard from the
standard clock by a local measurement of photon frequencies, a
photon would have to be kept alive for 30 years (or travel the
corresponding distance), assuming the rather optimistic precision
of measurement of 1 part in 10^{15}. The connection with photon
frequencies is however the road to an elucidation of the large-
scale physical differences between the standard clocks, and this
question in turn raises the issue of what is the energy? to which
I now turn.

7.   <u>Canonical energies for a given cosmos</u>. In non-relativistic
physics, quantum or classical, the energy may be specified rather
arbitrarily, as e.g. a self-adjoint operator, or a function on
phase space. This is no longer the case for relativistic physics,
or for a dynamics on a general cosmos, since the energy determines
the temporal evolution, which must be compatible with the causal
structure of the cosmos.

It appears that the only 'natural' definition of the energy,
i.e. one in which it is not simply given by external considerations
imposed on the given system, essentially as in non-relativistic
theory, is--as is also generally valid in non-relativistic theory,
but not always (as e.g. in Hoyle's steady-state model of the
universe)--as the generator of temporal evolution (also described
as the 'dual' of the time). This again defines both the classical
and the quantum energy.

What this means concretely may be illustrated by the treatment
of the energy of the highly simplified model for photons defined
by the two-dimensional wave equation $\phi_{x_0 x_0} - \phi_{x_1 x_1} = 0$. The
classical energy is the functional

$$E(\phi) = \int [\partial\phi/\partial x_0)^2 + (\partial\phi/\partial x_1)^2]dx_1 \; ;$$

the quantum energy is simply the operator H: $\phi \to \frac{\hbar}{i}(\partial\phi/\partial x_0)$,
which is self-adjoint in the Hilbert space of all normalizable
solutions of the wave equation. Now the wave equation is invariant
under all causal transformations, local or global, and in particular
under the non-standard temporal evolution group described earlier.
Accordingly, it will have a non-standard energy associated with
the alternative invariant clock. One finds for example the following which will be relevant later:

(1) The non-standard energy (say H' or E') always exceeds the
standard energy (H or E);

(2) While the standard quantum energy H depends only on the
immediate vicinity of the wave function ϕ--is 'intensive,' so to

speak--and is in fact simply the frequency for a wave function which
is locally a plane wave (within the attainable accuracy of physical
measurement)--the excess non-standard energy, H' - H, is independent
of the wave function in the immediate vicinity of any point--is
'extensive' so to speak--and varies <u>inversely</u> with the frequency
for a cutoff plane wave of a given number of oscillations; the
excess is thus 'delocalized' energy, intuitively speaking, and in
fact is too small to be conceivably measurable in the case of a
localized wave function.

Yet all of the general principles of theoretical physics,
including quantum field theory, apply equally to both the standard
and non-standard energies, as to the respective associated invariant
clocks. The same is equally true in 4 space-time dimensions, and
moreover applies also to Maxwell's equation, or any other relativistic equation left invariant by both standard and non-standard
temporal evolution.

Could the non-standard clock and energy conceivably be the
physically correct one? What sort of observational implications
would it have which might be checked against experiment? One
sees that the circumstance that the excess non-standard energy is
delocalized and vanishes within the limits of observation for a
localized wave function, and as if this were not enough, varies
inversely with frequency (for a given number of oscillations),
would make it extremely difficult to observe directly the excess
energy. But one sees also, that since the total energy is conserved
in the course of temporal evolution--trivially so, by virture of
the definition of the energy as the generator of temporal evolution--
the localized standard energy would decrease in the course of time,
and be transformed partially into the not directly observable
delocalized energy. This means that laboratory (i.e., local) observations on photons propagated over large distances would show a
redshift. Since the redshift of light emitted from apparently
distant galaxies is one of the striking facts of nature, an
investigation of the quantitative observable implications regarding
the redshift is thereby suggested.

On the other hand, conceivably there are many other non-
standard clocks; a priori in fact it is quite conceivable that
there exist such a multi-parameter family of non-standard clocks
that by adjustment of parameters one could fit an enormous variety
of observable relations involving the redshift. But this is not
the case; highly elementary and well-documented symmetry restric-
tions--'relativity,' as it is called, although 'invariance' would
have been preferred by Einstein, it seems, and is more apt--cut
down the (already quite limited) possibilities to the Minkowski
clock and the non-standard one already cited, leading to observable
redshift relations devoid of any free parameter, and thus highly
vulnerable to confrontation with observation.

8. <u>Generalized relativity</u>. Minkowski space as a causal manifold leads to the observed addition laws for large velocities, and otherwise agrees with local observations, confirming it as a model for the cosmos, but not necessarily uniquely affirming it. In principle, there has always been a possibility that some other model for the (empty-reference) cosmos could do as well or even better in certain respects. But is there any reasonably simple and unique way to restrict conceivable models that is both physically well-grounded, and mathematically cogent?

Physically, the most conservative restrictions involve observation at one point of the cosmos. Among such restrictions, supported by a great variety of quantitative evidence in the small as in the large, are spatial and temporal isotropy. Spatial isotropy means that there are no preferred space-like directions; that given any two space-like directions, at a given point of observation, there is a physically admissible transformation which carries one direction into the other; 'physically admissible' certainly involves causality, so that a highly conservative formulation of spatial isotropy is the requirement that for any two space-like directions, there exists a causal transformation on the cosmos mapping one into the other. Observationally, spatial isotropy is strongly indicated in the large by observations on the cosmic background radiation, and microscopically by conservation of angular momentum.

'Temporal isotropy' is the equivalence between observers in relative motion at the same point. In Minkowski space, it takes the form of Lorentz invariance. On a general cosmos, it becomes the equivalence of any two time-like directions. Just as in the case of spatial isotropy, this may be formulated conservatively as the requirement that given any two future directions at a point of the cosmos, there exist a causal transformation carrying one point into the other.

Adding to these two observable restrictions the philosophical one of anti-anthropocentrism, which originated the Copernican revolution and appears vital today, one obtains the requirement of spatial and temporal isotropy at every point of the cosmos. In spite of the conservatism, simplicity, and naturalness of these restrictions, they are mathematically extremely cogent, as shown by the important results of J. L. Tits and E. Vinberg. The work of Vinberg classifying homogeneous cones limits the causal cones C_p to cones definable in each tangent space by a quadratic equation, and thus implies that the causal structure must be induced from a Lorentzian metric. The work of Tits classifying Lorentzian manifolds enjoying various isotropy features shows that the cosmos must be locally Minkowskian, and in fact one of an explicitly enumerated set of possibilities. The absence of closed time-like paths earlier postulated then limits the possible cosmos to just 3: Minkowski

space, whose causal group is the Poincare group augmented by scale
transformations; the space $R^1 \times S^3$ earlier cited, whose causal
group is locally the conformal group on Minkowski space (i.e. the
group $O(4,2)$); and an hyperbolic space, whose causal group is
locally $O(3,2)$. All of these cosmos are imbeddable into $R^1 \times S^3$,
in such a way that their groups become subgroups of that of $R^1 \times S^3$,
and in fact they may be defined as orbits in $R^1 \times S^3$ under these
subgroups.

All three of these possible global cosmos are identical
locally, and so have the same local invariant clocks. If we
demand further of a physical clock that locally, at least, the
cosmos may be split into time and space components, in such a way
that temporal evolution affects only the first factor--in accordance
with standard physical thought, according to which a specific observation takes place at a particular time on objects distributed in
space--then there are only two possible local physical clocks,
namely those already cited. Corresponding to each of these there
is moreover a global separation into time and space of one of the
three models, in which temporal evolution changes only the time
component, and spatial transformations (i.e. causal transformations
leaving invariant the space component) act isotropically and
traisitively on space. (The hyperbolic space does not separate in
the large into time and space factors.)

In summary, then, with the one additional

Axiom 4. The cosmos admits an invariant clock relative to
which it can be separated locally into time and space components
(in the sense indicated above),

the cosmos is mathematically limited to 3 possibilities, all
locally Minkowskian; and the corresponding invariant clock is
limited locally to just two possibilities. Only one of these
possibilities gives rise to an intrinsic redshift arising from
the partial delocalization of the photon energy after propagation
through a large distance.

9. Cosmology and general relativity. The question of the
quantitative implications regarding the redshift of photons propagated over large distances, on the assumption that the correct
physical time is that derived from the local factorization as
$R^1 \times S^3$, is facilitated by the conformal invariance of Maxwell's
equations, and in fact the unitarity of the action of both standard
and non-standard time evolution groups. The predicted redshift z
after time t is then defined by the equation $(1+z)^{-1} =$
$\langle e^{itH'} He^{-itH'} \rangle / \langle H \rangle$, which remarkably is independent of the form
or frequency of the localized photon state with respect to which
the expectations are formed, leading to the redshift-distance
relation,

$$z = \tan^2(r/2).$$

The distance is not an observable quantity, but the purely geometric relations between apparent luminosity and distance (as well as other observed quantities, such an angular diameter) permit the distance to be eliminated, and purely observable stochastic relations derived. These relations may then be tested on the large samples of galaxies, quasars, and radio sources which are now available.

Comprehensive systematic tests of this nature have been conducted in accordance with contemporary statistical procedures, and remarkably good agreement between prediction and observation, especially by historical standards of observational cosmology, have been found. With rare exceptions, which may in fact arise from non-randomness of the relatively small samples involved in these cases, the redshift theory indicated (called the 'chronometric' theory, in view of its derivation from a general analysis of the nature of time) leads to a much better fit with observation than do Friedman-Lemaitre models with their two free parameter, i.e. q_0 and Λ. Counts of the numbers of galaxies, quasars, etc. below given redshifts or brighter than given luminosities are also found to be in very good agreement with the natural postulate of spatial homogeneity for their distribution, if the physical cosmos is assumed to take the form $R^1 \times S^3$.

In addition, a number of isolated anomalies within the Friedman-model cosmology are simply eliminated by the chronometric theory; among these are, for example, the apparent supperrelativistic lateral velocities of a number of sources, and the extraordinary luminosity and apparent evolution of quasars. The cosmic background radiation is predicted as the temporally homogeneous equilibrium photon gas established by the diffusion and scattering of electro-magnetic radiation around S^3 in accordance with energy conservation. The deviation from a pure Planck law detected in recent observations by Woody and Richards then becomes explicable as a consequence of a non-trivial level of isotropic angular momentum in the background radiation. The rough apparent coincidence of very large time scales is understandable from the relation between the Minkowski and chronometric times, $x_0 = 2 \tan (t/2)$ at a given point of space from which a uniform distribution in t implies a Cauchy distribution for x_0 whose median is of the order of the radius of the universe, S^3 (in units of c).

How is general relativity and its relation to cosmology affected? The postulated infinitesimal structure of space-time in general relativity, i.e. of reference or empty space-time, is changed from a Minkowski space, formed from the tangent space at the point of observation, to a chronometric space, $R^1 \times S^3$, invariantly attached to the point as the universal covering space of the conformal compactification of the tangent space with respect to the metric given in it. As far as is now known, the radius of the S^3 is too large (in conventional units; in natural units, the S^3 is of unit radius) to produce any presently observable effects in the small, and local observable aspects of general relativity are therefore unaffected.

In the large, because of the compactness of S^3 it is necessary, as Einstein proposed, to add the cosmological term to his equation. Overall, the resulting universe departs widely from the Friedman-Lemaitre model--any expansion, if present at all, must be slight-- but in its gross features is consistent with Einstein's original static conception. One is reminded of Einstein's original misgivings about Friedman's work; these now appear as another example of his prescience.

REFERENCES

1. A. Einstein (1917). Kosmologische Betrachtungen zur Allgemeinen Relativitatstheorie. Sitzungsberichte, Preuss. Akad. d. Wissenschaften.
2. J. Leray (1952). Hyperbolic partial differential equations. Ins. Adv. Study, Princeton, N.J.
3. I. E. Segal (1976). Mathematical cosmology and estragalactic astronomy. Academic Press, New York.
4. H. P. Jakobsen, M. Kon, and I. E. Segal (1979). Angular momentum of the cosmic background radiation. Phys. Rev. Lett. 42, pp. 1788-91.
5. H. P. Jackobsen, B. Orsted, I. E. Segal, B. Speh, and M. Vergne (1978). Symmetry and causality properties of physical fields. Proc. Nat. Acad. Sci. USA. 75, pp. 1609-11.
6. J. F. Nicoll and I. E. Segal (1978). Statistical scrutiny of the phenomenological redshift-distance square law. Ann. Phys. 113, pp. 1-28.
7. J. L. Tits. Lectures at College de France, 1978-79; see also Les espace isotropes de la relativite, Colloque sur la theorie de la relativite, 1959, pp. 107-119, Centre Belge Rech. Math., 1960.
8. E. Vinberg (1963). The theory of homogeneous convex cones. Trudy Moskov. Nat. Obsc. 12, pp. 303-358.
9. D. P. Woody and P. L. Richards (1979). Spectrum of the cosmic background radiation. Phys. Rev. Lett. 42, pp. 925-929.

PROJECTIONS OPERATORS FOR SEMISIMPLE COMPACT LIE GROUPS AND THEIR

APPLICATIONS

Yu. F. Smirnov

Institute of Nuclear Physics
Moscow State University
Moscow 117234; USSR

1. PROPERTIES OF THE PROJECTION OPERATORS FOR SIMPLE LIE GROUPS

When solving numerous quantum-mechanics problems, it becomes
necessary to expand the irreducible representations (IR) of groups
into irreducible components. The most straightforward way of solving
this problem is to use the projection operators (PO). The diverse
applications of the projection method to the nuclear theory are
reviewed in [1]. PO for group G will be understood henceforth to
be the operator $P_{mm'}^{[f]}$ which exerts the following action on the basis
vector $\psi_{[f']m''}$ belonging to the line m" of IR $D^{[f']}$ of group G:

$$P_{mm'}^{[f]}\psi_{[f']m''} = \delta_{m'm''}\delta_{[f][f']}\psi_{[f]m} \,. \tag{1}$$

Thereby, PO exhibits the properties

$$P_{mm'}^{[f]}P_{m''m'''}^{[f']} = \delta_{m'm''}\delta_{[f][f']}P_{mm'''}^{[f]} \,, \tag{2}$$

$$(P_{mm'}^{[f]})^{+} = P_{mm'}^{[f]} \,, \tag{3}$$

$$\hat{D}(g)P_{mm'}^{[f]} = \sum_{m''} D_{m''m}^{[f]}(g)P_{m''m'}^{[f]} \,, \tag{4}$$

$$P_{mm'}^{[f]}\hat{D}(g) = \sum_{m''} D_{m'm''}^{[f]}(g)P_{mm''}^{[f]} \tag{5}$$

where $\hat{D}(g)$ is the operator of the unitary representation of the
group G, $D_{mm''}^{[f]}(g)$ are the matrix elements of the transformation $g \in G$
in IR $D^{[f]}$.

Let us have some finite-dimensional reducible representation Δ of a group with basis ϕ_α ($\alpha = 1,2, \ldots, N$) which can be expanded into IR's as

$$\Delta = \sum_{[f]} \oplus \, \nu_f D^{[f]} \tag{6}$$

where ν_f is the multiplicity of IR $D^{[f]}$ in representation Δ. The vectors ϕ_α can be expanded into the basis vectors of IR's $D^{[f]}$ as

$$\phi_\alpha = \sum_{i,[f],m} C^\alpha_{i[f]m} \psi^i_{[f]m} \tag{7}$$

where the index $i = 1,2, \ldots, \nu_f$ labels the basis vectors of the multiple IR in the expansion (6). The action of PO $P^{[f]}_{mm'}$ on the vectors ϕ_α gives

$$P^{[f]}_{mm'}\phi_\alpha = \sum_i C^\alpha_{i[f]m'} \psi^i_{[f]m} \ , \tag{8}$$

i.e., PO $P^{[f]}_{mm'}$ projects the vectors ϕ_α onto the subspace of vectors $\psi^i_{[f]m'}$ ($i = 1,2, \ldots, \nu_f$) and then reduces that to the vectors which are transformed in correspondence with the line m of IR $D^{[f]}$. PO may be written in the Dirac notation as

$$P^{[f]}_{mm'} = \sum_i |\psi^i_{[f]m}\rangle\langle\psi^i_{[f]m'}| \ . \tag{9}$$

It can be seen from (4) and (8) that PO converts the arbitrary vector ϕ_α into a new vector which belongs to the definite line m of IR $D^{[f]}$ of the group. Practice often requires that new vectors with given symmetry properties (i.e., belonging to a definite line of some IR of group G) should be constructed out of the vectors which fail to exhibit a certain symmetry relative to group G. To illustrate the way of solving this problem through PO, we shall construct a new basis for the reducible representation (6) in which the representation matrices are of quasidiagonal "box-type" form. Consider the vectors

$$|[f]m\rangle_{m'\alpha} \equiv Q^{-1}P^{[f]}_{mm'}\phi_\alpha = \sum_\beta B_\beta \phi_\beta \ . \tag{10}$$

Normalization factors of these vectors (10) are of the form

$$Q^2 = \langle\phi_\alpha|P^{[f]}_{m'm'}|\phi_\alpha\rangle \tag{11}$$

while the coefficients B_β of their expansion into the initial basis ϕ_β are set by the formula

$$B_\beta = \frac{\langle\phi_\beta|P^{[f]}_{mm'}|\phi_\alpha\rangle}{\langle\phi_\alpha|P^{[f]}_{m'm'}|\phi_\alpha\rangle^{\frac{1}{2}}} \ . \tag{12}$$

By fixing the values of m' and α in (10) and giving m all the admissible values m = 1,2, ..., n_f (n_f is the dimension of IR $D^{[f]}$), we get the total set of the basis vectors of IR $D^{[f]}$. If the expansion (6) is nonmultiplicity free (i.e., $\nu_f > 1$ for some IR $D^{[f]}$) the problem of multiplicity may be solved in the following way. By varying the choice of the indices m' and α, we shall find for them such ν_f values which will give ν_f linearly independent vectors $|[f]m>_{m'\alpha}$. As a rule, such vectors will not be orthogonal to each other, but it will be easy to calculate the Gram matrix for them

$$<[f]m|[f]m>^{m'\alpha}_{m''\alpha'} =$$

$$\frac{<\phi_{\alpha'}|P^{[f]}_{m''m'}|\phi_\alpha>}{<\phi_\alpha|P^{[f]}_{m'm'}|\phi_\alpha>^{\frac{1}{2}}<\phi_{\alpha'}|P^{[f]}_{m''m''}|\phi_{\alpha'}>^{\frac{1}{2}}} \tag{13}$$

and to orthogonalize them using, for example, the Schmidt method. The set of the vectors $|[f]m>_i$ with different m, [f], i obtained after the above mentioned orthogonalization will be the sought basis in which the representation (6) of group G is of quasidiagonal form. Thus, PO can actually solve the problem of reducible representation expansion into irreducible components; the only necessary thing in this case is to know the explicit form of such operators and how to calculate their matrix elements.

PO for compact groups have been well known and may be presented in the integral form

$$P^{[f]}_{mm'} = \int dg\ D^{[f]}_{mm'}{}^*(g)\hat{D}(g) \tag{14}$$

where dg is the invariant measure on group G. However, this form of PO is not convenient since it is necessary to know the matrices $D^{[f]}_{mm'}(g)$ for finite transformations g which are not known now for various groups. In connection with this, another (infinitesimal) form of PO is of interest. The infinitesimal form was first proposed in [2] for the simplest group of three-dimensional rotations of SO(3) (or SU(2)). It has been shown in [2] that

$$P^J_{MM'} = \left[\frac{(J+M)!(J+M')!}{(2J)!(J-M)!(2J)!(J-M')!}\right]^{\frac{1}{2}}$$

$$\times \sum_{\tau=0}^{\infty} (-1)^\tau \frac{(2J+1)!}{\tau!(2J+\tau+1)!}\ J_-^{\tau+J-M}J_+^{\tau+J-M'} \tag{15}$$

where $J_\pm = J_x \pm iJ_y$ are the raising and lowering generators of the SO(3) group. It is assumed that the operator (15) acts on the arbitrary vectors of weight M', i.e., the eigenvectors of the generator $J_0 = J_z$

$$J_0 \psi_{M'} = M' \psi_{M'} \qquad (16)$$

which, however, cannot belong to a definite IR D^J of the SO(3) group; in other words, they are not, generally speaking, the eigenvectors of the Casimir operator $J^2 = J_x^2 + J_y^2 + J_z^2$. The operator (15) may be presented in the form

$$P_{MM'}^J = F_-(M) P^J F_+(M') \qquad (17)$$

where $P^J \equiv P_{JJ}^J$ is the operator of projection onto the highest weight vector of IR D^J and

$$F_{\mp}(M) = [(J + M)]^{\frac{1}{2}} \times [(2J)!(J - M)]^{-\frac{1}{2}} J_{\mp}^{J-M}$$

are the lowering (raising) operators which transform the highest weight vector $|JJ>$ of IR D^J into general basic vector of a lower weight of the same IR (or vice versa). Therefore, the main non-trivial part of PO is the operator of projection onto the highest weight vector. The result of [2] was generalized in [3-5] for all compact simple Lie groups. The operator $P^{[f]}$ of projection onto the highest weight vector of IR $D^{[f]}$ of group G_ℓ of rank ℓ was sought for in the form ($[f] = [f_1 f_2 \ldots f_\ell]$ is the highest weight, i.e., the IR signature):

$$P^{[f]} = \prod_{\alpha>0} P_\alpha \qquad (18)$$

where

$$P_\alpha = \sum_{\tau=0}^{\infty} C_\tau(\alpha) E_{-\alpha}^\tau E_\alpha^\tau , \qquad (19)$$

$E_{-\alpha}$, $E_\alpha = (E_{-\alpha})^+$ are the lowering and raising generators of group G_ℓ satisfying to the standard commutational Cartan-Weil relations and α is the root vector. In [3], the order of factors was set in a definite way and it was shown that the equations

$$E_\alpha P^{[f]} \psi_{(f)} = 0 \qquad \text{(for all positive roots),} \qquad (20)$$

$$P^{[f]} P^{[f]} \psi_{(f)} = P^{[f]} \psi_{(f)} \qquad (21)$$

which should govern the PO onto the highest weight vector, are satisfied if the factors $C_\tau(\alpha)$ are taken in the form

$$C_\tau(\alpha) = (-1)^\tau \frac{2^\tau [\frac{2(\alpha, f+\rho)}{(\alpha,\alpha)}]^{\frac{1}{2}}}{(\alpha,\alpha)^\tau \tau! [\frac{2(\alpha, f+\rho)}{(\alpha,\alpha)} + \tau]!} \qquad (22)$$

where ρ is the half-sum of positive roots. In expressions (20) and (21), $\psi_{(f)}$ is the arbitrary vector of weight $(f) = (f_1 f_2 \ldots f_\ell)$

which fails, generally speaking, to belong to a definite IR of the group G_ℓ. In other words, it has been assumed that the operator $P^{[f]}$ of the form (18), (19), (22) acts always on the vectors of definite weight (f) and it is why this operator fails to contain the diagonal Cartan generators H_1, H_2, ..., H_ℓ. For example, in case of IR of the U(3) group with the Young diagram $[f] = [f_1 f_2 f_3]$

$$P^{[f]} = P_{12} P_{13} P_{23} \qquad \text{where}$$

$$P_{ij} = \sum_\tau (-1)^\tau \frac{(f_i - f_j + j - i)!}{\tau!(f_i - f_j + j - i + \tau)!} A_{ji}^\tau A_{ij}^\tau .$$

The generators A_{ij} satisfy the standard commutation relations

$$[A_{ik}, A_{pq}] = \delta_{kp} A_{iq} - \delta_{iq} A_{pk} .$$

It is also possible to find the result of action of the operator (19-22) on an arbitrary (not necessarily weight) vector. It is necessary, for this purpose, to replace in (22) the 1-dimensional vector $f = \{f_1 f_2 \ldots f_\ell\}$ by the operator vector H with the components H_1, H_2, ..., H_ℓ; in this case, the resultant operator will project the general-form vectors ψ onto the space of all possible highest weight vectors relative to the group G_ℓ. An important property of PO should be noted, which can be obtained by means of Hermitian conjugation of (20):

$$P^{[f]} E_{-\alpha} = 0 . \tag{20a}$$

The relationship between the integral form (14) and infinitesimal form (18) of PO has been established in [6,7] by using the new universal parametrization for simple compact Lie groups. This parametrization has made it possible to factorize the invariant measure dg to find the explicit form of the matrix element

$$D_{ff}^{[f]}(g) = <[f](f)|\hat{D}(g)|[f](f)>$$

(where $|[f](f)>$ is the highest weight vector of IR $D^{[f]}$), and completely calculate the integral (14). The resultant PO has been obtained in the form (18), (19), (22). It follows from [6] (and has also been proved using the pure infinitesimal method in [4,5]) that the formula (18) admits any order of the factors P_α in which the factor P_β for each three positive roots α, β, γ satisfying the condition $\beta = \alpha + \gamma$ is located between the factors P_α and P_γ, i.e., the factors form the succession ... P_α ... P_β ... P_γ ... or ... P_γ ... P_β ... P_α ... . This means that there exist a great number of the equivalent factorized expressions (18) for PO $P^{[f]}$. The possibility of transition from one form of PO to another is very convenient for the simplifying of the various specific calculations involving the relevant PO. It will be noted that, later on,

Shapiro obtained PO for the scalar IR $D^{[00...0]}$ of an arbitrary simple compact Lie group. They can also be obtained naturally from (18) as a special case.

It is well known that an arbitrary semisimple Lie group is a direct product of a set of its simple subgroups. Therefore PO for semisimple groups is a product of projection operators for these subgroups.

Discussed above were only the operators of projection onto the highest weight vector. In order to construct PO of the general form

$$P^{[f]}_{mm'} = F_-(m) P^{[f]} F(m') \tag{23}$$

it is necessary to know the lowering and raising operators $F_\mp$. Finding the lowering operator $F_-(m)$ is essentially reduced to the construction of the total basis of IR $D^{[f]}$ out of the highest weight vector $|[f](f)\rangle$ using the lowering generators $E_{-\alpha}$. Such problem was examined for some groups [8-16] and has evoked the following remarks. First of all, there is an arbitrariness in choosing of the IR basis (and, hence of the operator F_+). The IR basis is usually constructed by the reduction of the group G_ℓ into a chain of subgroups $G_{\ell i}$ of lower ranks

$$G_\ell \supset G_{\ell 1} \supset G_{\ell 2} \supset \dots \quad . \tag{24}$$

This reduction may be canonical (i.e., the IR indices of subgroups $G_{\ell 1}$, $G_{\ell 2}$, ... are sufficient to label unambiguously the basis of IR $D^{[f]}$ of group G_ℓ) or noncanonical (most frequently in physical problems). Therefore, the problem of construction of the operators F_+ should be solved individually in each case. The examples of construction of some canonical and noncanonical bases through the lowering generators using the highest weight vectors have been discussed in [8-16]. Considered in those works were the reductions:

$$U(n) \supset U(n - 1) \supset \dots \supset U(2) \supset U(1) \quad [8],$$

$$O(n) \supset O(n - 1) \supset \dots \supset O(2) \quad [9]$$

(the canonical Gelfand-Tsetlin chains),

$$Sp(2n) \supset Sp(2n - 2) \supset \dots \supset Sp(2) \quad [10], \quad U(5) \supset SU_T(2) \times U(1)$$

(five dimensional quasispin [11] and the basis with pseudomoment in the method of generalized hyperspherical functions [12]), $U(5) \supset O(5) \supset O(3)$ (quadrupole oscillations in the unified model of nucleus [13]), $SU(3) \supset SO(3)$ [14] (the Elliott scheme in the oscillator shell model), $G_2 \supset SU(3)$ [15]. The same method was used to examine the expansion of the direct product of IR of the SU(3) group into irreducible components [16]. The structure and properties of the projected basis for the super-multiplet symmetry of SU(4) was discussed in [17].

The derived PO's (18) for the corresponding subgroups also prove to be useful in constructing such bases. The general principle of construction of the bases is as follows. First, the basis vectors of IR $D^{[f]}$ of group G_ℓ which are the highest weight vectors into the relation subgroup $\dot{G}_{\ell 1}$ from the chain (24) must be constructed. For example, the vector of the highest weight (f_1) for the subgroup $G_{\ell 1}$ is of the form

$$|[f_1](f_1)> \quad P^{[f_1]} E_{-\alpha_1}^{S_1} E_{-\alpha_2}^{S_2} \ldots E_{-\alpha_k}^{S_k} |[f](f)> \tag{25}$$

where $P^{[f_1]}$ is the PO for subgroup $G_{\ell 1}$. Considering 20a, the expression (25) comprises such lowering generators $E_{-\alpha i}$ which do not belong to the subgroup $G_{\ell 1}$ (a regular embedding of subgroup $G_{\ell 1}$ into group G_ℓ has been assumed). The nonnegative integers S_1, S_2, $\ldots$, S_k are selected so that the weight (f_1) will be obtained from the weight (f). If the IR $D^{[f_1]}$ of subgroup $G_{\ell 1}$ was contained several times in the IR $D^{[f]}$ of group G_ℓ, then the appropriate number of linearly independent vectors of type (25) should be selected and then orthonormalized. The equations have been formulated in [10-16] for the cases listed above, which make it possible to select the linearly independent vectors among the vectors (25) and to express the rest of the nonvanishing vectors of type (25) through them. The additional Hermitian operator Ω is often introduced to identify the linearly independent vectors $|[f_1](f_1)>$. The operator Ω is the scalar relative to subgroup $G_{\ell 1}$, which is composed of the generators of the group and, thereby, commutes with PO $P^{[f_1]}$. In such a way, we get

$$\Omega P^{[f_1]} E_{-\alpha_1}^{S_1} E_{-\alpha_2}^{S_2} \ldots E_{-\alpha_k}^{S_k} |[f](f)> =$$
$$= P^{[f_1]} \Omega E_{-\alpha_1}^{S_1} E_{-\alpha_2}^{S_2} \ldots E_{-\alpha_k}^{S_2} |[f](f)> . \tag{26}$$

After moving the raising operators in Ω to the right to the highest weight vector (where they are cancelled), replacing the diagonal generators H_i by their eigenvalues, and using the relation (20a), for the lowering generators of the subgroup $G_{\ell 1}$, we will be able to express vector (26) in the form of a linear combination of the vectors of type (25). In other words, this approach can give the matrix of operator Ω in the basis of vectors (25); in its turn, diagonalization of this matrix will make it possible to find the orthonormalized eigenvectors which will be characterized by the additional quantum number ω, i.e., by the eigenvalue of the operator Ω. After that, the same technique should be used

to examine the reduction of group $G_{\ell 1}$ into subgroup $G_{\ell 2}$, etc. As
a result, we will get the orthonormalized basis of IR $D^{[f]}$ of group
G_ℓ characterized by the total set of quantum numbers. Examples of
the construction of such bases for the cases where but a single
quantum number was missing have been given in [11-16]. In the presence of such bases, the operators $F_\pm$ become actually known and,
besides that, it becomes possible to solve numerous physical and
mathematical problems; for example, to use those bases in calculating
the matrix elements of the group generators. Namely this procedure
was used in [8,9] to derive the Gelfand-Tsetlin formulae for the
matrix elements of the generators of groups $U(n)$ [17] and $O(n)$ [18].

The above described approach based on PO properties is probably
the simplest way of deriving these important formulae. The matrix
elements of generators have also been obtained for some noncanonical
bases [11-16].

Thus, the PO method is an effective mean for solving numerous
problems in the theory of representations of semisimple compact Lie
groups which are of importance to physical applications. The natural
question arises, therefore, as to what extent this method can
be applied to the noncompact simple Lie groups. It is clear that
the results obtained for the compact groups may be directly applied
to the discrete series of unitary IR of noncompact groups for which
the highest weight (or lowest weight) vector exists. The problem of
continuous series needs additional treatment. However, the
analysis of the K-harmonic method made in [19] from the viewpoint
of the group $Sp(2R)$ (or $SU(1,1)$) has shown that the discrete series
are also an interesting object for physical applications. Examined
below will be the PO theory for the simplest noncompact group
$SU(1,1)$. This point can be also considered as an illustration of
the relation between integral and infinitesimal forms of PO.

2. PROJECTION OPERATORS FOR THE $SU(1,1)$ NONCOMPACT GROUP

It is known [20] that the generators J_0, $J_\pm$ of unitary IR of
the $SU(1,1)$ group, which make the bilinear form $|x_1|^2 - |x_2|^2$
invariant, satisfy the relations

$$[J_0, J_\pm] = \pm J_\pm \ ,$$

$$[J_+, J_-] = 2J_0 \ , \tag{27}$$

$$J_0^+ = J_0, \quad (J_\pm)^+ = -J_\mp \ , \tag{28}$$

i.e., the generators obey the same commutational relations as the
$SU(2)$ group generators; at the same time, in case of Hermitian conjugation the noncompact raising and lowering generators behave differently as compared with the $SU(2)$ generators. The eigenvectors

ψ_{jm} of the compact generator J_0

$$J_0\psi_{jm} = m\psi_{jm}, \quad m = 0, \pm\tfrac{1}{2}, \pm1, \ldots \tag{29}$$

will be used as the basis vectors of unitary IR D^j of the SU(1,1) group, while the IR's themselves will be labeled by index j which defines the eigenvalues of the Casimir operator $G(2) = J_-J_+ + J_0^2 + J_0$

$$G(2)\psi_{jm} = j(j + 1)\psi_{jm} . \tag{30}$$

This means that the SU(1,1) group is reduced to the compact subgroup O(2). The unitary IR's of the SU(1,1) group are subdivided (see, for example, [20]) into the positive discrete series

$$j = 0, \tfrac{1}{2}, 1, \ldots, \quad m = j+1, j+2, \ldots; \tag{31a}$$

the negative discrete series

$$j = 0, \tfrac{1}{2}, 1, \ldots, \quad m = -j-1, -j-2, \ldots; \tag{31b}$$

the first continuous series (integer)

$$j = -\tfrac{1}{2} + i\sigma, \quad \sigma > 0, \quad m = 0, \pm1, \pm2, \ldots; \tag{31c}$$

the second continuous series (semi-integer)

$$j = -\tfrac{1}{2} + i\sigma, \quad \sigma > 0, \quad m = \pm\tfrac{1}{2}, \pm\tfrac{3}{2}, \ldots; \tag{31d}$$

and the additional continuous series

$$j = -\tfrac{1}{2} + \rho, \quad 0 < \rho < \tfrac{1}{2}, \quad m = 0, \pm1, \pm2, \ldots . \tag{31e}$$

The conventional formulas

$$J_\pm\psi_{jm} = \sqrt{(j \mp m)(j \pm m + 1)}\ \psi_{jm\pm1} \tag{32}$$

hold for the unitary IR. An arbitrary finite transformation g of the SU(1,1) group is set by three Euler angles ϕ, τ, θ

$$(0 \le \phi \le 4\pi, \ 0 \le \theta \le 2\pi, \ 0 \le \tau < +\infty)$$

and can be described by the matrix

$$g(\phi,\tau,\theta) = \begin{pmatrix} e^{\frac{i}{2}(\phi+\theta)}\,\mathrm{ch}\ \tau/2 & e^{\frac{i}{2}(\phi-\theta)}\,\mathrm{sh}\ \tau/2 \\ e^{-\frac{i}{2}(\phi-\theta)}\,\mathrm{sh}\ \tau/2 & e^{-\frac{i}{2}(\phi+\theta)}\,\mathrm{ch}\ \tau/2 \end{pmatrix} . \tag{33}$$

The relevant operator of a unitary IR is of the form

$$\hat{R}(\phi,\tau,\theta) = e^{-i\phi J_0} e^{-\tau(J_+ + J_-)} e^{-i\theta J_0} \tag{34}$$

whence the matrix element of IR, D^j, may be written as

$$D^j_{mm'}(\phi\tau\theta) = \langle jm|\hat{R}(\phi,\tau,\theta)|jm'\rangle$$

$$= e^{-im\phi} d^j_{mm'}(\tau) e^{-im'\theta} \tag{35}$$

where

$$d^j_{mm'}(\tau) = \langle jm|e^{-\tau(J_+ + J_-)}|jm'\rangle \ .$$

The exponent in the last matrix element can be more conveniently
written as

$$e^{-\tau(J_+ + J_-)} =$$

$$= e^{\chi J_0} e^{-\lambda J_-} e^{-\lambda J_+} e^{\chi J_0}$$

$$= e^{-\chi J_0} e^{-\lambda J_+} e^{-\lambda J_-} e^{-\chi J_0} \tag{36}$$

where

$$\chi = \ln \operatorname{ch} \tau/2, \quad \lambda = \operatorname{sh} \tau/2 \ .$$

Then we get for $d^j_{mm'}(\tau)$:

$$d^j_{mm'}(\tau) = (-1)^{m-m'} (\operatorname{sh} \tau/2)^{m'-m} (\operatorname{ch} \tau/2)^{m+m'}$$

$$\times \sum_k \frac{(\operatorname{sh} \tau/2)^{2k}}{k!(k+m'-m)!} \langle jm|J_-^{k+m'-m} J_+^k|jm'\rangle$$

$$= (-1)^{m'-m} (\operatorname{ch} \tau/2)^{m+m'} (\operatorname{sh} \tau/2)^{m'-m}$$

$$\times \sqrt{\frac{\Gamma(j-m'+1)\Gamma(j-m+1)}{\Gamma(j+m'+1)\Gamma(j+m+1)}}$$

$$\times \sum_k \frac{\Gamma(j+m'+k+1)(\operatorname{sh} \tau/2)^{2k}}{k!\Gamma(k+m'-m+1)\Gamma(j-m'-k+1)} \ . \tag{37}$$

If the second equivalent form of the operator (36) is used, we
will similarly get

$$d^j_{mm'}(\tau) = (-1)^{m-m'}(\text{ch } \tau/2)^{-m-m'}(\text{sh } \tau/2)^{m-m'}$$

$$\times \sqrt{\frac{\Gamma(j+m'+1)\Gamma(j+m+1)}{\Gamma(j-m'+1)\Gamma(j-m+1)}} \; \sum_k \frac{\Gamma(j-m'+k+1)(\text{sh } \tau/2)^{2k}}{k!\,\Gamma(k+m-m'+1)\Gamma(j+m'-k+1)} \cdot \qquad (38)$$

Functions (35) are orthogonal to each other with the invariant measure $dg = \text{sh}\tau d\phi d\tau d\theta$. Therefore, the integral form of PO

$$P^j_{mm'} = \frac{2j+1}{16\pi^2} \int D^{j*}_{mm'}(\phi\tau\theta)\hat{R}(\phi\tau\theta)\text{sh}\tau d\tau d\phi d\theta \qquad (39)$$

exists for the SU(1,1) group exhibiting the properties

$$P^j_{mm'}\psi_{j'm''} = \delta_{m'm''}\psi_{jm} \begin{cases} \delta_{jj'} & \text{for discrete series} \\ \delta(j-j') & \text{for continuous series,} \end{cases} \qquad (40)$$

$$(P^j_{mm'})^+ = P^j_{m'm} \cdot \qquad (41)$$

It is of interest to find if an infinitesimal form of type (15) exists for the SU(1,1) group. According to [6,7], such an operator can be obtained from the integral expression (39) by substituting the operator (36) in (39), representing the operators in the right hand side of (39) in the form of Maclauren series and integrating over the Euler angles. These operations for IR, D^{j+}, belonging to the positive discrete series at $m = m' = j + 1$ will give

$$P^{j+}_{j+1\ j+1} =$$

$$\frac{2j+1}{2} \sum_{\tau=0}^{\infty} \int d\tau \text{sh}\tau(\text{ch } \tau/2)^{-4j-4}(\text{sh } \tau/2) \frac{2\tau J_+^\tau J_-^\tau}{\tau!\tau!} \cdot \qquad (42)$$

Use has been made here of the formula (38) for the matrix of finite transformation and of the second form (36) of the operator $\hat{R}(\phi\tau\theta)$. Had this operator been taken in the first form (36), we would have obtained nonconvergent integrals in the formula (42). Therefore, PO for the positive discrete series <u>does not exist</u> in the form

$$P^{j+}_{j+1\ j+1} = \sum_\tau c_\tau J_-^\tau J_+^\tau \cdot$$

It should be noted that, since the operator $P^{j+}_{j+1\ j+1}$ must commutate with J_0, only the terms with the same powers of J_+ and J_- remained in (42). Calculations of the integral in (39), which is obviously convergent at $\tau \leq 2j$, give the expression

$$P^{j+}_{j+1\ j+1} = \sum_{\tau=0}^{2j} (-1)^\tau \frac{(2j-\tau)!}{\tau!(2j)!} J_+^\tau J_-^\tau \qquad (43)$$

coinciding with the results of [19,21]. It should be borne in
mind, however, that in expression (43) the terms with $\tau > 2j$ were
omitted. These factors contain the nonconvergent integrals. This
means that expression (43) may be used only if the vector ψ_{j+1}
with weight $j + 1$ which does not belong to certain IR D^{j+} vanishes
under the action of the operator J^τ at $\tau > 2j$. This, however, is
not always the fact. For example, such situation takes place in case
of expansion into the irreducible component of Kronecker product of
two IR's when both of them belong to the discrete positive series
$D^{j_1+} \otimes D^{j_2+} = \sum_{j \geq j_1+j_2+1} D^{j+}$. In this case, according to [22], general basic vector Φ_{jm} for IR D^{j+} in this expansion can be obtained
through the action of PO

$$P^j_{mj+1} = (-1)^{\frac{m-j-1}{2}} \sqrt{\frac{(m-j-1)!\,(j+m)!}{(2j+1)!}} \; J^{m-j-1}_+ \; P^j_{j+1\;j+1} \qquad (44)$$

on the vector

$$\psi_{j_1 j_1+1}(x_1)\psi_{j_2 j-j_1}(x_2).$$

We obtain for this vector

$$[J_-(x_1) + J_-(x_2)]^\tau \psi_{j_1 j_1+1}(x_1)\psi_{j_2 j-j_1}(x_2) =$$

$$= \psi_{j_1 j_1+1}(x_1)[J_-(x_2)]^\tau \psi_{j_2 j-j_1}(x_2) = 0 \qquad (45)$$

if $\tau > j - j_1 - j_2 - 1$. Thus in this case PO can be used in the
form (43), so the results of [19] are correct. Consider now the
Kronecker product $D^{j_1+} \otimes D^{j_2-}$ where $j_1 > j_2$. According to [22]
this product comprises IR D^{j+} with $j \leq j_1 - j_2 - 1$. To construct
the basic vectors Φ_{jm} for such IR, the PO action must exert
again on the vector $\psi_{j_1 j_1+1}(x_1)\psi_{j_2 j-j_1}(x_2)$, in this case, however,
the vector $\psi_{j_2 j-j_1}(x_2)$ belonging to the negative discrete series
will not be turned into zero by any power of the operator $J_-(x_2)$,
so that the PO in form (40) cannot be used.

Thus, the example of the SU(1,1) group has shown that the
infinitesimal form of PO may be constructed for at least discrete
series of IR, but the scope of its applicability is limited as compared with that of the integral form of PO which may be used in
analyzing of arbitrary unitary IR [23]. Similar situation seems
to take place also for other noncompact (but locally compact) Lie
groups.

REFERENCES

1. N. McDonald, Adv. Phys., 19, 371 (1961).
2. J. Shapiro, J. Math. Phys., 6, 1680 (1965).
3. R. M. Asherova, Yu. F. Smirnov, Uspekhi Matematicheskih Nauk, 24, 227 (1969); Yu. F. Smirnov, in "Clustering Phenomena in Nuclei", IAE A, Vienna, 1969, p. 153; R. M. Asherova, Yu. F. Smirnov, V. N. Tolstoy, Theoreticheskaya i Matematicheskaya Fisika, 8, 255 (1971).
4. Yu. F. Smirnov, Thesis for Doctorate Degree, Inst. Nucl. Phys., Moscow State Univ., 1971.
5. R. M. Asherova, Yu. F. Smirnov, V. N. Tolstoy, Preprint FEI-397, Obninsk, 1973.
6. A. N. Leznov, M. V. Saveliev, Functional Analysis, 8, 87 (1974); Preprint IFVE STF-74-46, Serpuhkhov, 1974.
7. A. N. Leznov, M. V. Saveliev, Elem. Part. Atom. Nuclei, 7, 55 (1976); Theoreticheskaya i Matematicheskaya Fisika, 31, 273 (1977).
8. R. M. Asherova, Yu. F. Smirnov. V. N. Tolstoy, Theoreticheskaya i Matematicheskaya Fisika, 15, 107 (1973).
9. V. N. Tolstoy, Thesis for Candidate Degree, Inst. Nucl. Phys., Moscow State University, 1972; Yu. F. Smirnov, Preprint ITF-75-45, Kiev, 1975.
10. V. A. Mishukov, Graduation Thesis, Dept. Phys., Moscow State Univ., 1973.
11. Yu. F. Smirnov, V. N. Tolstoy, Rept. Math. Phys., 4, 105 (1973).
12. Yu. F. Smirnov, V. N. Tolstoy, Preprint ITF-75-90R, Kiev, 1975.
13. R. M. Asherova, Yu. F. Smirnov. V. N. Tolstoy, Preprint FEI-424, Obninsk, 1973; J. Phys. (G), 4, 205 (1978).
14. R. M. Asherova, Yu. F. Smirnov, Nucl. Phys. A144, 116 (1970); Rept. Math. Phys., 4, 83 (1973).
15. D. T. Sviridov, Yu. F. Smirnov, V. N. Tolstoy, Dokl. Akad. Nauk SSSR, 206, 53 (1972); Rept. Math. Phys., 7, 349 (1975); D. T. Sviridov, Yu. F. Smirnov, Theory of Optical Spectra of Transition Metals, Nauka, Moscow, 1977 (in Russian).
16. R. M. Asherova, Yu. F. Smirnov, Nucl. Phys. B4, 399 (1968). V. A. Knyr, Yu. F. Smirnov, V. N. Tolstoy, Rept. Math. Phys., 8, 343 (1975).
17. J. P. Draayer, J. Math. Phys., 11, 3225 (1970). E. Norvaisas, S. Alisauskas, Liet. fiz. Rink, 17, 457 (1977).
18. I. M. Gelfand, M. L. Tsetlin, Dokl, Akad. Nauk SSSR, 71, 825 (1970); 71, 1017 (1950).
19. V. A. Knyr, P. P. Pipiraite, Yu. F. Smirnov, Sov. Nucl. Phys. 22, 1063 (1975). Yu. F. Smirnov, K. V. Shitikova, Elem. Part. Atomic Nuclei, 8, 847 (1977).
20. N. Ya. Vilenkin, Special Functions and the Theory of Group Representation, Fizmatgiz, Moscow, 1965.
21. J. Patera, F. Stenger, Preprint CRM-199, Univ. Montreal, 1972.

22. N. Mukunda, B. Radhakrishnan, J. Math. Phys., 15, 1332, 1643,
        1656 (1974).
23. Yu. A. Smorodinsky, I. A. Verdiev, G. A. Kerimov, Sov. Nucl.
        Phys., 20, 827 (1974).

ORDERS IN NATURE: FROM QUANTUM TO CLASSICAL

H. Umezawa and H. Matsumoto

Department of Physics
University of Alberta
Edmonton, Alberta
Canada T6G 2J1

In this lecture a brief summary of a theory providing for the
deviation of quantum ordered states and for the creation of classically or quantum mechanically behaving macroscopic objects in these
ordered states is presented. This theory is called the boson theory.
Its practical applications are also mentioned. This list of contents
is as follows:

1. Microscopic Systems and Macroscopic Objects

2. Extended Objects in Quantum Systems and Soliton Solutions

3. The Boson Theory for Superconductivity

4. The Boson Theory for Crystals

5. The Dynamical Rearrangement of Symmetries

6. Topological Singularities, Dynamics of Extended Objects and
    Gravitational Theory.

1. <u>Microscopic Systems and Macroscopic Objects</u>

It is common to describe nature in terms of its strata structure. The crudest classification of this kind is given, for example, by the series of cosmological objects, everyday objects, molecules, atoms, and so on. One then hastily points out that the everyday objects consist of molecules, which consist of atoms, and so on,

and that phenomena belonging to different strata are usually described by different theories in physics. A particular distinction is made between the level of everyday objects and that of molecules. The objects which belong to the level of molecules or smaller are usually called microscopic, and other objects are called macroscopic.

However, it is obvious that nature is too complex to be grasped by such a "linear" view point. To point this out, let us recall the example of dislocations in a crystal. Dislocations are classically behaving macroscopic objects[1] which are created in a crystal and interact with various microscopic objects such as phonons. In one word, a crystal with dislocations presents an example of a system in which microscopic and macroscopic objects coexist and interact with each other[2,3]. In solid state physics we frequently meet similar situations: e.g. vortices in superconductors, magnetic domains in ferromagnets, etc. Even when one speaks of an object of cosmological scale, one faces the question of asking how the cosmological objects have been created out of very fundamental microscopic objects such as quarks or something more fundamental. One significant aspect of this question is to ask how the macroscopic objects come out of microscopic systems. A main purpose of this lecture is to formulate our answer to this question in a mathematical form which can be useful in the practical analysis of various physical phenomena. In solid state physics, one frequently catches a glimpse of the relation between micro and macroscopic objects. An example is given by the calculation of electric conductivity. This is a macroscopic quantity because it is the ratio of macroscopic current and macroscopic electric field. Linear response theory relates this macroscopic quantity to certain quantum fluctuation effects which are microscopic. However, the purpose of this paper is not to list known examples from solid state physics. Rather, we present a general formalism for the derivation of macroscopic theory from microscopic theory, and then, apply it to various phenomena to see whether it works well or not. According to this formalism, for example, the Josephson phenomenon, which is usually treated as the microscopic tunneling of Cooper pairs, is regarded simply as the creation of a macroscopic current[5,6] caused by the presence of a macroscopic surface singularity (the Josephson junction).

It is a remarkable fact that most classical objects manifest a certain order. To make this statement clearer, let us again consider the case of crystals. The basic equation for interacting molecules is translationally and rotationally invariant, as the equation for molecule gas should be. Under certain conditions, the molecule gas system manifests the crystal lattice order (creation of order) and the translational and rotational symmetry disappear from observation (spontaneous breakdown of symmetry). In this perfect crystal state we can excite phonon and many other quantum levels.

In other words, the perfect crystal state without boundaries is a quantum system of phonons and other excitations. However, we can modify the situation in such a way that there appear, in this quantum system, many kinds of extended objects (such as dislocations, grain boundaries, point defects, etc.), thus creating a situation in which classical and quantum objects coexist[2,3]. As will be shown later, even the boundary surface of a crystal of finite size can be regarded as a self-consistently maintained extended object[3]. In general terminology, we are concerned with the following processes: (a) creation of ordered states = spontaneous breakdown of symmetries, (b) creation of macroscopic objects in quantum systems.

Analysis of step (a) has shown that the phenomenon of spontaneous breakdown of symmetry is really a dynamical rearrangement[7,8,9] of symmetry, meaning that the observable manifestation of symmetry differs from the basic symmetry (i.e. the symmetry of basic equations); although the symmetry is never lost. In section 5 we briefly review the present status of the study of symmetry rearrangement. Analysis of step (b) leads to a systematic formulation of a theory for extended objects in quantum systems. This theory is called the boson theory. So far the boson theory has been applied to the analysis of superconductivity[5,10], itinerant electron ferromagnetism[11], crystal dislocations[3], the electron gas in solids[12] and surface phenomena in solids[3,6]. In this paper, we omit a description of these practical applications, although we make a brief sketch of the boson theory for superconductors and crystal dislocations in sections 3 and 4. In the next section we present an elementary model in order to elucidate some essential features of the boson theory. There it will also be shown that <u>the extended objects (or solitons) created in a quantum field system behave either as classical objects or as quantum mechanized objects, depending on their size and the experimental conditions.</u>

2. <u>Extended Objects in Quantum Systems and Soliton Solutions</u>

In this section we study the mechanism for creation of extended objects in quantum many body systems. Let ψ stand for the Heisenberg fields. Their equation is written as

$$\Lambda(\partial)\psi = F[\psi] \quad . \tag{2.1}$$

To make the essence of our argument transparent, we begin with a simple case, in which no composite particles appear and the perturbative expansion method is usable. We assume also that ψ is a boson field. Eq. (2.1) leads to the Yang-Feldman equation[13]

$$\psi = \varphi^{0} + (\Lambda(\partial))^{-1}F[\psi] \quad , \tag{2.2}$$

where φ^{0} is a free boson field satisfying

$$\Lambda(\partial)\varphi^O = 0 \quad . \tag{2.3}$$

It is important to note here that $F[\psi]$ in (2.1) is chosen in such a way that the mass in $\Lambda(\partial)$ is the physical mass, and therefore, that φ^O is the renormalized (or physical) free field. To understand this, we note that (2.2) should be read in the following way:

$$<a|\psi(x)|b> = <a|\varphi^O|b> + (-i) \int d^4y \quad \Delta(x-y)<a|F_y[\psi]|b>. \tag{2.4}$$

Here $|a>$ and $|b>$ are the vectors in the Fock space of φ^O, and $F_y[\psi]$ means $F[\psi]$ at the space-time position y. The function $\Delta(x)$ in (2.4) is the Green's function

$$\Lambda(\partial)\Delta(x) = i\delta(x) \quad . \tag{2.5}$$

The second term on the right hand side of (2.4) diverges unless $<a|F_y[\psi]|b>$ vanishes in a reasonable manner at the limit $t_y \to \pm \infty$. Such a divergence can be avoided when φ^O is renormalized, because the mass renormalization introduces a mass counter term which eliminates the divergence mentioned above. This argument implies that we regard the Heisenberg equation as an equation among matrix elements (i.e. the weak relation). All of the equations for Heisenberg operators in the following should be understood as weak relations. In the following φ^O will be called the physical field.

When we solve (2.2) by successive iteration, we are led to the usual perturbative expansion. The result is an expression of ψ in terms of φ^O:

$$\psi(x) = \psi(x;\varphi^O) \quad . \tag{2.6}$$

This is a linear combination of normal products of φ^O, and should be understood as a weak relation. Eq. (2.6) is called the dynamical map[7,9].

Let us now introduce a c-number function $f(x)$ which satisfies the equation for φ^O:

$$\Lambda(\partial)f = 0 \quad . \tag{2.7}$$

Then, we can generalize the Yang-Feldman equation as

$$\psi^f = \varphi^O + f + (\Lambda(\partial))^{-1}F[\psi^f] \quad . \tag{2.8}$$

Solving this by successive iteration, we obtain a new solution of the Heisenberg equation (2.1):

$$\psi^f(x) = \psi(x;\varphi^O + f) \quad . \tag{2.9}$$

Note that ψ^f is obtained from ψ by the replacement $\varphi^O \to \varphi^O + f$ (the boson transformation)[14]. The fact that both ψ and ψ^f satisfy the same Heisenberg equation (2.1) is the content of the <u>boson transformation theorem</u>[8,14]:

$$\Lambda(\partial)\psi^f = F[\psi^f] \quad . \tag{2.10}$$

Let us now introduce the c-number field ϕ^f as

$$\phi^f = <0|\psi^f|0> \quad . \tag{2.11}$$

When we ignore h (the Planck constant), ϕ^f will be denoted by ϕ_0^f:

$$\phi_0^f = \lim_{h \to 0} \phi^f \quad . \tag{2.12}$$

Now note that the difference between the vacuum expectation value of the product of ψ^f's and the product of vacuum expectation values of ψ^f is due to the contractions of physical fields, which create the loop diagrams in the course of successive iteration applied to (2.8). Since the contraction of physical fields create terms which vanish at h = 0, we can write as[15]

$$<0|F[\psi^f]|0> = F[\phi^f] + 0(h) \tag{2.13}$$

where $0(h)$ stands for those terms which vanish at h=0. Thus, (2.10) leads to the classical Eular equation[15]

$$\Lambda(\partial)\phi_0^f = F[\phi_0^f] \quad . \tag{2.14}$$

The above argument shows that ϕ_0^f is given by the tree diagrams only.

So far we have simplified the situation by several assumptions. Let us now reformulate our consideration in general terminology. Suppose that the Heisenberg equation (2.1) gives rise to a set of physical free fields $(\varphi_1^0, \varphi_2^0 \ldots \varphi_n^0 \ldots)$. Here φ_α^0 ($\alpha = 1 \ldots n$) are bosons and the other are fermions. The free field equations for the boson fields are

$$\lambda^\alpha(\partial)\varphi_\alpha^0 = 0 \tag{2.15}$$

where no summation over α is made. Since we require that ψ is realized in the Fock space of the physical fields, all of the matrix elements of ψ among the vectors in this Fock space should be well defined. This means that ψ can be expressed as a linear combination of normal products of the physical fields:

$$\psi(x) = \psi(x;\varphi_\alpha^0, \ldots) \quad . \tag{2.16}$$

This is the dynamical map[7,9] and should be understood as a weak relation.

Now we perform the substitution called the boson transformation[8,14]

$$\varphi^O_\alpha(x) \to \varphi^O_\alpha(x) + f_\alpha(x) \qquad\qquad (2.17)$$

in which $f_\alpha(x)$ ($\alpha = 1, \ldots, n$) are c-number functions which satisfy the equation for φ^O_α:

$$\lambda^\alpha(\partial) f_\alpha(x) = 0 \quad . \qquad\qquad (2.18)$$

Then, the boson transformation theorem states that the boson-transformed Heisenberg field

$$\psi^f(x) = \psi(x; \varphi^O_\alpha + f_\alpha, \ldots) \qquad\qquad (2.19)$$

also satisfies the Heisenberg equation (2.1). Note that the theorem holds true even when some members of the bosons φ^O_α are composite, and also even when the perturbative expansion method is not applicable.

Though the general proofs of the theorem available so far[8,14] have been very complicated, we have now a very simple proof. In the following we present a very brief sketch of this proof. We firstly show that the product and the boson transformation commute with each other, e.g.

$$\psi(x; \varphi^O + f)\psi(x; \varphi^O + f) = \left[\psi(x; \varphi^O)\psi(x; \varphi^O)\right]_{\varphi^O + f}$$

in which φ^O stands for the physical boson φ^O_α ($\alpha = 1,\ldots,n$). This leads to $F[\psi^f] = F^f[\psi]$, where F^f means the boson-transformed F. Since f and φ^O satisfy the same equation, the replacement $\dot\varphi^O \to \dot\varphi^O + \dot f$ is the same as $\omega(\vec\nabla)\varphi^O \to \omega(\vec\nabla)(\varphi^O + f)$ where $\omega(\vec p)$ is the energy of the φ^O-quantum. Thus, we can consistently perform the boson-transformation of φ^O carrying derivatives; $[\partial_\mu \varphi^O]^f = \partial_\mu(\varphi^O + f)$. Considering the fact that to solve the Heisenberg equation requires the combined uses of product, differentiation and integration, we can prove that

$$\Lambda(\partial)\psi^f = [\Lambda(\partial)\psi]^f = F^f[\psi] = F[\psi^f] \quad .$$

The c-number field ϕ^f is defined by (2.11). When we ignore h, ϕ^f becomes ϕ^f_o, i.e. Eq. (2.12). Following the argument which led to (2.13), we can prove[15] that ϕ^f_o satisfies the classical Euler equation (2.14).

The expression (2.17) for the boson transformation indicates that $f_\alpha(x)$ are created by the condensation of the bosons φ^O_α. The result of this condensation is the appearance of a certain extended object described by the c-number field ϕ^f. Note that, although $\phi^f(x)$ describes a classically behaving extended object, it implicitly contains quantum effects because we do not ignore h. Intuitively speaking, when the quantum fluctuation becomes much smaller than the macroscopic effect of the condensed bosons, the system behaves classically even though h=0 is not assumed. When we ignore h, ϕ^f becomes ϕ^f_o, which satisfies the classical Euler equation.

Eq. (2.18) for f_α admits various propagating wave solutions. A well-known example of wave-like extended objects in solid state physics are the ultrasonic waves in crystals.

We now consider a static extended object. Let $\omega_\alpha(\vec{p})$ denote the energy of the boson φ^o_α with momentum $\vec{p}$. We assume that $\omega_\alpha(\vec{p})$ is non-negative. Suppose that $\omega_\alpha(\vec{p}) = 0$ for $\vec{p} = \vec{p}^{(a)}$, $(a = 1,...)$. Assuming that $f_\alpha(x)$ is Fourier transformable, we write its Fourier amplitude by $f_\alpha(\vec{p})$. Then, Eq. (2.18) for static cases leads to the relation $\omega_\alpha(\vec{p})f_\alpha(\vec{p}) = 0$, implying that $f_\alpha(\vec{x})$ is a multi-periodic function of the form

$$f_\alpha(\vec{x}) = \sum_a c_a \exp[i\vec{p}^{(a)}\vec{x}] \ . \tag{2.20}$$

This static extended object is created by the condensation of the bosons in zero energy levels. All other static extended objects carry certain singularities[15] which prohibit the Fourier transform of f_α. <u>These singularities are either a divergent singularity or a topological singularity.</u> Here the divergent singularity means that $f_\alpha(x)$ diverges at $|\vec{x}| = \infty$ at least in a certain direction of $\vec{x}$. The topological singularity means that $f_\alpha(x)$ is not single valued, and therefore, is path-dependent. Obviously, $f_\alpha(x)$ <u>for time-dependent objects can also carry these singularities.</u>

A well-known example of an extended object with a topological singularity is the vortex in superconductivity. In the case of the straight line vortex the boson transformation parameter $f(x)$ is the cylindrical angle θ around the vortex line[5,9]. This $f(x)$ has a topological line singularity which makes θ multi-valued. The path-dependence of θ is manifested through the well-known relation $\vec{\nabla} \times \vec{\nabla}\theta = \delta(x_1)\delta(x_2)\vec{e}_3$ when the vortex line is along the x_3-axis. Here $\vec{e}_3$ is the unit vector in the x_3-direction. In general, when $f_\alpha(x)$ is path dependent, the topological singularity can be mathematically expressed by the relation

$$G^{\alpha\dagger}_{\mu\nu}(x) \neq 0 \quad \text{for certain x, } \mu, \nu \text{ and } \alpha \ , \tag{2.21}$$

where $G^{\alpha\dagger}_{\mu\nu}$ is defined by[3,14,16]

$$G_{\mu\nu}^{\alpha\dagger}(x) = [\partial_\mu, \partial_\nu] f_\alpha(x) \quad . \tag{2.22}$$

Furthermore, the existence of the path-dependent $f_\alpha(x)$ requires that $\partial_\rho f_\alpha$ should be single-valued:

$$[\partial_\mu, \partial_\nu] \partial_\rho f_\alpha(x) = 0 \quad . \tag{2.23}$$

Without loss of generality, we can put the free field equation (2.18) in the following form:

$$[D_\alpha^\mu \partial_\mu + m_\alpha^2] f_\alpha(x) = 0 \quad . \tag{2.24}$$

Here D_α^μ is a derivative operator and m_α^2 is a constant. It is important here that $D_\alpha^\mu \partial_\mu \exp(ip_\nu x_\nu) = 0$ for $p_\nu = 0$. Therefore, we have $m_\alpha^2 = 0$ when and only when the energy of the φ_α^o-quantum vanishes at $\vec{p} = 0$. When this happens, we say that the φ_α^o-quantum is a gapless boson.

Eqs. (2.22), (2.23) and (2.24) lead to

$$\partial_\nu f_\alpha(x) = \frac{1}{D_\alpha \partial + m_\alpha^2} D_\alpha^\mu G_{\mu\nu}^{\alpha\dagger}(x) \quad , \tag{2.25}$$

where $D_\alpha \partial$ means $D_\alpha^\mu \partial_\mu$. Since $G_{\mu\nu}^{\alpha\dagger}$ is Fourier transformable, $[1/(D_\alpha \partial + m_\alpha^2)]$ in (2.25) is well defined in terms of its Fourier representation. Note that (2.25) now leads to

$$D_\alpha^\nu \partial_\nu f_\alpha(x) = 0 \quad , \tag{2.26}$$

implying that the φ_α^o-quantum is a gapless boson. We thus conclude[15] <u>that extended objects associated with topological singularities (2.22) can be created only by the condensation of gapless bosons.</u> This is the reason why we find many kinds of extended objects in any ordered state, because the Goldstone bosons which maintain the order are gapless.

Let us now introduce the topological quantum number. To do this we consider a static extended object with the topological singularity of the property (2.21). When we define the currents $\vec{N}^\alpha$ by

$$N_i^\alpha(\vec{x}) = \frac{1}{2} \varepsilon^{ijk} G_{jk}^{\alpha\dagger}(\vec{x}) \quad , \tag{2.27}$$

these currents trivially conserve (i.e. $\vec{\nabla}\vec{N}^\alpha = 0$) according to (2.22) and (2.23). We thus have

$$0 = \int_V d^3x \vec{\nabla}\vec{N} = \int_S dS \, (\vec{n}\cdot\vec{N})$$

where S is the surface enclosing the domain V and $\vec{n}$ is the normal
vector on the surface. The above relations show that the quantities

$$B^{\alpha} \equiv \int_{Sc} dS(\vec{n} \cdot \vec{N}^{\alpha}) = \int_{Sc} dS(\vec{n} \cdot \vec{\nabla} \times \vec{\nabla}f) = \oint_{c} d\vec{s} \cdot \vec{\nabla}f \qquad (2.28)$$

are topological numbers in the following sense; B^{α} are independent
of change of the closed path c as far as the path does not cross the
singularity in the course of the modification of the path. Here
Sc is a surface whose circumference is the closed path c. In the
next two sections we show some examples in which the quantities B^{α}
can assume only certain discrete values. These B^{α} are called "topological quantum numbers". Extended objects with topological
singularities are usually not stable unless some of their topological numbers are quantized.

We have briefly described a method for the derivation of extended objects from quantum systems. This method has been used in
various problems in solid state physics, and has been called the
boson method. Recently, we applied[15] the boson method to the (1+1)-
dimensional model with the Heisenberg equation $[-\partial^2 - \mu^2]\psi(x) = \lambda\psi^3(x)$. We constructed the dynamical map $\psi(x; \varphi^0)$ by means of the
tree approximation and then performed the boson transformation
$(\varphi^0 \to \varphi^0 + f)$. Then, the choice f(x) = A exp(-mx), where m is the
physical mass of φ^0 and A is constant, led to the following form for
$\phi_o^f = <0|\psi^f|0>$:

$$\phi_o^f = v \tanh[(m/2)(x - a)] .$$

Here v and a are defined by the relation $A = -2ve^{ma}$. This result
for ϕ_o^f agrees exactly with the well-known static soliton solution
of the classical Euler equation $(-\partial^2 - \mu^2)\phi_o^f = \lambda[\phi_o^f]^3$. A group of
people at Strasbourg[17] have made a more extensive analysis of this
problem. They used the (1+1)-dimensional sine-Gordon Heisenberg
equation and showed that the choice

$$f(t, x) = \sum_{j=1}^{N} \exp[\alpha_i x + \beta_i t - \delta_i] .$$

with the conditions $\alpha_i^2 - \beta_i^2 = m^2$ leads to the N-soliton solutions[18]
of the classical sine-Gordon equation. We feel that this line of
study is going to clarify many features of the various kinds of
soliton solutions of non-linear equations.

Let us now clarify two questions. The first question is to ask
how the extended objects interact with quanta, while the second
question asks whether the extended objects always behave classically

or not. As will be shown in the following part of this section, study of the first question will lead us to an answer to the second question. To analyse the first question, we return to the Yang-Feldman equation (2.8) for the real boson field ψ and calculate the dynamical map ψ^f in (2.9) by means of successive iteration together with the tree approximation:

$$\psi^f(x) = \phi_o^f(x) + \int d^4y \; c(x; y)\varphi^o(y) + \ldots \; . \tag{2.29}$$

Here φ^o is the free field which satisfies (2.3). The second term on the right hand side of (2.29) will be called the linear boson term. The energy of the φ^o-quantum with momentum $\vec{k}$ will be denoted by ω_k.

For simplicity, we restrict our attention to the case of a static soliton in the tree approximation; the general consideration is found in ref. 19. A calculation[19] shows that the linear boson term of the dynamical map should satisfy the eigenvalue equation of the form

$$\{\Lambda(\partial) - F'[\phi_o^f(x)]\}u(x) = 0 \tag{2.30}$$

where $F' = \delta F/\delta\phi_o^f$. This equation shows that the φ^o-quantum feels the "self-consistent potential", $F'[\phi_o^f(x)]$, which is induced by the soliton.

The eigenvalue equation (2.30) admits, not only the eigen-solutions $u(x,\vec{k})$ with momentum $\vec{k}$ and energy ω_k, but also the solutions $u_i(x)$ with discrete energies ω_i; the latter solutions describing the bound states of the φ^o-quantum and the soliton. We choose $u(x, \vec{k})$ and $u_i(x)$ to form an orthonormalized complete set of the solutions of (2.30). Now note that the replacement $f(\vec{x}) \rightarrow f(\vec{x} + \vec{a})$ induces the spatial translation, $\vec{x} \rightarrow \vec{x} + \vec{a}$, in ϕ_o^f. Since $f(\vec{x} + \vec{a})$ also satisfies (2.3), we see that the Euler equation (2.14) is invariant under the translation. We therefore have

$$\{\Lambda(\partial) - F'[\phi_o^f]\}\partial_i\phi_o^f(\vec{x}) = 0 \quad , \tag{2.31}$$

which implies the presence of three zero-energy bound states. These states will be called the "translation modes". We define u_i in such a way that u_i with $i = 1,2,3$ correspond to the translation modes. Eq. (2.31) shows that the orthonormalized functions of the translation modes are linear combinations of $\partial\phi_o^f$; $u_i(\vec{x}) = n_{ij}\partial_j\phi_o^f(\vec{x})$ for $i = 1,2,3$.

The canonical commutation relation of the Heisenberg field requires that the linear boson term in the dynamical map (2.29) should be the boson field which contains all of the members of the complete set $[u_o(x, \vec{k}), u_i(x)]$. We therefore enlarge the Hilbert space in order to take into account all of the bound states. (See the Note added in Proof). To do this, we first build the Fock space, $\mathcal{H}$, of the quantum states of all of the modes <u>except the translation mode</u>. Let us denote by $\chi^o(\vec{x})$ the boson field which contains all of the wave functions, $u(x, \vec{k})$ and $u_i(x)$, except u_1, u_2 and u_3. Thus, $\mathcal{H}$ is the Fock space of the boson field χ^o. To take into account the translation modes, we introduce three sets of operators $[(q_i, p_i);\ i = 1,2,3]$, which satisfy the canonical commutation relation. Then, $[\chi^o(x) + u_i(\vec{x})q_i]$ is the boson field which contains all of the eigensolutions of (2.30). Note that $u_i(\vec{x})q_i = (\vec{Q}\cdot\vec{\partial})\phi_o^f$, where $Q_i = \eta_{ij}q_j$. Now the dynamical map (2.29) is modified as follows:

$$\psi^f(x) = \phi_o^f(x) + [\chi^o(x) + (\vec{Q} \cdot \vec{\partial})\phi_o^f(x)] + \ldots$$

$$= [1 + (\vec{Q} \cdot \vec{\partial}) + \ldots]\phi_o^f(\vec{x}) + \chi^o(x) + \ldots \qquad . \qquad (2.32)$$

Our new Hilbert space is the product of the Fock space $\mathcal{H}$ and the Hilbert space of the canonical variables (q_i, p_i). The operator $\vec{Q}$ will be called the "quantum coordinate".

The expression (2.32) suggests that the c-number coordinate $\vec{x}$ and the quantum coordinate $\vec{Q}$ appear only in the combination $(\vec{x} + \vec{Q})$. Our calculation [19] has indeed proved this; it showed that the dynamical map becomes

$$\psi^f(x) = \phi_o^f(\vec{x} + \vec{Q}) + \chi^o(\vec{x} + \vec{Q}, t)$$

$$+ \int d^4y \int d^4z\ \bar{c}(\vec{x} + \vec{Q}, t; y, z):\chi^o(y)\chi^o(z):$$

$$+ \ldots \qquad . \qquad (2.33)$$

Thus, the spatial translation $\vec{x} \to \vec{x} + \vec{a}$ is induced by $\vec{Q} \to \vec{Q} + \vec{a}$ (rearrangement of the translation symmetry). This implies that, as soon as the creation of an extended object breaks the translation and rotation symmetries, there appears the quantum coordinate $\vec{Q}$, which acts as the Goldstone mode and recovers the translational and rotational invariance. It can be shown[19] that ψ^f in (2.33) satisfies the Heisenberg equation (2.10). Although the tree approximation has been used in our consideration, the above conclusions are true quite generally[19].

When the size of the extended object is much larger than the quantum fluctuation of $\vec{Q}$, the extended object behaves as a classical object. When the quantum fluctuation of $\vec{Q}$ becomes as large as the size of the object, the extended object becomes a quantum mechanical object. Various small domains which appear in many kinds of ordered states in solid state physics may be these quantum extended objects. It may be that large nuclei also fall into this category. When we consider an extremely small extended object, we may find a variety of quantum extended objects as models for elementary particles. We come back to this point in the last section. (As for a recent development in this subject, see Note added in Proof).

3. The Boson Theory for Superconductivity

In order to apply the boson method to the calculation of many physical quantities associated with extended objects, we need to know the structure of the dynamical map $\psi(x; \varphi_\alpha^0 \ldots)$. It is common to make use of the Ward-Takahashi relations (W-T relations) in order to find those characteristic features of the dynamical map which are the results of the basic symmetry manifested in the original Heisenberg equation.

In this section let us illustrate the study of dynamical maps by using superconductivity as an example[10]. In this case we use the Lagrangian $\mathcal{L} = \mathcal{L}_1[\psi, A_\mu] - B\, D^\mu(\partial)A_\mu$, in which ψ and A_μ are the Heisenberg operators for the electron and the electromagnetic vector potential respectively. The role[20] of the field B is to act as the canonical conjugate of A_o. The term $\mathcal{L}_1[\psi, A_\mu]$ is the usual Lagrangian for the electron and electromagnetic fields, and therefore, is invariant under the gauge-transformation $\psi \to \exp[i(e/\hbar c)\lambda(x)]\psi$ and $A_\mu \to A_\mu + \partial_\mu\lambda$. Furthermore, the total Lagrangian is invariant under the global phase transformation $(\psi \to e^{i\theta}\psi, A_\mu \to A_\mu)$. We assume that the system is in an ordered state of the electric phase:

$$\Delta = <0|\psi_\uparrow(x)\psi_\downarrow(x)|0> \neq 0 \quad . \tag{3.1}$$

The gauge invariance, the global phase invariance and the condition (3.1) are the only assumptions which our consideration requires. The variational principle $\delta\mathcal{L}/\delta B = 0$ leads to the gauge condition $D^\mu(\partial)A_\mu = 0$. The derivative operator $D^\mu(\partial)$ is determined by the W-T relations.

The W-T relations lead us to the following set of physical free fields: the quasi-electron φ^0, the plasmon field U_μ^0, a gapless field χ^0 of positive norm and a gapless field b^0 of negative norm[10]. The gapless field χ^0 is the Goldstone field, and b^0 is called the ghost field which plays a role similar to the Gupta-Bleuler ghost in quantum electrodynamics. The χ^0 and b^0 satisfy the same free field equation. Use of the W-T relation leads us to the conclusion[10] that the derivative operator $D^\mu(\partial)$, which determines the gauge condition as $D^\mu(\partial)A_\mu = 0$, should be the one which appears in the field equation for χ^0 and b^0 in the following way:

$$D^\mu(\partial)\partial_\mu\chi^o = 0 \quad , \quad D^\mu(\partial)\partial_\mu b^o = 0 \quad . \tag{3.2}$$

Thus, we have $D^o(\partial) = \partial/\partial t$ and

$$i\vec{D}(i\vec{p}) \cdot \vec{p} = \omega_B^2(\vec{p}) \quad , \tag{3.3}$$

where $\omega_B(\vec{p})$ is the energy of the Goldstone boson and the ghost. It is common to introduce $v_B(\vec{p})$ by $\omega_B^2(\vec{p}) = (v_B/c)^2\vec{p}^2$. Further use of the W-T relations leads us to the following form for the dynamical maps[10]:

$$A_\mu(x) = a_\mu^o(x) + :\tilde{A}_\mu[x;\, \varphi^o,\, U^o - \frac{\hbar c}{e}\partial(\chi^o - b^o)]: \tag{3.4a}$$

$$\psi(x) = :\exp[i\chi^o(x)]\{\varphi^o(x) + \tilde{\psi}[x;\, \varphi^o,\, U^o - \frac{\hbar c}{e}\partial(\chi^o - b^o)]\}: \tag{3.4b}$$

$$B(x) = \alpha(\partial)[\chi^o(x) - b^o(x)] \tag{3.4c}$$

and

$$O_H(x) = :O[x;\, \varphi^o,\, U^o - \frac{\hbar c}{e}\partial(\chi^o - b^o)]: \quad . \tag{3.4d}$$

Here $O_H(x)$ stands for any gauge-invariant Heisenberg operator, and $\tilde{A}_\mu$ and $\tilde{\psi}$ do not contain any term which is linear in the physical free fields. The linear term $a_\mu^o(x)$ in (3.4a) is defined by

$$a_\mu^o(x) = U_\mu^o(x) + \frac{\hbar c}{e}\partial_\mu b^o(x) \quad . \tag{3.5}$$

It should be noted that the normalization of the free fields $(\varphi^o,\, U_\mu^o,\, \chi^o,\, b^o)$ is so chosen that the dynamical maps take simple forms. In other words, all the renormalization factors are absorbed into these free fields. In (3.4c), $\alpha(\partial)$ is a derivative operator which does not have any significance in the following consideration.

Note that the global phase transformation $(\psi \to e^{i\theta}\psi,\, A_\mu \to A_\mu)$ is induced by the following transformation of the Goldstone boson:

$$\chi^o(x) \to \chi^o(x) + \theta \quad . \tag{3.6}$$

This is the dynamical rearrangement of global phase symmetry[7,9,10]. The gauge transformation $(\psi \to \exp[i(e/\hbar c)\lambda]\psi,\, A_\mu \to A_\mu + \partial_\mu\lambda)$ is induced by the transformation

$$\chi^o(x) \to \chi^o(x) + (e/\hbar c)\lambda(x) \quad , \quad b^o(x) \to b^o(x) + (e/\hbar c)\lambda(x) \tag{3.7}$$

where $\lambda(x)$ is Fourier-transformable. This is the <u>dynamical</u>

<u>rearrangement of gauge symmetry</u>[10].

It is required that all of the observable states (say $|a\rangle$) should satisfy the Gupta-Bleuler type condition, i.e. $(\chi^o - b^o)^{(-)} |a\rangle = 0$. Now, the dynamical map (3.4d) implies that the Goldstone quanta χ^o and the ghost quanta b^o are unobservable (the Anderson-Higgs-Kibble mechanism[21,22]).

It can be shown[10] that the plasmon equation is of the form

$$\lambda_{\mu\nu}(\partial)U^o_\nu(x) = 0 \tag{3.8}$$

where

$$\lambda_{\mu\nu}(\partial) = \partial^2 g_{\mu\nu} - \partial_\mu \partial_\nu - m^2_{\mu\nu}(\partial) \quad . \tag{3.9}$$

Here $m^2_{\mu\nu}(\partial)$ is the proper self-energy of the photon (i.e. the plasmon) and has the form:

$$m^2_{\mu\nu}(\partial) = m^2(\partial)\eta_{\mu\nu}(\partial) - \delta m^T_{\mu\nu}(\partial) \quad . \tag{3.10}$$

Here $\delta m^T_{\mu\nu}(\partial)$ has transverse components only, i.e. $\delta m^T_{\lambda o} = \delta m^T_{o\nu} = 0$ and $\nabla_i \delta m^T_{ij} = \delta m^T_{\lambda i} \nabla_i = 0$. Its action on any vector, say $\vec{a}(x)$, is defined by

$$\delta m^T_{ij}(\partial)a_j(x) = [\vec{\nabla} \times \vec{M}(x)]_i \tag{3.11}$$

where

$$M_i = \left[\frac{4\pi\chi}{1 + 4\pi\chi}\right]_{ij} (\vec{\nabla} \times \vec{a})_j \quad . \tag{3.12}$$

Here the matrix $\chi = (\chi_{ij}(\vec{\nabla}))$ is defined as follows: $\chi_{ij}(i\vec{p})$ is the Fourier amplitude of the staggered susceptibility $\chi_{ij}(\vec{x} - \vec{y})$, i.e.

$$M_i(x) = 4\pi \int d^3y \; \chi_{ij}(\vec{x} - \vec{y})h_j(y) \tag{3.13}$$

where

$$h_j(x) = [1/(1 + 4\pi\chi)]_{ij} (\vec{\nabla} \times \vec{a})_j \quad . \tag{3.14}$$

When $\vec{a}$ is the vector potential, $\vec{h}$ is the magnetic field and $\vec{M}$ is the magnetic moment. This implies that the $\delta m^T_{\mu\nu}$-term in (3.10) appears only when the superconductor is magnetic.[23] The matrix elements $\eta_{\mu\nu}(\partial)$ have the properties $\eta_{oo} = -(c/v_B)^2$, $\eta_{ij} = \delta_{ij}$ and $\eta_{oi} = \eta_{io} = 0$. We have the relation $D_\mu(\partial) = (v_B/c)^2 \partial^\nu \eta_{\nu\mu}(\partial)$. Then, the plasmon equation (3.8) leads to $D^\mu(\partial)U^o_\mu = 0$. This is consistent with the gauge condition $D^\mu(\partial)A_\mu = 0$.

Let us now perform the following transformation[10]:

$$\chi^o \rightarrow \chi^o(x) + f(x) \tag{3.15a}$$

$$U_\mu^o(x) \rightarrow U_\mu^o(x) + a_\mu(x) \tag{3.15b}$$

which induces the change

$$a_\mu^o(x) \rightarrow a_\mu^o(x) + a_\mu(x) \quad . \tag{3.16}$$

Here $f(x)$ satisfies the equation for χ^o and b^o, i.e.

$$D^\mu(\partial)\partial_\mu f = 0 \quad . \tag{3.17}$$

In this case, the boson transformation theorem becomes slightly
more complicated because the invariance of the Gupta-Bleuler type
condition requires that the c-number a_μ should satisfy the relation

$$\lambda_{\mu\nu}(\partial)a_\nu(x) = - \frac{\hbar c}{e} m^2(\partial)\eta_{\mu\nu}(\partial)\partial_\nu f(x) \quad . \tag{3.18}$$

Since derivation of this equation is tedious and has been presented
in ref. 10, we do not repeat it here. Recalling (3.9) and (3.10),
we can rewrite (3.18) as

$$(\partial^2 g_{\mu\nu} - \partial_\mu\partial_\nu)a_\nu(x) = j_\mu(x) \tag{3.19}$$

with

$$j_\mu(x) = m^2(\partial)\eta_{\mu\nu}(\partial)[a_\nu(x) - \frac{\hbar c}{e}\partial_\nu f(x)] - \delta m^T_{\mu\nu}(\partial)a_\nu(x). \tag{3.20}$$

Here is an important note. Since $m^2_{\mu\nu}(p)$ is the photon self-
energy, it becomes the plasmon energy <u>only when</u> p_o is equal to the
plasmon energy $\omega_{p\ell}(\vec{p})$:

$$m^2_{ij}(p) = \omega^2_{p\ell}(\vec{p})\delta_{ij} \quad \text{for} \quad p_o = \omega_{p\ell}(\vec{p}).$$

Since the c-number parameter $a_\mu(x)$ does not satisfy the plasmon

equation (i.e. $\lambda_{\mu\nu}(\partial)a_\nu \neq 0$), we should use the off-the-energy-shell
$m^2_{\mu\nu}(\partial)$ in the above equations. When we consider a macroscopic phenomenon which changes slowly in time, we use $m^2_{\mu\nu}(\partial)$ with $\partial_o \simeq 0$.
It is common to express the current j_μ in terms of the boson characteristic function $c(\vec{V})$ and the London penetration depth λ_L which
are defined by $(1/\lambda_L)^2 \equiv m^2(0)$ and $c(\vec{p}) = \lambda_L^2 m^2(p)$ with $p_o = 0$. Note
that $(1/\lambda_L)$ is not necessarily equal to the zero-momentum plasmon
energy $m(\vec{p} = 0, p_o = \omega_{p\ell})$. Indeed we expect the relations

$(1/\lambda_L^2) = e^2 n_s/m_e c^2$ and $\omega_{p\ell}^2(\vec{p}=0) = e^2 n/m_e c^2$, where m_e is the electron mass, n the total electron density and n_s is the superconducting electron density.

Applying the boson transformation (3.15) to the dynamical maps in (3.4) and calculating the vacuum expectation values, we obtain

$$\hat{a}_\mu(x) = <0|A_\mu^f|0> \tag{3.21}$$

$$= a_\mu(x) + \tilde{A}_\mu[x; j_\mu(x)] \quad , \tag{3.22}$$

$$\hat{0}_H(x) = <0|0_H^f|0> \tag{3.23}$$

$$= 0[x, j_\mu(x)] \quad , \text{ etc.} \tag{3.24}$$

Here A_μ^f and 0_H^f mean the boson transformed A_μ and 0_H respectively. The c-number quantity $\tilde{A}_\mu$ is a linear combination of products of j_μ with powers higher than 1. Eq. (3.24) shows that all of the gauge-invariant quantities associated with the extended objects in super-conductors consist of $j_\mu(x)$. The quantity a_μ is the classical vector potential induced by the boson transformation. It was proved in the first paper in ref. 10 that the higher order term $\tilde{A}_\mu$ carries the factor ω_B^2/m^2, suggesting that $\tilde{A}_\mu$ is much smaller than a_μ. A careful calculation of the higher order term $\tilde{A}_\mu$ is now in progress. In the practical application of the boson theory to superconductivity, the linear approximation $\hat{a}_\mu \approx a_\mu$ has been made. The results have shown good agreement with experiments [10]. The calculation of the energy created by the boson transformation was also made in the first two papers in ref. 10. Since the results are complicated, we present here the result only for the static case, in which f(x) is independent of time. In the static case the energy induced by the boson transformation is given by

$$W_f = \int d^3x \ \vec{h}(\vec{x}) (\vec{\nabla} \times \vec{\nabla}) f \tag{3.25}$$

where $\vec{h}(\vec{x})$ is the magnetic field defined by (3.14), i.e. $\vec{h} = (1 + 4\pi\chi)^{-1} (\vec{\nabla} \times \vec{a})$.

It can be easily shown that when f has no topological singularities (i.e. $[\partial_\mu, \partial_\nu]f = 0$), then we have $j_\mu = 0$, implying that there are no extended objects without topological singularities[10]. This is a result of gauge invariance of the theory: when f(x) is Fourier-transformable, the boson transformation induces only a gauge transformation, the effect of which is unobservable. Eq. (3.25) also shows that $W_f = 0$ when f has no topological singularities.

When f(x) has a topological singularity, we find[10] that

$$\nu = \frac{1}{2\pi} \oint_c d\vec{s} \cdot \vec{\nabla} f \tag{3.26}$$

is a topological number according to (2.28).

In the case of superconductivity, the boson-transformed electron, ψ^f, carries the phase factor $\exp(if)$; other f's appearing in ψ^f carry derivatives. Thus, double-valuedness of the spinor ψ^f requires that $2\nu =$ integer, implying that ν is a topological quantum number[10]. This example illustrates the general fact that to answer the question asking which topological constants are quantized, we should know not only the symmetry of the original Heisenberg equation, but also the observable (rearranged) symmetry[24]. Since Eq. (3.19) leads to $a_\mu = (\hbar c/e)\partial_\mu f$ at $|\vec{x}| \to \infty$, (3.26) shows that

$$\int (d\vec{S} \cdot \vec{\nabla} \times \vec{a}_\mu) = \frac{2\pi \hbar c}{e} \nu \tag{3.27}$$

which implies the flux quantization[10].

Common theoretical analysis of superconductivity is based on use of the Gor'kov equation[25] for the order parameter $\Delta(x) = <0|\psi^f_\uparrow \psi^f_\downarrow + \psi^{f\dagger}_\downarrow \psi^{f\dagger}_\uparrow|0>$ where ψ^f is the boson transformed field. According to the dynamical map (3.4b), $\Delta(x)$ has the structure: $\Delta(x) = \exp[2if(x)]\bar{\Delta}(x; j_\mu)$. We can regard the Gor'kov equation as classical Euler equation for $\Delta(x)$. When $\Delta(x)$ is small, the Gor'kov equation becomes the Ginsburg-Landau equation[26]. However, the G-L equation is useful only when $T \approx T_c$ or $H \approx H_{c2}$ (i.e. the cases of small order parameter). On the other hand, practical application of the boson theory is based on the use of Eq. (3.19) which is easily handled. The only tedious part is the calculation of $m^2(\partial)$ and $\eta_{\mu\nu}(\partial)$. These computations were made in several previous articles[10]. The strength of the boson theory lies in the fact that the theory can describe the electromagnetic properties in the entire domain of the magnetic field and temperature by means of a relatively simply computation. The theory can give a detailed account of the structure of the interactions among vortices. The boson theory for superconductivity has been applied to the study of the first order phase transition in Nb and Va at $H = H_{c1}$, anisotropic properties[27] of cubic superconductors such as Nb and Pb-Tl, and the A-15 structure superconductors[10]. The results have shown reasonable agreement with experiments. A recent application is a study of magnetic superconductors[23], i.e. the superconducting state of the rare earth compounds, $(RE)Rh_4B_4$, $(RE)Mo_6S_8$ and $(RE)Mo_6Se_8$, an experimental study which has become extremely active in the last two years. According to our analysis, many features of these magnetic superconductors originate from the following features[23,28]. The superconductivity persistent current strongly shields the effects of spin moments[28]. In the mixed states the vertex current is modified in such a way that the flux quantization is not violated by the presence of spin moments. This effect creates a strong attractive interacion among vortices[23]. When the spin-moment effects and the effects of the persistent current are in balance, the magnetic state seems to be able to coexist with the superconducting state in the form of a spin-spiral phase[28]. In the boson theory the temperature effects are treated by Thermo Field Dynamics[5,29] which is a purely quantum-field theoretical formulation of quantum statistical mechanics.

4. The Boson Theory for Crystals

We assume that the Lagrangian for the Heisenberg field $\psi(x)$ for molecules is translationally and rotationally invariant. It is also assumed that $v(\vec{x}) = \langle 0|\psi^{\dagger}(x)\psi(x)|0\rangle$ has the crystal periodicity: $v(\vec{x}) = v(\vec{x} + \vec{a}_i)$ $(i = 1, 2, 3)$. An extensive use of the W-T relations leads to the following dynamical map[2]:

$$0_H(x) = \sum_{\lambda} : \varphi_{\lambda}(\vec{x} + \vec{\chi}^o(x))\hat{0}_{\lambda}(x; \partial\vec{\chi}^o, \varphi^o) : . \qquad (4.1)$$

Here $\vec{\chi}^o$ denotes the three phonon fields which are the Goldstone bosons and φ^o is the quasi-molecule. The symbol $0_H(x)$ stands for any Heisenberg operator, and $\{\varphi_{\lambda}(\vec{x})\}$ is a complete set of the periodic functions; $\varphi_{\lambda}(\vec{x} + \vec{a}_i) = \varphi_{\lambda}(\vec{x})$. The momenta of $\vec{\chi}^o$ and φ^o are confined to the first Brillouin zone. The phonon equation has the form $\lambda_{ij}(\partial)\chi^o_j(x) = 0$, where $\lambda_{ij}(\partial) = - \rho_{ij}(\vec{\nabla}) \cdot (\partial/\partial t)^2 + C^{\ell m}_{ij}(\vec{\nabla})\partial_{\ell}\partial_m$. The matrix $\rho_{ij}(\vec{k})$ is the effective density, and the quantities $C^{\ell m}_{ij}(\vec{k})$ are called the elastic constants although they depend on $\vec{k}$.

The dynamical map (4.1) shows that the spatial translation $0_H(\vec{x} + \vec{\theta}, t)$ is induced by the following transformation of phonons and quasi-molecules: $\vec{\chi}^o(\vec{x}, t) \rightarrow \vec{\chi}^o(\vec{x} + \vec{\theta}, t) + \vec{\theta}$ and $\varphi^o(\vec{x}, t) \rightarrow \varphi^o(\vec{x} + \vec{\theta}, t)$. This is the dynamical rearrangement of translational symmetry in crystal[2,3].

The boson transformation $\vec{\chi}^o(x) \rightarrow \vec{\chi}^o(x) + \vec{f}(x)$ with $\vec{f}(x)$ satisfying $\lambda_{ij}(\partial)f_j(x) = 0$ creates various kinds of extended objects in crystals[3]. The boson transformed Heisenberg operator is

$$0_H(x) = \sum_{\lambda} : \varphi_{\lambda}(\vec{x} + \vec{f}(x) + \vec{\chi}^o(x))\hat{0}_{\lambda}(x; \partial\vec{\chi}^o + \partial f, \varphi^o) : . \quad (4.2)$$

Then $\langle 0|0_H(x)|0\rangle$ gives various physical quantities associated with extended objects created by the boson transformation. When $\vec{f}(x)$ is single-valued it gives the classical sound wave. When $\vec{f}(x)$ is multi-valued, the periodic nature of the functions $\varphi_{\lambda}(\vec{x})$ implies that the topological constants defined by (2.28), i.e.

$$B_i^c = \oint_c d\vec{S} \cdot \vec{\nabla} f_i, \qquad (i = 1, 2, 3) \qquad\qquad (4.3)$$

should be quantized[2,3] as $\vec{B}_c = \sum n_i^c \vec{a}_i$ (n_i^c: integers). The vector $\vec{B}^c$ is called the Burgers vector. Line singularities associated with $f(x)$ create crystal dislocations, plane surface singularities create grain boundaries, enclosed surface singularities create crystal point defects, and oscillating boundary-surface singularities create surface sound wave[3] (Rayleigh wave). The temperature effects are treated by Thermo Field Dynamics[5,29].

5. <u>The Dynamical Rearrangement of Symmetries</u>

In the last two sections we have seen some examples of dynamical rearrangement of symmetry. In general, when the basic Heisenberg equation is invariant under the transformation $\psi \to \psi' = G[\psi]$ (G-symmetry), and when this transformation of Heisenberg fields is induced by transformation of the physical fields $\varphi^0 \to \tilde{G}[\varphi^0]$ ($\tilde{G}$-symmetry), then we say that the G-symmetry is rearranged[7,8,9] into the $\tilde{G}$-symmetry. The creation of any ordered state (i.e. spontaneous breakdown of any symmetry) is caused by a certain symmetry rearrangement. Furthermore, it has been shown[30] that in gauge theories such as quantum electrodynamics, there always appear certain Goldstone bosons and a symmetry rearrangement takes place even when no ordered state is created.

It has been proven that[31] $G = SU(2)$ with $H = U(1) \to \tilde{G} = E(2)$ and that[32] $G = SU(2) \times SU(2)$ with $H = $ isospin $SU(2) \to \tilde{G} = E(3)$. Here H means the stability group. These examples suggest[33] the rule that the rearrangement of a simple Lie group is a group contraction[34]. However, an exception to this rule has been presented[35] in the case $G = SU(2)$ with no stability group; here the $\tilde{G}$-transformations do not form a group. A study of general structure of symmetry rearrangement was made in ref. 36, where the reason for the many exceptions to the above-mentioned rule was explained. The group theoretical study of symmetry rearrangement is far from complete. A general rule is that when a symmetry rearrangement takes place, the $\tilde{G}$-transformations contain the translation of Goldstone bosons; $\chi_i^0 \to \chi_i^0 + \theta_i$. A necessary condition for the rearrangement to become a group contraction is that the stability group H is a maximum subgroup of G. The origin of the symmetry rearrangement lies in the fact that the dynamical maps $\psi(x; \varphi_\alpha^0 \ldots)$ are weak relations. Furthermore, when G_i stands for the generators of G-symmetry, the matrix elements $\langle a|G_i|b\rangle$ have the form:

$$\langle a|G_i|b\rangle = \int d^3x \langle a|\rho_i(x)|b\rangle \quad . \qquad\qquad (5.1)$$

When $\langle a|\rho_i(x)|b\rangle$ contains a term of order $1/V$ (V: volume), such a term does not contribute to $<a|G_i|b>$, because the limit $V \to \infty$ should be performed before the spatial integration is made. The limiting process $(1/V) \to 0$ induces the symmetry rearrangement[31] (such as group contraction, etc.). Recently, some people have been studying the mechanism of the reverse process, $G \to G$, by means of deformation theory[37].

6. Topological Singularities, Dynamics of Extended Objects and Gravitational Theory

It was shown in section 2 that a condensation of gapless bosons(such as Goldstone bosons) can create extended objects with topological singularities which are defined by

$$G_{\mu\nu}^{\alpha\dagger}(x) \neq 0 \quad \text{for certain x, } \mu, \nu \text{ and } \alpha \tag{6.1}$$

with $G_{\mu\nu}^{\alpha\dagger} = [\partial_\mu, \partial_\nu]f_\alpha$. As shown by (2.25), a path-dependent f_α satisfying $D_\alpha^\nu \partial_\nu f_\alpha = 0$ is determined as[3,16]

$$\partial_\nu f_\alpha(x) = \frac{1}{D_\alpha^\rho \partial_\rho} \, D_\alpha^\mu G_{\mu\nu}^{\alpha\dagger}(x) \tag{6.2}$$

where no summation over α is made. It has been proven that the choice of $G_{\mu\nu}^{\alpha\dagger}$ is conditioned only by the continuity equation $\partial_\lambda G_\alpha^{\lambda\rho} = 0$, in which $G_\alpha^{\lambda\rho}$ is the dual conjugate of $G_{\mu\nu}^{\alpha\dagger}$, i.e. $G_\alpha^{\lambda\rho} = (\tfrac{1}{2})\varepsilon^{\lambda\rho\mu\nu}G_\nu^{\dagger\alpha}$. The continuity equation leads **us** to conclude that the singular domain at each time is either a line without end points (such as superconducting vortex or a crystal dislocation) or a surface which consists of endless lines (e.g. Josephson junction, crystal grain boundaries). An enclosed surface is also permitted because it consists of rings. In cases of line and surface singularities we have[3,16]

$$G_\alpha^{\lambda\rho}(x) = \nu_\alpha \iint d\sigma d\tau \partial[y^\lambda, y^\rho]/\partial[\tau, \sigma] \cdot \delta^{(4)}(x - y(\sigma,\tau)) \tag{6.3}$$

and

$$G_\alpha^{\lambda\rho}(x) = \iint d\sigma_1 d\sigma_2 d\tau \, \partial[y^\lambda, y^\rho, F_\alpha]/\partial[\tau, \sigma_1, \sigma_2]$$
$$\times \, \delta^{(4)}(x - y(\sigma_1, \sigma_2, \tau)) \,, \tag{6.4}$$

respectively. Here the ν_α is constant, τ is the time-like parameter and F_α is a function of σ, σ_1 and σ_2. An enclosed surface singularity creates point defects in crystals[3] and MIT-type[38] bag in a relativistic theory[39].

In section 2 we have seen that the boson transformation creates classical objects which are solutions of classical Euler equations. We now see that we can associate a variety of topological singularities with these classical objects, because the shape of the singular domain and its time-dependence is extremely arbitrary. This demonstrates the richness of the classical Euler equations. Even when we consider a Lorentz-covariant Euler equation, there are solutions, the behaviour of whose singular domains does not follow the laws of special relativity. This leads us to expect that even when the basic equation is Lorentz-covariant, the presence of extended objects may easily induce the dynamics of a curved space-time geometry.

This consideration leads us to the following speculation for the gravitational theory. Construct the gauge invariant quantum theory of a massless field with spin two. This equation, which may be regarded as the Einstein equation for Minkovsky geometry, is the Heisenberg equation, with which our consideration starts. We expect that this equation leads to the graviton as a member of the set of physical fields. When certain extended objects are created by the boson transformation, a Fourier transformable $f(x)$ may give rise to a classical gravitational wave. We are particularly interested in the extended objects whose $f(x)$-functions carry a variety of topological singularities, which may introduce curved space-time geometry. It has been speculated[24] that the black hole may be an extended object with an enclosed surface singularity.

Acknowledgement

We wish to thank Mr. G. Semenoff for valuable discussions and careful reading of the manuscript. This work was supported by the Natural Sciences and Engineering Research Council, Canada and by the Dean of the Faculty of Science of the University of Alberta.

References

1. J.F. Nye, "Physical Properties of Crystals", Oxford University Press, Oxford (1957).
   L.D. Landau and E.M. Lifshitz, "Theory of Elasticity", Pergamon Press, New York (1959).
2. M. Wadati, H. Matsumoto, Y. Takahashi and H. Umezawa, Phys. Lett. $\underline{62A}$,(1977) 255, 258.
   M. Wadati, H. Matsumoto, Y. Takahashi and H. Umezawa, Fortsch. Phys. $\underline{26}$, (1978) 357.
3. M. Wadati, H. Matsumoto and H. Umezawa, Phys. Rev. $\underline{B18}$, (1978) 4077.
4. R. Kubo, J. Phys. Soc. Japan $\underline{12}$, (1957) 570.

5. L. Leplae, H. Umezawa and F. Mancini, Physics Reports 10C,
    No. 4 (1974).
6. F. Mancini, H. Matsumoto, H. Umezawa and M. Wadati, Prog.
    Theor. Phys. (Kyoto) 62 (1979) 12.
7. H. Umezawa, Nuovo Cimento 40 (1965) 450.
    L. Leplae, R.N. Sen and H. Umezawa, Suppl. Prog. Theor. Phys.
    (Kyoto) (Commerative issue for 30th Anniversary of the Meson
    Theory by Dr. H. Yukawa) (1965) 637.
8. H. Matsumoto, N.J. Papastamatiou and H. Umezawa, Nucl. Phys.
    B82 (1974) 45.
9. H. Umezawa, "Self-Consistent Quantum Field Theory and Symmetry
    Breaking", from Renormalization and Invariance in Quantum
    Field Theory, Edited by E.R. Caianiello, Plenum Publishing
    Corporation, New York (1973).
10. H. Matsumoto and H. Umezawa, Fortsch. Phys. 24, (1976) 357.
    H. Matsumoto, M. Tachiki and H. Umezawa, Fortsch. Phys. 25
    (1977) 273.
    M. Tachiki and T. Koyama, Phys. Rev. B15 (1977) 3339.
    M. Tachiki and H. Umezawa, Phys. Rev. B15 (1977) 3332.
    F. Mancini, M. Tachiki and H. Umezawa, Physica 94B (1978) 1.
11. H. Matsumoto, H. Umezawa, S. Seki and M. Tachiki, Phys. Rev.
    B17 (1978) 2276.
12. H. Matsumoto, G. Semenoff, M. Tachiki and H. Umezawa, (to be
    published in Fortsch. Phys.).
13. C.N. Yang and D. Feldman, Phys. Rev. 79 (1950) 772.
14. L. Leplae, F. Mancini and H. Umezawa, Phys. Rev. B2 (1970)
    3594.
    H. Matsumoto, N.J. Papastamatiou and H. Umezawa, Nucl. Phys.
    B97 (1975) 90.
15. H. Matsumoto, P. Sodano and H. Umezawa, Phys. Rev. D19 (1979)
    511.
16. M. Wadati, H. Matsumoto and H. Umezawa, Phys. Rev. D18 (1978)
    520.
17. G. Oberlechner, M. Umezawa and Ch. Zenses, Lettere al Nuovo
    Cimento 23 (1978) 641.
18. R. Hirota, J. Phys. Soc. Japan 33 (1972) 1459.
19. H. Matsumoto, G. Oberlechner, H. Umezawa and M. Umezawa, J.
    Math. Phys. 20 (1979) 2088.
    As for the study of the quantum coordinate based on other
    methods, see
    J.L. Gervais and B. Sakita, Phys. Rev. D11 (1975) 2943.
    J.L. Gervais and A. Jevicki, Nucl. Phys. B110 (1976) 93.
    C.G. Callan, Jr. and D.J. Gross, Nucl. Phys. B93 (1975) 29.
20. N. Nakanishi, Prog. Theor. Phys. (Kyoto) 49 (1973) 640;
    50 (1973) 1388.
21. M.P. Anderson, Phys. Rev. 110 (1958) 827, 1900.
    P. Higgs, Phys. Rev. 45 (1960) 1156.
    T.W.B. Kibble, Phys. Rev. 155 (1967) 1554.
22. H. Matsumoto, N.J. Papastamatiou, H. Umezawa and G. Vitiello,

Nucl. Phys. $\underline{B97}$ (1975) 61.

23. M. Tachiki, H. Matsumoto and H. Umezawa, Phys. Rev. $\underline{B20}$ (1979) 1915.

24. M. Wadati, H. Matsumoto and H. Umezawa, Phys. Rev. $\underline{D15}$ (1978) 520.

25. L.P. Gor'kov, Zh. Eksperim. i Teor. Fiz. $\underline{36}$ (1959) 1918; $\underline{37}$ (1959) 1407 [Sov. Phys. JETP $\underline{9}$ (1959) 1364; $\underline{10}$, (1960) 998].

26. V.L. Ginzburg and L.P. Landau, Zh. Eksperim. i Teor. Fix. $\underline{20}$ (1950) 1064.

27. T. Koyama, M. Tachiki, H. Matsumoto and H. Umezawa, Phys. Rev. $\underline{B20}$ (1979) 1915.

28. H, Matsumoto, H. Umezawa and M. Tachiki, Solid State Comm. $\underline{31}$ (1979) 157.
M. Tachiki, A. Kotani, H. Matsumoto and H. Umezawa, Solid State Comm. $\underline{31}$ (1979) 927.

29. Y. Takahashi and H. Umezawa, Collective Phenomena $\underline{2}$ (1975) 55.
H. Matsumoto, Fortsch. Phys. $\underline{25}$ (1977) 1.

30. H. Matsumoto, P. Sodano and H. Umezawa, Phys. Rev. $\underline{D15}$ (1977) 2192.

31. H. Matsumoto, H. Umezawa, G. Vitiello and J.K. Wyly, Phys. Rev. $\underline{D9}$ (1974) 2806.

32. Y. Fujimoto and N.J. Papastamatiou, Nuovo Cimento $\underline{40}$ (1977) 468.

33. J. Joos and E. Weimar, Nuovo Cimento $\underline{32A}$ (1976) 283.
C. de Concini and G. Vitiello, Nucl. Phys. $\underline{B116}$ (1976) 141; Phys. Lett. $\underline{B70}$ (1977) 355.

34. E. Inonii and E.P. Wigner, Proc. Natl. Acad. Sci. U.S. $\underline{39}$ (1953) 510.

35. H. Matsumoto, N.J. Papastamatiou and H. Umezawa, Phys. Rev. $\underline{D13}$ (1976) 1054.

36. M. Hongoh, H. Matsumoto and H. Umezawa, "The Dynamical Rearrangement of Symmetry and the Group Contraction", Preprint, University of Alberta (1978).

37. E. Weimar, Acta Phys. Austrica $\underline{48}$ (1978) 201.

38. A. Chodos, R. Jaffe, K. Johnson, C. Thorn and V. Weisskopf, Phys. Rev. $\underline{D9}$ (1974) 3471; $\underline{10}$ (1974) 2599.

39. M. Wadati, H. Matsumoto and H. Umezawa, Phys. Rev. $\underline{D18}$ (1978) 1192.

Note Added in Proof 1

Note added in Proof (1): Let $\alpha(\vec{k})$ and $\alpha_i (i \neq 1, 2$ or $3)$ denote the annihilation operators in the physical field χ^o; $\alpha(\vec{k})$ for the scattering states and α_i for the bound states except the translation modes. The comment here is that $\alpha(\vec{k})$ is not the annihilation operator (say $a(\vec{k})$) in φ^o which is the physical field without extended objects. Indeed, $a(\vec{k})$ is a linear combination of $\alpha(\vec{k})$, $\alpha^\dagger(\vec{k})$ α_i, $\alpha_i^\dagger$ and $\vec{Q}$ $(i \neq 1, 2, 3)$, with coefficients being complicate functionals of the boson transformation parameter f.

Note Added in Proof (2)

Note added in Proof (2): The following results have been recently obtained in a study of quantum corrections of solitons (H. Matsumoto, G. Semenoff, H. Umezawa and M. Umezawa, preprint, University of Alberta, 1979).

(1)   When a theory without extended objects is renormalizable, so is the theory with extended objects. This had been proven in the approximation of one-loop correction. The recent proof does not need any approximation.

(2)   Though it had been known that the quantum coordinate $\vec{Q}$ and the position $\vec{x}$ always appear through the combination $\vec{x} + \vec{Q}$, nothing has been known about the appearance of $\vec{Q}$. Recently, the appearance of $\vec{Q}$ and $\dot{\vec{Q}}$ in the Heisenberg field ψ^f has been fully analysed.

(3)   It can be proven that the specific choice of f (and therefore, of the solution ϕ_o^f of the classical Euler equation) uniquely determines the dynamical map of the boson-transformed Heisenberg field ψ^f with all of the quantum corrections.

SYSTEMATIC METHODS FOR DETERMINING THE CONTINUOUS

TRANSFORMATION GROUPS ADMITTED BY DIFFERENTIAL EQUATIONS

Carl E. Wulfman

Department of Physics
University of the Pacific
Stockton, CA 95211

INTRODUCTION

This talk will present some recent results obtained by using Lie's systematic methods to uncover transformation groups admitted by several types of differential equations. We begin by sketching the methods.

1. Lie Algebras Admitted by Differential Equations

In order to be able to treat a variety of differential equations on the same footing let us consider a general set of equations

$$F^\nu(z) = 0, \ \nu = 1,,V \ \text{and} \ z = (z^1, , z^N),\qquad (1)$$

in which the F^ν are differentiable functions of the variables z^r. The variables z^r will be classified into "independent" variables, z^j, $j = 1, , n$ and <u>potential</u> "dependent" variables z^Λ, $\Lambda > n$. The potential "dependent" variables will be further classified into <u>potential</u> derivatives of order 0, 1, , M according to the following scheme and definitions:

Potential "dependent" variables of order zero:

$$z^\lambda, \ \lambda = n+1,,n+\eta;\qquad (2a)$$

Potential first derivatives $z^{\lambda j}$ defined by the Pfaffian equations

$$dz^\lambda = z^{\lambda j}dz^j \ \text{(summation convention)};\qquad (2b)$$

Potential m^{th} order derivatives $z^{\Lambda j}$ defined by the Pfaffian equations

$$dz^{\Lambda} = z^{\Lambda j} dz^{j}, \tag{2c}$$

where

$\Lambda = \lambda$, if $m = 1$, and otherwise $\Lambda = \lambda j^{m-1} . . j^{i}$, $m = 2, , M,$
with each j^{i} having the same range as j.
Equations (1) define V relations among the N variables z, and so
define a manifold or set of manifolds in an N-dimensional space.
Equations (2) define $\eta(n^{M}-1)/(n-1)$ relations between the variables
z and infinitesimal displacements dz and thereby define an Mth
order jet space. They restrict displacements on the manifold(s)
defined by (1). Any suitably smooth system of differential
equations (and many differentio-integral equations) may be written
in the form (1,2). If the equations possess sufficiently differentiable solutions the potential "dependent" variables and derivatives will take on fixed values when fixed values are assigned to
the "independent" variables and the initial and/or boundary value
data are fixed. In such cases functions g^{Λ} will exist such that

$$z^{\Lambda} \rightarrow g^{\Lambda}(z^{1}, ,z^{n}) \tag{3a}$$

and

$$z^{\Lambda j} \rightarrow \partial g^{\Lambda}/\partial z^{j} \equiv g^{\Lambda}_{j} . \tag{3b}$$

When the initial and/or boundary values are not fixed assigning
values to the z^{j}, $j = 1, ,n$, does not fix the value of a g^{Λ} and
even if the g^{Λ} are <u>assigned</u> values, the value of any $\partial g^{\Lambda}/\partial x^{j}$
remains indeterminate. The methods sketched here for determining
the Lie algebras admitted by the equations (1), (2) do not require
any investigation of the integrability of (1,2), but integrability
or differentiability--and other--conditions may be included in (1).

We shall however suppose that equations (1), (2) are such
that when they possess solutions the functions g^{Λ} take on real
values when real values are assigned to the z^{j}.[1] We shall also
suppose that equations (1) contain no repeated factors.[2] Both
those restrictions may be removed with little mathematical difficulty, but doing so would require extensive verbal elaboration of
the following discussions.[3]

Consider a one-parameter group of infinitesimal transformations of the space of the variables z into itself, and suppose it
defined in Lie's standard form

$$z^r \to \bar{z}^r = z^r + \delta a Z^r(z), \quad r = 1, \ ,N. \tag{4}$$

The Lie generator of the group is then

$$U = Z^r(z) \ \partial/\partial z^r. \tag{5}$$

Taken by themselves equations (1) are invariant under, or admit, the transformation (4) iff

$$F^v(\bar{z}) = 0, \quad v = 1, \ ,V \tag{6}$$

for all values of the variables z that satisfy (1). Substituting (4) into (6) one easily finds (6) holds iff for all such z one has

$$UF^v(z) = 0, \quad v = 1, \ , \ V. \tag{7}$$

Equations (2) taken by themselves are invariant under the transformation (4) iff

$$d\bar{z}^\Lambda = \bar{z}^{\Lambda j} d\bar{z}^j \tag{8a}$$

for all values of the variables satisfying

$$dz^\Lambda = z^{\Lambda j} dz^j. \tag{8b}$$

It is shown in the appendix that this requires

$$Z^{\Lambda k} = D_k(Z^\Lambda) - \sum_j z^{\Lambda j} D_k(Z^j), \quad j,k = 1, \ , \ n, \tag{9a}$$

where

$$D_k \equiv \partial/\partial z^k + \Sigma z^{\Lambda k} \ \partial/\partial z^\Lambda. \tag{9b}$$

Let us use the summation convention and also use the symbol $|$ to signify the imposition of the conditions required by the equations

$$F^v(z) = 0, \quad v = 1, \ ,V. \tag{10a}$$

Then it is easily seen that equations (1), (2) taken together are invariant under the transformation (4) iff

$$UF^v(z) \ | = 0, \quad v = 1, \ ,V \ , \tag{10b}$$

and for all Λ, k

$$\{Z^{\Lambda k} - D_k(Z^\Lambda) + z^{\Lambda j} D_k(Z^j)\}| = 0. \tag{10c}$$

These determining equations are always <u>linear</u> partial differential equations, and they always have the trivial solutions $Z^r = 0$, r = 1, , N, which defines the generator U of identity transformations. They may also yield a finite, a denumerably infinite, or a non-denumerably infinite number of linearly independent generators U.

If U_α and U_β are solutions of the determining equations then

$$U_\gamma = aU_\alpha + bU_\beta \quad \text{(a,b parameters)} \tag{11a}$$

and

$$U_\gamma = [U_\alpha, \; U_\beta] \tag{11b}$$

are also solutions. The generators admitted by equations (1), (2) consequently are elements of a Lie algebra of either finite or infinite dimension.

The U's possess the derivative property

$$U \, A(z) \, B(z) = (UA)B + A(UB). \tag{12}$$

2. Infinite-Order Contact Transformations

When the equations (1) involve more than one independent variable one generally finds that differentiating them yields more new variables $z^{\Lambda j}$ than new equations $F^{v'} = 0$, $v' > V$, so that the manifold(s) defined by the equations increase(s) in dimension. If the original set of V equations possesses solutions that are sufficiently differentiable, then these solutions also satisfy the consequent equations obtained by differentiation. There may therefore exist transformations admitted by the enlarged set and

such that all variables $\bar{z}$ are dependent upon the new variables $z^{\Lambda j}$ as well as the original variables z^r, $r = 1,$,N. This process of developing differential consequences of the original equations (1) and enlarging the space of the variables z may be continued indefinitely. The set of equations (2) then also becomes infinite.

One defines <u>infinite order contact transformations</u> as transformations

$$z^r \to \bar{z}^r \, (z),$$

$$r = j, \lambda, \Lambda; \quad j = 1, \,n; \quad \lambda = n{+}1, \, , \, n{+}\eta; \tag{13a}$$

$$\Lambda = \lambda j^{m-1} . \; . \; j^1; \quad m = 2,,, \; \infty$$

and

$$\Lambda = \lambda \, , \; m = j;$$

which have the property that if

$$dz^\Lambda = z^{\Lambda j} dz^j, \tag{13b}$$

then

$$d\bar{z}^\Lambda = \bar{z}^{\Lambda j} d\bar{z}^j. \tag{13c}$$

The equations define an infinite order jet space: the transformations have also been called Lie-Backlund.[4] For $U = Z^r (z)$ $\partial/\partial z^r$ to be the generator of an infinite order contact transformation it is necessary that for all Λ , k (cf appendix)

$$z^{\Lambda k} = D_k(Z^{\Lambda}) - z^{\Lambda j}D_k(Z^j). \tag{14}$$

A group of infinitesimal infinite-order contact transformations will leave invariant the set of equations

$$F^v(z) = 0, \; v = 1, \, , \, \infty \tag{15a}$$

iff for all v

$$UF^v \; \big| \; = 0 \tag{15b}$$

and for all Λ,k

$$\{z^{\Lambda k} - D_k(z^{\Lambda}) + z^{\Lambda j}D_k(Z^j)\}\big| = 0. \tag{15c}$$

As in the case of finite-order contact transformations, a set of differential equations that admits a set of generators U_α, admits the Lie algebra based upon the U_α.

3. Admission of Groups and Semi-Groups of Finite Transformations

Lie proved that a system of equations S will admit a one-parameter group or semi-group of finite transformations only if it admits the corresponding group of infinitesimal transformations.[5]
An operator $U = Z^r(z) \; \partial/\partial z^r$ that anhillates every member of the set of equations for all values of the variables allowed by the equations is always the generator of a local one-parameter group of infinitesimal transformations admitted by the equations. However it may not be the generator of a global one-parameter group of finite transformations admitted by the equations. To determine whether it is or is not such, one must carry out an investigation of the integrability of the transformation and of the differential equations of the set S. As such investigations depend critically and subtly upon the particular form of the set we will not here enter into them. Also there is as yet no well developed theory of the integration of infinite-order contact transformations. Hence we will sometimes proceed formally and suppose, unless information to the contrary is noted, that with each generator U admitted by our equations is associated a one-parameter group with operator exp aU, and that with each finite dimensional Lie algebra is associated a finite dimensional Lie group with operator $\exp a^i U_i$. In dealing with equations defined on Banach or Hilbert spaces it is possible to be more definite as the theory of one-parameter semi-group and group operations in such spaces is well developed.

When a set of differential equations admits an infinite dimensional Lie algebra, this may only generate a Lie pseudo-group. Infinite dimensional algebras may however contain finite dimensional subalgebras which generate finite-dimensional Lie groups of transformations.

4. Specializations

If one considers transformations which act on the dependent variables but do not change the independent variables then all $z^j = 0$, and

$$z^{\Lambda k} = D_k (z^{\Lambda}). \tag{16}$$

If, on the other hand, one considers transformations of the independent variables only, then all $z^{\lambda} = 0$, and

$$z^{\Lambda k} = z^{\Lambda j} D_k (z^j). \tag{17}$$

In any case, when $z^{\Lambda} \rightarrow g^{\Lambda}$, D_k becomes the total derivative operator for the independent variable z^k whether or not g^{Λ} is a unique $g(z)$.

Let $z^{\lambda} \rightarrow g^{\lambda}$, $z^{\Lambda k} \rightarrow g^{\Lambda}_k$, and let $F = z^{\lambda} - g^{\lambda}$. Also let

$U = U^o + Z^{\Lambda k} \partial/\partial z^{\Lambda k}$ with $Z^{\Lambda k}$ satisfying (14), and with

a.) $U^o \Rightarrow U^{oa} = z^j \partial/\partial z^j + z^{\lambda} \partial/\partial z^{\lambda}$; b.) $U^o \Rightarrow U^{ob} = \tilde{z}^{\lambda} \partial/\partial z^{\lambda}$

$$\tag{18}$$

where

$$\tilde{z}^{\lambda} = z^{\lambda} - z^{\lambda j} z^j.$$

Then $U^{oa} F = U^{ob} F$ and the generators U^a, U^b act equivalently on the space of all solutions of the differential equations. If one knows all $\tilde{z}^{\lambda}$ admitted by a system, one can use this result to obtain <u>all</u> U's admitted by it.[6]

If one has a set of linear differential equations with a single dependent variable z^{λ} then the Z^{Λ} (including the $Z^{\Lambda k}$) must be linear in the dependent variables z^{Λ}, including the potential derivatives. When the $z^{\Lambda} \rightarrow g^{\Lambda}$, whether g^{Λ} is a fixed function or not, it is possible to replace the generator U which acts on all the variables by a linear operator Q of the form

$$Q = q^o(z^1,,z^n) + q^j(z^1,,z^n) \, \partial/\partial z^j + q^{jk}(z^1,,z^n) \, \partial/\partial z^j \, \partial/\partial z^k$$

$$+ + + \tag{19}$$

The correspondence between the U and Q operators is not 1:1, for the family of U's with

$$U^o = (z^j + \Theta^j)\partial/\partial z^j + (z^{\lambda} - \Theta^j z^{\lambda j})\partial/\partial z^{\lambda} , \quad \Theta^j \text{ arbitrary}, \tag{20}$$

all give rise to the same Q.

When there is more than one dependent variable, the Q of (19) must be replaced by a matrix Q whose elements are differential operators.[7]

The Q's admitted by a set of linear differential equations are elements of a Lie algebra. However they do not in general act as derivations:

$$QAB \neq (QA)B + A(QB).$$

5. Algebras and Groups Admitted by Hamilton's Equations; A Theorem.[8]

Consider a classical mechanical system in the space of states and energy, PQET. Let z_c^i, $i = 1, 2, n$; $(z_c) = (PQET)$, be the cartesian coordinates of the system point, and let the Hamiltonian manifold M have defining equations $W^\sigma(z_c) = 0$, $\sigma = 1, 2, , S$. These might simply be $H - E = 0$.[9] We shall only concern ourselves with manifolds that are defined by functions that are at least twice continuously differentiable (C^2) functions.

The coordinates z_c may, as usual, be separated into two classes: "position" coordinates x_c^r, $r = 1, 2 \ldots n/2$, with $x_c = q, t$; and "momentum" coordinates y_c^r, $r = 1, 2, \ldots n/2$, with $y_c = p, - E$. The Poisson bracket of two functions $A(z)$ and $B(z)$ is defined by

$$\{A, B\}_{yx} = \sum_r \left(\frac{\partial A}{\partial y_c^r} \frac{\partial B}{\partial x_c^r} - \frac{\partial B}{\partial y_c^r} \frac{\partial A}{\partial x_c^r} \right) . \tag{21}$$

One then has the usual canonical relations

$$\{y_c^r, x_c^s\}_{yx} = \delta^{rs}, \quad \{y_c^r, y_c^s\}_{yx} = 0 = \{x_c^r, x_c^s\}_{yx}. \tag{22}$$

When no confusion is likely to result we will drop the subscript yx.

Consider now an arbitrary one-parameter group of diffeomorphisms $T(a)$, with a parameter a, defined by

$$z^i \rightarrow \bar{z}^i = F^i(\underset{\sim}{z},a), \tag{23}$$

$$z^i = F^i(\underset{\sim}{z},0) \, ,$$

and suppose that these are of class C^k, $k \geq 2$. We no longer require that the z's are cartesian coordinates. Let

$$Z^i(\underset{\sim}{z}) = \partial F^i(z,a)/\partial a \Big|_{a\,=\,0} \, . \tag{24}$$

Then the generator of the transformation T(a) is

$$U^i = Z^i(\underset{\sim}{z}) \, \partial/\partial z^i, \tag{25}$$

and the transformation T(a) is defined locally by

$$z^i \rightarrow \bar{z}^i = z^i + \delta a(\delta z^i/\delta a), \tag{26a}$$

where

$$\delta z^i/\delta a = Z^i(\underset{\sim}{z}) \, . \tag{26b}$$

The manifold M is invariant under the transformation T(a) if and only if

$$0 = UW^\sigma \Big|_M \, , \quad \sigma = 1, \, , \, s, \tag{27a}$$

that is to say, if and only if the tangency condition

$$Z^i(z)W^\sigma_i(z) = 0, \quad W^\sigma_i = \partial W^\sigma/\partial z^i, \tag{27b}$$

is satisfied everywhere on M. Letting $W = c_\sigma W^\sigma$ with the c_σ being arbitrary multipliers we write (27a) as

$$UW \Big|_M = 0 \tag{27c}$$

and will write (27b) as

$$Z^i(z) \, W_i(z) \Big|_M = 0 \, . \tag{27d}$$

Corresponding to the division of the coordinates into two classes we write,

$$U = \xi^r \partial/\partial x^r + \Pi^r \partial/\partial y^r \, , \quad r = 1, \, 2, \, .., \, n/2 \, . \tag{28}$$

A transformation will be canonical on M if and only if

$$\{\bar{z}^i, \bar{z}^j\}_{yx}\Big|_M = \{z^i, z^j\}_{yx}\Big|_M \tag{29}$$

Inserting (26) into (29) one finds immediately that T(a) is canonical on M, if and only if everywhere on M, and for all r and s,

$$\partial\xi^r/\partial x^s = -\,\partial\Pi^s/\partial y^r\ ,$$

$$\partial\xi^r/\partial y^s = \partial\xi^s/\partial y^r\ , \tag{30}$$

$$\partial\Pi^r/\partial x^s = \partial\Pi^s/\partial x^r\ .$$

Similarly one finds T(a) is canonical throughout PQET if equations
(30) are satisfied throughout PQET.

The trajectories of the system in PQET, all of which lie on M,
may be parameterized by a real variable ω. Hamilton's equations
then read[9]

$$\dot{z}^i - \{W, z^i\} = 0, \tag{31a}$$

where

$$\dot{z}^i = dz^i/d\omega\ . \tag{31b}$$

Let us define a <u>locally stable</u> transformation T(a) of M as one
satisfying the general requirement (27) and the further, highly
restrictive requirement,

$$0 = \partial/\partial z^j\ (Z^i(\underset{\sim}{z})W_i(\underset{\sim}{z}))\Big|_M\ ,\quad j = 1,2\ldots, m \tag{32}$$

If VB is C^∞ and δz is infinitesimal one then has

$$Z^i(\underset{\sim}{z} + \delta\underset{\sim}{z})W_i(\underset{\sim}{z} + \delta\underset{\sim}{z}) = 0\quad \text{if } W(\underset{\sim}{z}) = 0\ . \tag{33}$$

Thus, if U is the generator of a locally stable transformation of
M, the transformation carries not only the exact M, but also a
"thin shell" about M, into itself.

We now establish the following THEOREM:

A continuous group T(a) of diffeomorphisms of class C^k, $k > 2$,
canonical on a Hamiltonian defining manifold M of class $C^{k'}$, k'
> 2, leaves invariant Hamilton's equations of motion if and only
if T(a) is locally stable on M.

In supplying the proof, we note first that a transformation

in z space induces a transformation in the extended space $z,\dot{z}$ (c.f. appendix) whose generator $\hat{U}$, the prolongation of U, is

$$\hat{U} = U + \hat{Z}^i \partial/\partial \dot{z}^i \ . \tag{34}$$

As the transformation does not depend explicitly on ω one has

$$\hat{Z}^i = dZ^i/d\omega \ , \tag{35}$$

and if $\dot{z}$ satisfies (31) this becomes

$$\hat{Z}^i = \{W,z^i\} \quad . \tag{36}$$

Furthermore, for the equations of motion to be left invariant by T(a) it is necessary and sufficient that for all $\underset{\sim}{z},\underset{\sim}{\dot{z}}$ satisfying the equations of motion one has

$$\hat{U}(\underset{\sim}{\dot{z}} - \{W,\underset{\sim}{z}\} \ ) = 0 \ . \tag{37}$$

This condition is equivalent to the following conditions on U itself:

$$UW\Big|_M = 0 \ , \tag{38}$$

and

$$\{W,Z^i\} - U\{W,z^i\}\Big|_M = 0 \tag{39}$$

The first condition simply expresses the fact that the transformation must carry M into itself. The second condition, (39), follows from substituting (36) into (37). Now on writing equations (39) out in detail one finds they can be put in the following form

$$\Sigma_s \ ([\partial\xi^r/\partial x^s + \partial\Pi^s/\partial y^r]\partial W/\partial y^s - [\partial\xi^r/\partial y^s - \partial\xi^s/\partial y^r \ \partial W/\partial x^s$$

$$- \partial(UW)/\partial y^r)\Big| = 0, \quad r = 1,2,\ldots,n/2; \tag{40a}$$

$$\Sigma_s ([\partial\Pi^r/\partial x^s - \partial\Pi^s/\partial x^r]\partial W/\partial y^s - [\partial\Pi^r/\partial y^s + \partial\xi^s/\partial x^r]\partial W/\partial x^s$$

$$+ \partial(UW)/\partial x^r)\Big|_M = 0, \quad r = 1,2,\ldots,n/2 \ . \tag{40b}$$

The content of each square bracket vanishes for transformations canonical on M. Consequently, one sees that if U is the generator

of a transformation canonical on M, the transformation will be
an invariance transformation of the equations of motion if and
only if

$$Z^i(\underset{\sim}{z})W_i(\underset{\sim}{z})\Big|_M = 0, \tag{41a}$$

and

$$\partial/\partial z^j(\underset{\sim}{z})W_i(\underset{\sim}{z}))\Big|_M = 0, \quad j = 1,2,..,\eta .$$

This proves the theorem.

The physical and computational importance of the theorem
arises because the restriction (32) is a stability requirement.
As a one-parameter group of canonical transformations is associated
with a constant of motion G by the relations

$$\xi^r = \partial G/\partial y^s , \qquad \Pi^r = - \partial G/\partial x^r, \tag{42}$$

it follows from the theorem that one can determine the functions
G even when one is uncertain of the coordinates of the Hamiltonian
manifold by amounts δz^i.

6. Groups Of Infinite Order Contact Transformations Admitted By

    Schroedinger Equations.

    The Schroedinger equation

$$(H - i\partial/\partial t)\psi = 0 \tag{43}$$

is invariant under the infinitesimal transformation with generator
Q iff

$$(H - i\partial/\partial t)Q\psi\big| = 0. \tag{44}$$

We shall suppose the only boundary conditions are the usual ones
"at infinity." If an operator Q satisfies (44), then also

$$Q(H - i\partial/\partial t\psi\big| = 0, \tag{45}$$

and

$$(\partial/\partial t\, Q - i[H,Q])\psi\big| = 0. \tag{46}$$

Thus Q is the quantal analog of a classical constant of the motion
if it is a self-adjoint operator. In any case, (46) implies that

$$d/dt \int dv\, \psi_a^* Q\psi_b = 0. \tag{47}$$

If a Q satisfying (44) is not a function of t, then it corresponds to a classical first integral;

$$[H,Q]\psi\big| = 0. \tag{48}$$

Such a Q is a generator of a degeneracy group of the Schroedinger
equation.

When a Q satisfying (44) is an explicit function of t it
corresponds to a classical second integral, or "time-dependent
constant of motion."

If Q is self-adjoint under the usual Schroedinger scalar
product, then exp iaQ, a real, is unitary and converts a normalized solution of (43) into a normalized solution. One then knows
that the Q is actually a generator of a one-parameter Lie group
admitted by the Schrodinger equation.

Because Schroedinger's equation is a partial differential
equation one must in general expect that the Lie generators U
that it admits are those of infinite-order contact transformations. Consequently one must expect the operator Q to take
derivatives of arbitrarily large order, and one might expect
that determining the form of the Q's admitted by a Schroedinger
equation would be a hopeless task.

However both the quantal free particle and the quantal harmonic oscillator admit spectrum generating algebras whose operators
Q contain derivative operators of order two at most. In classical
physics one uses Hamilton-Jacobi techniques to map separable
integrable dynamical systems onto oscillator and free-particle
systems - bound trajectories being homeomorphic to oscillator
trajectories, and non-bound trajectories being homeomorphic to
free particle trajectories if one works with the phase space PQ.
Now quantal oscillator and free-particle systems have in common
the coordinate-independent property of possessing energy spectra
that are linear in quantum numbers that take on only non-negative
values:

$$E_{osc} = e^i n_i + e^o \equiv n + e^o, \quad n_i = 0,1,2, \, , \,$$

$$E_{fp} = e^i n_i \equiv n, \quad n_i = k_i, \quad -\infty < k_i < \infty.$$

We have used this fact in developing an algorithm that maps the
quantal analogs of separable integrable systems into quantal
systems in which $i\partial/\partial t$ has a linear spectrum even when the systems
possess both bound and unbound states.[10] In every case investigated we have found that the mapping yields a system that

admits a state generating algebra of Q's containing derivative
operators of order no higher than those in the differential equation obtained by the mapping, i.e., second order. Once the determining equations for these Q's have been solved, applying the
inverse mapping yields the Q's that provide a state generating
algebra for the Schroedinger equation of interest. The transformation algorithm is as follows:

Given a Schroedinger equation

$$(H - i\partial/\partial t)\psi = 0 \tag{50}$$

in which H has a known spectrum $E(n)$, $n = \vec{f}(n_1, n_2, \ ,\ )$, and

$$\Psi = \oint c(n_i)\Phi(n_i, x, t), \quad \Phi = \phi(n_i, x)\exp\text{-}iE_n t, \quad x = (x_1, x_2, \ ,\ ) \tag{51}$$

we seek a transformation

$$\Psi \rightarrow \widetilde{\widetilde{\Psi}} = \oint c(n_i)\widetilde{\widetilde{\Phi}}(n_i, x, t), \quad \widetilde{\widetilde{\Phi}} = \exp\ iV\ \Phi(n_i, x, t) \tag{52}$$

such that

$$i\partial/\partial t\ \widetilde{\widetilde{\Phi}}(n_i, x, t) = (n+e^o)\widetilde{\widetilde{\Phi}}, \quad \widetilde{\widetilde{\Phi}} = \widetilde{\widetilde{\phi}}(n, x)\exp -i(n+e^o)t \tag{53}$$

and such that

$$e^{iV}(H - i\partial/\partial t)e^{-iV} = F^1 F^2 \tag{54a}$$

$$F^2 \widetilde{\widetilde{\Psi}}\big| = 0. \tag{54b}$$

Here the operator F^2 contains no derivatives of negative order,
and the sign of evaluation, $\big|$, indicates that (54b) is to hold
true for all Ψ satisfying (50)

Now if Q is such that

$$e^{iV}(H - i\partial/\partial t)e^{-iV}\widetilde{\widetilde{Q}}\widetilde{\widetilde{\Psi}}\big| = 0 \tag{55}$$

one knows that

$$Q = e^{-V}\widetilde{\widetilde{Q}}e^{iV} \tag{56}$$

satisfies

$$(H - i\partial/\partial t)Q\Psi\big| = 0. \tag{57}$$

We let

$$\exp\ iV = \exp\ iV_s \exp\ iV_t \tag{58}$$

where V_t is of the form

$$V_t = -i\{\ln\left[\frac{\tilde{n}}{E(\tilde{n})}\right]\}t\partial/\partial t. \tag{59}$$

The operator $\tilde{n}$ is a function of H which satisfies

$$(\tilde{n}(H) - n)\Phi(n_i,x) = 0. \tag{60}$$

Writing

$$\tilde{\Psi} = e^{iV}t\Psi \tag{61}$$

and using the fact that

$$(\tilde{n} - i\partial/\partial t)\tilde{\Psi} = 0 \tag{62}$$

one finds that (50) becomes

$$\{H - E(\tilde{n}(i\partial/\partial t))\tilde{\Psi}| = 0. \tag{63}$$

If this equation contains no derivatives of negative order, then one sets $V_s = 0$, and $\tilde{\tilde{\Psi}} = \tilde{\Psi}$, and then proceeds directly to the determination of the Q's it admits. Otherwise one seeks a V_s of the form

$$V_s = -i\{\ln\tilde{n}\{g(\tilde{n}(i\partial/\partial t))^{-1}\}\}r\partial/\partial r + \text{const.} \tag{64}$$

and such that (63) takes on the form

$$F^1(x,\partial/\partial x,t, \partial/\partial t)F^2(x,\partial/\partial x,t,\partial/\partial t)\tilde{\tilde{\Psi}} = 0, \tag{64}$$

where F^2 has no negative order derivative operators, and the set of solutions Ψ of the equation

$$F^2\tilde{\tilde{\Psi}} = 0 \tag{66}$$

yields all the solutions

$$\Psi = e^{-iV}\tilde{\Psi} \tag{67}$$

of the equation (50).

The algorithm sketched here has been used to find state generating algebras admitted by the hydrogen-like atom in one, two, and three dimensions, by one-dimensional systems with potentials kx, kx^{-2}, cx^2, $cx^{-2} + kx^2$, by Gegenbauer's equation, by the two-dimensional harmonic oscillator, and by the rigid

rotator and symmetric top.$^{(10a,b,c,d)}$ The algebra of the top
was not previously known, and the realizations obtained for
the other systems were, with the exception of the oscillator and
free particle, previously unknown – for in every case the operators obtained as generators of invariance transformations of
Schroedinger's equation are the quantal operators representing
classical constants of motion.

7. Limit Cycles As Invariant Functions Of Lie Groups $^{(11)}$

Consider a set of autonomous ordinary differential equations
that define a differentiable flow:

$$\dot{x}^i - f^i(x) = 0 \ , \ i - 1, \ ,N \tag{68}$$

$$\dot{x}^i = dx^i/dt, \ -\infty < t < \infty \ .$$

Let a solution S_o of (68) be defined by

$$g_o^i(x,t) = 0, \ i = 1, \ ,N \tag{69}$$

it being supposed that the functions g_o^i are differentiable. If
there exists a set of solutions of (68) definable by

$$g^i(x,t;\delta a) \equiv g_o^i(x,t) + \delta a g_1^i(x) = 0, \ i = 1, \ ,N \tag{70}$$

where δa is an arbitrary infinitesimal parameter, then we may say
that S_o is a member of a local one-parameter family of solutions
$S(\delta a)$. A local one-parameter family of trajectories, $T(\delta a)$, perhaps corresponging to $S(\delta a)$, may be defined on the phase space $\{x\}$
by

$$h^i(x;\delta a) \equiv h^i(x) + \delta a h_1^i(x) = 0, \ i = 2, \ , \ N. \tag{71}$$

Again assuming h^i are differentiable we may write (70) as

$$g^i = (1 + \delta a W)g^i, \ \ W = \Omega^j(x,t)\partial/\partial x^j + T(x,t)\partial/\partial t \tag{72}$$

and in (71)

$$h^i = (1 + \delta a X)h^i, \ \ X = \chi^j(x)\partial/\partial x^j. \tag{73}$$

Thus we may always suppose that the families (70), (71) are
obtained by the action of a one-parameter family of first-order
differential operators acting on one of their members. However,
though the solution curves $S(\delta a)$ are all diffeomorphic to the
real line, the corresponding trajectories $T(\delta a)$ need not be
diffemorphic to each other. Let us make the following DEFINITION:

A differentiable trajectory T_o is an <u>exceptional</u> trajectory if (68) admits no local one-parameter group of diffeomorphisms of phase space that converts T_o into a one parameter family of trajectories of (68).

If a trajectory is an exceptional one there are no neighbouring trajectories diffeomorphic to it. Because equations (68) determine a flow, if one establishes that a T_o is an exceptional trajectory, one immediately knows that T_o is either a limit cycle or a "limit line". T_o will be a limit cycle iff the one-parameter group generated by U_o is a compact transformation group. It will be a limit line iff the group generated by U_o is a non-compact transformation group.

Now equations (68) will admit a group of diffeomorphisms of $\{x\}$ with generator U iff the determining equations for U are satisfied. These may be written in the form

$$[U,V] = 0 \tag{74}$$

where

$$U = \xi^i(x)\partial/\partial x^i, \tag{75}$$

and

$$V = f^i(x)\partial/\partial x^i \tag{76}$$

is the generator of the evolution group of (68).

Suppose U_o, U_1 satisfy equations (74). U_o will have T_o as an invariant function iff $\underset{\sim}{\xi}_o = (\xi_o^1, \xi_o^2, , \xi_o^N)$ is everywhere on T_o tangent to T_o. U_1 will generate a one-parameter family of trajectories containing T_o iff $\underset{\sim}{\xi}_1 = (\xi_1^1, \xi_1^2, , \xi_1^N)$ is not tangent to T_o. This will be true iff there is no function $\phi(x)$ s.t.

$$\underset{\sim}{\xi}_1(x) = \phi(x)\underset{\sim}{\xi}_o(x). \tag{77}$$

As every trajectory of (68) is an invariant function of the evolution operator, we may set $U_o = V$ and thereby establish the following THEOREM:

Let U be the generator of a one-parameter group of diffeomorphisms that is admitted by a set S of autonomous ordinary differential equations defining a flow with evolution operator V. Let T be
a trajectory defined by these equations. Then T is an exceptional
trajectory if and only if the determining equations derived from
S have no solution yielding a U that is functionally independent
of V on T.

The exceptional trajectory will be a limit cycle if and only
if the transformation group with generator U is compact.

To exemplify the theorem consider the equations

$$\dot{\theta} = 1, \quad \dot{r} = r - r^3 \tag{78}$$

which have the limit cycle trajectory $r = 1$ when r, θ, are polar
coordinates. The general solution of the determining equations is
an arbitrary linear combination of U's with

$$z^\theta = r^\alpha (r^2 - 1)^{-\frac{1}{2}\alpha} \exp -\alpha\theta \quad , \quad \alpha \text{ an arb. const.,} \tag{79}$$

$$z^r = r^{\beta+1} (r^2 - 1)^{1-\frac{1}{2}\beta} \exp -\beta\theta \quad , \quad \beta \text{ an arb. const.} \tag{80}$$

The requirement that they define a real local diffeomorphism reduces
U to the form

$$U = a \, \partial/\partial\theta + br(r^2 - 1)\partial/\partial r \quad , \quad a, b \text{ real constants.} \tag{81}$$

When $r = 1$, the only non vanishing U is the rotation generator, in
agreement with the theorem and the observation that $r = 1$ is an
exceptional trajectory and a limit cycle.

8. Applications To Non-linear Partial Differential Equations With

    Soliton Solutions

Kumei has pioneered the systematic analysis of the group
properties of non-linear partial differential equations which
have soliton solutions. He investigated the sine-Gordon equation,[12a]
the Korteweg-deVries equation,[12b] and the cubic Schroedinger
equation.[12c] In each case he found solitons to be invariant
solutions of one-parameter groups of infinite order contact transformations. He showed that the series of known conservation laws
obeyed by these equations exist as a direct consequence of the
existence of these groups, and he obtained new conservation laws
for the sine-Gordon equation by uncovering their associated invariance groups. Finally, Kumei was able to give a general, _and_

<u>purely</u> <u>local</u>, discussion of the relation between the generators of
groups of Lie-Backlund transformations admitted by PDE's that may
be put in Hamiltonian form, and conservation laws obeyed by the
equations.(12c) Ibragimov has given a discussion of these relations
from a Lagrangian standpoint using Noether's theorem.(13)

We will conclude this talk with an example taken from Kumei[12a]
which nicely illustrates the application of the general theory:
we consider the sine-Gordon equation

$$y_{xt} = \sin y .\tag{82}$$

Using the notation of Section 1 we let $j = 1,2$; $\lambda = 3$, and
make the identifications $z^1 = x$, $z^2 = t$, $z^3 = y$. We then replace
(82) by the equations

$$dz^3 = z^{31}dz^1 + z^{32}dz^2 \tag{83}$$

$$dz^\Lambda = z^{\Lambda 1}dz^1 + z^{\Lambda 2}dz^2$$

and

$$z^{312} = \sin z^3 . \tag{84}$$

Since we are interested in Lie-Backlund transformations we supple-
ment this last equation with its differential consequences:

$$z^{3121} = z^{31}\cos z^3 \tag{85}$$

$$z^{3122} = z^{32}\cos z^3$$

$$\vdots$$

It is possible to impose the integrability conditions

$$z^{\Lambda ij} = z^{\Lambda ji} . \tag{86}$$

Let the generator of the transformation of interest be

$$U = U^o + Z^{\Lambda k}\partial/\partial z^{\Lambda k}$$

where

$$\tag{87}$$

$$U^o = Z^3 \partial/\partial z^3 .$$

Then

$$Z^{\Lambda k} = D_k(Z^\Lambda) . \tag{88}$$

Taking into account (85) and (86) one sees that one will be dealing with the most general possible U^O depending upon potential derivatives of order no greater than three if one lets Z^3 be a function of z^3, z^{31}, z^{32}, z^{311}, z^{322}, z^{3111}, z^{3222}. Finally, one sees the equations (83), (84), (85) admit the generator U iff

$$z^{312} - z^3 \cos z^3 \Big| = 0 \tag{89}$$

and

$$D_1 D_2 z^3 - z^{312} \Big| = 0 . \tag{90}$$

Kumei showed that these equations have the four linearly independent solutions:

$$Z^3_a = z^{31}, \qquad Z^3_b = z^{32}$$

$$Z^3_c = z^{3111} + \tfrac{1}{2}(z^{31})^3 \tag{91}$$

$$Z^3_d = z^{3222} + \tfrac{1}{2}(z^{32})^3.$$

He then established that they give rise to conservation laws

$$\partial f^1/\partial z^1 + \partial f^2/\partial z^2 = 0 \tag{92}$$

with the following fluxes $\underset{\sim}{f} = (f^1, f^2)$:

$$\underset{\sim}{f}_a: \quad (z^{32}z^{312}, \, -z^{31}\sin z^3), \quad (z^{32}z^{322}, \, -z^{32}\sin z^3) \tag{93}$$

$$\underset{\sim}{f}_b: \quad (\{z^{31112} + \tfrac{3}{2}(z^{31})^2 z^{312}\}z^{32}, \, -\{z^{3111} + \tfrac{1}{2}(z^{31})^3\}\sin z^3)$$

$$\underset{\sim}{f}_c: \quad (\{z^{32222} + \tfrac{3}{2}(z^{32})^2 z^{322}\}z^{32}, \, -\{z^{3222} + \tfrac{1}{2}(z^{32})^3\}\sin z^3) .$$

Kumei was not only the first to establish the relationship between conservation laws such as (92) and the transformation groups admitted by the governing PDE--he was also the first to show that the solutions governed by these laws were invariant functions of the group and consequently easily calculable.

In so doing Kumei also established that a particularly simple class of infinite order contact transformations is of fundamental importance in the study of many non-linear PDE's: these are transformations whose generators depend explicitly on only a few derivatives of order higher than those in the original PDE. Such generators are particularly easy to uncover by the systematic means pioneered by Lie.

Kumei showed how to exponentiate such generators to iteratively determine the action of the corresponding group operator. However, the problem of obtaining the finite transformations in closed form by integration of the characteristic equations of the group is still unsolved, primarily because the equations are infinite in number.

There is now much interest in translating the study of infinite order contact transformations from the language of Lie into the language of differential topology (jet spaces, tangent bundles, cohomologies, etc.). It is to be hoped that this will prove of aid in solving the fundamental problem of integrating the characteristic equations of groups of these transformations.

APPENDIX 1

The Transformation Law for Derivatives and Ratios of Arbitrary Displacements: Contact Transformations of Arbitrary Order.

We wish to establish the relation between the transformation law for variables z and the transformation law for

$$\frac{dz^\Lambda}{dz^j} \Bigg|_{dz^{j'} = 0, \; j' \neq j}. \tag{A1}$$

When a functional relation is established between z^Λ and z^j this ratio is just the partial derivative $\partial z^\Lambda/\partial z^j$. We anticipate this possibility in our notation by dividing the z variables into classes: potential dependent variables z^Λ, $\Lambda > n$, and independent variables z^j, $j \leq n$. The z^Λ may themselves become derivatives of arbitrary order, and we will allow the transformation law to depend both upon the z^j and the z^Λ. In the derivation which follows we modify the argument of Ibragimov and Anderson[4] so as to establish that the law of transformation of (A1) does not depend upon the ratio being a derivative. This enables us to determine the algebras admitted by differential equations without establishing integrability and uniqueness conditions for the differential equations.

Consider the equation

$$dz^\Lambda - \sum_j z^{\Lambda j} dz^j = 0, \; j = 1, 2, , , n; \; \Lambda > n. \tag{A2}$$

We suppose the variables $z^{\Lambda j}$ to be members of the set of variables z^Λ. Let us carry out a transformation of the variables z: $z \to \overline{z}$, and require that the transformation leaves (A2) invariant, i.e., that

$$\left. d\overline{z}^{\Lambda} - \Sigma_j \overline{z}^{\Lambda j} d\overline{z}^j \right| = 0. \tag{A3}$$

If we let

$$z^r \rightarrow \overline{z}^r = z^r + \delta a Z^r(\underset{\sim}{z}), \qquad r = j, \Lambda \tag{A4}$$

then

$$dz^r \rightarrow d\overline{z}^r = dz^r + \delta a Z^r_s \, dz^s, \tag{A5}$$

where

$$Z^r_s = \partial Z^r / \partial z^s. \tag{A6}$$

Equations (A3), (A5) then yield:

$$\left. \{ Z^{\Lambda}_s dz^s - \Sigma_j (z^{\Lambda j} Z^j_s dz^s + z^{\Lambda j} dz^j) \} \right| = 0. \tag{A7}$$

Separating the range of the variable s into $s \leq n$, $s > n$, and label-ing $s = k$ in the first range and $s = \ell$ in the second, we obtain, after making a change in dummy indices in the last term on the right of (A7):

$$\left. \Sigma_k \{ (Z^{\Lambda}_k - \Sigma_j z^{\Lambda j} Z^j_k - z^{\Lambda k}) dz^k + (Z^{\Lambda}_\ell - \Sigma_j z^{\Lambda j} Z^j_\ell) dz^\ell \} \right| = 0. \tag{A8}$$

Now since $\ell > n$ we must have by (A2)

$$dz^\ell = \Sigma_k z^{\ell k} \, dz^k.$$

Consequently (A8) becomes

$$\Sigma_k \{ -z^{\Lambda k} + (Z^{\Lambda}_k + Z^{\Lambda}_\ell z^{\ell k}) - \Sigma_j z^{\Lambda j} (Z^j + Z^j_\ell z^{\ell k}) \} dz^k = 0. \tag{A9}$$

Making use of the independence of the dz^j we find for each value of Λ and each value of j

$$z^{\Lambda j} = Z^{\Lambda}_j + Z^{\Lambda}_\ell z^{\ell j} - \Sigma_k z^{\Lambda k} (Z^k_j + Z^k_\ell z^{\ell j}). \tag{A10}$$

Defining the operator

$$D_j = \partial / \partial z^j + z^{\ell j} \partial / \partial z^\ell, \quad j = 1, \dots, n; \; \ell = n+1, \, n+2, \, \dots \tag{A11}$$

we have

$$z^{\Lambda j} = D_j(Z^{\Lambda}) - \Sigma_k z^{\Lambda k} D_j(Z^k), \quad j = 1, \, \dots, \, n. \tag{A12}$$

Thus when

$$z^j \rightarrow \bar{z}^j = z^j + \delta a Z^j, \quad j = 1, \ldots, n \qquad\qquad\text{(A13)}$$

$$z^\Lambda \rightarrow \bar{z}^\Lambda = z^\Lambda + \delta a Z^\Lambda, \quad \Lambda > n,$$

and

$$z^{\Lambda j} \rightarrow \bar{z}^{\Lambda j} = z^{\Lambda j} + \delta a Z^{\Lambda j},$$

we must allow $Z^{\Lambda j}$ to be restricted by (A12).

Finally we note that when the Z^j, Z^λ are functions only of
the variables z^j, z^λ the contact transformation defined by (A12),
(A13) is termed an extended point transformation, the space of the
variables z^j, z^λ, z^Λ is termed the extension or prolongation of
the space of z^j, z^λ, and the generator U is said to be the prolongation of the generator.

$$U^o = Z^j\, \partial/\partial z^j + Z^\lambda\, \partial/\partial z^\lambda. \qquad\qquad\text{(A14)}$$

REFERENCES

1. The variables z^r take on hyperreal values as well, since
    $z^r + dz^r$ is also a z^r: c.f.e.g., K. Stroyan, W. A. J.
    Luxemburg, Introduction to the Theory of Infinitesimals,
    Academic Press, NY, 1976.
2. c.f. A. Cohen, An Introduction to the Lie Theory of One-parame-
    ter Groups, Stechert, NY, 1931, pp. 16-23.
3. For Lie-Backlund transformations of PDE's with complex varia-
    bles see S. Kumei, J. Math. Phys., 18, 256, (1977).
4. N. Ibragimov, R. L. Anderson, J. Math. Anal. & Appl., 59, 145,
    (1977).
5. S. Lie, Differentialgleichungen, Leipzig, 1891, reprinted,
    Chelsea, NY, 1967; pp. 299-305.
6. This seems to have first been recognized by Kumei (unpub. 1974).
7. For an example see T. Shibuya, C. Wulfman, Rev. Mex. Fis. 22,
    171 (1973).
8. C. Wulfman, T. Sumi in Atomic Scattering Theory, J. Nuttall,
    ed., U. of Western Ontario, London, Ont., 1978; pp. 197-
    202. See Also C. Wulfman, Dynamical Groups in Atomic and
    Molecular Physics, in Recent Advances in Group Theory and
    Their Application to Spectroscopy, J. Donini, ed., Plenum,
    NY, 1979.
9. c.f. J. L. Synge, Classical Physics, in Encyclopedia of Physics,
    S. Flugge, ed., Vol. III/1, Springer, Berlin, 1960.
10. R. L. Anderson, S. Kumei, C. Wulfman; a.) Phys. Rev. Lett., 28,
    988, 1972; b.) Rev. Mex. Fix. 21, 1, (1972); c.) Rev. Mex.
    Fis. 21, 35, (1972); d.) J. Math. Phys. 14, 1527 (1973).

11. C. Wulfman, J. Phys. $\underline{A12}$, L73, (1979).
12. S. Kumei, a.) J. Math. Phys., $\underline{16}$, 2461, (1975); b.) ibid., $\underline{18}$, 256, (1977); c.) ibid. $\underline{19}$, 195, (1978).
13. N. H. Ibragimov, Lett. in Math. Phys. $\underline{1}$, 423, (1977).

SYMMETRY BREAKING IN EMBRYOLOGY AND IN NEUROBIOLOGY

J. D. Cowan

Department of Biophysics and Theoretical Biology
The University of Chicago
Chicago, IL 60637

INTRODUCTION

As Sattinger (1980) has emphasized, there is an intimate
connection between the symmetries of a dynamical system, and the
nature of the solutions manifest at a point of bifurcation. Many
aspects of symmetry breaking in biology can be interpreted in
such terms. In this paper I shall confine myself to differing
aspects of symmetry-breaking in embryology and in neurobiology.

The English mathematician Turing (1952) first formulated such
a process of biological symmetry-breaking in terms of the properties of diffusion coupled chemical reactions. The analysis of
symmetry-breaking in such systems can be carried out via bifurcation theory (Matkowsky, 1970; Auchmuty and Nicolis, 1975; and
Nicolis and Prigogine, 1977). Consider the nonlinear diffusion
coupled reaction,

$$\partial_t c - D \cdot \nabla\nabla c = \mu R(c) \tag{1}$$

in which c is a vector of concentrations, R(c) a vector of reaction
rates with coupling parameters μ, and D the diffusion coefficient,
a tensor spatially, but a matrix with respect to concentration
indices. In general there is also a convection term $v \cdot \nabla c$, but I
shall ignore this. Given appropriate initial and boundary conditions, eqn. (1) is to be solved for R(c) of the general form
shown in fig. 1.

For example, consider the problem on a finite rectangular
enclosure of sides L_x and L_y respectively, with Neumann boundary
conditions $\nabla c = 0$, and an initial perturbation $c(\underline{r}, 0 : \varepsilon)$

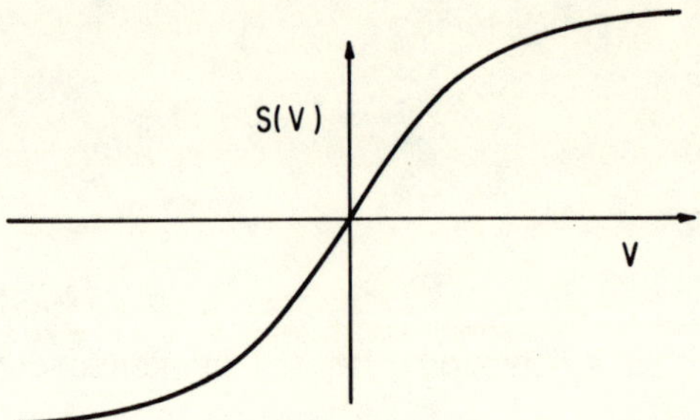

Fig. 1. A typical nonlinear reaction rate vs. chemical
concentration relation.

asymptotically equivalent to $\rho_o + \varepsilon\rho(\underline{r})$, where $\nabla\rho = 0$ on the
boundaries of the enclosure. For these conditions the stationary
solution $c = c^o$ is unique and stable for values of $\mu < \mu_c$, but
becomes unstable for $\mu \geq \mu_c$, and is displaced by nonuniform solutions
Thus $\mu = \mu_c$ is said to be a point of bifurcation. A stability
analysis of the linearized problem (Cohen, 1971; Lacalli and
Harrison, 1978),

$$\partial_t c - D\cdot\nabla\nabla c = \mu R^{(1)}(c^o)\cdot c \tag{2}$$

leads to the asymptotic solution

$$c(\underline{r},t;\varepsilon) - c^o \sim \varepsilon \sum_n \rho_n e^{\sigma_n t} \cos(\underline{k}_n \cdot \underline{r}) \tag{3}$$

where ρ_n are the Fourier coefficients of $\rho(\underline{r})$, where $\det[\sigma - R^{(1)}$
$(c^o)(\mu-\mu_n)] = 0$, $\mu_n = [R^{(1)}(c^o)]^{-1}D(\underline{k}_n \cdot \underline{k}_n)$, $\underline{k}_n \cdot \underline{k}_n = |\underline{k}_n|^2 = \pi^2$
$[(n_x/L_x)^2 + (n_y/L_y)^2]$. It follows that the stationary solution c^o
becomes unstable at the point $\mu = \mu_c = \min \mu_n > 0$, $= [R^{(1)}(c^o)]^{-1}$
$\min[D(\underline{k}_n \cdot \underline{k}_n)]$, and that the subsequent perturbation grows at a rate
proportional to $(\mu-\mu_c)$, until the nonlinear nature of $R(c)$ begins
to have an effect: thereafter the perturbation grows at some rate

$(\mu-\mu_c)^\alpha$ until the neighborhood of a new stationary solution is reached. Thus there are several time scales in the process, and one of the main problems is to find the correct scales.

These can be obtained as follows (Matkowsky, 1970; Gordon and Hoppenstaedt, 1975). Let $\varepsilon = (\mu-\mu_c)^\alpha$, $\tau = (\mu-\mu_c)t$, and assume that:

$$c(\underline{r},t;\varepsilon) - c^o \sim \sum_i w_i(\underline{r},t;\tau)\varepsilon^i \tag{4}$$

On substituting (4) into (1) and comparing powers of ε, a hierarchy of _linear_ equations is obtained. The form of this hierarchy depends upon α. The correct choice of α depends in turn on the nature of the solutions to the stationary bifurcation problem (Sattinger, 1980; Nicolis and Prigogine, 1977):

$$D\cdot\nabla\nabla c + \mu R(c) = 0 \tag{5}$$

which can be solved by assuming that (Busse, 1978):

$$c(\underline{r},\infty) - c^o \sim \sum_i \bar{w}_i(\underline{r})\varepsilon^i, \tag{6}$$

and

$$\mu - \mu_c \sim \sum_i \mu_i\varepsilon^i. \tag{7}$$

For the given boundary conditions, solutions of the form $\cos(\underline{k}_n\cdot\underline{r})$ obtain for certain $\underline{k}_n$, and $\alpha = 1/2$, whence the hierarchy,

$$[\partial_t - D\cdot\nabla\nabla - \mu_c R^{(1)}(c^o)]w_1 \equiv L(\mu_c)w_1 = 0, \tag{8}$$

$$L(\mu_c)w_2 = \frac{\mu_c}{2}R^{(2)}(c^o)w_1w_1, \tag{9}$$

$$L(\mu_c)w_3 = -\partial_\tau w_1 + R^{(1)}(c^o)w_1 + \mu_c R^{(2)}(c^o)w_1w_2$$

$$+ \frac{\mu_c}{6}R^{(3)}(c^o)w_1w_1w_1, \tag{10}$$

$$\cdot$$

$$\cdot$$

$$L(\mu_c)w_i = r_i(w_1,\ldots,w_{i-2}), (i>2), \tag{11}$$

with initial and boundary conditions

$$w_1(\underline{r},o) = \rho(\underline{r}), \quad w_i(\underline{r},o) = 0 \quad (i>2), \tag{12}$$

$$\nabla w_i = 0, \text{ on all boundaries.} \tag{13}$$

The solution to eqn. (8) can be written as,

$$w_1(\underline{r},t;\tau) = \sum_n U_n a_{1n}(\tau) e^{\sigma_n t} \cos(\underline{k}_n \cdot \underline{r}) \tag{14}$$

where $\sigma_n = 0$ for $n = n_c$, the critical wave number corresponding to $\min[D(\underline{k}_n \cdot \underline{k}_n)]$, and $\sigma_n < 0$ otherwise. Thus (14) can be written as

$$w_1 = U_{n_c} a_{1n_c}(\tau) \cos(\underline{k}_{n_c} \cdot \underline{r}) + \text{decaying modes.} \tag{15}$$

To solve for $a_{1n_c}(\tau)$ it suffices to note that eqns. (8)-(11) are linear, and possess bounded solutions iff (Matkowsky, 1970),

$$\lim_{T \to \infty} \frac{1}{T} \int_0^T \int_A dt d\underline{r} w_1 r_i \equiv \overline{\langle w_1, r_i \rangle} = 0, \tag{16}$$

the well-known Fredholm alternative property (Sattinger, 1980).
On applying this solvability criterion to eqns. (9)-(11), the
bifurcation equations,

$$\frac{\mu_c}{2} R^{(2)}(c^o) \overline{\langle w_1, w_1 w_1 \rangle} = 0, \tag{17}$$

$$\overline{\langle w_1, \partial_\tau w_1 \rangle} = R^{(1)}(c^o) \overline{\langle w_1, w_1 \rangle} + \mu_c R^{(2)}(c^o) \overline{\langle w_1, w_1, w_2 \rangle}$$
$$+ \frac{\mu_c}{6} R^{(3)}(c^o) \overline{\langle w_1, w_1 w_1 w_1 \rangle} \tag{18}$$

$$.$$

$$.$$

appear, subject to the boundary conditions (12) and (13). Eqn.
(17) is satisfied identically and does not determine $a_{1n_c}(\tau)$, but
it does show that all the decaying modes give zero contribution.
This is the origin of the adiabatic elimination method introduced
by Stuart and Watson (1960) in fluid convection problems, and

(independently) by Haken (1975) in laser physics, and formalized by Matkowsky (1970). Eqns. (9), (15) and (18) then lead to an equation for $a_{1n_c}(\tau)$,

$$\partial_\tau a_{1n_c}(\tau) = \lambda a_{1n_c}(\tau) + a_o \left| a_{1n_c} \right|^2 a_{1n_c}(\tau)$$

$$+ \sum_m b_{mn_c} \left| a_{1m_c} \right|^2 a_{1n_c}(\tau), \qquad (19)$$

where $\quad \lambda = R^{(1)}(c^o) \qquad (20)$

$$a_o = \frac{1}{8}\mu_c U_n U_n \left[2\mu_c (R^{(2)}(C^o))^2 \left(\frac{1}{\sigma(0)} + \frac{1}{\sigma(2k_n)} \right) \right.$$

$$\left. + \frac{1}{3} R^{(3)}(c^o) \right] \qquad (21)$$

$$b_{mn_c} = \frac{1}{8}\mu_c U_m U_n \left[\mu_c (R^{(2)}(c^o))^2 \frac{2}{\sigma(k_m + k_n)} \right.$$

$$\left. + \frac{1}{3} R^{(3)}(c^o) \right] \qquad (22)$$

Eqn. (19) will be recognized as a form of the Landau-Ginzburg equation. It is interesting to examine the solution of this equation in the case of single plane-waves or <u>rolls</u>, in which case the sum over k_{m_c} vanishes, and eqn. (19) can be solved exactly as,

$$a_{1n_c}(\tau) = \pm \left[\rho_{n_c}^{-2} e^{-2\lambda\tau} - \frac{a_o}{\lambda}(1 - e^{-2\lambda\tau}) \right]^{-\frac{1}{2}} \qquad (23)$$

where ρ_{n_c} is the Fourier coefficient of the perturbation at the critical wavelength k_{n_c}. It follows that as τ tends to ∞, $a_{1n_c}(\tau)$ tends to the constant $\pm\sqrt{(\lambda/-a_o)}$, and w_1 to the constant solution,

$$\pm U_{n_c} \sqrt{(\lambda/-a_o)} \cdot \cos(k_{n_c} \cdot r) \qquad (24)$$

a stationary state whose dependence on the initial perturbation is one of sign only (Matkowsky, 1970). Similar conclusions obtain for higher order terms $a_{1n_c}(\tau)$ in (4): in fact each a_{in} approaches a constant which depends only upon $a_{1n_c}(\infty) = \pm\sqrt{(\lambda/-a_o)}$, so as τ tends to ∞, the entire expansion describes a stationary state that is independent of the initial perturbation. It follows also from eqn. (19) that for $a_{1n_c}(\tau) \lessgtr a_{1n_c}(\infty)$, $d_\tau a_{1n_c}(\tau) \gtrless 0$, so the stationary state is stable, and takes the form,

$$c(\underline{r},\infty;\varepsilon)-c^{o} \sim \pm\, U_{n_c}\, \sqrt{(\lambda(\mu-\mu_c)/-a_o)}\cos(\underline{k}_{n_c}\cdot\underline{r}) \qquad (25)$$

$$+\, 0(\varepsilon^2)$$

This result corresponds to that obtained by stationary bifurcation theory (Sattinger, 1980; Nicolis and Prigogine, 1977).

Similar conclusions obtain for the existence of <u>cross-rolls</u>, i.e., squares and rectangles satisfying the boundary conditions, in which case eqn. (19) obtains in its full form. Although the time-dependent case cannot be solved explicitly, steady state solutions can be obtained as in the case of simple rolls, in the form,

$$c(\underline{r},\infty;\varepsilon)-c^{o} \sim \pm\, U_{n_c}\, \sqrt{(\lambda(\mu,\mu_c)/-a_o-b_{mn})}\cdot[\cos(\underline{k}_{n_c}\cdot\underline{r})$$

$$+\, \cos(\underline{k}_{m_c}\cdot\underline{r})]\, +\, 0(\varepsilon^2) \qquad (26)$$

where $\underline{k}_n\cdot\underline{k}_{m_c} = 0$, and where b_{mn} is given by eqn. (22). It can be shown that the relative stability of rolls vs. cross-rolls is determined by the magnitudes of the coefficients a_o and b_{mn}. Thus rolls are stable and cross-rolls unstable, whenever $b_{mn} < a_o < 0$; whereas cross-rolls are stable and rolls unstable whenever $a_o \pm b_{mn} < 0$.

It is of interest to compare such stability properties with those of the solutions on an infinite planar domain. In such a case the Fredholm alternative property does not generally hold (Sattinger, 1980), but it will be assumed here. In such a case solutions of the form $\cos(\underline{k}_n\cdot\underline{r})$ again obtain for certain $\underline{k}_n$, but α may take values different from 1/2. In particular the case $\alpha = 1$ can occur, leading to a Landau-Ginzburg equation of the form

$$\partial_\tau a_{1n_c}(\tau) = \lambda a_{1n_c}(\tau) + ca_{1\ell_c}\, a_{1m_c}(\tau)\cdot\delta(\underline{k}_{\ell_c}+\underline{k}_{m_c}-\underline{k}_{n_c}) \qquad (27)$$

where $\lambda = R^{(1)}(c^{o})$ as before, and where $c = (1/2)\mu_c U_n R^{(2)}(c^{o})$. Such an equation does not generate bounded stable solutions. Thus although there are <u>hexagonal</u> solutions to eqn. (1) of the form:

$$c^O_{n_c} \pm U_{n_c} |\lambda/-c| (\mu-\mu_c) \cdot [\cos(\underline{k}_{n_c} \cdot \underline{r}) + \cos(\underline{k}_{m_c} \cdot \underline{r})$$

$$+ \cos(\underline{k}_{\ell_c} \cdot \underline{r})] \delta(\underline{k}_{\ell_c} + \underline{k}_{m_c} - \underline{k}_{n_c}), \quad (28)$$

such solutions are not stable.

Stable hexagons arise from eqn. (18), which generates a Landau-Ginzburg equation of the form,

$$\partial_\tau a_{1n_c}(\tau) = \lambda a_{1n_c}(\tau) + a_o |a_{1n_c}|^2 a_{1n_c}(\tau) + \sum_\ell b'_{\ell_c n_c} |a_{1\ell_c}|^2 a_{1n_c}$$

$$+ \sum_{m_c} b'_{m_c n_c} |a_{1m_c}|^2 a_{1n_c} \quad (29)$$

where $\underline{k}_{m_c} \cdot \underline{k}_{n_c} = \underline{k}_{\ell_c} \cdot \underline{k}_{n_c} = |k_o|^2 \cos(\pm\pi/3) = |k_o|^2/2$. λ and a_o are again given by (20) and (21) respectively, but $b'_{\ell_c n_c}$ and $b'_{m_c n_c}$ differ slightly from (22), reflecting the hexagonal lattice rather than a rectangular one. It follows that hexagons of the form

$$c^O_{n_c} \pm U_{n_c} \sqrt{(\lambda(\mu-\mu_c)/-a_o - 2b'_{in_c})} [\cos(\underline{k}_{\ell_c} \cdot \underline{r}) + \cos(\underline{k}_{m_c} \cdot \underline{r})$$

$$+ \cos(\underline{k}_{n_c} \cdot \underline{r})] \cdot \delta(\underline{k}_{\ell_c} + \underline{k}_{m_c} - \underline{k}_{n_c}) \quad (30)$$

can exist, and are stable whenever $a_o < b'_{1n_c}$ and $a_o + a'_{in_c} < 0$, $i = \ell_c, m_c$ whereas roll solutions of (29) are stable whenever $b'_{in_c} < a_o$, $i = \ell_c, m_c$. It should be mentioned, however, that stability, as I have used it above, means stability with respect to disturbances within the same lattice class (Sattinger, 1980); it is not yet clear if rolls and cross-rolls are stable with respect to hexagonal perturbations and vice-versa. The reader should also keep in mind that the analysis applies only to small amplitude perturbations, and does not necessarily apply to large amplitude perturbations or solutions. Nevertheless there is some hope that for reactions of the type depicted in Fig. 1, in which reaction rates are monotonic functions of concentration, the large amplitude solutions do in fact grow from the small amplitude ones described above, with little change (Keener, 1978).

EMBRYOLOGY

Such a conclusion has been partly confirmed in numerical
computations by Gierer and Meinhardt (1972) of various versions of
eqns. (1), which they used to model various aspects of morphogenesis
and regeneration in invertebrates and vertebrates. They and others
(Wolpert, 1971) have, for example, modelled the morphogenesis and
regeneration of the freshwater polyp Hydra in terms of such reaction
pairs. Hydra comprises no more than 10^5 cells, and is a few mm long.
It is essentially a tube with tentacles and a foot. The tube regions
comprise a head, gastric region, budding area and foot. If the head
is removed, after several hours a new one will grow out of the
gastric region (Wolpert, 1971). Similarly a new foot will regenerate
from the stump. A two gradient model comprising separate and
independent head and foot morphogens, each consisting of an activator-
inhibitor pair, reproduces very well the observed phenomena. For
example, the activator-inhibitor equations (Gierer and Meinhardt,
1972):

$$\partial_t \begin{pmatrix} c_1 \\ c_2 \end{pmatrix} = \begin{pmatrix} c\rho c_1/c_2 - \mu & \cdot \\ c'\rho'c_1 & -\nu \end{pmatrix} \begin{pmatrix} c_1 \\ c_2 \end{pmatrix}$$

$$+ \begin{pmatrix} D_1 & \cdot \\ \cdot & D_2 \end{pmatrix} \partial_{xx} \begin{pmatrix} c_1 \\ c_2 \end{pmatrix} + \begin{pmatrix} \rho_o\rho \\ \cdot \end{pmatrix} \tag{31}$$

replicate very well a series of head and foot regeneration experiments. The reason for this can be seen very easily from the small
amplitude bifurcation analysis (Matkowsky, 1970; Auchmuty and
Nicolis, 1975; Haken and Olbrich, 1978) of such equations. As I
have already discussed, with zero flux boundary conditions the
steady solutions are:

$$c(x,\infty) - c^o \sim U_{n_c} \sqrt{(\lambda(\mu-\mu_c)/-a_o)} \cos(k_{n_c} x) \tag{32}$$

where

$$k_{n_c} = \pi/L$$

If one now assumes a process of gene-switching to occur as the
cell and tissue responses to such a distribution of morphogen,
then one can easily see how the various hydroid regions might be
encoded in terms of the concentrations of such morphogens. If a
head or foot is now removed, corresponding to excision of one end
of the domain, the remaining morphogen distribution is no longer
stable and it will reorganize until a new stable state is reached

of the form:

$$c(x,\infty) - c^{O} \sim U_{n_c} \sqrt{(\lambda(\mu-\mu_c)/-a_o)} \cos(k'_{n_c} x)$$

$$k'_{n_c} = \pi/L' \tag{33}$$

where L' is the length of the remainder. Evidently there are limits
to the ability of the system to reorganize into a normal, or nearly
normal hydra. These limits have been studied within the format
described above (Cohen, 1971; Lacalli and Harrison, 1978; Gierer and
Meinhardt, 1972).

More complex, two and three-dimensional problems have been
investigated within the same framework. For example, in the larval
stage of Drosophila there exist many segments, some of which contain
groups of epidermal cells, called imaginal discs (Garcia-Bellido, et
al., 1973) that are destined to become specific adult structures.
Thus there are a number of head discs, several thoracic discs,
including wing and leg discs, and several abdominal discs. Each disc
comprises a group of cells derived from just a few precursors, that
is each disc is a polyclone (Crick and Lawrence, 1975) formed from a
few clonal lines. Examination of hundreds of mutant Drosophilae has
uncovered a process of symmetry-breaking associated with the specifi-
cation of adult structures within such discs. Initially cells from
various clones can appear anywhere within such discs. Initially
cells from various clones can appear anywhere within a disc. However,
after a certain developmental time such clones are found to be
restricted to either anterior or else posterior regions of any disc.
In effect a clonal boundary appears which separates regions determined to become anterior from those determined to become posterior.
It turns out that there is a sequence of progressive restrictions
on the fate of clonal tissue, as shown in Fig. 2. Thus the anterior-
posterior boundary is followed by the appearance of a dorsal-ventral
boundary, a wing-notum boundary, a proximo-distal boundary, and so
on, until each disc ends up as a number of compartments. Exactly
how this takes place is as yet unknown, but it probably takes place
by a process of gene-switching similar to that in Hydra, following
the appearance of non-uniform two-dimensional morphogenetic distributions (Kauffman et al., 1978).

Cellular tissues or aggregates can undergo symmetry-breaking
configurational changes even in the absence of any underlying change
of morphogen or prepattern. For example, an aggregate of dissociated
heart cells will coalesce into a tightly coupled ball of cells, and
an aggregate of dissociated embryonic retinal and heart cells will
"sort out" into balls of heart cells partly or completely surrounded
by retinal cells (Steinberg, 1963). This effect occurs because heart
and retinal cells have differing adhesivities, by virtue of the
differing sets of contact adhesion molecules they display on their

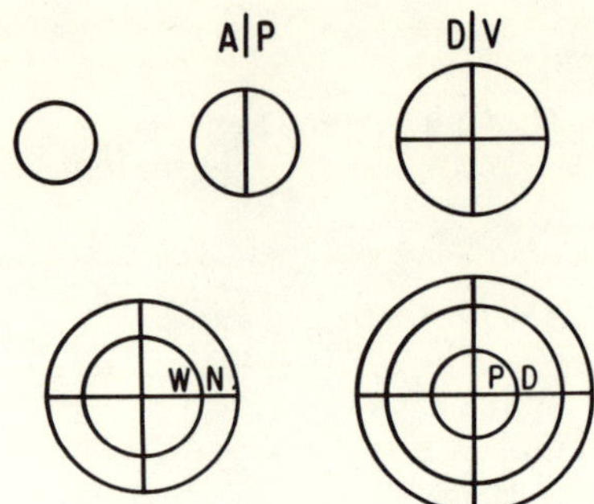

Fig. 2.   The sequence of polyclonal compartments appearing
in the early _Drosophila_ wing disc.

surfaces. The final stable configuration is one that minimizes the
surface free-energy associated with cell-cell contacts. Such processes, and many others, play key roles in the development and
differentiation of the vertebrate nervous system. It has recently
been discovered, for example, that the cell lineage of the _Xenopus_
nervous system unfolds in a fashion similar to that depicted in
Fig. 2 for insect imaginal discs, and it therefore seems likely
that there are underlying distributions of morphogens or prepatterns associated with differing neuronal structures. It follows
that the problem of neuronal specificity, the wiring-up of the
nervous system into highly specific circuits, can be thought of in
terms of the matching-up of _maps_ provided by prepatterns. An
interesting example of this is to be found in the way in which nerve
fibers from the _Xenopus_ retina make precise topographic connections
in a central region of the brain, the optic lobes or tectum. In
many cases, individual groups of nerve fibers can find the appropriate group of target cells in the optic tectum, even under abnormal
conditions: for example after sectioning the optic nerve bundle and
rotating either the eye or the tectum (Sperry, 1963). This can be
understood in terms of the matching of retinal and tectal map
coordinates, encoded in terms of prepatterns of contact adhesion
molecules, together with a kind of Darwinian competition on the
part of retinal fibers and tectal cells, for complimentary cells
and fibers (Prestige and Willshaw, 1975; Whitelaw and Cowan, 1980).
Once again the notions of symmetry-breaking can be applied to such
a process, in which an initially disordered retino-tectal map sorts
out into a precise topographically ordered map, as depicted in Fig. 3

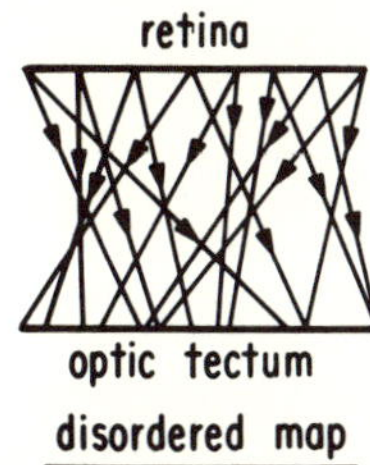
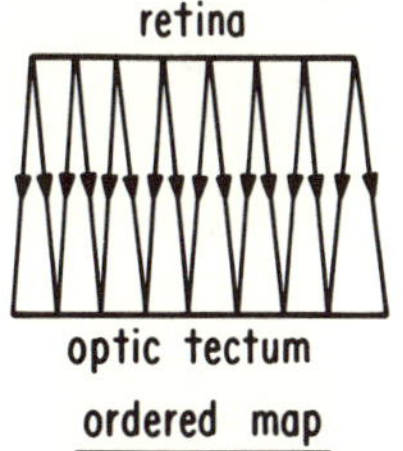

Fig. 3. The sorting out of an initially disordered retino-tectal
map into an ordered map.

NEUROBIOLOGY

     Many other examples of symmetry-breaking are to be found in
neurobiology. Consider for example the properties of aggregates
of coupled nerve cells or neurons. Neurons in the vertebrate
central nervous system are either excitatory or inhibitory in their
action on other neurons, or on muscles or glands. What then are
the properties of nets of such neurons? It turns out that they
are very similar to the diffusion-coupled chemical reactions discussed in §1. Thus a sheet of excitatory and inhibitory neurons,
coupled together in all possible ways, can support many differing
kinds of stable activity. Let the ground-state correspond to, on
the average, zero activity: assumed to be stable to small random
perturbations. If the excitability of the sheet is, however,
increased beyond a critical value, symmetry-breaking bifurcations
to new states will occur, exactly as discussed in §1. Depending
upon the way in which the neurons within the sheet are connected,
such new states comprise either temporal oscillations, travelling
waves, or else standing spatial patterns (Wilson and Cowan, 1973).
A particularly interesting example occurs in the production of
drug-induced visual hallucination patterns (Siegel and West, 1975).
Visual hallucinations appear in many conditions: migraine, epilepsy,
hypnagogic hallucinosis, etc. I shall concentrate on those hallucination patterns seen in the earliest stages of drug induced
hallucinosis, which appear as simple geometrical forms (Siegel, 1977).
Klüver (1967) classified these forms as follows: grating or lattice,
cobweb, funnel or tunnel, and spiral. Fig. 4 shows such form constants.

     Of course, one has to relate such patterns seen in the visual
field to corresponding patterns of cortical activity. To do this
it suffices to note that there is a conformal projection of the
visual field onto the visual cortex (Schwartz, 1977; Cowan, 1977).

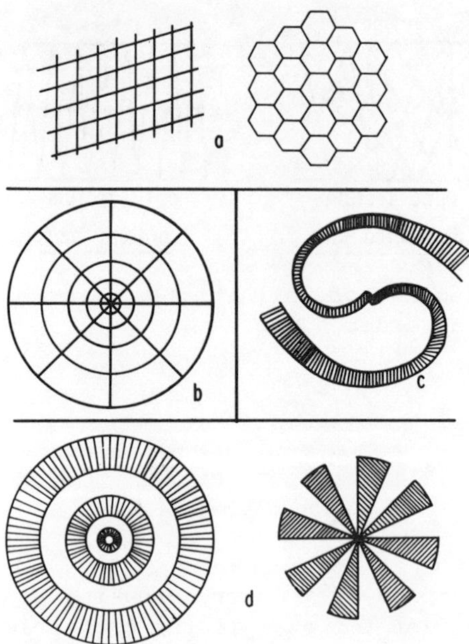

Fig. 4.   Typical form constants.   a. Lattice, b. Cobweb,
          c. Spiral, d. Tunnel and Funnel.

This transformation has been shown to take the detailed form
(Cowan, 1977):

$$x = \sqrt{\left(\frac{4k}{\pi\varepsilon}\right)} \cdot \ln\left[(\sqrt{\varepsilon}\phi + \sqrt{(w_o^2 + \varepsilon\phi^2)})/w_o\right],$$

$$(34)$$

$$y = \sqrt{\left(\frac{4k}{\pi}\right)} \cdot \phi\theta \cdot \sqrt{(w_o^2 + \varepsilon\phi^2)}^{-1}$$

where (x,y) are cortical coordinates, (ϕ,θ) the corresponding visual
field ones, and (w_o,ε,k) are constants reflecting anatomical parameters.   It will be seen that close to the center of the visual field
(ϕ small), $x = (4k/\pi) \cdot \phi/w_o$, $y = \sqrt{(4k/\pi)} \cdot \phi\theta/w_o$, i.e., visual field
coordinates in disguise; whereas far from the center, $x = \sqrt{(4k/\pi\varepsilon)} \cdot$
$\ln[\sqrt{\varepsilon}\phi/w_o]$, $y = \sqrt{(4k/\pi\varepsilon)} \cdot \theta$.   This is the complex logarithm (Fischer,
1973; Schwartz, 1977).   That is, a point in the visual field may be
represented by the complex variable $z = \sqrt{\varepsilon} \cdot \phi/w_o \cdot \exp[i\theta]$, and the
corresponding cortical point by $w = \sqrt{(4k/\pi\varepsilon)} \cdot \ln z$, for sufficiently
large ϕ.   It follows that lattice form constants, which are usually
small and central, are not changed very much by the above transformation, but the other form constants: cobwebs, tunnels and funnels,
and spirals, are usually larger, and are therefore subject to the
complex logarithmic transformation described above.   Fig. 5 shows
their cortical images.   It will be seen that these images are <u>rolls</u>
of various orientations.   It follows immediately from §1 that such

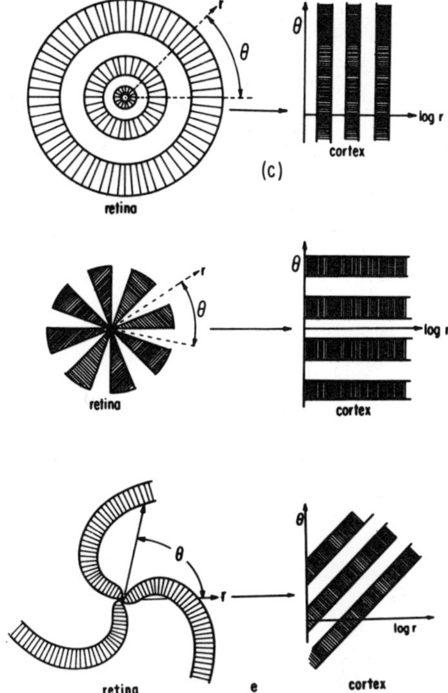

Fig. 5.   Cortical images of visual form constants.

rolls can be obtained as the bifurcating small amplitude solutions
to some system of nonlinear equations for large-scale neuronal
activity in two-dimensional sheets of neuronal tissue, exactly as
in the case of diffusion-coupled chemical reactions previously
considered.

To demonstrate this, G. B. Ermentrout and I (Ermentrout and
Cowan, 1979) have made use of a nonlinear field theory for neuronal
activity, developed in analogy with eqn. (1). Let $v(\underline{r},t)$ be a vector
representing the mean voltage built-up in the membranes of excitatory
and inhibitory cells at the point $\underline{r}$, in a sheet of tissue. Then v
satisfies the nonlinear integro-differential equation:

$$\partial_t v = -v(\underline{r},t) + \int_{-L/2}^{L/2} k(\underline{r},\underline{r}')s[v(\underline{r}',t)]d\underline{r}'$$
$$+ \rho_0 \rho(\underline{r})$$

(35)

where $k(\underline{r})$ is a matrix of coupling kernels, $s[v]$ a smooth nonlinear
functional of v, and where the excitatory stimulus is $\rho_0\rho(\underline{r})$. With
a few modifications, eqn. (35) can be analyzed in exactly the same

fashion as eqn. (1), with similar results. It turns out that to produce stable rolls rather than cross-rolls, a four-component equation of the form:

$$\partial_t \begin{pmatrix} v_1 \\ v_2 \\ v_3 \\ v_4 \end{pmatrix} = - \begin{pmatrix} v_1 \\ v_2 \\ v_3 \\ v_4 \end{pmatrix} + \begin{pmatrix} \cdot & \cdot & k_{31} & k_{41} \\ k_{12} & \cdot & \cdot & \cdot \\ \cdot & \cdot & \cdot & k_{43} \\ k_{14} & k_{24} & \cdot & \cdot \end{pmatrix} s\left[\begin{pmatrix} v_1 \\ v_2 \\ v_3 \\ v_4 \end{pmatrix} \right]$$

$$+ \begin{pmatrix} \rho_o \rho \\ \cdot \\ \cdot \\ \rho_o \rho \end{pmatrix} \tag{36}$$

is required, where $\int_{-L/2}^{L/2} k_{ij}(r-r') s[v_j(r')] d\underline{r}' = k_{ij} s[v_j]$, and where $k_{ij}(\underline{r}) = b_{ij} \exp[-|\underline{r}|^2/\lambda_{ij}^2]$. b_{ij} represents the strength of coupling between the ith and jth cell populations, and λ_{ij} represents the space-constant of such coupling. Such equations are analogous to a similar system of diffusion-coupled reaction equations proposed recently by Meinhardt and Gierer (1980) for similar purposes. The uniform ground state is destabilized by the external stimulus $\rho(\underline{r})$, exactly as in the case of eqn. (31), and stable rolls result, with or without boundary conditions. Corresponding to this, the various form constants are then presumably seen in the visual field.

What is the physiological basis for such a model? The neuro-pharmacology of psychomimetric drug action is not yet sufficiently understood to permit the pinpointing of the pathways leading from such drugs as LSD, marijuana, cocaine or peyote to cortical excitability, but such pathways must exist. One possibility is that a brain-stem nucleus is involved, possibly the so-called Raphé nucleus (Aghajanian et al., 1970), but this remains to be determined. The stimulus $\rho(\underline{r})$ may be in fact a disinhibition of the cortex resulting from the effects of drugs on such a nucleus. In any event, I have given the conditions under which a very general class of neuronal interactions will generate small amplitude organized cooperative activity in the form of rolls, and I have outlined how the conformal transformation embodied in the retino-cortical map acts to convert such cortical forms into visual apparitions.

CONCLUDING REMARKS

It will no doubt be evident to the reader that I have barely scratched the surfaces of the fields of embryology and of neurobiology. Nevertheless, I hope I have given some indication of how bifurcation theory can be used to obtain information about some of the nonlinear phenomena that underlie development and perception. It remains to be seen how far one can take such an approach, before the problems get too complicated. The British biologist J.B.S. Haldane once said of biological problems that "they may be stranger than we can know". I reject this view, but it will clearly take a very long time indeed before a scientific understanding develops of embryos and brains, of an intellectual depth that would have appealed to Albert Einstein.

Aghajanian, G. K., Foot, W. F., and Sheard, M. H., 1970, J. Pharmacol. Exp. Ther., 171: 178.

Auchmuty, J. F. G. and Nicolis, G., 1975, Bifurcation analysis of nonlinear reaction-diffusion equations I, Bull. Math. Biol. 37: 323.

Busse, F. H., 1978, Nonlinear properties of thermal convection, Rep. Prog. Phys. 41: 1929.

Cohen, M. H., 1971, Models for the control of development, Symp. Soc. Exptl. Biol. XXV: 455.

Cowan, J. D., 1977, in Neuronal Mechanisms of Visual Perception, ed. E. Pöppel et al., NRP Bull. 15: 3.

Crick, F. H. C. and Lawrence, P. A., 1975, Compartments and polyclones in insect development, Science 189: 340.

Ermentrout, G. B. and Cowan, J. D., 1979, A mathematical theory of visual hallucination patterns, Biol. Cybernetics 34: 137.

Fischer, B., 1973, Overlap of receptive field centers and representation of the visual field in the cat's optic tract, Vision Res. 13: 2113.

Garcia-Bellido, A., Ripoli, P., and Morata, G., 1973, Nature New Biol., 245: 251.

Gierer, A. and Meinhardt, H., 1972, A theory of biological pattern formation, Kybernetik 12: 30.

Haken, H., 1975, Generalized Ginzburg-Landau equations for phase transition-like phenomena in lasers, nonlinear optics, hydrodynamics, and chemical reactions, Z. Physik. B 21: 105.

Haken, H. and Olbrich, H., 1978, Analytical treatment of pattern formation in the Gierer-Meinhardt model of morphogenesis, J. Math. Biol. 6: 317.

Hoppenstaedt, F. and Gordon, N., 1975, Nonlinear stability analysis of static states which arise through bifurcation, Comm. Pure Appl. Math. XXVIII: 355.

Kauffman, S. A., Shymko, R. M., and Trabert, K., 1978, Control of sequential formation in Drosophila, Science 199: 259.

Keener, J. P., 1978, Activators and inhibitors in pattern formation, Studies in Applied Math., 59: 1.

Klüver, H., 1967, "Mescal and the Mechanisms of Hallucination," Chicago: Univ. of Chicago Press.

Lacalli, T. C., and Harrison, L. G., 1978, The regulatory capacity of Turing's model for morphogenesis, with applications to slime molds, J. theor. Biol. 70: 273.

Matkowsky, B. J., 1970, A simple nonlinear dynamic stability problem, Bull. Amer. Math. Soc. 76: 3.

Meinhardt, H. and Gierer, A., preprint.

Nicolis, G. and Prigogine, I., 1977, "Self-Organization in Non-equilibrium Systems," Wiley, New York.

Prestige, M. C. and Willshaw, D. J., 1975, On a role for competition in the formation of patterned neural connections, Proc. Roy. Soc. Lond. B, 190: 77.

Sattinger, D. H., this volume.

Schwartz, E., 1977, Spatial mapping in the primate sensory projection, Biol. Cybernetics 25: 181.

Siegel, R. K., 1977, Hallucinations, Scientific Amer. 237: 4.

Siegel, R. K. and West, L. J., 1975, "Hallucinations: Behavior, Experience, and Theory," Wiley, New York.

Sperry, R. W., 1963, Chemoaffinity in the orderly growth of nerve fiber patterns and connections, PNAS 50: 703.

Steinberg, M. S., 1963, Reconstruction of tissues by dissociated cells, Science 141: 401.

Stuart, J. T., 1960, On the nonlinear mechanics of wave disturbances in stable and unstable parallel flows I, J. Fluid. Mech. 9: 353.

Turing, A. M., 1952, The chemical basis of morphogenesis, Phil. Trans. Roy. Soc. (Lond.) B: 641: 37.

Watson, J., 1960, On the nonlinear mechanics of wave disturbances in stable and unstable parallel flows II, J. Fluid. Mech. 9: 371.

Whitelaw, G. and Cowan, J. D., in preparation.

Wilson, H. R. and Cowan, J. D., 1973, A mathematical theory of the functional dynamics of cortical and thalamic nervous tissue, Kybernetik 13: 55.

Wolpert, L., 1971, Adv. Morphogenesis 6: 183.

ON GLOBAL PROPERTIES OF QUANTUM SYSTEMS

H. D. Doebner* and J. Tolar**

Technische Universität*
and
Czech Technical University**

1. The Physical Arena

1.1 Localized Systems

Consider a smooth connected manifold M and a physical
system which is "based" on M (configuration space),
i.e. one can localize the system in a sufficiently
large class of regions U M and observe how the
localization region moves. The properties of such a
system will depend on the geometry of M. Its description
may be based in a natural manner on those objects on M,
which are characteristic for its structure, like com-
pactly supported functions, n-forms, vector fields,
jets, etc. Special examples are Hamiltonian systems [1],
non-relativistic quantum systems on homogeneous spaces
[2], [3] and Maxwell fields on manifolds [4], [5].

If the system does not behave like a "standard free"
system an interaction can be defined and it is tempting
and successful in particle physics and gravitation to
understand at least certain types of interactions through
a supplementary (internal) dynamic attributed to the
standard free system and coupled to its (external)
dynamic on M such that the interacting system is repro-
duced. The internal dynamic is assumed to act on a
smooth connected or on a discrete manifold F_U, which
may depend on the localization region U. It can be
observed only indirectly through its influence or
"shadow" on the physical quantities measured on M.

To introduce the coupling we construct out of M and

F_U a generalized configuration space A (the arena).

We assume for the system a pointlike localization $U = \{m\}$, $m \in M$, we use $\{m\} \times F_m$ as arena for $U = \{m\}$, we take $F_m \sim F$ and have in general $U \times F$ as arena for $U \subset M$. To couple the dynamic on M and F smoothly, choose an open covering $\{U_\alpha\}$ of M and glue the strips

$U_\alpha \times F$ smoothly together to construct A.

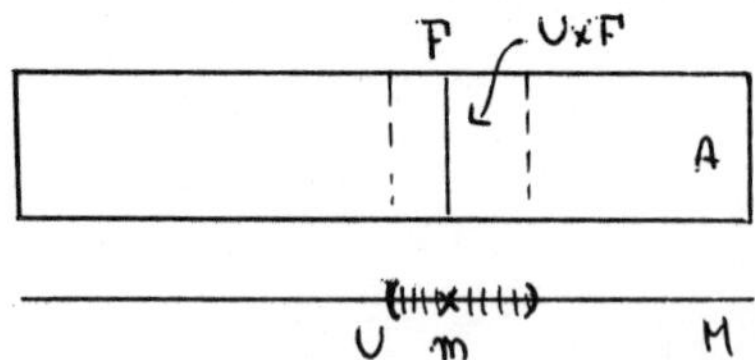

Fig.1: Generalized configuration space (fibre bundle)

The mathematical model for this situation is a fibre bundle $E = (A, \pi, M, F)$ with A as total space, M as basis, F as fibre; π is a continuous surjective map $\pi: E \to M$. The local structure of A as $U_\alpha \times F$ is imposed by a family $\{\psi_\alpha\}$ of diffeomorphisms $\psi_\alpha: U_\alpha \times F \to \pi^{-1}(U_\alpha)$ such that $\pi \circ \psi_\alpha(m, y) = y$, $m \in U_\alpha$, $y \in F$, i. e. the internal space F "grows" locally on the external one by application of π^{-1}. As mentioned above, the description of the system on A is based on certain objects on E, but the physical accessible quantities are on M. So one needs some additional information how to project the system from E to M, i. e. how to construct the system's shadow on M. For mathematical details see [6].

1.2 Examples

Locally any configuration space M, as a smooth n-dimensional manifold is isomorphic to an open set in $\mathbb{R}^n$, i. e. there is a smooth map $\varphi_\alpha: U_\alpha \to \mathbb{R}^n$. Globally that is not true in general. M can have non-trivial global properties. The same holds for generalized configuration spaces A; the glueing of $U_\alpha \times F$ may lead to global properties of A, say a twist, depending on the topology of M. Certainly, the root of such global structure is a physical one. We give now examples [7], [8], for globally non-trivial configuration spaces and discrete fibre bundles.

The Moebius strip as configuration space

Take two pointlike particles on a circle S^1. The configuration space of the system is $S^1 \times S^1 = \{(\varphi_1, \varphi_2) \mid 0 \leq \varphi_{1,2} < 2\pi\}$ which is the torus T^2 (fig. 2). Suppose the particles cannot sit in the same point, i. e. the configurations $D = \{\varphi_1, \varphi_1\}$ are forbidden and the configuration space is $S^1 \times S^1 - D$. Suppose furthermore the particles to be indistinguishable, then the configurations (φ_1, φ_2) and (φ_2, φ_1) have to be identified. To do this we use the symmetric group S_2 and have as a configuration space $(S^1)^{[2]} = (S^1 \times S^1 - D)/S_2$.

It is a smooth manifold, because S_2 acts properly discontinuous on $(S^1 \times S^1 - D)$ (this is not the case for $S^1 \times S^1$) and it is diffeomorphic to the Moebius strip with its obvious global twist property. Note, that $(S^1)^{[2]}$ is even a fibre bundle $((S^1)^{[2]}, \pi, S^1, I)$, with I being a finite interval in $\mathbb{R}$, but then the basis and the fibres are external spaces; the localization of position, not only on the basis S^1 but on the total space, is an observable.

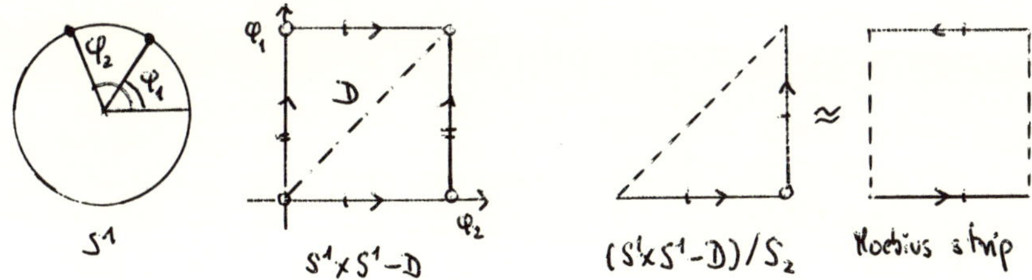

Fig. 2. Moebius Strip as Configuration Space

A projective space as configuration space

The global property of the configuration space above was obviously the result of the indistinguishability of the two systems on the (compact) S^1. A corresponding result should be valid also on the (non-compact) $\mathbb{R}^3$. Here take two pointlike indistinguishable particles on $\mathbb{R}^3$, which cannot sit on the same point. The configuration space of the systems is $(\mathbb{R}^3)^{[2]} = (\mathbb{R}^3 \times \mathbb{R}^3 - D)/S_2$ with $D = \{(\vec{x}_1, \vec{x}_2) \mid \vec{x}_1 = \vec{x}_2\}$. To see its structure, introduce coordinates $\vec{X} = \vec{x}_1 + \vec{x}_2$ and $\vec{X}' = \vec{x}_1 - \vec{x}_2$. Then $(\mathbb{R}^3)^{[2]} = \mathbb{R}^3_X \times (\mathbb{R}^3_{X'} / S_2) =$

$\mathbb{R}^3_x \times \mathbb{R}^3_r \times \mathbb{R}P^2$ with $\mathbb{R}^3_{x_2} = \mathbb{R}^3_x - \{0\}$, $\mathbb{R}_r = \{r \,|\, r = |x|\}$ and $\mathbb{R}P^2$ being the two-dimensional projective space, i. e. the sphere $S^2 = \{\vec{x} \,|\, |\vec{x}| = 1\}$ with points $\vec{x}$ and $-\vec{x}$ being identified because of S_2. $\mathbb{R}P^2$ is non-orientable, as the Moebius strip, and its first homotopy group is $\pi_1(\mathbb{R}P^2) = S_2$, (see [6]).

Coverings as configuration spaces

Take a pointparticle on M with non-trivial $\pi_1(M)$. Such M have regular coverings M^c which are geometrically (principal) fibre bundles $(M^c, \pi, M, \pi_1(M)/N)$ with N being a normal subgroup of $\pi_1(M)$ and $\pi_1(M)/N$ being its discrete fibre. A nice example is $M = \dot{\mathbb{R}}^2 = \mathbb{R}^2 - \{0\}$ (see fig. 3) with $\pi_1(\dot{\mathbb{R}}^2) = \mathbb{Z}$ with $\mathbb{Z}$ as a group of integers. The homotopy class of $r \in \mathbb{Z}$ contains loops circulating r-times around $\{0\}$ in positive $(r \geq 0)$ or negative $(r < 0)$ directions. If we choose $N = p \cdot \mathbb{Z}$, p integer, then we find a p-fold covering $(\dot{\mathbb{R}}^{2(p)}, \pi, \dot{\mathbb{R}}^2, \mathbb{Z}/p \cdot \mathbb{Z})$, i. e. a Riemannian surface with p leaves (Fig. 3 for $p = 2$). A non trivial $\pi_1(M)$ provides us with a canonical internal space associated with the topology of M. This space can carry a discrete internal motion attributed to a system on M, if a covering M^c of M is used as generalized configuration space. And we have the technical advantage that there exists a canonical lift of differential operations etc. from M to M^c.

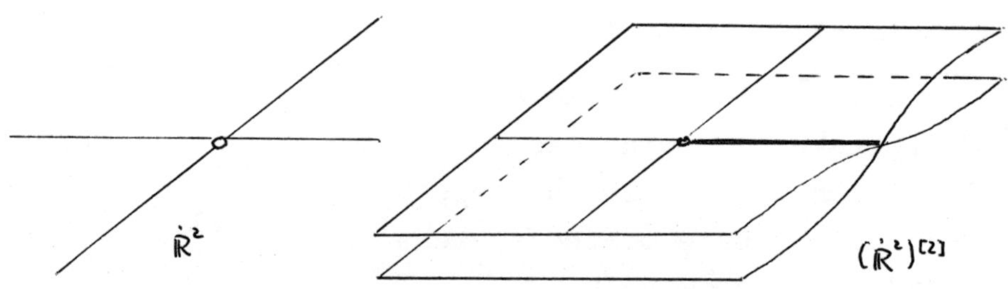

Fig. 3: $\dot{\mathbb{R}}^2$ and a twofold covering $(\dot{\mathbb{R}}^2)^{[2]}$

If the geometry of M is significant for all systems localized on M, one may argue that there are physical examples with such a discrete internal dynamic and one can at least describe the system on M and on M^c and see what the difference is (see section below). Referring to the configuration space $(\mathbb{R}^3)$ [2] we note that M/G_D with G_D being a discrete group acting on M properly discontinuously, the quotient space M/G_D is an "undercovering" of M.

1.3   Global Physical Properties

The examples given above show that the (external) configuration space M itself or the generalized configuration space A can have global properties which cannot be seen locally. The question is now : Are there observable consequences of such global properties of M or of A for a system localized on $U \subset M$, i. e. one can "hear" the "shape" of M or A — like the shape of a drum — through specific physical properties of the system on $U \subset M$ [9]. There are systems for which the answer is positive. We refer to gauge theories on principle bundles [10] and to spin structures on non simply connected manifolds and their use in many body physics [11] . A further example is non relativistic quantum mechanics on M with non trivial $\pi_1(M)$. This is because of the following reasons: The selfadjointness of operators representing observables and acting on suitably defined wave functions on M feel M not only locally; if all vector fields on M correspond to observables, the properties of the set of these operators will depend on $\pi_1(M)$. This will be discussed in the next chapter. For a simple example consider $\dot{\mathbb{R}}^2$ and define [8] wave functions on the 2-fold covering $(\dot{\mathbb{R}}^2)$ [2] as sections in the associated vector bundle through a representation of $\mathbb{Z}/2\mathbb{Z}$ or as equivariant functions on $(\dot{\mathbb{R}}^2)$ [2]. Then one can formulate quantum mechanics on $(\dot{\mathbb{R}}^2)$ [2] and compare it with the standard formulation on $\dot{\mathbb{R}}^2$. It turns out e. g. that the representations of the Heisenberg Lie algebra of (local) position and momentum operators $[\vec{p},\vec{q}]=-i\hbar\,\mathbb{1}$ via essentially skew adjoint operators on an invariant common dense domain obtained by a quantization on $\dot{\mathbb{R}}^2$ and $(\dot{\mathbb{R}}^2)$ [2] are unitarily inequivalent and that the corresponding harmonic oscillators have different spectra.

2.   Borel kinematics and Segal Quantisations
2.1 Borel kinematics

Take a classical particle-like system on M without
internal motion. For convenience we specify the
localization regions to be Borel sets from a Borel field
$\mathcal{L}$(M) on M. To mimic the motion of such a region U,
we shift U through the flow F(X) of a smooth vector
field $X \in \mathcal{V}$(M) to some region U'

$$U' = \left\{ m' \mid m' = F^X(m,t), \; m \in U, \; (m,t) \in D(X,t) \right\}$$

as indicated in figure 4 with flow

$$F^X: (m, t) \in D(X) \subset M \times \mathbb{R} \longrightarrow F^X(m, t) \in M$$

defined on its (open) domain $D(X, t) = \left\{ m \mid (m, t) \in D(X) \right\}$.
The Borel sets and the vector fields are obviously
some generalizations of positions and momenta of a
system moving on M, the flow parameter may be interpreted
as time. The pair $(\mathcal{L}$(M)$, \mathcal{V}$(M)$)$ is called a Borel
kinematic on M [12] .

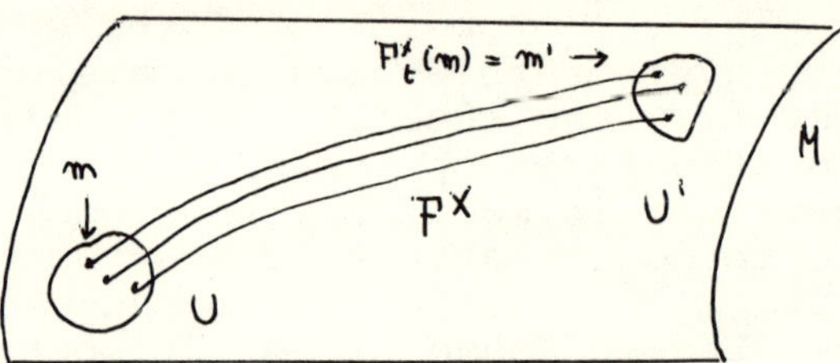

Fig. 4: On the Borel kinematics

2.2 Segal Quantisations

A suitable method to quantize the kinematic on M is to
construct a realization (Segal Quantisation) $\mathcal{Q}$ of
$(\mathcal{L}$(M)$, \mathcal{V}$(M)$)$ as operators in some Hilbert space
and to specify $\mathcal{Q}$ from physical requirements.

1

Consider $U \in \mathcal{L}$(M). In our approach, the localization
of the system in U is an observable also on the
quantum level. Hence $\mathcal{Q}$ maps U onto a selfadjoint
operator denoted by $\mathbb{B}$ (U). We require $\mathbb{B}$ (U) to be a
projection operator (position projection), as it is
for $M = \mathbb{R}^n$ or as it can be justified in an axiomatic
approach to a quantisation on M [17], and we construct

a realization of $\mathcal{L}(M)$ via a σ-additive mapping into the set $P(\mathcal{H})$ of the projection operators on $\mathcal{H}$

$$\mathcal{B} : U \in \mathcal{L}(M) \longrightarrow P(\mathcal{H}) \ni \mathbb{B}(U)$$

with $\mathbb{B}(\emptyset) = 0$, $\mathbb{B}(M) = \mathbb{1}$. The probability for the system in a state $\varphi \in \mathcal{H}$ to be localized in U is given by $\| \mathbb{B}(U)\varphi \|^2$. A probability density can be defined with respect to a measure class on M. We choose the class $[\mu_L]$ obtained from the Lebesgue class on $\mathbb{R}^n$ through a partition of unity, i. e. $\mu = K(m) \mu_o$ with k being a smooth positive real function on M and μ_o a standard measure in $[\mu_L]$. To have further properties of $\mathcal{B}$, observe, that any U can be a localization region, i. e. for any U with $\mu(U) \neq 0$ there exists a $\varphi \in \mathcal{H}$ such that $\| \mathbb{B}(U)\varphi \|^2 \neq 0$ and that the probability $\| \mathbb{B}(U)\varphi \|^2$ vanishes for a given U for all $\varphi \in \mathcal{H}$ if and only if $\mu(U) = 0$. Then $\mathcal{H}$ can be realized as a space of complex vector-(r)-valued functions $\Psi : M \rightarrow \mathbb{C}^{(r)}$, $r = r$ (m), square integrable with respect to μ with r depending in general on $m \in M$. We choose for simplicity the scalar case r = 1, i. e. $L^2(M, d\mu)$; here the position projection acts as $\mathbb{B}(U)\Psi(m) = \chi(U)\Psi(m)$ with χ as characteristic function of U.

   Different choices $\mu_i \in [\mu]$, i = 1, 2 yield physically equivalent descriptions of the system. The spaces $\mathcal{H}^{(i)} = L^2(M, \mu_i)$, are unitarily related through $U_{21} : \Psi^{(1)}(m) \longrightarrow S(m)\Psi^{(2)}(m))$, $S(m) = R(m)p(M)$, $R(m) = (d\mu_1/d\mu_2)^{1/2}$, $|p(m)|^2 = 1$, with $S(m)$ smooth and nonvanishing on M. For the position projection $\mathbb{B}^{(i)}(U) = \mathbb{B}^{(2)}(U)$ holds.

<u>2</u>

Consider $X \in \mathcal{V}(M)$. Similarly as the localization, also at least some classical momenta are observables on the quantum level. The corresponding selfadjoint operators are $\hat{P}(X)$. We construct first a linear map $\mathcal{P}$ of $\mathcal{V}(M)$ into the space $\mathcal{L}(\mathcal{H})$ of linear operators in $\mathcal{H}$.

$$\mathcal{P} : X \in \mathcal{V}(M) \longrightarrow \mathcal{L}(\mathcal{H}) \ni \mathbb{P}(X)$$

and restrict $\mathcal{P}$ such that $\mathbb{P}(X)$ is skew symmetric on some dense domain $\mathcal{D}$ in $\mathcal{H}$. In a second step the skew adjoint operators are selected or constructed via a selfadjoint extension of $\mathbb{P}(X)$.

   To specify $\mathcal{P}$ we translate the classical shift of U through F_t^X to the quantum level: Take complex smooth functions with compact support, $\Psi(m) \in C_c^\infty(M)$, define

for X and $t_X > 0$ a linear shift operator V^X_t via ($t \in R$, $|t| < t_X$). $V^X_t \Psi(m) = K^X_t(m) \Psi(F^X_t(m))$ on $C^\infty_o(M)$, supp $\Psi = U_\Psi \subset D(X, t)$, i. e. V^X_t maps $\Psi \in H$ with support U_Ψ into $(V^X_t \Psi)$ with support $U'_\Psi = \{m' | F^X_t(m) = m',\ m \in U_\Psi\}$, in complete agreement with the situation described in figure 3. $K^X_t(m)$ is a given smooth complex non vanishing function on M, depending on X and t, introduced because the shift may change the probability
for the localization: on one hand the measure μ may be not X-invariant, on the other, there might be an inbuilt structure on M simulating an interaction. For t_X small enough V^X_t is densely defined and $V^X_0 = \mathbb{1}$ and $(V^X_t)^{-1} = V^X_{-t}$ holds. We assume $V^X_t \Psi$ to be smooth in t around t = 0.

On the classical level the momentum was the differential of the flow. We do the same here and define (the factor $\hbar$ is introduced for convenience)

$$\mathbb{P}(X) \equiv \hbar \lim_{t \to 0} \frac{1}{t} (V^X_t - \mathbb{1}) \quad \text{on} \quad C^\infty_o(M)$$

This implies $\mathbb{P}(X) = \hbar\, \theta(X) + \alpha(X)$ $\qquad$ (∗)

with $\theta(X)$ as Lie derivative and $\alpha(X) = \frac{d}{dt}\Big|_{t=0} K^X_t(m)$

as a complex 1-form on M, i. e. $\alpha \in A^1(M, \mathbb{C})$.
So far we have used the linear structure of $\mathcal{V}(M)$; to have also its Lie-algebraic structure on the quantum level, we assume $\mathcal{P}$ to be a Lie-algebra homomorphism (up to domain questions), i. e.

$$\mathbb{P}([X, Y]) = [\mathbb{P}(X), \mathbb{P}(Y)]$$

This restricts the 1-form α to a closed one, i. e.

$$\alpha \in Z^1(M, \mathbb{C}) \subset A^1(M, \mathbb{C}).$$

To enforce now the skew symmetry of $\mathbb{P}(X)$ as a requirement on $\mathcal{P}$, we have to fix the measure $\mu \in [\mu_L]$ and we find $\mathbb{P}(X)$ skew symmetric on some $\mathcal{T}_x$ on $L^2(M, d\mu)$ if $\text{Re}\, \alpha(X) = \frac{\hbar}{2}\, \text{div}_\mu X$ or

$$\mathbb{P}(X) = \hbar \cdot \theta(X) + \frac{\hbar}{2} \cdot \text{div}_\mu X + ia(X) \quad ,$$

i. e. the real part of $\alpha \in Z^1(M, \mathbb{C})$ depends on the measure, the imaginary part $a_2 \in Z^1(M, \mathbb{C})$ is arbitrary. In unitarily related spaces $L^2(M, d\mu_i)$, $\mu_i \in [\mu_1]$, i = 1, 2, the momentum operators are physically equivalent. Their difference $\mathbb{P}^{(1)}(X) - U^* \mathbb{P}^{(2)}(X) U = \frac{dS}{S} =: \beta(X)$ is a socalled logarithmically exact complex 1-form, i. e. there exists a non vanishing smooth $S : M \longrightarrow \mathbb{C}$ such that $\beta = \frac{dS}{S}$. The 1-form β is called, furthermore, induced if $|S(m)|^2 = 1$, i. e. $\mu_1 = \mu_2$. Properties of this form are discussed in section 2.3.1.

Summing up, we define a mapping $\mathcal{Q} : (U, X) \longrightarrow (\mathbb{B}(U), \mathbb{P}(X))$ as a scalar Segal quantisation of $(\mathcal{L}(M), \mathcal{W}(M))$ if $\mathcal{H}$ is realized as $L^2(M, d\mu)$, $\mu \in [\mu_L]$ and with $\mathbb{B}(U) = \chi(U) \cdot$ on $\mathcal{H}$, $\mathbb{P}(X) = \hbar \cdot \theta(X) + \alpha(X)$ on $C_c^\infty(M) \subset \mathcal{H}$ with a closed complex 1-form α and $\mathrm{Re}\ \alpha(X) = \frac{\hbar}{2} \mathrm{div}_\mu X$ (see [13]).

The quantizations are characterized by α and two different $\mathcal{Q}_i : (U, X) \longrightarrow (\mathbb{B}^{(i)}(U), \mathbb{P}^{(i)}(X))$ in $L^2(M, d\mu_i)$, $\mu_i \in [\mu_L]_2$, $i = 1, 2$, are unitarily equivalent if $\mathbb{B}^{(1)}(U) = \mathbb{B}^{(2)}(U)$, $\mathbb{P}^{(1)}(X) - \mathbb{P}^{(2)}(X) = \alpha^1(X) - \alpha^2(X) = \beta(X)$ with β being a logarithmically exact complex 1-form.

3

We add the remark, that the shift operators and the position projections are related through $|t| < t_x$

$$V^x_{-t}\, \mathbb{B}(U')\, V^x_t = \mathbb{B}(U) \quad \text{on}\quad C_c^\infty(M)$$

which is a local version of Mackey's imprimitivity relation, i. e. a local 1-dimensional symmetry with classical generator X. This local imprimitivity relation may be used also as a defining relation for V^x_t. Its infinitesimal form on $C_c^\infty(M)$ is

$$[\mathbb{P}(X), \mathbb{Q}(g)] = \hbar\, \mathbb{Q}(\theta(X)g) \qquad (**)$$

with $\mathbb{Q}(g) \varphi(m) = g(m) \varphi(m)$, $g \in C_c^\infty(M)$. Viewed as an operator equation for $\mathbb{P}(X)$ its solution is again (*). Momentum operators and position functions with (**) are known as local Heisenberg systems [13], [14], [15]. For a more detailed discussion see [12].

2.3 Parametrisation of SEGAL Quantisations

There are unitarily inequivalent Segal quantizations $\mathcal{Q}$ in $L^2(M, d\mu)$. We parametrize them, using global properties of M, especially its first de Rham cohomology.

1

The set $L^1(M,\mathbb{C})$ of logarithmically exact complex
1-forms β form an additive group. The vector space
and hence additive group $B_1^1(M,\mathbb{C})$ of exact complex
1-forms is a subgroup of $L^1(M,\mathbb{C})$. The elements of
the quotient group $K^1(M,\mathbb{C}) = L^1(M,\mathbb{C})/B^1(M,\mathbb{C})$ are
logarithmically exact cohomology classes $[\beta]$. If β is
decomposed into real and imaginary parts $\beta = \mathrm{Re}\,\beta +$
$i\,\mathrm{Im}\,\beta$, then $\mathrm{Re}\,\beta$ is exact and $i\,\mathrm{Im}\,\beta$ is induced and
any $[\beta]$ contains induced logarithmically exact 1-forms.
One checks that any $\beta \in L^1(M,\mathbb{C})$ is also closed, i. e.
$L^1(M,\mathbb{C})$ is a subgroup of $Z^1(M,\mathbb{C})$.

Inequivalent $\mathcal{O}$ are given through $\alpha \in Z^1(M,\mathbb{C})$ up to
an additive $\beta \in L^1(M,\mathbb{C})$, i. e. by the elements of the
quotient group $H_L^1(M,\mathbb{C}) = Z^1(M,\mathbb{C})/L^1(M,\mathbb{C})$, which is
a subgroup of the first de Rham cohomology space
$H^1(M,\mathbb{C}) = Z^1(M,\mathbb{C})/B^1(M,\mathbb{C})$, and $H_L^1(M,\mathbb{C}) =$
$H^1(M,\mathbb{C})/K^1(M,\mathbb{C})$ holds.

2

With this result inequivalent $\mathcal{O}$ can be parametrized [13]
(see [6] and the results of R. N. Palais quoted
therein.).

$H^1(M,\mathbb{C})$ is a vector space over $\mathbb{C}$ with dimension
equal to the first Betti number $b^1(M)$, depending on
the topology of M (see 2.3.3.). Suppose b^1 to be finite.
According to the de Rham existence theorem there are
1-forms $\beta_i \in L^1(M,\mathbb{C})$, $i = 1,\ldots,b_1$, which can be
chosen even to be induced $\beta_i = \beta_{ind,i}$, such that
their logarithmically exact cohomology classes $[\beta_{ind,i}]$,
$i = 1,\ldots, b^1$, are a basis in $H^1(M,\mathbb{C})$. Furthermore,
$H_L^1(M,\mathbb{C})$ can be considered as a vector space over the
integers $\mathbb{Z}$; it has dimension b^1 and $[\beta_{ind,i}]$, $i =$
$1,\ldots, b^1$, are a basis in $H_L^1(M,\mathbb{C})$. This implies, that
inequivalent $\mathcal{O}$ can be parametrized through b^1
parameters $0 \leq P_i < 1$, $i = 1,\ldots, b^1$, with respect
to a suitably chosen basis of induced logarithmically
exact 1-forms. Physically these $\beta_{ind,i}$ can be
interpreted as some kind of potentials, living on M,
which influence the system similarly as a vector
potential, a charge. A corresponding result is valid
in gauge theories.

<u>3</u>

To connect $b^1(M)$ more directly with topological
properties of M, consider $\pi_1(M)$ and assume $\pi_1(M)/C(\pi_1(M))$
to be finitely generated. Then (Hurewicz-theorem) [6]
$\pi_1(M)/C(\pi_1(M)) = F(M) \oplus T(M)$ holds; F(M) is a free
abelian group (1. Betti group) with $b^1(M)$ generators
and $T(M) = T_1(M) \oplus ... \oplus T_n(M)$ is the 1. torsion group
with cyclic groups $T_j(M)$ of finite order. We give some
examples [16]

M	b(M)	T(M)
S^1	1	O
Moebius strip	1	O
$\mathbb{R}P^2$	O	S_2
Klein bottle	1	S_2

Therefore, non equivalent Segal quantisations are
related to $\pi_1(M)$ and hence to (certain) coverings M^c
of M and to discrete internal motions as indicated in
section 1.3. A more detailed discussion with worked
examples will be given in [12].

2.4 A Remark on Indistinguishable Systems

Extending the above method to bundles as generalized
configuration spaces in a way, similarly to that
described in [8], one can show, that non equivalent
quantisations of the kinematics on bundles with
discrete or compact groups G as fibres are given
through non equivalent unitary representations of G.
This result is related to the classification of Segal
quantisations. The physically interesting case M = $(\mathbb{R}^3)^{[2]}$
has $b^1(M) = O$, i. e. there is only one Segal quantisation.
But $\pi_1(M) = S_2$ implies the existence of a non trivial
covering with the group S_2 as discrete fibre. Hence we
find two quantisations on the covering of M and it is
easily shown, that they correspond to symmetric and
antisymmetric wave functions. So the Pauli principle
results as a consequence of the non trivial topology
of the configuration space. This is not surprising,
since the particular topology was enforced on the space
by the indistinguishability of the two particles. But
it shows the possibility of retracing the Pauli principle
to the classical level and of formulating indistinguishability in a way such that a subsequent quantisation
gives the usual results. It also gives a simple example
of how a physically well justified assumption yields

a non trivial topological structure on which different
quantisations exist and are used in nature.

Acknowledgements

We acknowledge fruitful discussions with W. Greub,
E, Binz, J. Hennig, F. Pasemann and B. Angermann.

References

1. R. Abraham and J.E. Marsden, "Foundation of Mechanics",
   2 nd edition, Benjamin (1978)
2. R.W. Machey, "Induced Representations and Quantum
   Mechanics", Benjamin, New York (1968)
3. H.D. Doebner and J. Tolar, J. Math. Phys.,16:975 (1975)
4. C.M. Misner and J.A. Wheeler, Ann. Phys. (N.Y.),
   2:525 (1957)
5. L.L. Henry, J. Math. Phys., 18:662 (1977)
6. W. Greub, St. Halperin and R. Vanstone, "Connections,
   Curvature and Cohomology", Academic Press, New York
   (1972)
7. J.M. Leinaas, Nuovo Cim., 47A:19 (1978)
8. H.D. Doebner and J.E. Werth, J. Math. Phys., 20:1011
   (1979)
9. M. Kac, see e.g. B. Booß, "Topologie und Analysis",
   Springer, Berlin (1977)
10. M.F. Atiyah, N.J. Hitchin and I.M. Singer, Proc.
    Roy. Soc. London, A362:425 (1978)
11. J. Petry, "Proceedings of the Conference on Differen-
    tialgeometric Methods in Mathematical Physics,
    Clausthal, 1978, to be published in "Lecture Notes
    in Physics", Springer, Berlin (1980)
12. H.D. Doebner, J. Tolar in preparation
13. I.E. Segal, J. Math. Phys., 1:468 (1960)
14. F. Pasemann, Proceedings of the Conference on
    Differentialgeometric Methods in Mathematical
    Physics, Clausthal, 1978, to be published in
    "Lecture Notes in Physics", Springer, Berlin (1980)
15. H. Snellman, Ann. Inst. Henri Poincaré, A24:393
    (1976)
16. S.T. Hu, "Homology Theory", Holden-Day, San Francisco
    (1970)
17. J.M. Jauch, "Foundations of Quantum Mechanics",
    Addison-Wesley, Reading, Mass. (1968)

INVITED SPEAKERS AND INVITED WORKSHOP PARTICIPANTS

Mehmet Abak, Karadeniz Universitesi, Trabzon, Turkey
Ali Attiya Abdulla, University of Baghdad, Baghdad, Iraq
Rafal Ablamowicz, Southern Illinois University at Carbondale
Hadi H. Aly, Southern Illinois University at Edwardsville
Asim O. Barut, University of Colorado, Boulder, Colorado
Brian L. Beers, Science Application Inc., McLean, Virginia
L. C. Biedenharn, Duke University, Durham, North Carolina
Arno Böhm, University of Texas at Austin, Austin, Texas
Subir Bose, Southern Illinois University at Carbondale
Philip H. Butler, University of Canterbury, Christchurch, New Zealand
Jack D. Cowan, University of Chicago, Chicago, Illinois
Robert M. Delaney, Saint Louis University, St. Louis, Missouri
P. A. M. Dirac, Nobel Laureate, Florida State University,
    Tallahassee, Florida
H. D. Doebner, Universität Clausthal, Clausthal, West Germany
John L. Gammel, Saint Louis University, St. Louis, Missouri
Murray Gell-Mann, Nobel Laureate, California Institute of Technology,
    Pasadena, California
Robert Gilmore, University of South Florida, Tampa, Florida
John Gregory, Southern Illinois University at Carbondale
Bruno Gruber, Southern Illinois University at Carbondale
Khidir A. A. Hamza, Nuclear Research Center, Baghdad, Iraq
Walter C. Henneberger, Southern Illinois University at Carbondale
B. R. Judd, The John Hopkins University, Baltimore, Maryland
Anatoli U. Klimyk, Academy of Sciences of the Ukrainian SSR, Kiev,
    USSR
Peter Kramer, Universität Tübingen, Tübingen, West Germany
Melvin Lax, City College of New York, New York

F. A. Matsen, University of Texas at Austin, Austin, Texas
Willard Miller, Jr., University of Minnesota, Minneapolis, Minnesota
Richard S. Millman, Southern Illinois University at Carbondale
Marcos Moshinsky, Universidad Nacional Autonoma de Mexico, Mexico
Lochlainn O'Raifeartaigh, Dublin Institute for Advanced Studies,
    Dublin, Ireland
Peter Ortoleva, Indiana University, Bloomington, Indiana
Jiri Patera, Universite de Montreal, Montreal, Canada
Ryszard Raczka, Institute for Nuclear Research, Warsaw, Poland
Alladi Ramakrishnan, Institute for Mathematical Sciences, Madras,
    India
Paul Roman, State University of New York, Plattsburg, New York
Jan Rzewuski, Wroclaw University, Wroclaw, Poland
Frank C. Sanders, Southern Illinois University at Carbondale
T. S. Santhanam, The Australian National University, Canberra,
    Australia
David S. Sattinger, University of Minnesota, Minneapolis, Minnesota
I. E. Segal, Massachusetts Institute of Technology, Cambridge,
    Massachusetts
Robert T. Sharp, McGill University, Montreal, Canada
Yu. F. Smirnov, Moscow State University, Moscow, USSR
E. G. C. Sudarshan, University of Texas at Austin, Austin, Texas
M. Samuel Thomas, Royal Military College of Canada, Kingston,
    Ontario, Canada
Hiroomi Umezawa, Killam Memorial Professor, University of Alberta,
    Edmonton, Alberta, Canada
R. Vasudevan, Institute for Mathematical Sciences, Madras, India
Mahmoud Abd El-Wahab Khalil, Alexandria University, Egypt
Richard E. Watson, Southern Illinois University at Carbondale
Eugene P. Wigner, Nobel Laureate, Princeton University, Princeton,
    New Jersey
Carl E. Wulfman, University of the Pacific, Stockton, California

INDEX